Solution of Differential Equation Models by Polynomial Approximation

PRENTICE-HALL INTERNATIONAL SERIES
IN THE PHYSICAL AND CHEMICAL ENGINEERING SCIENCES

NEAL R. AMUNDSON, EDITOR, *University of Minnesota*

ADVISORY EDITORS

ANDREAS ACRIVOS, *Stanford University*
JOHN DAHLER, *University of Minnesota*
THOMAS J. HANRATTY, *University of Illinois*
JOHN M. PRAUSNITZ, *University of California*
L. E. SCRIVEN, *University of Minnesota*

AMUNDSON *Mathematical Methods in Chemical Engineering: Matrices and Their Application*
AMUNDSON AND ARIS *Mathematical Methods in Chemical Engineering: Vol. II, First Order Partial Differential Equations with Applications*
ARIS *Elementary Chemical Reactor Analysis*
ARIS *Introduction to the Analysis of Chemical Reactors*
ARIS *Vectors, Tensors, and the Basic Equations of Fluid Mechanics*
BALZHISER, SAMUELS, AND ELIASSEN *Chemical Engineering Thermodynamics*
BRIAN *Staged Cascades in Chemical Processing*
CROWE et al. *Chemical Plant Stimulation*
DENN *Stability of Reaction and Transport Processes*
DOUGLAS *Process Dynamics and Control: Vol. I, Analysis of Dynamic Systems*
DOUGLAS *Process Dynamics and Control: Vol. II, Control System Synthesis*
FOGLER *The Elements of Chemical Kinetics and Reactor Calculations: A Self-Paced Approach*
FREDRICKSON *Principles and Applications of Rheology*
FRIEDLY *Dynamic Behavior of Processes*
HAPPEL AND BRENNER *Low Reynolds Number Hydrodynamics with Special Applications to Particulate Media*
HIMMELBLAU *Basic Principles and Calculations in Chemical Engineering*, 3rd edition
HOLLAND *Fundamentals and Moulding of Separation Processes: Absorption, Distillation, Evaporation, and Extraction*
HOLLAND *Multicomponent Distillation*

HOLLAND *Unsteady State Processes with Applications in Multicomponent Distillation*
KOPPEL *Introduction to Control Theory with Applications to Process Control*
LEVICH *Physicochemical Hydrodynamics*
MEISSNER *Processes and Systems in Industrial Chemistry*
MYERS AND SEIDER *Introduction to Chemical Engineering and Computer Calculations*
NEWMAN *Electrochemical Systems*
OHARA AND REID *Modeling Crystal Growth Rates from Solution*
PERLMUTTER *Stability of Chemical Reactors*
PETERSEN *Chemical Reaction Analysis*
PRAUSNITZ *Molecular Thermodynamics of Fluid-Phase Equilibria*
PRAUSNITZ, ECKERT, ORYE, AND O'CONNELL *Computer Calculations for Multicomponent Vapor-Liquid Equilibria*
RUDD et al. *Process Synthesis*
SCHULTZ *Polymer Materials Science*
SEINFELD AND LAPIDUS *Mathematical Methods in Chemical Engineering: Vol. III, Process Modeling, Estimation and Identification*
VILLADSEN AND MICHELSEN *Solution of Differential Equation Models by Polynomial Approximation*
WILDE *Optimum Seeking Methods*
WILLIAMS *Polymer Science and Engineering*
WOODS *Financial Decision Making in the Process Industry*

Solution of Differential Equation Models by Polynomial Approximation

JOHN VILLADSEN
MICHAEL L. MICHELSEN

Instituttet for Kemiteknik
Denmark

PRENTICE-HALL, INC.
Englewood Cliffs, New Jersey 07632

Library of Congress Cataloging in Publication Data

VILLADSEN, JOHN.

Solution of differential equation models by polynomial approximation.

(Prentice-Hall international series in the physical and chemical engineering sciences)

Includes bibliographies and index.

1. Differential equations—Numerical solutions. 2. Approximation theory. 3. Polynomials. 4. Chemical engineering—Mathematical models. I. Michelsen, M. L., joint author. II. Title.

QA.371.V52 515'.35 77–4331

ISBN 0–13–822205–3

10 9 8 7 6 5 4 3 2 1

Printed in the United States of America

Prentice-Hall International, Inc., *London*
Prentice-Hall of Australia Pty. Limited, *Sydney*
Prentice-Hall of Canada, Ltd., *Toronto*
Prentice-Hall of India Private Limited, *New Delhi*
Prentice-Hall of Japan, Inc., *Tokyo*
Prentice-Hall of Southeast Asia Pte. Ltd., *Singapore*
Whitehall Books Limited, *Wellington, New Zealand*

Contents

Preface

The principal aim of this book is to support the engineer—specifically the chemical engineer—who is interested in quantitative treatment of physical models. Engineering models often appear in the guise of differential equations and we present some tools for solving these models by polynomial approximation.

Many years of teaching experience have made it abundantly clear that chemical engineering students and graduates are by no means less mathematically gifted than their colleagues working in other engineering sciences. Once they understand that even very complex models can be attacked with the aid of only a few, basic methods and that these methods are available as computer codes, many of the students develop a voracious appetite for mathematical studies and may have to be pushed gently away from their newly acquired hobby of mathematical modeling before they become estranged from their true vocation of chemical engineering: inventiveness in the field of chemical processes, the actual large- or small-scale synthesis of chemical compounds, and the interplay between the chemical industry and society as a whole.

Scandinavian technical universities have a long tradition of giving their students a solid background in classical mathematics—at least with respect to the number of required courses. Still it is an undeniable fact that their lack of preparedness for numerical manipulations with models is often lamentable. Since the same observation is also made by colleagues in American universities, it may express some rather fundamental misconception in our way of teaching mathematics to young people even before the university level.

The graduate student or engineer working on a mathematical project is interested in the solution of the model and in the physical insight that the solution can give him. Unless he feels that the mathematics course helps him understand the physical background of his own problem, he will approach it as just another required course that stands between him and his true interests. Not only may a successful course in mathematics improve the student's understanding of physical sciences, but it may also prompt him at some later stage of his development to study specific subjects of mathematics in depth.

Countless books exist on applied mathematics and quite a few on numerical methods. Hildebrand's series is an example of excellent publications from the first category, while Lapidus' *Digital Computation for Chemical Engineers* has taught many generations of students how to talk sense with a computer.

Our text tries to incorporate model understanding and numerical model analysis into a single book. It does not deal with numerical methods per se (classical methods such as finite differences are hardly touched); neither is it a book about applied (or engineering) mathematics. Classical methods for solving linear differential equations are reviewed in chapter 1 and well-known techniques such as approximation by perturbation series occur frequently. Yet there are much better and more complete treatises on these subjects.

On reflection, the book seems to owe much to Lanczos' beautifully written *Applied Analysis.* He tried to enamor the student with the philosophy of numerical work rather than presenting the subject as either a string of theorems or a stack of digested subroutines on punched cards. Each physical problem must be treated on its own merits—at least if it is worth any research effort—and a careful analysis of the problem is more than half its solution. Once the ephemeral "feeling" for numerical work has been instilled into the student, he can produce remarkable results with only a few tools.

The first chapter presents what we regard as the basic units of mathematical models in chemical engineering: fluid flow, diffusion, and chemical reaction. In developing the model for a fixed bed reactor, the intimate coupling between transport processes that leads to an almost untractable mathematical complex is clearly seen. Much emphasis is put on the simplifications that are possible by judicious model approximation. The student who understands how the model can be simplified without sacrificing the desired essential features will save himself many hours of fruitless computer work, and he is likely to emerge from his analysis with a far better understanding of the physical process.

The personal interests of the authors in chemical reactor simulation as an academic research subject and in industrial practice is apparent not only in the first chapter but in the entire book. We make no *amende honorable* for this preoccupancy with chemical reactions and transport phenomena, but hopefully the last two examples of chapter 1 will give some inspiration to the reader who is interested in fluid mechanics and rheology. The numerical methods of the following chapters can certainly also be applied to these subjects.

Chapters 2 to 4 give a systematic treatment of weighted residual methods applied to linear problems. These chapters may be regarded as the core of the book, and the student who becomes familiar with the application of the algorithms in chapters 3 and 4 will be able to solve many models of practical interest. The orthogonal collocation method is our preferred choice of weighted residual method, for reasons that are stated clearly in chapters 2 and 4. We are not convinced, however, that this very popular method is a vademecum for the solution of differential equations. We believe that collocation is a convenient, mechanical method that may give even more desirable results in unison with quite different techniques such as perturbation methods. There are numerous examples of this combination of methods in chapters 2, 4, 6, 8, and 9.

Nonlinear problems are treated in chapters 5 and 8. Here simplifications of the numerical procedure—similar to the model simplifications of chapter 1—are pursued whenever possible. To reduce the numerical work, many different tricks have to be played. In chapters 5 and 8, many of these tricks are advocated, most notably the sensitivity analysis that makes tracing of the solution as a function of model parameters much easier.

Specific difficulties of the problem may require special techniques. Chapter 7, which describes global spline collocation as a method of solving entry length problems, is specifically noteworthy. Part of chapter 8 treats an efficient procedure for integration of sets of stiff equations, a numerical problem of tremendous interest in current research. The procedure is based on semi-implicit Runge–Kutta methods that are related to the collocation methods of chapter 4 but are easier to apply.

The techniques of chapter 6—one-point collocation methods—are eminently suited for qualitative study of the problem. The one-point collocation method is mentioned in numerous papers from the last few years. Different variants of the method are reported, and usually the object is to study the influence of model parameters on the solution.

In chapter 9 we report results on three larger research projects. The problems are all from very active areas of research, and their solution might be interesting in itself, but the intention of the chapter is to show how closely all the methods of the previous chapters work together. Perhaps this will give the reader determination to analyze his own problems in a similar fashion.

Finally, in the appendix we have collected a number of subroutines that should help the reader to solve a majority of computational problems in differential equations. The routines have been used for over 6 years at this university and in some large Danish engineering companies. The routines are available in the libraries of computer centers at the University of Houston and at the University of California at Berkeley and are used at a number of other American universities; in Salford, Great Britain; at the Technion in Haifa; and in various Scandinavian universities. We believe our long experience with the programs have made them almost failproof.

The readers of this text are usually from various fields of engineering or science. The problems may be stated in chemical engineering terms but they are usually of a quite general nature. In our university (with a 5-year curriculum for the degree of MSc in engineering) it is used in the fourth year as a textbook in a one-semester course (4 credit hours). In American universities it is probably best suited for a first-year graduate course, although some of the material can certainly be taught at an undergraduate level. Sections 1.1 to 1.4, 1.7, and 8.2, and chapters 2, 3, 4, and 6 contain material for approximately one semester. Very brief continuing-education courses, supplemented by one week of intense problem solving, have been given from the appendix alone with good results.

Exercises are given in each chapter. Many of these have been selected to support the understanding of a specific point in the text. Another group of exercises points to applications of the methods that are not immediately obvious from the text. Finally, a number of exercises are formulated as small research problems—either as an analysis of a recent paper from the literature or as an extension of previously reported results. This last group is the most important to us because we wish to equip the student for individual research in numerical treatment of mathematical models. Naturally some basic computer experience should be required as a prerequisite to the course, but in many of the computer exercises, no more than a straightforward application of the algorithms of the appendix is needed.

The introductions to each chapter should help focus the attention on the major ideas of the chapter and after each chapter there is a brief list of references (with comments about the most important ones). No attempt has been made to present a complete set of references for each subject.

In finishing a project of this size the authors have incurred debts to many persons. J. Villadsen wishes to thank Professor Warren E. Stewart of the University of Wisconsin for an invigorating experience more than 10 years ago. It was in cooperation with this dedicated scientist that the first version of orthogonal collocation was conceived.

Thanks are also due to the many chemical engineering departments in different parts of the world who have provided funds for short courses on part of the material; most recently a full-semester course was offered at the University of Houston, where the graduate students helped to draw attention to weak passages.

Drs. Bruce A. Finlayson and Bruno van den Bosch read the manuscript and suggested many valuable corrections. The authors graciously acknowledge their contributions.

Finally, the secretaries at Instituttet for Kemiteknik have labored over several years to produce the editions of notes that have now finally reached a measure of completeness. Mrs. Bente Hansen and Mrs. Johanne Nielsen bore the brunt of this work. For their patience we owe them our sincere thanks.

Lyngby, Denmark

John Villadsen
Michael L. Michelsen

Solution of Differential Equation Models by Polynomial Approximation

A Preliminary Study of Some Important Mathematical Models from Chemical Engineering

1

Introduction

In this chapter, we present the main items of the book:

1. Define the models that are to be solved by approximate methods and, in a purely qualitative way, discuss the expected behavior of the mathematical solution.
2. Present a few important mathematical tools that have been used to obtain "exact" or "closed-form" solutions of various differential equation models.

The physical systems to be considered are composed of one or several phases, each of which is supposed to be continuous on the microscopic scale. Differential mass balances and heat balances for each phase and equilibrium relations between the phases are our main modeling tools. If the system is a flowing fluid, both mass and heat balance are influenced by the velocity field of the fluid, and this is described by a momentum balance. Our starting point is equations (1) to (3) of section 1.1, and these equations are used as far as possible in an attempt to give a unified treatment of the model formulation. The equations are certainly not an exact description of the physical systems, and the reader should note the restrictions that the application of a simple set of basic equations imposes on the treatment. Model formulation must necessarily be based on actual physical systems, and sections 1.2 to 1.6 are a catalog of models that we have chosen to use as examples. The choice of examples is directed by the interests of the authors, and the reader who finds that too much emphasis has been put on models for chemical reactors can hopefully be convinced that models for these systems exhibit most of the important features of the vaguely defined model building concept. The examples are organized as shown in table 1.1.

TABLE 1.1
ORGANIZATION OF MODELS DESCRIBED IN CHAPTER 1

Section	Description of model
1.1	General model: momentum–energy–and mass balances. Explicit models for velocity distribution in unidirectional flow.
1.2	Models for a homogeneous fluid phase with or without reaction. No interaction fluid-bed packing.
1.3	Steady state and transient models for a solid phase on which a chemical reaction occurs.
1.4	The interplay between the flowing fluid phase and the stationary solid phase. Steady state and transient fixed bed reactor models.
1.5	Purification of a fluid by reverse osmosis in hollow fibers. The example shows how complicated models can sometimes be treated gradually by suitable assumptions.
1.6	Two-dimensional flow and heat transfer to a non-Newtonian fluid. Material beyond the level of section 1.1 is used in the model formulation. The final model has an unusual mathematical appearance.

The models of sections 1.2 to 1.4 are all characterized by rather simple flow patterns: The convective fluid flow is one-dimensional in the axial direction of a cylinder. Diffusion gradients inside solids are considered only across a thin plate, in the axial and radial directions of a cylinder and radially in a sphere. With these restrictions added to the assumptions of the basic model (1) to (3), an apparently complete treatment of these frequently occurring systems, all of which are treated in subsequent chapters by numerical methods, can be given.

The layout of the chapter would give the reader too bright a picture of model formulation as an easily systematized subject if no more examples were included. Some feeling for the complexity of the subject may be obtained through a study of the last two examples in sections 1.5 and 1.6. These examples are not very complicated, although they include a few features such as a two-dimensional velocity field that was not present in the previous examples.

Results from classical mathematical analysis occur throughout the book interspersed with the approximative methods that will be our main tools for solving models.

Two subjects, however, are given separate treatments. These are the solution of coupled linear differential equations with constant coefficients and the Fourier series solution of linear partial differential equations.

As further argued in the introduction to section 1.7, these specific mathematical topics are used in almost any type of numerical treatment of much more complicated nonlinear differential equations.

1.1 The General Mathematical Model

A differential mass balance for component A in a dilute binary mixture takes the form

$$\frac{\partial c_A}{\partial t} + (\mathbf{v} \cdot \nabla c_A) = (\nabla \cdot D\, \nabla c_A) - R_A \tag{1}$$

A corresponding thermal energy balance is almost identical to (1):

$$\rho c_p \left[\frac{\partial T}{\partial t} + (\mathbf{v} \cdot \nabla T) \right] = (\nabla \cdot k \nabla T) + Q \tag{2}$$

The velocity field $\mathbf{v}$, which must be known in order to solve these two equations, is given by a momentum balance for the fluid.

For a constant-density, constant-viscosity Newtonian fluid, one obtains the following vector differential equation, the so-called Navier–Stokes equation:

$$\rho \left[\frac{\partial \mathbf{v}}{\partial t} + (\mathbf{v} \cdot \nabla \mathbf{v}) \right] = \mu \nabla^2 \mathbf{v} + \rho \mathbf{G} - \nabla p \tag{3}$$

Equation (3) introduces a new variable, the pressure gradient ∇p. In the examples of sections 1.2 to 1.4, the pressure is known at every point of the system. Equation (3) is supplemented by the equation of continuity

$$\frac{\partial \rho}{\partial t} = -[\nabla \cdot (\rho \mathbf{v})]$$

or for a flow at constant density, simply

$$\nabla \cdot \mathbf{v} = 0 \tag{4}$$

The reader who is familiar with the subject of transport phenomena will have no difficulty in recognizing this set of four equations as the source of innumerable mathematical models of interest to chemical engineers. He will also know that these leave some physical phenomena and systems untouched. If he is in doubt about their applicability in a specific situation, he should consult a standard text such as Bird (1960, chapters 3, 10, and 18) for an almost complete guide to the model building process.

We shall never need a more general basis for the models solved in the present text. If any phenomenon that is not covered by the assumptions of (1) to (3) appears, the system will be very simple otherwise and the appropriate differential balances can be set up from first principles.

The two terms of the left-hand side of the equations describe the accumulation of mass, thermal energy, and momentum per unit time in a unit volume of the fluid as seen by an observer who travels through the system, passively drifting along the streamlines. This total accumulation term—the substantial derivative or D/Dt in the notation of Bird (1960, p. 73)—may also be conceived as the sum of two perhaps more well-known terms. The first term is the conventional accumulation term in a volume that is fixed in space, and the second term is the rate of change of the property by the convective flow through the element.

The right-hand sides of the equations also consist of two terms. The first term is the change of molecular diffusion flow across the element. Fick's law, Fourier's law, and Newton's viscosity law are used to correlate the flux q of an extensive property (mass, heat, or momentum) to the gradients of concentration, temperature, and fluid velocity by simple semi-empirical proportionality relations with mass diffusivity D, thermal heat conductivity k, and viscosity μ as proportionality constants.

As written in (1) and (2), D and k are molecular properties of the fluid, and these quantities are well known either from experiments or from theoretical calculations. We shall often use equations (1) and (2) for systems that are much too complex to allow a complete solution of (1) to (3). An empirical velocity distribution is inserted in (1) and (2), but now the transport parameters are empirical quantities without relation to k and D as they stand in (1) and (2): A turbulent axial mixing coefficient, a mass transport coefficient inside the pore system of a catalyst, and a radial heat conductivity in a fixed bed reactor will all be called D and k with only a brief reference to their physical significance. The reason that this violation of the assumptions of the general model does not invalidate the results is that the mechanisms of the more complicated transport phenomena are empirically found to be similar to those leading to the diffusion terms of (1) and (2), and consequently the mathematical models look alike although their foundation is different.

The last terms—R_A, Q, and $(\rho \mathbf{G} - \nabla p)$—are volumetric production terms. R_A is the rate of disappearance of A per unit volume by a chemical reaction. Q is a corresponding heat of reaction by the chemical reaction or another continuous heat source in the fluid. Internal friction, absorption of radiant heat, or dielectric heating all contribute to Q, while external heating from a steam jacket surrounding the fluid has to be treated differently in an accurate model, although we shall see that an approximation of a boundary heat source by an average volumetric heat source is quite often used.

Finally, $\rho\mathbf{G}$ and $-\nabla p$ in equation (3) describe the production of momentum by body forces proportional to the fluid density (the action of gravity is an example) or by a pressure gradient ∇p in the flowing system.

Equations (1) and (2) are scalar equations, and their solutions are the scalar quantities c_A and T as functions of the space coordinates and time t. Equation (3) is a vector equation with three components, one equation for each of the three unknown scalar components of the fluid velocity $\mathbf{v}$.

We denote the coordinates of the normal rectangular coordinate system as either (x, y, z) or (x_1, x_2, x_3), whichever is more convenient. The axial coordinate in cylinder geometry is called z, and the radial coordinate in cylinder and spherical geometry is called r. We shall never treat problems in the angular coordinates of these last two geometries—the dependent variables will be constant in these directions. This restriction makes a complete description of the relevant vectorial quantities of equations (1) to (3) for all three geometries quite easy.

The three components of $\mathbf{v}$ are v_{x1}, v_{x2}, v_{x3} or, in a more convenient notation, v_1, v_2, v_3. The gradient of the scalar quantities c_A and T is a vector with the partial derivatives in the 1-, 2-, and 3-directions as components. In rectangular coordinates, the scalar products $\mathbf{v} \cdot \nabla c_A$ and $\mathbf{v} \cdot \nabla T$ of equations (1) and (2) are

$$\sum_1^3 v_i \frac{\partial c_A}{\partial x_i} \quad \text{and} \quad \sum_1^3 v_i \frac{\partial T}{\partial x_i}$$

In equation (3), the quantity $\nabla\mathbf{v}$—the gradient of the vector $\mathbf{v}$—is a second-order tensor, which is a (3×3) matrix $\mathbf{M}$ with element $m_{ij} = \partial v_j/\partial x_i$ in row i and column j.

The scalar product $\mathbf{v} \cdot \nabla\mathbf{v}$ is the product of the (row) vector $\mathbf{v} = (v_1, v_2, v_3)$ and $\mathbf{M}$. It is thus a (1×3) vector, where each component is the scalar product of $\mathbf{v}$ and a column of $\mathbf{M}$.

$$\mathbf{v} \cdot \nabla\mathbf{v} = (v_1, v_2, v_3)\begin{pmatrix} \dfrac{\partial v_1}{\partial x_1} & \dfrac{\partial v_2}{\partial x_1} & \dfrac{\partial v_3}{\partial x_1} \\[2ex] \dfrac{\partial v_1}{\partial x_2} & \dfrac{\partial v_2}{\partial x_2} & \dfrac{\partial v_3}{\partial x_2} \\[2ex] \dfrac{\partial v_1}{\partial x_3} & \dfrac{\partial v_2}{\partial x_3} & \dfrac{\partial v_3}{\partial x_3} \end{pmatrix} \tag{5}$$

$$= \left(\sum_{j=1}^3 v_j \frac{\partial v_1}{\partial x_j}, \quad \sum_{j=1}^3 v_j \frac{\partial v_2}{\partial x_j}, \quad \sum_{j=1}^3 v_j \frac{\partial v_3}{\partial x_j}\right)$$

Assume for the moment that D and k are constant; i.e., they are the same in all directions and they are independent of c_A and T. In this case, the first term on the right-hand sides of (1) and (2) can be written $D\nabla^2 c_A$

and $k\nabla^2 T$. ∇^2 is called the *Laplacian operator* and, in rectangular coordinates, the diffusion terms contribute with the following scalars in equations (1) and (2) and in the ith equation of (3):

$$D \sum_{j=1}^{3} \frac{\partial^2 c_A}{\partial x_j^2}, \quad k \sum_{j=1}^{3} \frac{\partial^2 T}{\partial x_j^2}, \quad \mu \sum_{j=1}^{3} \frac{\partial^2 v_i}{\partial x_j^2}$$

The scalars $\nabla^2 v_i$ ($i = 1, 2, 3$) that appear, one in each of the three scalar equations (3), are, of course, only the correct contributions for the viscous forces when the fluid is Newtonian (and has a constant density: $\nabla \cdot \mathbf{v} = 0$). For more general fluids the correct viscous term $-\nabla \cdot \boldsymbol{\tau}$ must be used in the momentum balance in place of $-\mu\nabla^2\mathbf{v}$. $\boldsymbol{\tau}$ is the shear stress tensor, a (3×3) matrix with components τ_{ji}, and the vector-matrix product $-\nabla \cdot \boldsymbol{\tau}$ is similar in form to (5) and gives a contribution $-\sum_{j=1}^{3} \frac{\partial \tau_{ji}}{\partial x_j}$ to the ith equation of (3).

The curvilinear expressions for the components of ∇ and $-\nabla \cdot \boldsymbol{\tau}$ and for the scalar ∇^2 are summarized in table 1.2 for the r- and z-directions of a cylinder ($v_\phi = 0$ and v_r, v_z independent of ϕ) and for the r-direction of a sphere. ($v_\theta = v_\phi = 0$ and v_r is independent of θ and of ϕ.)

TABLE 1.2

EXPLICIT EXPRESSIONS FOR SOME TERMS THAT OFTEN APPEAR IN EQUATIONS (1) TO (3) IN CYLINDRICAL AND SPHERICAL GEOMETRY

	Components of ∇	The scalar ∇^2	Components of $-\nabla \cdot \boldsymbol{\tau}/\mu$
Cylinders			
r-direction	$\frac{\partial}{\partial r}$	$\frac{1}{r}\frac{\partial}{\partial r}\left(r\frac{\partial}{\partial r}\right) + \frac{\partial^2}{\partial z^2}$	$\frac{\partial}{\partial r}\left[\frac{1}{r}\frac{\partial}{\partial r}(rv_r)\right] + \frac{\partial^2 v_r}{\partial z^2}$
z-direction	$\frac{\partial}{\partial z}$		$\frac{1}{r}\frac{\partial}{\partial r}\left(r\frac{\partial v_z}{\partial r}\right) + \frac{\partial^2 v_z}{\partial z^2}$
Spheres			
r-direction	$\frac{\partial}{\partial r}$	$\frac{1}{r^2}\frac{\partial}{\partial r}\left(r^2\frac{\partial}{\partial r}\right)$	$\frac{1}{r^2}\frac{\partial}{\partial r}\left(r^2\frac{\partial v_r}{\partial r}\right) - \frac{2}{r^2}v_r$

Examples:

1. The energy balance in spherical geometry with no variation in the θ- or ϕ-directions:

$$\rho c_p\left(\frac{\partial T}{\partial t} + v_r\frac{\partial T}{\partial r}\right) = \frac{1}{r^2}\frac{\partial}{\partial r}\left(r^2 k\frac{\partial T}{\partial r}\right) + Q \tag{6}$$

2. The momentum balance in cylinder geometry for a constant-property Newtonian fluid. Only the z-component and the r-component are shown. It is assumed that $v_\phi = 0$ and that v_z and v_r do not change in the ϕ-direction.

$$
\begin{aligned}
&\rho\left(\frac{\partial v_z}{\partial t} + v_r\frac{\partial v_z}{\partial r} + v_z\frac{\partial v_z}{\partial z}\right)\\
&\qquad = \mu\left[\frac{1}{r}\frac{\partial}{\partial r}\left(r\frac{\partial v_z}{\partial r}\right) + \frac{\partial^2 v_z}{\partial z^2}\right] + \rho G_z - \frac{\partial p}{\partial z}\\
&\rho\left(\frac{\partial v_r}{\partial t} + v_r\frac{\partial v_r}{\partial r} + v_z\frac{\partial v_r}{\partial z}\right)\\
&\qquad = \mu\left\{\frac{\partial}{\partial r}\left[\frac{1}{r}\frac{\partial}{\partial r}(v_r r)\right] + \frac{\partial^2 v_r}{\partial z^2}\right\} + \rho G_r - \frac{\partial p}{\partial r}
\end{aligned}
\tag{7}
$$

In the flow problems of the following sections we shall assume that (3) can be solved independently of (1) and (2). This leads to an enormous simplification of the computational work since the three coupled equations in (3) can be solved separately for the components of $\mathbf{v}$ and the solution $\mathbf{v}(\mathbf{x}, t)$ can afterward be inserted into the two coupled equations (1) and (2), which are finally solved for c_A and T. The assumptions implied in this computational scheme are that $\mathbf{v}$ and the parameters ρ and μ of (3) are independent of c_A and T and that the chemical reaction must occur without volume change. These assumptions—except perhaps the temperature independence of ρ and μ—are generally quite reasonable for diluted reacting mixtures.

It is convenient to present a few classical solutions of Navier–Stokes equations since these explicit solutions will be used throughout the text without reference to equation (3).

For laminar flow in a cylinder with inpenetrable walls, v_r and v_ϕ are zero and the axial velocity v_z is determined by the axial pressure gradient and by the fluid friction. Since the fluid is considered incompressible and any volume change due to chemical reaction is neglected, v_z is independent of z. It does change in the r-direction since a dissipation of z-momentum is necessary to maintain the fluid flow in the z-direction and the r-component of the z-momentum vector is a function of $\partial v_z/\partial r$. Consequently for steady state laminar flow of a constant density and viscosity, Newtonian fluid $v_z(r)$ is obtained by solution of the z-component of (3) with

$$v_\phi = v_r = \frac{\partial v_z}{\partial z} = \frac{\partial v_z}{\partial \phi} = G_z = 0$$

$$\mu\frac{1}{r}\frac{d}{dr}\left(r\frac{dv_z}{dr}\right) = \frac{dp}{dz} = \text{constant}$$

With a no-slip condition $v_z = 0$ at the tube wall $r = R$, the solution, which is finite at $r = 0$, is

$$\begin{aligned} v_z(r) &= \left(-\frac{dp}{dz}\right)\frac{R^2}{4\mu}\left(1 - \frac{r^2}{R^2}\right) \\ &= v_{\max}\left(1 - \frac{r^2}{R^2}\right) = 2v_{\mathrm{av}}\left(1 - \frac{r^2}{R^2}\right) \end{aligned} \tag{8}$$

where either the maximum velocity $v_{\max}$ at $r = 0$ or the average velocity v_{av} defined by

$$\begin{aligned} v_{\mathrm{av}} &= \frac{1}{A}\int_A v_z\,dA = \frac{1}{R^2}\int_0^R \left(-\frac{dp}{dz}\right)\frac{R^2}{4\mu}\left(1 - \frac{r^2}{R^2}\right)dr^2 \\ &= \frac{1}{2}\left(-\frac{dp}{dz}\right)\frac{R^2}{4\mu} = \frac{1}{2}v_{\max} \end{aligned}$$

may be used to normalize the velocity profile.

If the fluid is non-Newtonian, the elements τ_{ji} of the shear stress tensor must be used in the viscous terms of (3) rather than the velocity gradients.

For cylinder symmetric flow in the z-direction ($v_r = v_\phi = 0$), we need only the change of z-momentum in the r-direction or, from table 1.2, $(1/r)(d/dr)(r\tau_{rz})$. We shall now insert an empirical expression for τ_{rz} in the third component of (3):

$$\frac{1}{r}\frac{d}{dr}(r\tau_{rz}) = -\frac{dp}{dz} = \text{constant}$$

One example is the *power law fluid* described by the constants m and n [Bird (1960), p. 11]:

$$\tau_{rz} = -m\left|\frac{dv_z}{dr}\right|^{n-1}\frac{dv_z}{dr} \tag{9}$$

In tube flow, dv_z/dr is always negative and the absolute value sign in (9) can be eliminated:

$$\tau_{rz} = m\left|-\frac{dv_z}{dr}\right|^{n-1}\left(-\frac{dv_z}{dr}\right) = m\left(-\frac{dv_z}{dr}\right)^n \tag{10}$$

Now the third equation of (3) becomes

$$m\frac{1}{r}\frac{d}{dr}\left[r\left(-\frac{dv_z}{dr}\right)^n\right] = -\frac{dp}{dz}$$

which is integrated twice with $v_z(r = R) = \left.\dfrac{dv_z}{dr}\right|_{r=0} = 0.$

$$
\begin{aligned}
v_z &= \frac{n}{n+1}\left(\frac{1}{2m}\right)^{1/n}\left|-\frac{dp}{dz}\right|^{1/n} R^{(n+1)/n}\left[1-\left(\frac{r}{R}\right)^{(n+1)/n}\right] \\
&= v_{\max}\left[1-\left(\frac{r}{R}\right)^{(n+1)/n}\right] = \frac{3n+1}{n+1}v_{\text{av}}\left[1-\left(\frac{r}{R}\right)^{(n+1)/n}\right] \qquad (11)\\
&= \left(1+\frac{2}{M}\right)v_{\text{av}}\left[1-\left(\frac{r}{R}\right)^{M}\right]
\end{aligned}
$$

where $v_{\max}$ and v_{av} have the same meaning as in equation (8) and $M = (n+1)/n$. In laminar flow of a pseudo-plastic material ($n < 1$ or $M > 2$) the velocity profile is steeper than the parabolic velocity profile (8) for a fluid with Newtonian properties. A fluid with dilatant behavior is characterized by $n > 1$.

Fluid flow in the turbulent regime is described empirically by a relation similar to (11) but with $M \gg 1$. v_z is independent of r except close to the wall $r = R$. In packed beds we shall assume the velocity profile to be flat since the effect of an r-dependent axial velocity is probably small in comparison with other effects—at least for a large length-to-diameter ratio.

With the velocity profile given by (8), (11), or v_z = constant, we now treat the steady state solution of (1) and (2) for a homogeneous fluid-phase reaction in a tubular reactor.

k and D are assumed to be independent of c_A and T, but not necessarily the same in the r- and z-directions. D and k will usually be empirical quantities describing axial mixing in a packed bed or a mixed solid-fluid heat conductivity. The conditions for $\mathbf{v}$ are those used above to solve the trivial versions of Navier–Stokes equation for unidirectional flow. With these remarks the steady state mass and heat balances for sections 1.2 to 1.4 can be set up.

$$
v_z\frac{\partial c_A}{\partial z} = D^r\frac{1}{r}\frac{\partial}{\partial r}\left(r\frac{\partial c_A}{\partial r}\right) + D^z\frac{\partial^2 c_A}{\partial z^2} - R_A \qquad (12)
$$

$$
\rho c_p v_z\frac{\partial T}{\partial z} = k^r\frac{1}{r}\frac{\partial}{\partial r}\left(r\frac{\partial T}{\partial R}\right) + k^z\frac{\partial^2 T}{\partial z^2} + Q \qquad (13)
$$

where superscripts r and z are used for radial and axial transport properties, respectively. v_z is based on the empty tube velocity whether the bed is packed with solid particles or not and R_A is conversion per unit volume packed bed.

1.2 Steady State Homogeneous Flow Model for a Reacting System

1.2.1. Derivation of a one-dimensional model

Dimensionless independent variables $\zeta = z/L$ and $x = r/R$ are introduced in (12) and (13):

$$\frac{v_z}{v_{av}}\frac{\partial c_A}{\partial \zeta} = \frac{LD^r}{R^2 v_{av}}\frac{1}{x}\frac{\partial}{\partial x}\left(x\frac{\partial c_A}{\partial x}\right) + \frac{D^z}{Lv_{av}}\frac{\partial^2 c_A}{\partial \zeta^2} - \frac{L}{v_{av}}R_A \tag{14}$$

$$\frac{v_z}{v_{av}}\frac{\partial T}{\partial \zeta} = \frac{Lk^r}{R^2 v_{av}\rho c_p}\frac{1}{x}\frac{\partial}{\partial x}\left(x\frac{\partial T}{\partial x}\right) + \frac{k^z}{Lv_{av}\rho c_p}\frac{\partial^2 T}{\partial \zeta^2} + \frac{LQ}{v_{av}\rho c_p} \tag{15}$$

These equations are extremely simple compared to (1) to (3): The momentum balance has been solved independently of the mass and energy balances, the time dependence of c_A and T has been dropped, and the transport properties are assumed constant. Even so, the resulting model consists of two coupled nonlinear partial differential equations (R_A and Q depend nonlinearly on c_A and T); its complete solution is a major numerical enterprise that is beyond the scope of this text.

Simplifications of the two equations are indeed desirable and often possible in practice. Here we shall discuss how these simplifications can be brought about by discarding small terms and by suitable averaging methods that transform (14) and (15) into two coupled nonlinear first-order *ordinary* differential equations (25) and (26). In subsection 1.2.3, the axial dispersion terms on the right-hand sides are reintroduced—but the model still consists of ordinary differential equations. Finally, in subsection 1.2.4, the total set of side conditions of (14) and (15) are discussed and a particularly simple variant of the equations is treated a little further.

In equations (14) and (15) the coefficients of the second-order derivatives with respect to ζ are frequently several orders of magnitude smaller than the coefficients of the first-order derivatives. Consequently these terms can frequently be neglected.

On the other hand, the coefficients of the radial diffusion terms are quite large since L/R is usually $\gg 1$. Thus, except for unusually rapid reactions, one may assume that the radial variation of c_A and T is of minor importance; this indicates that the mean values of c_A and T over the cross section of the reactor may be used rather than the x- and ζ-dependent variables in (14) and (15) if the purpose of the calculations is to compute the conversion of a given feed in a reactor of given length L.

Integrating (14) and (15) over the cross section, neglecting the axial diffusion terms and imposing a condition of zero flux across the cylinder axis, yields

$$\int_0^1 \frac{v_z}{v_{av}} \frac{\partial c_A}{\partial \zeta} dx^2 = 2\frac{LD^r}{R^2 v_{av}}\left(\frac{\partial c_A}{\partial x}\right)_{x=1} - 2\int_0^1 \frac{L}{v_{av}} R_A x\, dx \tag{16}$$

$$\int_0^1 \frac{v_z}{v_{av}} \frac{\partial T}{\partial \zeta} dx^2 = 2\frac{Lk^r}{R^2 v_{av}\rho c_p}\left(\frac{\partial T}{\partial x}\right)_{x=1} + 2\int_0^1 \frac{L}{v_{av}\rho c_p} Q x\, dx \tag{17}$$

For an impermeable wall the concentration gradient at $x = 1$ is zero, while by the film theory the temperature gradient is proportional to the difference between the reactor wall temperature T_W and the fluid temperature $T_{x=1}$ just inside the wall.

$$\frac{k^r}{R}\left(\frac{\partial T}{\partial x}\right)_{x=1} = U(T_W - T_{x=1}) \tag{18}$$

Inserting in (16) and (17) yields

$$2\int_0^1 \frac{v_z}{v_{av}} \frac{\partial c_A}{\partial \zeta} x\, dx = -2\int_0^1 \frac{L}{v_{av}} R_A x\, dx \tag{19}$$

$$2\int_0^1 \frac{v_z}{v_{av}} \frac{\partial T}{\partial \zeta} x\, dx = 2\frac{LU}{R v_{av}\rho c_p}(T_W - T_{x=1}) + 2\int_0^1 \frac{L}{v_{av}\rho c_p} Q x\, dx \tag{20}$$

The average concentration and temperature in the cross section of the reactor are defined by

$$\bar{c}_A v_{av} = \int_0^1 2 v_z c_A(\zeta, x) x\, dx \tag{21}$$

$$\bar{T} v_{av} = \int_0^1 2 v_z T(\zeta, x) x\, dx \tag{22}$$

and for small radial concentration and temperature gradients the right-hand side integrals of (19) and (20) are well approximated by

$$2\int_0^1 \frac{L}{v_{av}} R_A x\, dx = \frac{L}{v_{av}} R_A(\bar{c}_A, \bar{T}) \tag{23}$$

$$2\int_0^1 \frac{L}{v_{av}\rho c_p} Q x\, dx = \frac{L}{v_{av}\rho c_p} Q(\bar{c}_A, \bar{T}) \tag{24}$$

Finally we would like to substitute $T_W - T_{x=1}$ in (20) by a corresponding temperature driving force in terms of the average temperature $\bar{T}$ of (22). Now $|T_W - \bar{T}|$ is necessarily greater than $|T_W + T_{x=1}|$ and in order to compensate for this a modified heat transfer coefficient $\bar{U}$ (which is smaller than U) is introduced at the same time.

Our model (14) and (15) now reduces to a one-dimensional form in terms of the variables $\bar{c}_A$ and $\bar{T}$ defined by (21) and (22):

$$\frac{d\bar{c}_A}{d\zeta} = -\frac{L}{v_{av}} R_A(\bar{c}_A, T_A) \tag{25}$$

$$\frac{d\bar{T}}{d\zeta} = 2\frac{L\bar{U}}{Rv_{av}\rho c_p}(T_W - \bar{T}) + \frac{L}{v_{av}\rho c_p} Q(\bar{c}_A, \bar{T}) \tag{26}$$

A suitable choice of $\bar{U}$ is obtained by

$$\frac{1}{\bar{U}} = \frac{1}{U} + \alpha\frac{R}{k'} \tag{27}$$

where the best choice of α depends on the velocity profile as discussed in chapter 6. For a flat velocity profile, $\alpha = \frac{1}{4}$ is best.

Let us consider (25) and (26) for an nth-order irreversible reaction where the rate constant k is given by an Arrhenius form:

$$R_A = kc_A^n = a \exp\left(-\frac{E}{R_G T}\right) c_A^n \tag{28}$$

and the heat production term Q arises due to the heat of reaction

$$Q = (-\Delta H) R_A \tag{29}$$

In (28), a is a temperature-independent constant, E is the activation energy, and R_G the universal gas constant.

Let the reactor inlet temperature and concentration be $\bar{T} = T_0$ and $\bar{c}_A = c_0$, respectively. R_A can be rewritten into a dimensionless form:

$$\begin{aligned} R_A &= a \exp\left(-\frac{E}{R_G T_0}\right) \exp\left[\frac{E}{R_G T_0}\left(1 - \frac{T_0}{T}\right)\right] c_0^n y^n \\ &= k_0 c_0^n y^n \exp\left[\gamma\left(1 - \frac{1}{\theta}\right)\right] \end{aligned} \tag{30}$$

where $\theta = \bar{T}/T_0$, $y = \bar{c}_A/c_0$, $k(T_0) = k_0$, and $\gamma = E/R_G T_0$. Equations (25) and (26) now become

$$\frac{dy}{d\zeta} = -\,\mathrm{Da}\, y^n \exp\left[\gamma\left(1 - \frac{1}{\theta}\right)\right] \tag{31}$$

$$\frac{d\theta}{d\zeta} = \beta \,\text{Da}\, y^n \exp\left[\gamma\left(1 - \frac{1}{\theta}\right)\right] - H_W(\theta - \theta_W) \tag{32}$$

$$\text{Da} = \frac{Lk_o}{v_{av}} c_0^{n-1}$$

$$\beta = \frac{c_0(-\Delta H)}{\rho c_p T_0}$$

$$H_W = \frac{2\bar{U}}{R} \frac{L}{c_p \rho v_{av}}$$

The Damköhler number, Da, is a measure of rate of reaction at inlet conditions, γ is a dimensionless activation energy, β is the adiabatic temperature rise relative to inlet temperature T_0 if all the reactant is consumed, and H_W is a dimensionless heat transfer coefficient (the number of heat transfer units in the reactor).

Thus the resulting reactor model is a set of two nonlinear first-order ordinary differential equations that are much easier to solve than (14) and (15). The major model simplification lies in the averaging process (16) and (17), which can also be used if the axial diffusion (or dispersion) terms of (14) and (15) are retained.

In certain cases (31) and (32) can be solved in closed form. Thus for an adiabatic reactor ($\bar{U} = 0$), multiplication of (31) by β and addition to (32) yields

$$\frac{d}{d\zeta}(\theta + \beta y) = 0$$

or $\theta + \beta y$ is equal to a constant, which is determined from the inlet conditions $\theta = y = 1$ to be $1 + \beta$.

Now θ can be eliminated from (31):

$$\frac{dy}{d\zeta} = -\text{Da}\, y^n \exp\left[\frac{\gamma\beta(1-y)}{1+\beta(1-y)}\right] = -F(y)$$

or

$$\zeta = \int_y^1 \frac{du}{F(u)} \tag{33}$$

For each value of y the integral (33) must be found by a numerical method (except for $\gamma\beta = 0$) but tabulation of ζ as a function of y is nevertheless a trivial problem. If y is desired at a fixed grid of ζ-values, (33) is interpreted as an algebraic equation in y. Inverse interpolation in a table of ζ versus y may give sufficiently accurate values for y at the grid points.

The final step in the solution of (31) and (32) for an adiabatic reactor is to determine the temperature profile from the algebraic equation $\theta = 1 + \beta[1 - y(\zeta)]$, which, of course, does not present any problem once $y(\zeta)$ is known.

It is of some interest to note that an analytical solution of (31) and (32) is possible also for a nonadiabatic reactor if $E \cong 0$, i.e., if the rate constant is temperature independent as is the case in the high temperature SO_2-converter beds. After solving the mass balance and inserting $y(\zeta)$ in the energy balance, this becomes a first-order inhomogeneous equation with constant coefficients. A "hot spot," i.e., a maximum in $\theta(\zeta)$, may well occur also when $E = 0$. After the hot spot, all reactant has been burnt off and the reactor acts as a simple heat exchanger. Exercise 1.1 treats a reactor where this simple solution is applicable.

1.2.2. Limiting solutions to the full fluid-phase model

The maximum effect of the simplifications leading from (14) and (15) to (31) and (32) can be estimated by a study of two limiting cases—either infinite or zero radial diffusivity. This will be done for an isothermal first-order reaction where (14) and (15) as well as (31) and (32) are particularly simple to solve.

The full model for laminar Newtonian flow with first-order isothermal reaction in an empty tube is

$$2(1 - x^2)\frac{\partial y}{\partial \zeta} = \frac{D'L}{R^2 v_{av}} \frac{1}{x} \frac{\partial}{\partial x}\left(x\frac{\partial y}{\partial x}\right) - \text{Da}\, y \tag{34}$$

when the axial diffusion is neglected. The side conditions of (34) are

$$y = \frac{c_A}{c_0} = 1 \quad \text{at } \zeta = 0$$

$$\frac{\partial y}{\partial x} = 0 \quad \text{at } x = 0 \text{ and } x = 1 \quad \text{for } \zeta \geq 0$$

Two limiting cases are considered.

1. $\dfrac{D'L}{R^2 v_{av}} \to \infty$ or $y(\zeta, x) = y(\zeta)$ for all ζ

   With no radial gradients the solution of (34) and (31) are identical and the average outlet concentration

$$\bar{y}(\zeta = 1) = \int_0^1 4(1 - x^2)y(1, x)x\, dx$$

is found by solution of (31) with $\theta = 1$ and $n = 1$:

$$\bar{y}(\zeta = 1) = \exp(-\text{Da})$$

2. In the absence of radial diffusion ($D' = 0$), a fluid element retains its identity throughout its passage of the reactor and

$$\bar{y}(\zeta = 1) = 4\int_0^1 x(1 - x^2)\exp\left(-\frac{\text{Da}}{2(1 - x^2)}\right) dx$$

The average concentration at the reactor outlet obtained from (34) for a finite D' lies between the solution of these limiting cases that are compared in table 1.3 for various values of the Damköhler number.

TABLE 1.3
AVERAGE OUTLET CONCENTRATION FOR INFINITE AND ZERO RADIAL DIFFUSIVITY

Da	$D' \to \infty$	$D' = 0$
0.1	0.905	0.910
0.2	0.819	0.832
0.5	0.607	0.649
1.0	0.368	0.443
2.0	0.135	0.219

It is seen that the model simplification is without practical consequences when the reaction is slow since both limiting models give the same result for small values of Da. It is shown at the end of the next subsection that a finite radial dispersion can be treated by means of an equivalent axial dispersion term; even though this introduces a second-order derivative in (31), the resulting model is still considerably simpler than the partial differential equation (34).

1.2.3 Axial dispersion of mass and energy

The averaging process of subsection 1.2.1 can be performed even though the axial diffusion terms are retained. One obtains the following set of second-order nonlinear equations:

$$\frac{1}{\text{Pe}_M}\frac{d^2y}{d\zeta^2} - \frac{dy}{d\zeta} - \text{Da}\, y^n \exp\left[\gamma\left(1 - \frac{1}{\theta}\right)\right] = 0 \tag{35}$$

$$\frac{1}{\text{Pe}_H}\frac{d^2\theta}{d\zeta^2} - \frac{d\theta}{d\zeta} + \beta\, \text{Da}\, y^n \exp\left[\gamma\left(1 - \frac{1}{\theta}\right)\right] - H_W(\theta - \theta_W) = 0 \tag{36}$$

The two dimensionless groups Pe_M and Pe_H are axial Peclet numbers defined by

$$\mathrm{Pe}_M = \frac{v_{\mathrm{av}}L}{D} \quad \text{and} \quad \mathrm{Pe}_H = \frac{v_{\mathrm{av}}L\rho c_p}{k}$$

The transport coefficients D and k which enter into the Peclet numbers are empirical quantities which bear no relation to the molecular properties. Experiments for a packed bed [Gunn (1971)] show that Pe_M is between 0.8 and 2 times the bed length per particle diameter ratio L/d_p for various packings. Pe_H is smaller, but of the same order of magnitude. Since L/d_p is very large in most industrial reactors, the numerical coefficient of the second-order terms are small. Since the second derivatives are of the same order of magnitude as the first derivatives, it is certainly reasonable to neglect the axial dispersion terms in a steady state calculation as we did in subsection 1.2.1.

Models (35) and (36) are interesting, however, from a mathematical point of view; as we shall see at the end of the subsection, inclusion of an axial diffusion term may lead to a useful interpretation of the often far more important radial diffusion terms.

Solution of the set of two second-order differential equations (35) and (36) requires two side conditions for each dependent variable.

The correct specification of these side conditions is by no means an easy task and it might be helpful to consider the slightly simplified model where $H_W = 0$. Here $y(\zeta)$ is a monotonically decreasing function of ζ and $\theta(\zeta)$ is an increasing function when the reaction is exothermic. Mass or energy is carried toward $\zeta = 1$ by convective transport. The diffusion mechanism acts in the direction of decreasing y and θ and consequently by this mechanism energy is carried toward $\zeta = 0$ while mass is carried toward $\zeta = 1$.

Both effects tend to decrease y and increase θ near the reactor entrance and it would certainly be incorrect to take the side condition $y = \theta = 1$ at $\zeta = 0$ from section 1.2 since obviously $y < 1$ and $\theta > 1$ at $\zeta = 0$.

Bischoff (1961) gives a very clear presentation of the side condition problem, and his procedure will be briefly reviewed for the mass balance since the resulting side conditions apply to the present example as well as to a more complicated example in the next subsection.

Assume that the reaction section $0 \le \zeta \le 1$ is preceded by an entrance section $-\infty < \zeta < 0$ and followed by an exit section $1 < \zeta < \infty$. The Peclet numbers are, respectively, Pe_- and Pe_+ in these sections, and no chemical reaction occurs except in $0 \le \zeta \le 1$—a situation that may be visualized by a radiation-catalyzed chemical reaction where shielding prevents reaction for $\zeta < 0$ and $\zeta > 1$.

Mass balances for the two outer sections are

$$\frac{1}{\text{Pe}_-}\frac{d^2y}{d\zeta^2} - \frac{dy}{d\zeta} = 0 \quad \text{for } \zeta < 0 \quad \text{with } y = 1 \text{ } at\, \zeta \to -\infty$$

$$\frac{1}{\text{Pe}_+}\frac{d^2y}{d\zeta^2} - \frac{dy}{d\zeta} = 0 \quad \text{for } \zeta > 1 \quad \text{with } y \text{ finite at } \zeta \to \infty$$

The solutions of these equations are

$$\begin{aligned} y &= 1 + A_1 \exp(\text{Pe}_- \zeta) \quad \text{for } \zeta < 0 \\ y &= A_2 + A_3 \exp(\text{Pe}_+ \zeta) \quad \text{for } \zeta > 1 \end{aligned} \tag{37}$$

where $A_3 = 0$ to satisfy the condition at $\zeta \to \infty$.

The solutions for the outer sections tell us that y decreases from 1 to $1 + A_1$ when ζ increases from $-\infty$ to 0 and that y is constant ($= A_2$) for $\zeta > 1$.

Mass balances for differential volumes around $\zeta = 0$ and $\zeta = 1$ are

$$\begin{aligned} y_- - \frac{1}{\text{Pe}_-}\frac{dy_-}{d\zeta} &= y_+ - \frac{1}{\text{Pe}_M}\frac{dy_+}{d\zeta} \\ Y_- - \frac{1}{\text{Pe}_M}\frac{dY_-}{d\zeta} &= Y_+ - \frac{1}{\text{Pe}_+}\frac{dY_+}{d\zeta} \end{aligned} \tag{38}$$

where y_+ and Y_- are values of y inside the reactor, $y_- = 1 + A_1$, $Y_+ = A_2$, and the derivatives in the outer sections are found by differentiation of (37):

$$\frac{dy_-}{d\zeta} = \left.\frac{dy}{d\zeta}\right|_{\zeta=0_-} = A_1\,\text{Pe}_- \quad \text{and} \quad \frac{dY_+}{d\zeta} = \left.\frac{dy}{d\zeta}\right|_{\zeta=1_+} = 0$$

Inserting these results into (38) one obtains

$$1 = y_+ - \frac{1}{\text{Pe}_M}\frac{dy_+}{d\zeta} \tag{39}$$

$$Y_- - \frac{1}{\text{Pe}_M}\frac{dY_-}{d\zeta} = A_2 \tag{40}$$

We finally apply that the concentration profile is continuous across the boundaries at $\zeta = 0$ and $\zeta = 1$. Thus $y_+ = y_- = 1 + A_1$ and $Y_- = Y_+ = A_2$ whereby (40) is reduced to

$$\frac{1}{\text{Pe}_M}\frac{dY_-}{d\zeta} = 0 \quad \text{or} \quad \frac{dY_-}{d\zeta} = 0 \quad \text{for any finite Pe}_M \tag{41}$$

Equations (39) and (41) are the required set of boundary values for (35).

The boundary conditions are independent of the rate expression in (35) and also independent of the specific values of Pe_+ and Pe_-, i.e., on the conditions in the outer sections.

An identical derivation for the heat balance shows that the side conditions for (36) are completely analogous to (39) to (41) with (θ, Pe_H) instead of (y, Pe_M).

If $\mathrm{Pe}_- \to \infty$, y very rapidly attains the value 1 for $\zeta < 0$ – which is the reason why (39) is often described as implying a discontinuity of y (or θ) at this point. The dependent variables are both continuous of course but the first derivatives are discontinuous except when $\mathrm{Pe}_- = \mathrm{Pe}_M$, i.e., when the diffusivity is the same in the entrance sector and in the reactor.

It is also noted that the gradient at $\zeta = 1$ is zero only when Pe_M is finite and in fact nothing may be concluded in advance about the gradient at $\zeta = 1$ when $\mathrm{Pe}_M \to \infty$. This is in full agreement with our expectations: (39) degenerates for $\mathrm{Pe}_M \to \infty$ to $y=1$ at $\zeta = 0$ and (35) degenerates to the first-order equation (31) for which only one side condition is necessary.

It is finally noted that direct integration of (35) and (36) for $H_W = 0$ and insertion of the side conditions derived above yields

$$1 - y(\zeta = 1) = \mathrm{Da} \int_0^1 \exp\left[\gamma\left(1 - \frac{1}{\theta}\right)\right] y^n \, d\zeta$$

$$\theta(\zeta = 1) - 1 = \beta \, \mathrm{Da} \int_0^1 \exp\left[\gamma\left(1 - \frac{1}{\theta}\right)\right] y^n \, d\zeta$$

or

$$\theta = 1 + \beta(1 - y) \quad \text{at } \zeta = 1$$

Furthermore if Pe_M and Pe_H both approach zero, the model degenerates to the adiabatic stirred tank model for which θ and y are constant and equal to their exit values $\theta(\zeta = 1)$ and $y(\zeta = 1)$.

Consequently the model with side conditions (39) to (41) and $H_W = 0$ correctly reduces to the adiabatic stirred tank model in the limiting case $(\mathrm{Pe}_M, \mathrm{Pe}_H) = 0$:

$$\begin{aligned} 1 - y &= \mathrm{Da} \exp\left[\gamma\left(1 - \frac{1}{\theta}\right)\right] y^n \\ \theta &= 1 + \beta(1 - y) \end{aligned} \tag{42}$$

Axial mass diffusion in an isothermal reactor implies that $y(\zeta = 0) < 1$ and that the concentration gradient is less steep than in an ideal plug flow reactor. The plug flow reactor gives a higher exit conversion than any reactor with axial dispersion and $y(\zeta)$ must cross the plug flow profile exactly one time in $0 < \zeta < 1$.

The nonisothermal reactor might behave quite differently since the transport of energy toward $\zeta = 0$ greatly enhances the chemical reaction in the inlet section of the reactor and the conversion may in fact become higher than in a plug flow reactor.

It is well known that the limiting case (42) of an adiabatic stirred tank reactor may exhibit three steady states even for first-order reactions. The other limiting case (31) and (32) is an initial value problem with no singularities in the derivatives and it has one and only one solution whatever the values of the parameters.

Consequently one suspects that for a small value of the Peclet number and large enough values of the heat sensitivity parameters β and γ, several solutions of the adiabatic reactor model with axial mixing may exist in a certain range of Da-values.

Over the last 10 years numerous articles from chemical engineering literature have treated the two equations (35) and (36) and many interesting phenomena of multiple solutions have been found also for $H_W = 0$. A short list of references to some interesting papers on this subject is given at the end of the chapter.

In table 1.3 we considered two limiting solutions of the partial differential equation (34), which describes first-order isothermal reaction for laminar flow of the reactant in an empty tube. The case $D^r \to \infty$ corresponds to a plug flow reactor while the case $D^r = 0$ corresponds to a strong apparent axial dispersion of reactant: The fluid that reaches $\zeta = 1$ along an x-coordinate close to 1 has had an extremely long residence time in the reactor while the fluid in the center of the tube reaches the outlet after $L/2v_{av}$ time units. The result of the uneven distribution of residence times is clearly a much smaller conversion for large values of Da.

The interpretation of a less than perfect radial mixing in terms of an axial dispersion coefficient was originally given by Taylor (1953). For flow in empty tubes a one-dimensional model approximation for a system with laminar flow requires the use of an "effective" axial diffusivity given by

$$D^z_{\text{eff}} \simeq D^z + \frac{1}{48}\frac{R^2 v_{av}{}^2}{D^r} \quad \text{or} \quad \frac{1}{\text{Pe}_{M,\text{eff}}} \simeq \frac{1}{\text{Pe}_M} + \frac{1}{48}\frac{v_{av}R^2}{D^r L} \tag{43}$$

If we consider the effect of the last term alone, we may compare the average concentration at the reactor outlet as calculated by (34) and by (35) for various values of the dimensionless group $D^r L/v_{av}R^2$. Table 1.4 shows the result of a calculation by the complete model (34) and by the "improved" averaged model (35) for Da = 1.

TABLE 1.4
IMPROVED ONE-DIMENSIONAL REACTOR MODEL

$D^r L/v_{av} R^2$	$\bar{y}_{\zeta=1}$ (full model)	$\bar{y}_{\zeta=1}$ [by (35) and (43)]
0.1	0.4038	0.4179
0.2	0.3931	0.3982
0.5	0.3809	0.3818
1	0.3750	0.3752

It is seen that the Taylor dispersion concept permits a much better model approximation than the crude averaging process of subsection 1.2.1 that leads to either $\bar{y}_{\zeta=1} = 0.3679$ or to $\bar{y}_{\zeta=1} = 0.443$ for all entries of table 1.4 depending on whether complete mixing or no mixing at all is assumed.

Wicke (1975) reviews the application of Taylor dispersion for packed beds where numerical coefficients are selected that are different from those of (43).

1.2.4 Boundary conditions for two-dimensional model—the extended Graetz problem

After having discussed some important and (from a practical point of view) very reasonable simplifications of the complete model for the fluid phase, we shall briefly return to (14) and (15) and review the boundary conditions of these equations.

For laminar flow in an empty tube, D and k can be taken to be equal in the radial and axial directions and they are molecular transport properties. The dimensionless model is

$$v_z(x)\frac{\partial y}{\partial \zeta} = \frac{1}{\mathrm{Pe}_M}\left[\frac{\partial^2 y}{\partial \zeta^2} + \frac{L^2}{R^2}\frac{1}{x}\frac{\partial}{\partial x}\left(x\frac{\partial y}{\partial x}\right)\right] - \mathrm{Da}\, y^n \exp\left[\gamma\left(1 - \frac{1}{\theta}\right)\right] = 0$$

$$v_z(x)\frac{\partial \theta}{\partial \zeta} = \frac{1}{\mathrm{Pe}_H}\left[\frac{\partial^2 \theta}{\partial \zeta^2} + \frac{L^2}{R^2}\frac{1}{x}\frac{\partial}{\partial x}\left(x\frac{\partial \theta}{\partial x}\right)\right] + \beta\, \mathrm{Da}\, y^n \exp\left[\gamma\left(1 - \frac{1}{\theta}\right)\right] = 0 \tag{44}$$

The assumption of $D^z = D^r$ and $k^z = k^r$ is not correct for flow in a packed bed and the resulting model will become considerably more complex than (44).

Boundary conditions at $\zeta = 0$ and at $\zeta = 1$ are taken from the discussion in subsection 1.2.3.

$$\begin{aligned} \zeta = 0: \quad & y - \frac{1}{\mathrm{Pe}_M}\frac{\partial y}{\partial \zeta} = 1 \quad \text{and} \quad \theta - \frac{1}{\mathrm{Pe}_H}\frac{\partial \theta}{\partial \zeta} = 1 \\ \zeta = 1: \quad & \frac{\partial y}{\partial \zeta} = \frac{\partial \theta}{\partial \zeta} = 0 \end{aligned} \tag{45}$$

Symmetry across the cylinder axis gives

$$\frac{\partial y}{\partial x} = \frac{\partial \theta}{\partial x} = 0 \quad \text{at } x = 0 \tag{46}$$

The tube wall is assumed to be impenetrable for the reactant

$$\frac{\partial y}{\partial x} = 0 \quad \text{at } x = 1 \tag{47a}$$

The side conditions at $x = 1$ may be much more complicated for the energy balance. Assume that the tube wall is insulated from $\zeta = 0$ to $\zeta = \zeta_1$ and that heat transfer to an exterior medium of temperature $T_W(\zeta)$ is admitted for $\zeta_1 \le \zeta \le 1$:

$$\begin{aligned} \frac{\partial \theta}{\partial x} &= 0 \quad \text{at } x = 1 \quad \text{for } 0 \le \zeta < \zeta_1 \\ \frac{\partial \theta}{\partial x} &= -\frac{UR}{k}(\theta - \theta_W) \quad \text{for } \zeta_1 \le \zeta \le 1 \end{aligned} \tag{47b}$$

θ_W may be a constant or it may vary with ζ. For a given mass flow of coolant and a given inlet temperature $T_W(\zeta = 0)$ or $T_W(\zeta = 1)$, a total heat balance over part of the reactor plus heat exchanger gives an ordinary differential equation to determine $T_W(\zeta)$ simultaneously with the solution of the partial differential equation model for the reactor. This type of side condition is discussed in chapter 4 for a considerably simpler model.

It should be noted that (44) with its side conditions (45) to (47) is a boundary value problem in ζ as well as in x. A considerable simplification is obtained if the axial diffusion terms, which are often quite insignificant, are neglected; the equations then become of first order in ζ and can be solved by a marching technique from $\zeta = 0$. The somewhat complicated boundary condition (47b) at $x = 1$ presents no difficulty, the adiabatic and the nonadiabatic reactor differ only in the side condition for the energy balance and not in the structure of the differential equations.

It may finally be observed that the two equations (44) as well as the side conditions can be decoupled if $\mathrm{Pe}_M = \mathrm{Pe}_H$. This leads to another significant simplification; however, this is not based upon a true physical description of the system since Pe_M is usually considerably greater than Pe_H.

We shall repeatedly have occasion to discuss the solution of the linear partial differential equation that appears when the heat generation term is dropped from the heat balance (15).

$$2v_{\mathrm{av}}(1 - x^2)\frac{\partial T}{\partial z} = \frac{k}{R^2 \rho c_p}\left[\frac{1}{x}\frac{\partial}{\partial x}\left(x\frac{\partial T}{\partial x}\right) + R^2 \frac{\partial^2 T}{\partial z^2}\right] \tag{48}$$

A detailed analysis of this problem may give some insight into the difficulties of solving the complete model (14) and (15) without neglecting the axial diffusion terms.

As a reference for the numerical solution of (48), which will be started in chapter 4 and continued in chapter 9, we shall here give an outline of the dimensionless quantities that can be derived from the equation.

The boundary conditions are imposed at $z \to -\infty$ and at $z \to \infty$ and we shall assume that the tube is insulated from $z = -\infty$ to $z = 0$ and that the wall temperature is constant, T_W for $z \geq 0$. The Newtonian fluid enters the tube at $z \to -\infty$ with $T = T_0$, flows in fully developed laminar flow through the tube, and attains the temperature $T = T_W$ when $z \to \infty$. The absence of a natural scale for the z-direction permits a small simplification of the model in comparison with (44) if the dimensionless variables are chosen as follows:

$$\zeta = \frac{kz}{2\rho c_p v_{\mathrm{av}} R^2} = \frac{z}{\mathrm{Pe}\, R}$$

$$x = \frac{r}{R} \quad \text{and} \quad \theta = \frac{T - T_W}{T_o - T_W}$$

$$(1 - x^2)\frac{\partial \theta}{\partial \zeta} = \frac{1}{x}\frac{\partial}{\partial x}\left(x\frac{\partial \theta}{\partial x}\right) + \frac{1}{\mathrm{Pe}^2}\frac{\partial^2 \theta}{\partial \zeta^2} \tag{49}$$

$$\theta = 1 \quad \text{for } \zeta \to -\infty \quad \text{and } \theta \text{ is finite } (\to 0) \text{ for } \zeta \to \infty \tag{50}$$

$$\frac{\partial \theta}{\partial x} = 0 \quad \text{at } x = 0 \qquad \text{for all } \zeta \tag{51}$$

$$\frac{\partial \theta}{\partial x} = 0 \quad \text{at } x = 1 \qquad \zeta < 0 \tag{52a}$$

$$\theta = 0 \quad \text{at } x = 1 \qquad \zeta \geq 0 \tag{52b}$$

The linear model (49) with boundary conditions (50) to (52) may be called the *extended Graetz problem* since (in 1885) Graetz was the first to solve the model with negligible axial conduction, $\text{Pe} \to \infty$. The equation is a representative model for a large number of systems such as tracer dispersion and gas absorption; this explains its frequent appearance in the chemical engineering literature.

The quantity of primary interest is the Nusselt number for $\zeta \geq 0$; it is defined in the usual way by the first relation in (53):

$$\begin{aligned} \text{Nu}(\zeta) &= \frac{2Rh}{k} = -\frac{2(\partial T/\partial x)|_{x=1}}{\bar{T} - T_W} = -\frac{2(\partial \theta/\partial x)|_{x=1}}{\bar{\theta}} \\ &= \frac{-\frac{1}{2}(d\bar{\theta}/d\zeta) + (2/\text{Pe}^2)(d^2/d\zeta^2)\int_0^1 x\theta\, dx}{\bar{\theta}} \end{aligned} \tag{53}$$

h is a rigorously defined average heat transfer coefficient based on a driving force $\bar{T} - T_W$:

$$\begin{aligned} \bar{T} &= \frac{1}{v_{\text{av}}A} A \int v_z T\, dA \\ \bar{\theta} &= \frac{\bar{T} - T_W}{T_0 - T_W} = 4\int_0^1 x(1 - x^2)\theta\, dx \end{aligned} \tag{54}$$

The last relation in (53) is found by integration of the energy balance (49) over the cross section. The boundary condition at $x = 0$ and the definition of $\bar{\theta}$ are inserted and the numerator of the last expression appears.

For $\text{Pe} \to \infty$ a very convenient expression for $\text{Nu}(\zeta)$ is obtained:

$$\text{Nu}(\zeta) = -\frac{(d\bar{\theta}/d\zeta)}{2\bar{\theta}} \tag{55}$$

1.3 Steady State and Transient Models for Solids

1.3.1 Catalyst pellet model

Whenever we have referred to a packed bed in the previous section, we might have thought of an ion exchange column, a dehumidifier, a heat recuperator, a fixed bed reactor, or a number of other important chemical units. The solid component of the bed is, of course, an active partner in the process that we wish to simulate but mass and energy balances have been written in section 1.2 for the fluid phase alone, as if there was no coupling between the phases. It has implicitly been assumed that the temperature and concentration at any point in the solid at tubular position (r, z) is equal to the temperature and concentration in the fluid at the same position. This is certainly not true: The concentration of adsorbent is lower in the fluid phase inside the pore system of the adsorbent than in surrounding bulk fluid phase—otherwise the fluid phase would not be purified during its passage of the adsorption column. In steady state operation the temperature on the surface of a catalyst particle is higher than in the fluid at the same tubular position when the reaction is exothermic, and the external surface is again cooler than the interior of the particle since the major part of the conversion takes place on the large interior surface of the catalyst pellet.

In this section we shall review the most common mass and heat transfer models for steady state and transient behavior of porous particles. This is a necessary preliminary step if the reactor model of the previous section is to be refined—and the arguments for this refinement seem pretty strong. It is, however, important to strike a balance between degree of refinement and computational work. Model (44) for the fluid phase is quite complicated even without axial dispersion, and the occurrence of a complicated coupling mechanism between fluid phase and solid phase may push the model beyond the capability of even a large computer—especially for transient studies or in steady state optimization of a reactor operation.

We shall visualize the catalyst particle as a porous solid with twisting pores of different diameter and length, with crevices, dead-end pores, and perhaps a micropore system imbedded in a system of larger pores. These large pores serve as main passages for the fluid reactant that penetrates the solid and reacts on the pore walls.

The heat of reaction $Q = (-\Delta H)R_A$—which may be positive or negative—is conducted through the pore system and the solid matrix to the pellet surface and from there into the surrounding bulk fluid phase.

Formally the mass and energy balances for pure diffusional flow of one reactant in a large surplus of inert are derived from (1) and (2) when the convective terms are dropped.

$$\begin{aligned} \varepsilon\frac{\partial c_A}{\partial t} &= \frac{D_e}{R^2}\nabla^2 c_A - R_A \\ \rho c_p\frac{\partial T}{\partial t} &= \frac{k_e}{R^2}\nabla^2 T + Q \end{aligned} \tag{56}$$

The two transport coefficients D_e and k_e, which are considered to be independent of spatial position in the pellet (either directly or implicitly through c_A and T), are composite properties that reflect the different transport mechanisms inside the complex pore structure. Experimental values for D_e and k_e are found in Satterfield (1970) pp. 65 and 171.

∇^2 is the Laplacian operator for the solid phase. It may contain one, two, or at most three spatial coordinates in the pellet. For one-dimensional diffusion,

$$\nabla^2 = \frac{1}{x^s}\frac{\partial}{\partial x}\left(x^s\frac{\partial}{\partial x}\right) \tag{57}$$

where $s = 0, 1$, and 2 for planar, cylindrical, and spherical geometry, respectively, and x is the ratio of the distance from the center measured relative to the pellet radius or (for slabs) the half thickness of the pellet.

The accumulation term on the left-hand side of the mass balance contains the pellet porosity ε since it is assumed that the accumulation occurs in the gas phase only. Correspondingly, ρ and c_p are pellet density and heat capacity, respectively, and not properties of the solid matrix alone.

A number of other physical processes of considerable interest to chemical engineering, e.g., leaching of solids, adsorption of vapors on solids, and thermal treatment of solids, are described by one or both equations in (56). The detailed treatment that we give these equations in several chapters of this text is well justified by their importance to engineering design and also by the importance of linear variants of the equations in theoretical numerical analysis.

The production terms $-R_A$ and Q are, of course, not always caused by a chemical reaction. In a physical adsorption process, R_A is the rate of increase of adsorbed material on the solid and Q is the heat of adsorption per unit adsorbed material. The physical adsorption may be described as an instantaneous reaction—the immobilized material on the solid surface is always in equilibrium with material in the fluid phase at the same position.

For one spatial variable x the side conditions at $x = 0$ and $x = 1$ are

$$\frac{\partial c_A}{\partial x} = \frac{\partial T}{\partial x} = 0 \quad \text{at } x = 0 \tag{58}$$

$$\left.\begin{aligned} \frac{\partial c_A}{\partial x} &= \frac{Rh_M}{D_e}[c_b - c_A(x = 1)] \\ \frac{\partial T}{\partial x} &= \frac{Rh}{k_e}[T_b - T(x = 1)] \end{aligned}\right\} \quad \text{at } x = 1 \tag{59}$$

T_b and c_b are properties of the bulk fluid phase. They are generally functions of position in the reactor and of time but in the present section we shall regard them as known quantities. h_M and h are mass and heat transfer coefficients for the film that surrounds the pellets. The two dimensionless groups

$$\mathrm{Bi}_M = \frac{Rh_M}{D_e} \tag{60}$$

and

$$\mathrm{Bi} = \frac{Rh}{k_e} \tag{61}$$

are called the Biot numbers for mass and heat transfer, respectively.

The pellet model is seen to be a set of coupled nonlinear partial differential equations, which are again coupled to the fluid phase mass and energy balances through the boundary conditions (59). The fluid phase balances are themselves coupled partial differential equations with time and one or two spatial coordinates as independent variables.

The solution of the complete set of balances for the coupled phases is a considerable task even in the steady state case, and development of a simplified pellet model is certainly warranted.

Our objective is (1) to obtain a solution of the pellet model for the steady state problem, and (2) to discuss model simplifications for the transient calculations.

One dimensionless form of (56) is

$$\frac{\partial y}{\partial \tau} = \nabla^2 y - \Phi^2 \exp\left[\gamma\left(1 - \frac{1}{\theta}\right)\right] y^n \tag{62}$$

$$\mathrm{Le}\frac{\partial \theta}{\partial \tau} = \nabla^2 \theta + \beta\Phi^2 \exp\left[\gamma\left(1 - \frac{1}{\theta}\right)\right] y^n \tag{63}$$

for an nth-order irreversible reaction.

$$y = \frac{c_A}{c_0} \quad \text{and} \quad T = \frac{T}{T_0}$$

where c_0 is a reference concentration and T_0 a reference temperature. τ is a dimensionless time given by

$$\tau = \frac{D_e}{R^2 \varepsilon} t \tag{64}$$

The dimensionless groups are the Thiele modulus Φ, the Lewis number Le, β, and γ; β and γ are the dimensionless heat generation and activation energy, respectively.

$$\begin{aligned}
\Phi^2 &= \frac{R^2}{D_e}\left[a \exp\left(-\frac{E}{R_G T_0}\right)\right] c_0^{n-1} \\
\text{Le} &= \frac{D_e \rho c_p}{k_e \varepsilon} \\
\beta &= \frac{D_e(-\Delta H)c_o}{k_e T_0} \\
\gamma &= \frac{E}{R_G T_0}
\end{aligned} \tag{65}$$

Finally, the dimensionless spatial boundary conditions are

$$\frac{\partial y}{\partial x} = \frac{\partial \theta}{\partial x} = 0 \quad \text{at} \quad x = 0 \tag{66}$$

$$\left(\frac{\partial y}{\partial x}\right)_{x=1} = \text{Bi}_M\,(y_b - y_{x=1}) \tag{67}$$

$$\left(\frac{\partial \theta}{\partial x}\right)_{x=1} = \text{Bi}\,(\theta_b - \theta_{x=1}) \tag{68}$$

1.3.2 The steady state solution of the pellet model

We shall first consider the steady state solution of (62), (63), (66), (67), and (68). For convenience (c_0, T_0) are chosen as (c_b, T_b) and the steady state model is

$$\nabla^2 y = \Phi^2 y^n \exp\left[\gamma\left(1 - \frac{1}{\theta}\right)\right] = \Phi^2 \text{ rate}\,(y, \theta) \tag{69}$$

$$\nabla^2 \theta = -\beta \Phi^2 y^n \exp\left[\gamma\left(1 - \frac{1}{\theta}\right)\right] = -\beta \nabla^2 y \tag{70}$$

$$\begin{aligned}
x = 0&: \qquad \frac{dy}{dx} = \frac{d\theta}{dx} = 0 \\
x = 1&: \qquad \frac{dy}{dx} = \text{Bi}_M\,(1 - y) \quad \text{and} \quad \frac{d\theta}{dx} = \text{Bi}\,(1 - \theta)
\end{aligned} \tag{71}$$

In particular we are interested in the so-called effectiveness factor η, the volume averaged reaction rate relative to the rate at bulk phase temperature and concentration.

$$\eta = \frac{\int_0^1 \Phi^2 \text{ rate } (y, \theta)\, dx^{s+1}}{\int_0^1 \Phi^2 \text{ rate } (1, 1)\, dx^{s+1}} = \int_0^1 \text{rate } (y, \theta)\, dx^{s+1} \tag{72}$$

In section 1.4 we use η wherever possible to represent the solution of the pellet problem when the combined fluid-phase pellet model is solved.

For industrially important problems, Bi is quite often small and the intraparticle temperature gradient may be negligible. This leads to a significant simplification of (69) and (70):

Integrate (70) over the pellet volume to obtain

$$(s + 1)\,\overline{\text{Bi}}(1 - \bar{\theta}) \sim -\beta\Phi^2 \int_0^1 y^n \exp\left[\gamma\left(1 - \frac{1}{\bar{\theta}}\right)\right] dx^{s+1}$$
$$= -\beta\Phi_1^2 \eta(\theta = \bar{\theta}) \tag{73}$$

In (73) a driving force $(1 - \bar{\theta})$ has been used rather than $(1 - \theta_{x=1})$ which contains an additional unknown temperature $\theta_{x=1}$. We compensate for the use of a larger temperature driving force by using a smaller value $\overline{\text{Bi}}$ for the transport coefficient, quite in analogy with (27).

$$\frac{1}{\bar{h}} = \frac{1}{h} + \alpha_p \frac{R}{k_e} \quad \text{or} \quad \overline{\text{Bi}} = \frac{\text{Bi}}{1 + \alpha_p \text{Bi}}$$

where α_p is chosen as $1/(s + 3)$ as shown in chapter 6.

The mass balance becomes

$$\nabla^2 y = \Phi^2 y^n \exp\left[\gamma\left(1 - \frac{1}{\bar{\theta}}\right)\right] \tag{74}$$

in terms of y and the constant average temperature $\bar{\theta}$.

For a first-order reaction, (74) can be solved analytically to give, e.g., for spherical geometry,

$$\eta = \frac{3\,\text{Bi}_M}{\Phi_1^2} \frac{\Phi_1 \coth \Phi_1 - 1}{\Phi_1 \coth \Phi_1 + \text{Bi}_M - 1} \tag{75}$$

η is a function of $\bar{\theta}$ since $\Phi_1^2 = \Phi^2 \exp\left[\gamma\left(1 - \frac{1}{\bar{\theta}}\right)\right]$. Equation (75) is inserted into (73) to give an algebraic equation in $\bar{\theta}$:

$$\overline{\text{Bi}}\,(1 - \bar{\theta}) = -\beta\, \text{Bi}_M \frac{\Phi_1 \coth \Phi_1 - 1}{\Phi_1 \coth \Phi_1 + \text{Bi}_M - 1}$$

The solution $\bar{\theta}$ is inserted into (75) to give the value of η.

An important feature of this greatly simplified pellet model is that it may have more than one steady state solution, just as the true model (69) and (70). For $n \neq 1$, (74) is easily solved numerically, perhaps in parallel with (73), and quite accurate values of η are obtained even for rather large values of β.

Numerical solution of the complete model (69) and (70) is, however, not in principle more complicated. Multiplying (69) by β and adding the result to (70) yields

$$\nabla^2(\theta + \beta y) = 0 \quad \text{or} \quad \frac{d\theta}{dx} + \beta\frac{dy}{dx} = c_1 = 0 \quad \text{[from (71)]}$$

A further integration yields

$$\theta + \beta y = c_2 \tag{76}$$

where the integration constant is related to the surface concentration and temperature by

$$c_2 = \theta_{x=1} + \beta y_{x=1} = 1 - \beta\frac{\text{Bi}_M}{\text{Bi}}(y_{x=1} - 1) + \beta y_{x=1}$$

or

$$\theta = \beta(y_{x=1} - y) + 1 - \beta\frac{\text{Bi}_M}{\text{Bi}}(y_{x=1} - 1) \tag{77}$$

Equation (77) is used to eliminate θ from (69), which can now be solved with its two boundary conditions. Finally $\theta(x)$ is found from (77) and the solution of (69).

Equation (77) is independent of the surface concentration in two cases:

1. $\text{Bi} = \text{Bi}_M$
2. $\text{Bi}_M \gg 1$ and $\text{Bi} \gg 1$ in which case $\theta_{x=1} = y_{x=1} = 1$

In both situations (77) simplifies to

$$\theta = 1 + \beta(1 - y) \tag{78}$$

The solution of the steady state model is consequently seen to present only minor numerical problems. Whether the full model or a fairly accurate approximate model (73) and (74) is used, one obtains the effectiveness factor η as a function of the bulk fluid phase properties (c_b, T_b), the two dependent variables of the fluid-phase steady state model that are again functions of one or two spatial variables in the reactor.

The transient equations (62) and (63) for the pellet are much more difficult to solve; when these are coupled to the transient, mass and energy balances for the fluid phase, an almost unsurmountable numerical task is at hand. Thus simplifications of the dynamic pellet model are certainly desired.

1.3.3 Linearized dynamic model for the pellet

For small deviations from a steady state solution, the rate expression in (62) and (63) may be linearized around this steady state.

Let (y_{ss}, θ_{ss}) be the solution to (69) to (71) and define deviation variables

$$\begin{aligned}\hat{y}(x, \tau) &= y(x, \tau) - y_{ss}(x)\\ \hat{\theta}(x, \tau) &= \theta(x, \tau) - \theta_{ss}(x)\end{aligned} \tag{79}$$

$$\Phi^2 y^n \exp\left[\gamma\left(1 - \frac{1}{\theta}\right)\right] \simeq \Phi^2 y_{ss}^n \exp\left[\gamma\left(1 - \frac{1}{\theta_{ss}}\right)\right] + \hat{y}R_y + \hat{\theta}R_\theta$$

where

$$R_y = \Phi^2 \frac{n}{y_{ss}(x)} y_{ss}^n \exp\left[\gamma\left(1 - \frac{1}{\theta_{ss}}\right)\right] \tag{80}$$

$$R_\theta = \Phi^2 \frac{\gamma}{\theta_{ss}^2} y_{ss}^n \exp\left[\gamma\left(1 - \frac{1}{\theta_{ss}}\right)\right] \tag{81}$$

The linearized transient equations are

$$\begin{aligned}\frac{\partial \hat{y}}{\partial \tau} &= \nabla^2 \hat{y} - R_y \hat{y} - R_\theta \hat{\theta}\\ \mathrm{Le}\frac{\partial \hat{\theta}}{\partial \tau} &= \nabla^2 \hat{\theta} + \beta R_y \hat{y} + \beta R_\theta \hat{\theta}\end{aligned} \tag{82}$$

with boundary conditions

$$\begin{aligned}&\frac{\partial \hat{y}}{\partial x} = \frac{\partial \hat{\theta}}{\partial x} = 0 \quad \text{at } x = 0\\ &\left.\begin{aligned}\frac{\partial \hat{y}}{\partial x} + \mathrm{Bi}_M \hat{y} = 0\\ \frac{\partial \hat{\theta}}{\partial x} + \mathrm{Bi}\, \hat{\theta} = 0\end{aligned}\right\} \quad \text{at } x = 1\end{aligned} \tag{83}$$

Substitution of an exponential time dependence for $\hat{y}$ and $\hat{\theta}$ yields an eigenvalue problem. If all eigenvalues have negative real parts, the solution $(\hat{y}, \hat{\theta})$ of (82) will decrease to zero for all x when $\tau \to \infty$. Under these circumstances the steady state (y_{ss}, θ_{ss}) is said to be asymptotically stable.

This linear stability analysis is treated in detail in chapters 5 and 9.

1.3.4 Simplifications of the nonlinear transient model

If the effective heat conductivity k_e of the solid is sufficiently large in comparison with the film transfer coefficient h (i.e., if Bi is small), the whole temperature increase occurs over the film, and the energy balance in (56) can be integrated over the system volume to give an ordinary differential equation in the average pellet temperature $\bar{T}$:

$$\rho c_p \frac{d\bar{T}}{dt} = (s+1)\frac{\bar{h}}{R}(T_b - \bar{T}) + \int_0^1 (-\Delta H) R_A \, dx^{s+1} \tag{84}$$

where the replacement of $T_b - T_{x=1}$ by $T_b - \bar{T}$ may as in the steady state model be compensated for by using a smaller value $\bar{h}$ of the heat transfer coefficient

$$\frac{1}{\bar{h}} = \frac{1}{h} + \alpha_p \frac{R}{k_e} \quad \text{or} \quad \bar{h} = \frac{h}{1 + \alpha_p \,\mathrm{Bi}}$$

The mass balance cannot be similarly simplified since $\mathrm{Bi}_M \gg 1$:

$$\mathrm{Bi}_M = \frac{h_M R}{D_e} = \frac{1}{2}\left(\frac{2h_M R}{D_G}\right)\frac{D_G}{D_e} = \frac{1}{2}\,\mathrm{Sh}\,\frac{D_G}{D_e}$$

The Sherwood number (based on the reactant diffusion coefficient D_G through the film) is never less than 2. D_G may be several orders of magnitude larger than D_e, the "effective" diffusion in the porous pellet, and consequently $\mathrm{Bi}_M \gg 1$.

Simplification of the transient mass balance is, however, possible but for quite different reasons.

The parameter Le is generally very large, of the order of 10 to 1000, for most industrial reactions. This means that the concentration changes much more rapidly than the temperature of the pellet.

Under these circumstances $(\rho c_p/k_e)R^2$ is a more natural time scale than $(\varepsilon/D_e)R^2$ that was used in (64), and (62) and (63) will appear as

$$\frac{1}{\mathrm{Le}}\frac{\partial y}{\partial \tau_1} = \nabla^2 y - \Phi^2 y^n \exp\left[\gamma\left(1 - \frac{1}{\theta}\right)\right] \tag{62a}$$

$$\frac{\partial \theta}{\partial \tau_1} = \nabla^2 \theta + \beta \Phi^2 y^n \exp\left[\gamma\left(1 - \frac{1}{\theta}\right)\right] \tag{63a}$$

$$\tau_1 = \frac{t}{R^2}\frac{k_e}{\rho c_p}$$

Our assumption that the concentration profile is instantaneously established—the so-called quasi-stationary assumption for the mass balance—means that the left-hand side of (62a) can be equated to zero.

The energy balance (63a) is handled by the previous simplification (84) and we obtain the following set of equations:

$$0 = \nabla^2 y - \Phi^2 y^n \exp\left[\gamma\left(1 - \frac{1}{\bar{\theta}}\right)\right] \tag{85}$$

$$\frac{d\bar{\theta}}{d\tau_1} = (s + 1)\,\overline{\text{Bi}}\,(\theta_b - \bar{\theta}) + \beta\Phi^2 \exp\left[\gamma\left(1 - \frac{1}{\bar{\theta}}\right)\right]\int_0^1 y^n\, dx^{s+1} \tag{86}$$

Solution of (85) analytically (for $n = 1$) or numerically gives the concentration profile as a function of $\bar{\theta}$ just as in (74) and (75). The integral in (86) is evaluated and the dynamics of the pellet is described by a single ordinary differential equation in $\bar{\theta}$.

This extremely simplified—and yet realistic—model retains several important characteristics of the complete model (62) and (63). For example, it allows a "quenched" state with a small reaction rate to slide into an "ignited" state with a large reaction rate and a significant temperature drop across a boundary film.

1.3.5 Pellet parameters for a sulfuric acid catalyst

A numerical example will indicate typical values of the pellet dimensionless groups for a gaseous reactant.

Villadsen (1970) and Livbjerg and Villadsen (1972) present the following data for a silica-based SO_2 oxidation catalyst:

Pellet: $\varepsilon = 0.4$, $D_e = 5.2 \times 10^{-2}\ \text{cm}^2/\text{s}$,
$k_e = 7 \times 10^{-4}$ cal/s cm °K,
$(\rho c_p) = 0.4$ cal/cm^3 °K, $R = 0.3$ cm
Reaction conditions: Inlet of an industrial converter.
Pressure 1 atm, $c_{b0} = c_{SO_2} = 1.6 \times 10^{-6}$ mol/cm^3 for 10 vol % SO_2 in air,
$T_0 = 773$ °K, superficial gas velocity $v_0 = 100$ cm/s,
and bed porosity $\varepsilon_b = 0.4$, heat of reaction, 23,100 cal/mol
Gas-phase properties at 500 °C: $\rho_b = 4.57 \times 10^{-4}$ g/cm^3,
$(c_p)_b = 0.24$ cal/g °K,
$D_b = 0.655$ cm^2/s, $\mu_b = 3.58 \times 10^{-4}$ g/cm s.
$k_b = 9.19 \times 10^{-5}$ cal/cm s °K

With these data one obtains

$$\mathrm{Re} = \frac{2 \cdot v_0 R \rho_b}{\mu_b} = 77, \qquad \mathrm{Pr} = \left(\frac{\mu c_p}{k}\right)_b = 0.94, \qquad \mathrm{Sc} = \left(\frac{\mu}{\rho D}\right)_b = 1.26$$

$$j_D \text{ [Satterfield (1970), p. 82]} = \frac{0.357}{\varepsilon_b \, \mathrm{Re}^{0.359}} = 0.19$$

$$\frac{j_H}{j_D} = 1.37 \quad \text{for most gaseous reactants [Satterfield (1970), p. 83]*}$$

$$h = j_H (\rho c_p)_b v_0 \, \mathrm{Pr}^{-2/3} = 2.97 \times 10^{-3} \text{ cal/cm}^2 \text{ s °K}$$

$$\mathrm{Bi} = \frac{Rh}{k_e} = 1.27$$

$$\frac{\mathrm{Bi}_M}{\mathrm{Bi}} = \frac{h_M}{h}\frac{k_e}{D_e} = \left(\frac{\mathrm{Pr}}{\mathrm{Sc}}\right)^{2/3} \frac{j_D}{j_H} \frac{k_e}{(\rho c_p)_b D_e} = 5450 \frac{k_e}{D_e} = 73$$

$$\beta = \frac{c_{b0}(-\Delta H) D_e}{k_e T_0} = 3.6 \times 10^{-3}$$

$$\mathrm{Le} = \frac{D_e \rho c_p}{k_e \varepsilon} = 74$$

The value of β is so small that in the absence of heat and mass film resistance the pellet will be at substantially the fluid-phase temperature T_0 ($\Delta T = 2.8$ °K if all reactant is converted in the pellet). For spherical pellets and Bi = 1.27 we predict an $\bar{h}$ value of $0.80h$, while the value of $\bar{h}/h$ that is obtained from the full model is 0.78 (see section 6.2). Thus our model simplification (84) is fully justified. Bi_M is large and the simpler boundary condition $y = 1$ at $x = 1$ could be used rather than (67)—the resulting reduction in complexity of the model is, however, small.

If the catalyst is extremely active, the reaction occurs on the pellet surface and $y \sim 0$ at any point in the pellet, including $x = 1$. $\bar{\theta}$ is given by (77) and it can be as large as $1 + \beta(\mathrm{Bi}_M/\mathrm{Bi}) = 1 + 73\beta = 1.26$—or $\Delta T = 200$ °K across the film.

This maximum temperature increase is seen to be

$$\beta \frac{\mathrm{Bi}_M}{\mathrm{Bi}} = \frac{j_D}{j_H}\left(\frac{\mathrm{Pr}}{\mathrm{Sc}}\right)^{2/3}\left(\frac{c_{b0}(-\Delta H)}{(\rho c_p)_{b0} T_{b0}}\right) \sim 0.6 \beta_{b0}$$

where β_{b0} is the dimensionless adiabatic temperature rise for a pure gas-phase conversion of the reactant.

In practice this large temperature gradient across the film is, of course, never obtained since the reaction proceeds at a finite rate.

* Bird (1960, p. 647) uses $j_H = j_D$.

The parameter Le for the pellet dynamics is very large and our assumption of quasi-stationarity for the mass balance (62a) is certainly justified.

Many theoretical studies of the dynamic behavior of catalyst pellets have been made with Le < 1 where—as introduced in chapter 9—most of the mathematically interesting phenomena occur. If for a moment we assume that Le $\ll 1$, the energy balance (63) instantaneously follows even rapid variations in the concentration profile. Now, consider a pellet whose pore volume is filled with reactant at concentration c_{b0}. The pellet is suddenly ignited and an instantaneous temperature rise

$$\Delta\theta = \frac{c_{b0}\varepsilon(-\Delta H)}{\rho c_p T_{b0}} = \frac{\beta}{\mathrm{Le}}$$

is registered if the rate of reaction is very fast.

The transient temperature rise may be much larger than β when Le is close to zero; but as we have seen from the numerical example, the Lewis number is in reality much larger than unity and the energy balance reacts sluggishly to changes in concentration.

1.4 Heterogeneous Model for a Reacting System

In section 1.2 we derived homogeneous models for a reactor, that is, models where pellet reactant concentration and temperature are the same as in the fluid outside the catalyst pellet. In section 1.3, steady state and dynamic models for the pellet were discussed. A combination of results from the two sections leads to a heterogeneous reactor model in which pellet phase and fluid phase may have different reactant concentration and temperature at the same position in the reactor. The two phases are coupled through a set of boundary conditions.

1.4.1 Parameters and variables in heterogeneous model

The dynamics of the reactor bed are characterized by two time constants t_M and t_H. The first time constant t_M is the ratio of the free reactor volume to the volumetric inlet flow rate, i.e., a measure of fluid residence time. The second time constant t_H is the ratio of the reactor heat capacity to the heat capacity of the inlet stream. t_H is consequently a measure of response time to thermal disturbances and is called the thermal residence time.

$$t_M = \frac{L\varepsilon_b}{v_0}$$

$$t_H = \frac{L\varepsilon_b(\rho c_p)_{b0} + L(1-\varepsilon_b)(\rho c_p)}{v_0(\rho c_p)_{b0}} = t_M\left[1 + \frac{1-\varepsilon_b}{\varepsilon_b}\frac{\rho c_p}{(\rho c_p)_{b0}}\right]$$

Subscript b refers to the fluid phase and subscript 0 to inlet fluid conditions. When the same quantities are used for both fluid and solid phases, they are shown without subscript for the solid phase.

The bed porosity ε_b is between 0.3 and 0.7 and with a reactor pressure of the order of 1, the last term in t_H is several orders of magnitude larger than 1. Hence the thermal residence time is much larger than the fluid residence time t_M and in practice all capacitance terms except the thermal capacitance of the solid phase may be neglected. We shall see that this leads to a tremendous model simplification for transient calculations.

It will be assumed that axial dispersion can be neglected and that radial gradients can be accounted for by using an effective wall heat transfer coefficient $\bar{U}_b$ defined by (27) with $\alpha = \frac{1}{4}$.

The fluid phase is taken to be a nearly ideal gas and the pressure drop in the reactor is neglected. For an equimolar chemical reaction the linear fluid velocity at reactor temperature T_b is

$$v = v_0\frac{\rho_{b0}}{\rho_b} = v_0\frac{T_b}{T_{b0}}$$

1.4.2 Models for the two coupled phases

For spherical pellets the fluid-phase mass and energy balances are

$$-\pi R_b^2\frac{\partial}{\partial z}(vc_b) = \varepsilon_b\pi R_b^2\frac{\partial c_b}{\partial t} + (1-\varepsilon_b)\pi R_b^2\frac{3}{R}h_M(c_b - c_s)$$

$$-\pi R_b^2\frac{\partial}{\partial z}[v(\rho c_p T)_b] = \varepsilon_b\pi R_b^2\frac{\partial(\rho c_p T)_b}{\partial t} + (1-\varepsilon_b)\pi R_b^2\frac{3}{R}h(T_b - T_s)$$
$$+2\pi R_b\bar{U}_b(T_b - T_W)$$

(T_s, c_s) are temperature and concentration, respectively, of the reactant on the surface of the pellets. (T_b, c_b), the fluid-phase variables, are assumed to be constant in the cross section of the bed.

Dimensionless variables $y_b = c_b/c_{b0}$, $\theta_b = T_b/T_{b0}$, and $\zeta = z/L$ are introduced. $v = v_0(T_b/T_{b0})$; $(c_p)_b$ is independent of temperature and reactant concentration.

$$-\theta_b \frac{\partial y_b}{\partial \zeta} - y_b \frac{\partial \theta_b}{\partial \zeta} = \frac{\varepsilon_b L}{v_0} \frac{\partial y_b}{\partial t} + \frac{(1-\varepsilon_b)L}{v_0} \frac{3h_M}{R}(y_b - y_s)$$

$$-\frac{\partial \theta_b}{\partial \zeta} = \frac{(1-\varepsilon_b)L}{v_0(\rho c_p)_{b0}} \frac{3h}{R}(\theta_b - \theta_s) + \frac{L\bar{U}_b}{v_0(\rho c_p)_{b0}} \frac{2}{R_b}(\theta_b - \theta_W)$$

The solid-phase balances are (56):

$$\varepsilon \frac{\partial c}{\partial t} = \frac{D}{R^2}\nabla^2 c - R_A(c, T)$$

$$\rho c_p \frac{\partial T}{\partial t} = \frac{k}{R^2}\nabla^2 T + (-\Delta H)R_A(c, T)$$

The subscript e that was used to denote effective transport coefficients k_e and D_e has been dropped here to avoid confusion of nomenclature. Boundary conditions for the solid-phase balances are

$$\left(\frac{\partial c}{\partial x}\right)_{x=1} = \frac{h_M R}{D}(c_b - c_s)$$

$$\left(\frac{\partial T}{\partial x}\right)_{x=1} = \frac{hR}{k}(T_b - T_s)$$

where x is the dimensionless pellet coordinate.

The model simplification (85) and (86) is introduced into the solid balances:

$$\frac{D}{R^2}\nabla^2 c = R_A(c, \bar{T})$$

$$\rho c_p \frac{\partial \bar{T}}{\partial t} = \frac{3\bar{h}}{R}(T_b - \bar{T}) + (-\Delta H)\int_0^1 R_A(c, \bar{T})\, dx^3$$

The mass balance can be solved first, and the concentration profile $c(x)$ is obtained as a function of c_b and $\bar{T}$.

An effectiveness factor η is defined by the following relation:

$$\eta R_A(c_{b0}, T_{b0}) = \eta R_{A0} = \int_0^1 R_A(c, \bar{T})\, dx^3$$

Now the mass balance can be formally integrated over the pellet volume

$$3\frac{D}{R^2}\frac{h_M R}{D}(c_b - c_s) = \eta R_{A0}$$

and the energy balance becomes

$$\rho c_p \frac{\partial \bar{\theta}}{\partial t} = \frac{3\bar{h}}{R}(\theta_b - \bar{\theta}) + \frac{(-\Delta H)}{T_{b0}} \eta R_{A0}$$

or

$$\frac{1-\varepsilon_b}{\varepsilon_b} \frac{\rho c_p}{(\rho c_p)_{b0}} \frac{\partial \bar{\theta}}{\partial t} = \frac{3\bar{h}}{R} \frac{1-\varepsilon_b}{\varepsilon_b} \frac{1}{(\rho c_p)_{b0}}(\theta_b - \bar{\theta}) + \frac{1-\varepsilon_b}{\varepsilon_b} \frac{(-\Delta H)c_{b0}}{T_{b0}(\rho c_p)_{b0}} \eta \frac{R_{A0}}{c_{b0}}$$

The factor $(-\Delta H)c_{b0}/T_{b0}(\rho c_p)_{b0}$ is the dimensionless adiabatic temperature rise β_{b0} defined in subsection 1.3.5. The factor of $\partial\bar{\theta}/\partial t$ was defined in subsection 1.4.1 as $(t_H/t_M) - 1$. Note that $\bar{\theta}$ is a function of t as well as of the bed axial variable ζ, which explains why the pellet energy balance is still a partial differential equation even after the averaging process over the pellet variable x.

We shall finally introduce H_b and H_W, two dimensionless groups that contain the number of heat transfer units, fluid to pellets and fluid to reactor wall, for the reactor of length L:

$$H_b = t_M \frac{1-\varepsilon_b}{\varepsilon_b} \frac{1}{(\rho c_p)_{b0}} \frac{3\bar{h}}{R} = \frac{(1-\varepsilon_b)L}{v_0(\rho c_p)_{b0}} \frac{3\bar{h}}{R}$$

$$H_W = t_M \frac{1}{\varepsilon_b(\rho c_p)_{b0}} \frac{2\bar{U}_b}{R_b} = \frac{2L\bar{U}_b}{v_0(\rho c_p)_{b0} R_b}$$

The combined fluid-phase and pellet model is now

$$-\theta_b \frac{\partial y_b}{\partial \zeta} - y_b \frac{\partial \theta_b}{\partial \zeta} = t_M \frac{\partial y_b}{\partial t} + \frac{1-\varepsilon_b}{\varepsilon_b} t_M \eta \frac{R_{A0}}{c_{b0}} = t_M \frac{\partial y_b}{\partial t} + \mathrm{Da}\, \eta$$

$$-\frac{\partial \theta_b}{\partial \zeta} = H_b(\theta_b - \bar{\theta}) + H_W(\theta_b - \theta_W) \tag{87}$$

$$(t_H - t_M)\frac{\partial \bar{\theta}}{\partial t} = H_b(\theta_b - \bar{\theta}) + \beta_{b0}\, \mathrm{Da}\, \eta$$

where $\mathrm{Da} = [(1-\varepsilon_b)/\varepsilon_b] t_M (R_{A0}/c_{b0})$ is the Damköhler number as defined in section 1.2 but based on the catalyst volume of the bed and the inlet superficial gas velocity v_0.

The steady state model that is obtained by dropping the time derivatives in (87) is

$$-\theta_b \frac{dy_b}{d\zeta} - y_b \frac{\partial \theta_b}{\partial \zeta} = \mathrm{Da}\, \eta$$

$$-\frac{d\theta_b}{d\zeta} = H_b(\theta_b - \bar{\theta}) + H_W(\theta_b - \theta_W) \tag{88}$$

$$0 = H_b(\theta_b - \bar{\theta}) + \beta_{b0}\, \mathrm{Da}\, \eta$$

The model consists of two coupled first-order ordinary differential equations coupled to an algebraic equation. The pellet steady state model is hidden in $\eta = \eta(\bar{\theta}, y_b)$, which is obtained by (73) and (74) or for large values of Bi from the solution of (69) to (71).

If the reactor is adiabatic, further simplification of (88) is possible similar to (33) in section 1.2. Since $H_W = 0$ for an adiabatic reactor, $\bar{\theta}$ can be eliminated between the second and third equation to give

$$-\frac{d\theta_b}{d\zeta} = -\beta_{b0}\,\mathrm{Da}\,\eta$$

The first equation is multiplied by β_{b0} and the two equations are added:

$$\frac{d(\beta_{b0}\theta_b y_b + \theta_b)}{d\zeta} = 0 \quad \text{or} \quad \theta_b = \frac{1 + \beta_{b0}}{1 + \beta_{b0} y_b}$$

and θ_b can be eliminated to give a final model that consists of one ordinary differential equation for y_b and two algebraic equations for θ_b and $\bar{\theta}$.

The coupling of a set of ordinary differential equations with algebraic equations or the coupling of partial differential equations with ordinary differential equations in the boundary conditions as seen in subsection 1.2.4 is a very common feature of chemical engineering models.

The transient model (87) is frequently—and with full justification according to the arguments of subsection 1.4.1—simplified by dropping the accumulation terms with factor t_M. The resulting model is

$$\begin{aligned} -\theta_b\frac{\partial y_b}{\partial \zeta} - y_b\frac{\partial \theta_b}{\partial \zeta} &= \mathrm{Da}\,\eta \\ -\frac{\partial \theta_b}{\partial \zeta} &= H_b(\theta_b - \bar{\theta}) + H_W(\theta_b - \theta_W) \\ \frac{\partial \bar{\theta}}{\partial t/t_H} &= H_b(\theta_b - \bar{\theta}) + \beta_{b0}\,\mathrm{Da}\,\eta \end{aligned} \tag{89}$$

The model is still a set of three coupled partial differential equations with y_b, θ_b, and $\bar{\theta}$ as dependent variables and with the pellet dynamics hidden in η, which is given by the solution of (85) and (86).

The time constant t_p for transport of heat from the pellet to the fluid phase is obtained if (86) is divided by $3\,\overline{\mathrm{Bi}}$ for spherical pellets:

$$\frac{R\rho c_p}{3\bar{h}}\frac{d\bar{\theta}}{dt} = (\theta_b - \bar{\theta}) + \frac{\beta\Phi^2}{3\,\overline{\mathrm{Bi}}}\exp\left[\gamma\left(1 - \frac{1}{\bar{\theta}}\right)\right]\int_0^1 y^n\,dx^3$$

or

$$t_p = \frac{R\rho c_p}{3\bar{h}} = \frac{t_H}{H_b}$$

as also seen from the last equation of (89).

t_p may well be much smaller than t_H if H_b is large but it is certainly much larger than the internal pellet time constant, which has been neglected in our previous averaging process (84).

If t_p is much smaller than t_H, that is, if we assume $\bar{\theta}$ to follow θ_b instantaneously—the so-called "chromatographic assumption" of Aris and Amundson (1973, p. 35), because it corresponds to an instantaneous local equilibrium between the solution and adsorbed species in a chromatographic column—then additional simplification of the model (89) is possible. Subtraction of the middle equation of (89) from the last equation yields

$$\frac{\partial \theta_b}{\partial \zeta} + \frac{\partial \theta_b}{\partial (t/t_H)} = \beta_{b0}\,\mathrm{Da}\,\eta - H_W(\theta_b - \theta_W) \tag{90}$$

The consequence of this last simplification is best seen by consideration of a nonreactive system (Da = 0) where (90) is a first-order linear partial differential equation in θ_b. This is solved by the method of characteristics as described in Aris and Amundson (1973) to give a temperature front that marches with constant velocity $1/t_H$ through the reactor without changing shape.

Thus by neglecting the heat transfer resistance between pellets and fluid phase, our dynamic model becomes unable to predict the broadening of sharp temperature disturbances that is observed in experimental reactor dynamic studies. Slowly changing temperature disturbances may still be accurately simulated but the model is unable to cope with high-frequency disturbances.

A dispersive element may be included in the transient reactor model even when the pellet dynamics is neglected. Most industrial reactors are built of high heat capacity material and the dynamics of the reactor wall may well be much more important (large time constant) than the pellet dynamics [see, e.g., Hoiberg and Foss (1971)]. This coupling element has not been considered in our models but it may easily confound a laboratory study of reactor dynamics.

It should finally be mentioned that axial dispersion can be included in the model without increasing its complexity. Some investigators, e.g., Wæde Hansen (1974) and Sørensen (1976), collect all dispersive effects (pellet fluid, fluid wall, etc.) into an effective axial heat dispersion term in the same way as real gradients are handled in subsection 1.2.3. If a suitable Pe-number can be derived from experiments to describe this lumping process, the method leads to the same dynamic results as in the present treatment; this does, however, have the benefit of operating with real physical properties.

1.5 A Model for Hollow-Fiber Reverse-Osmosis Systems

The flow systems of the previous sections were all characterized by a unidirectional convective flow. This greatly simplified the solution of Navier–Stokes equations and the complications arose due to diffusional effects in the fluid or in a solid phase that was coupled to the fluid phase.

In this and the next section we deviate somewhat from the main stream of our text, which by and large deals with solution of models taken from the previous sections. The excursion into reverse osmosis here or into extrusion of polymers in the next section is not motivated by a desire to treat these subjects on their own merits. Rather we wish to stress that model building is certainly more complicated than we have shown so far.

1. *Description of the process in a hollow-fiber membrane:* Heavy-walled hollow cylindrical membranes with an internal diameter of 50–20 microns are spun from a semipermeable material and imbedded in an epoxy tube sheet that is housed in an exterior shell. Pressurized feed solution flows over the fibers in the shell. The product water permeates the fibers and flows inside them toward the open ends, which are at atmospheric pressure. A very large membrane area per unit volume is provided and concentration polarization—a major source of efficiency loss in reverse osmosis (or ultrafiltration)—is negligible due to the permissible small product flux. These features of the hollow-fiber membrane make it popular in a variety of situations: desalination and waste water treatment in general—as well as in biological separations where the treatment of uremia by hemodialysis is only one example.

The setting up and analysis of a model for the highly complicated flow pattern in a hollow-fiber agglomerate is important to establish rational design criteria for wall thickness, outside fiber radius, and fiber length and to determine the capacity at given feed flow rate, operating pressure, and inlet feed concentration. Optimal values of outside fiber radius r_0 and fiber length L can be determined for given operating conditions and this has been the object of the study by Gill (1973). His model looks considerably more complicated than the one that is derived below, but when the various assumptions that he makes through his derivation are introduced immediately, the results become identical.

2. *Assumptions in the mathematical model:* The arrangement of fibers is shown in figure 1-1(a). Variations in concentration and

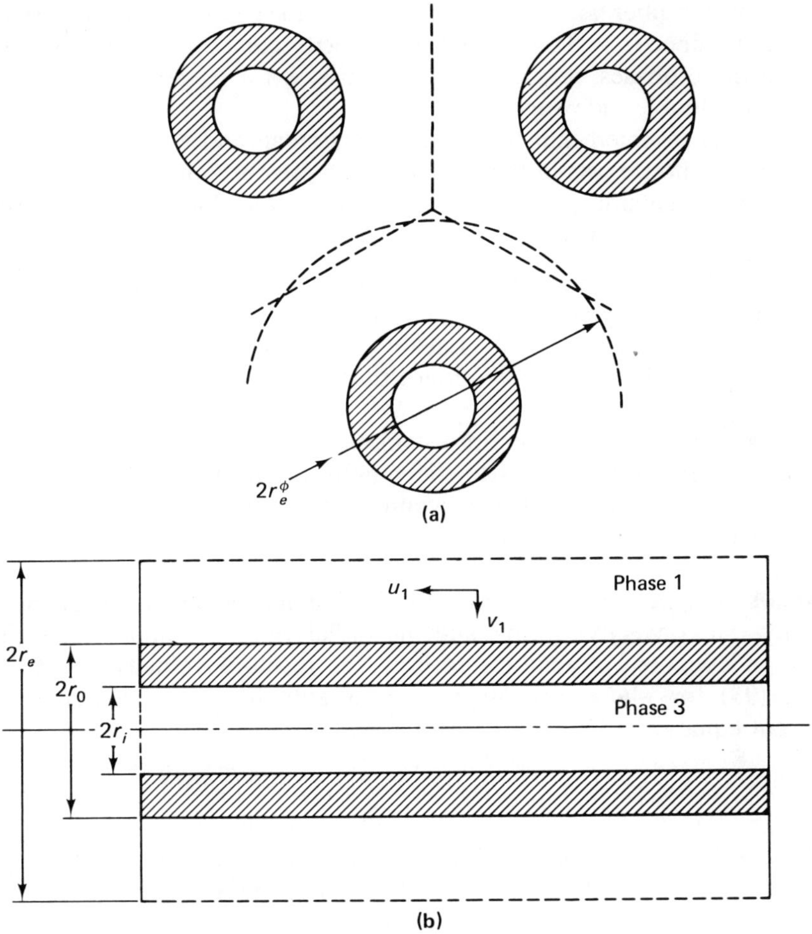

Figure 1-1. (a) Hollow fiber reverse osmosis. (b) Arrangement of fibers and fluid flow directions through phases 1–3.

pressure are small around the periphery of the imaginary hexagonal boundary that surrounds each fiber, and this boundary is replaced by an equivalent circle having the same annular area between it and the outside fiber wall.

The radius r_e of the equivalent annulus is determined by the porosity of the bed ε and the outside fiber radius r_0:

$$r_e \equiv \left(\frac{1}{1-\varepsilon}\right)^{1/2} r_0 \tag{91}$$

The velocity vector **v** for the fluid flow through the system shown in figure 1-1(b) has two components v_r and v_z. Since the system is composed of three phases, (1) outside the fibers ($i = 1$), (2) in the fiber wall ($i = 2$), (3) in the passage on the product side of the fiber ($i = 3$), and the two velocity components are used in all three phases, it is convenient to use u_i for v_{zi} and v_i for v_{ri} to avoid double indexing.

It is assumed that the fluid is Newtonian and that steady state laminar flow exists in the system. Both pressure and concentration decrease significantly in the axial z-direction due to friction loss and product removal.

The simultaneous solution of nine coupled nonlinear partial differential equations (two equations for **v** and one mass balance for each phase) is an almost impossible task. Luckily a number of simplifying circumstances exist that drastically reduce the computational work.

Assumption 1: The equation of motion can be solved separately. The present system admits to the simplification discussed in section 1.1 since **v** is independent of concentration and the process is assumed to be carried out isothermally.

Assumption 2: Axial diffusion terms are neglected in comparison with axial convective terms, and the radial velocity component v_1 is thus independent of axial distance z. The resulting velocity field (92) and (93) for phase 1 contains considerably fewer terms than the parent equations (7):

$$u_1\frac{\partial u_1}{\partial z} + v_1\frac{\partial u_1}{\partial r} = -\frac{1}{\rho}\frac{\partial p_1}{\partial z} + \nu\frac{1}{r}\frac{\partial}{\partial r}\left(r\frac{\partial u_1}{\partial r}\right) \tag{92}$$

$$v_1\frac{\partial v_1}{\partial r} = -\frac{1}{\rho}\frac{\partial p_1}{\partial r} + \nu\frac{\partial}{\partial r}\left[\frac{1}{r}\frac{\partial}{\partial r}(rv_1)\right] \tag{93}$$

Assumption 3: The small flow v_{1w} out of phase 1 probably rules out the possibility of concentration polarization that would set up radial concentration gradients. Thus c_1 is assumed to be independent of r and a mass balance over the annular volume $V_1 = \pi(r_e^2 - r_0^2)\,dz$ will give a relation between the loss of solute from the bulk phase 1 and the amount that has penetrated the membrane at radius r_0:

$$\pi(r_e^2 - r_0^2)\frac{d(u_{1\mathrm{av}}c_{1w})}{dz} = -(1-K)2\pi r_0 v_{1w}c_{1w} \tag{94}$$

K is called the rejection coefficient of the membrane. If K is close to unity, almost no solute enters the fiber wall; a value of K close to zero means that the solute is strongly retained (absorbed) by the fiber material. K is clearly an empirical constant for the fiber membrane. It is found to vary only a few percent with the main design objective, which is the productivity Φ, defined by the area-averaged linear velocity in phase 1,

$$\Phi = \frac{u_{1av}(z) - u_{1av}(z = 0)}{u_{1av}(z = 0)} \tag{95}$$

and the representation of c_{1w} by (94) with a constant K is reasonable for Φ less than approximately 0.5–0.6.

The area-averaged axial velocity u_{1av} is defined similarly to (94) by a fluid balance from $z = 0$ to z using the wall radial velocity v_{1w}, which is independent of z by assumption 2.

$$u_{1av}(z) = \frac{1}{A_1}\int_{A_1} u_1(r, z)\, dA_1 = u_{1av}(z = 0) - \frac{2r_0}{r_e^2 - r_0^2} v_{1w} z \tag{96}$$

Equation (96) shows that $du_{1av}/dz = -[2r_0/(r_e^2 - r_0^2)]v_{1w}$, which is inserted into (94) to give

$$u_{1av}\frac{dc_{1w}}{dz} = -K\frac{du_{1av}}{dz}c_{1w} \tag{97}$$

Integration of (97) from $z = 0$ to z with the initial condition $c = c_0$ and $u_{1av} = u_{1av}(z = 0)$ at $z = 0$ immediately gives the concentration of solute at the wall as a function of z:

$$\frac{c_{1w}(z)}{c_0} = \left[\frac{u_{1av}(z = 0)}{u_{1av}(z)}\right]^K \tag{98}$$

The wall velocity v_{1w} can also be represented in terms of the difference in fluid pressure across the fiber wall, $p_{1w} - p_{3w}$, modified by the corresponding difference is osmotic pressure, $\pi_{1w} - \pi_{3w}$:

$$v_{1w} = -A[(p_{1w} - p_{3w}) - (\pi_{1w} - \pi_{3w})] \tag{99}$$

where A is a permeability coefficient for the membrane.

Assumption 4: The flow through the membrane is in the radial direction only—a reasonable assumption considering its low permeability. Consequently, by a simple geometric argument,

$$v_{3w} = v_{1w}\frac{r_0}{r_i} \tag{100}$$

$$c_{3w} = (1 - K)c_{1w} \tag{101}$$

Assumption 5: The variable osmotic pressure difference in (99) is eliminated by the assumption that osmotic pressure is proportional to concentration of solute or

$$\pi = \frac{c}{c_0}\pi(z = 0) = \frac{c}{c_0}\pi_0$$
$$v_{1w} = -A\left(p_{1w} - p_{3w} - \frac{\pi_0}{c_0}Kc_{1w}\right) \tag{102}$$

We can now take an overview of the resulting computational work.

Phase 1: To obtain the velocity field $[u(r, z), v(r)]$, it is necessary to solve equations (92) and (93) using the following boundary conditions:

$$\frac{\partial u_1}{\partial r} = 0 \quad \text{at } r = r_e \quad \text{[no shear at ``free surface'']}$$
$$u_1 = 0 \quad \text{at } r = r_0 \quad \text{[no slip at fiber wall]}$$
$$u_1 = u_{1av}(z = 0) \cdot h(r) \quad \text{at } z = 0 \quad \text{[where } h(r) \text{ is the inlet velocity distribution in the annulus]} \tag{103}$$
$$v_1 = 0 \quad \text{at } r = r_e \quad \text{[no material flow across } r = r_e\text{]}$$
$$v_1 = v_{1w} \quad \text{at } r = r_0$$

The pressure p is eliminated by introducing the equation of continuity (4) in the system of equations.

For constant density $\nabla \cdot \mathbf{v} = 0$ or from Bird (1960, p. 83),

$$\frac{\partial u_1}{\partial z} + \frac{1}{r}\frac{\partial}{\partial r}(rv_1) = 0 \tag{104}$$

The mass balance equation (2) will not be set up since the important quantity $c_{1w}(z)$ has already been found in (98).

Phase 2: Assumptions 4 and 5 make a closer study of this phase superfluous since v_{3w}, c_{3w}, and p_{3w} are all known by (100) to (102) as functions of phase 1 variables.

Phase 3: The flow in the hollow fibers may be countercurrent $[u_3(L, r) = 0]$ or concurrent $[u_3(0, r) = 0]$, but the flow field inside the membrane need not be calculated since the product concentration in the fluid that leaves the system at the fiber end can be calculated from c_{3w} and v_{1w} and these are already known.

In conclusion, the design procedure of Gill (1973) reduces to the solution of (92), (93), and (104), with boundary conditions given by (103). All remaining variables are found by simple manipulations using the various expressions (98) to (102) that were derived when the assumptions of the model were cataloged.

1.6 Flow of Polymer Melts in Extruders

The purpose of an extruder is to force molten polymer at a controlled rate through a die. Solid polymer is fed through a hopper at the rear of the screw and conveyed forward in a channel of gradually decreasing depth. Heat is transferred to the material from the barrel wall, and in the forward section of the screw the pressure increases due to the channel geometry. The polymer starts to melt and internal friction generates further heat. The last section of the extruder, the metering section, where all polymer is molten, is of particular interest in the control of the extruder operation. Zamodits (1969) has set up a model for this metering section and by solution of the equations of motion and energy he is able to predict the effect of geometrical parameters, operating conditions, and melt rheology on the extruder performance. Trowbridge (1973), in a study of criteria for multiplicity of the solution to nonlinear boundary value problems, makes an almost successful attempt to restate this model and he computes approximate regions of uniqueness for the solution. We shall generally follow the development of Zamodits (and Pearson), which brings out some very interesting aspects of model formulation, especially with respect to the final problem statement in a form which is different from the explicit differential equations that we have encountered previously.

A screw and barrel system with constant helix angle α is shown in figure 1–2. The screw diameter is D, the largest distance between screw and barrel is h, and the flight walls of height $(h - \delta)$ almost touch the barrel. The distance δ is small enough to exclude any flow over the top of the flight wall. Consequently the polymer flows in a shallow channel of breadth B and depth $\sim h$, which can be unrolled to a long straight channel when $h \ll D$ (or B), in which case curvature effects in the (x, z)-plane are neglected.

The components of the velocity vector are called u, v, and w in the x-, y-, and z-directions. v is zero (except very near the flight edges) and the transversal (u) and down-channel (w) velocity components are independent of x and z in the unrolled channel of constant depth h and constant pressure gradients $\partial p/\partial x = p_x$ and $\partial p/\partial z = p_z$ in the x-, and z-directions. Thus the velocity field (u, w) is a function of y alone. It is described by the relative motion of the top and bottom surfaces of the channel and

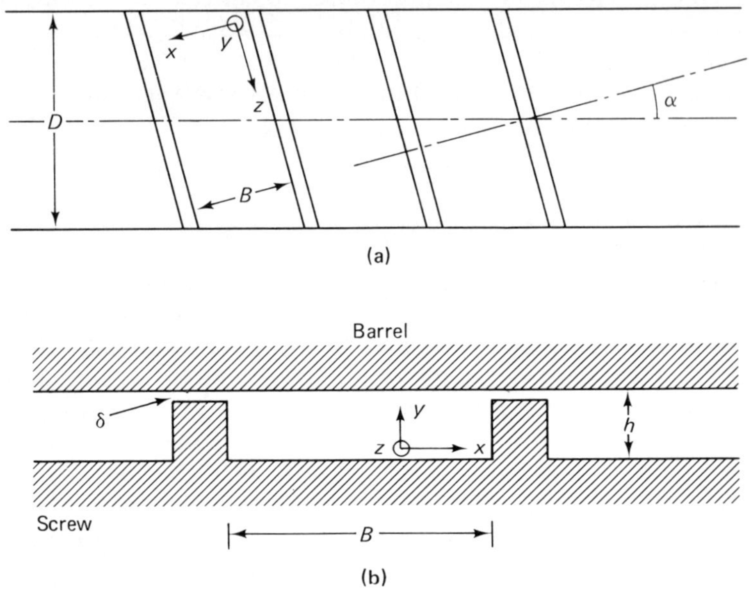

Figure 1-2. Screw and barrel system for polymer extrusion.

by the local pressure gradients, the constants p_x and p_z of which only p_z is known from pressure measurements along the channel.

Inertia forces (centrifugal forces in this case) are small and in the unrolled channel model they are neglected. Now the screw may be regarded as stationary and the barrel as rotating (N revolutions per time unit) without altering the problem. With no slip at the screw and barrel surfaces, the velocity boundary conditions are

$$u = 0, \quad w = 0 \quad \text{at } y = 0$$

$$u = \pi DN \sin \alpha, \quad w = \pi DN \cos \alpha \quad \text{at } y = h$$

or in terms of the dimensionless variables $Y = y/h$, $U = u/(\pi DN \sin \alpha)$, and $W = w/(\pi DN \cos \alpha)$ that we shall use later,

$$U = W = 0 \quad \text{at } Y = 0, \qquad U = W = 1 \quad \text{at } Y = 1 \tag{105}$$

Furthermore,

$$\int_0^h u\,dy = \int_0^1 U\,dY = 0 \tag{106}$$

which expresses that no material is transferred perpendicular to the flight walls.

We shall consider an insulated screw and a barrel with constant temperature T_0—a situation likely to occur in practice. The flow of heat in the shallow unrolled channel is in the y-direction only. The assumption of no heat transfer over or through the flight walls is consistent with the approach to the flow of melt in the channel.

$$\frac{dT}{dY} = 0 \quad \text{at } Y = 0, \qquad T = T_0 \quad \text{at } Y = 1 \tag{107}$$

With our present assumptions on the flow field (u and w are only functions of y and $v = 0$) it may be checked from the expression for $\mathbf{v} \cdot \nabla \mathbf{v}$ in (3) that the equation of motion only contains viscous terms and that we only need to consider the change in the y-direction of x- and z-momentum flux in the y-direction (τ_{yx} and τ_{yz}):

$$\begin{aligned} p_x &= -\frac{d\tau_{yx}}{dy} \\ p_z &= -\frac{d\tau_{yz}}{dy} \end{aligned} \tag{108}$$

The equation of energy (2) is also free of convective terms and Q is the (often appreciable) heat generated by internal friction. Bird (1960), p. 731, gives the following expression for the heat generation term $Q = -(\boldsymbol{\tau} : \nabla \mathbf{v})$ for our velocity field:

$$-Q = \tau_{yx}\frac{du}{dy} + \tau_{yz}\frac{dw}{dy} = \tau_{yx}u_y + \tau_{yz}w_y$$

Apart from this heat generation term the equation of energy only contains the conductive term in the y-direction. With a temperature-independent heat conductivity k one obtains

$$k\frac{d^2T}{dy^2} + Q = 0 \tag{109}$$

with boundary conditions (107).

Zamodits used a two-dimensional form of the power law expression (9) for the nonzero components of the shear stress:

$$\begin{aligned} \tau_{yx} &= -C[(u_y^2 + w_y^2)^{1/2}]^{n-1}u_y \\ \tau_{yz} &= -C[(u_y^2 + w_y^2)^{1/2}]^{n-1}w_y \end{aligned} \tag{110}$$

We are now able to write the balances (108) and (109) in terms of the velocity gradients u_y and w_y:

$$p_x = -\frac{d\tau_{yx}}{dy} = \frac{d}{dy}[C(u_y^2 + w_y^2)^{(n-1)/2}u_y] \tag{111}$$

$$p_z = -\frac{d\tau_{yz}}{dy} = \frac{d}{dy}[C(u_y^2 + w_y^2)^{(n-1)/2} w_y] \tag{112}$$

$$k\frac{d^2T}{dy^2} + C(u_y^2 + w_y^2)^{(n-1)/2}(u_y^2 + w_y^2) = 0 \tag{113}$$

where C may be a function of T—and thus indirectly a function of y since T is a function of y.

u_y and w_y are first calculated from (111) and (112). The result is (117) and (118). Next, with u_y and w_y formally expressed as functions of y, (113) can be solved to give (120). Unfortunately the solution (117) and (118) of the equation is very untidy. It contains three unknown parameters: $P_1 = p_x/p_z$ (p_z can be measured but p_x is unknown), y_1, and y_2. The last two parameters are the values of y where, respectively, u_y and w_y are zero. The three unknown parameters in the end are expressed through the boundary conditions (105) and the total mass balance (106) but the result (121) to (123) is not a neat one.

The stress neutral surfaces $u_y = 0$ and $w_y = 0$ deserve a physical interpretation. $u_y = 0$ is uninteresting—it occurs for some value $y_1 \in (0, h)$. $w_y = 0$, which occurs at y_2, is more interesting.

For a large N, a small value of p_z, or a small value of h, the extremum of w is outside the channel ($y_2 < 0$). But for small N (large p_z or h), w_y may change sign for $y \in (0, h)$. In the first case, w decreases from $\pi DN \cos\alpha$ at $y = h$ to zero at $y = 0$. In the second case, w passes through a minimum and since $w = 0$ at $y = 0$, it is negative close to the channel bottom. This flow reversal is of course unwanted since the volumetric output q from the extruder is directly related to w:

$$q = B\int_0^h w\, dy \tag{114}$$

Direct integration of (111) and (112) from y_1 to y and from y_2 to y gives

$$p_x(y - y_1) = C(u_y^2 + w_y^2)^{(n-1)/2} u_y$$
$$p_z(y - y_2) = C(u_y^2 + w_y^2)^{(n-1)/2} w_y$$

or in dimensionless form

$$\frac{hp_x}{C}(Y - Y_1)\left(\frac{\pi ND}{h}\right)^{-n} = (U_Y^2 \sin^2\alpha + W_Y^2\cos^2\alpha)^{(n-1)/2} U_Y \sin\alpha \tag{115}$$

$$\frac{hp_z}{C}(Y - Y_2)\left(\frac{\pi ND}{h}\right)^{-n} = (U_Y^2 \sin^2\alpha + W_Y^2\cos^2\alpha)^{(n-1)/2} W_Y \cos\alpha \tag{116}$$

Each of the equations is squared. The squared equations are added and the quantity V is found:

$$V = U_Y^2 \sin^2 \alpha + W_Y^2 \cos^2 \alpha = \left(\frac{hp_z}{C}\right)^{2/n}\left(\frac{\pi DN}{h}\right)^{-2}[P_1^2(Y - Y_1)^2 + (Y - Y_2)^2]^{1/n}$$

Now from (115)

$$\begin{aligned} U_Y \sin \alpha &= \frac{hP_1}{C} p_z(Y - Y_1)\left(\frac{\pi ND}{h}\right)^{-n} V^{-(n-1)/2} \\ &= P_1(Y - Y_1)\frac{h}{\pi DN}\left(\frac{hp_z}{C}\right)^{1/n}[P_1^2(Y - Y_1)^2 \\ &\quad + (Y - Y_2)^2]^{(1-n)/2n} \end{aligned} \tag{117}$$

Insertion of V in (116) gives

$$W_Y \cos \alpha = (Y - Y_2)\frac{h}{\pi DN}\left(\frac{hp_z}{C}\right)^{1/n}[P_1^2(Y - Y_1)^2 + (Y - Y_2)]^{(1-n)/2n} \tag{118}$$

Finally V is inserted in the energy balance (109) and a temperature dependence for C is introduced:

$$C = C_0 \exp[-A(T - T_0)] = C_0 \exp(-\theta) \tag{119}$$

$$\begin{aligned} &\frac{d^2\theta}{dY^2} + \frac{h^2}{k}AC_0 \exp(-\theta)(u_y^2 + w_y^2)^{(n+1)/2} \\ &= \frac{d^2\theta}{dY^2} + \frac{Ah^{(3n+1)/n}p_z^{(n+1)/n}}{k}C_0^{-1/n} \exp\left(\frac{\theta}{n}\right) \\ &\quad \times [P_1^2(Y - Y_1)^2 + (Y - Y_2)^2]^{(n+1)/2n} = 0 \end{aligned} \tag{120}$$

Integration of (117) and (118) from $Y = 0$ to Y gives U and W as functions of Y. When $Y = 1$, $U = W = 1$ and two equations in the unknown quantities Y_1, Y_2, and P_1 result.

$$\begin{aligned} \sin \alpha = \frac{hP_1}{\pi DN}\left(\frac{hp_z}{C_0}\right)^{1/n} \int_0^1 \exp\left(\frac{\theta}{n}\right)(Y - Y_1) \\ \times [P_1^2(Y - Y_1)^2 + (Y - Y_2)^2]^{(1-n)/2n}\, dY \end{aligned} \tag{121}$$

$$\begin{aligned} \cos \alpha = \frac{h}{\pi DN}\left(\frac{hp_z}{C_0}\right)^{1/n} \int_0^1 \exp\left(\frac{\theta}{n}\right)(Y - Y_2) \\ \times [P_1^2(Y - Y_1)^2 + (Y - Y_2)^2]^{(1-n)/2n}\, dY \end{aligned} \tag{122}$$

Finally from (106)

$$0 = \int_0^1 \int_0^Y \exp\left(\frac{\theta}{n}\right)(Y - Y_1)[P_1^2(Y - Y_1)^2 + (Y - Y_2)^2]^{(1-n)/2n}\, dY\, dY \tag{123}$$

If $\theta = A(T - T_0)$ is zero for $0 \le Y \le 1$, the problem is completely defined by the three definite integrals (121) to (123) to determine Y_1, Y_2, and P_1, and the integrals of (117) and (118) to give U and W as explicit functions of Y.

θ may be zero because $T = T_0$, in which case the energy balance is unnecessary. If $A = 0$ (a temperature-independent viscosity), (120) is reformulated in terms of $T - T_0$, and the temperature increase from $Y = 1$ to $Y = 0$ can be found from the resulting linear differential equation.

The real difficulty appears when the effect of a temperature-dependent viscosity ($A \neq 0$) is desired.

In this case the system of definite integrals (121) to (123) cannot be calculated unless the nonlinear boundary value problem (120), (107) has been solved for $\theta(Y)$; since (120) contains all three unknowns P_1, Y_1, and Y_2, the final model formulation looks like a minor version of the Gordian knot.

Zamodits has guessed P_1, Y_1, Y_2 and has solved (120) by finite differences. The problem is excellently suited for the numerical methods presented in this text. A reasonably accurate solution is obtained by the one-point collocation method of chapter 6 in which $\theta(Y)$, the solution of (120), is expressed by means of a parabola in Y. The integrals are evaluated by quadrature formulas and finally the whole problem is resolved in terms of four nonlinear algebraic equations.

1.7 Solution of Linear Differential Equations

The mathematical models that occur in the current chemical engineering literature are by and large nonlinear. It is the nonlinear character of the models that is interesting, and the research potential of linear models has largely been exhausted, especially by an intensive effort in the field of linear control theory.

Very few of the models of the previous sections have been tractable by the analytical methods that constitute the main body of typical courses in mathematics or "applied" mathematics. The idea of extracting useful information from a simplified, linear version of the model should, however, always be kept in mind. In section 1.2 the case of zero activation energy produced a linear model with all the interesting features of the nonlinear parent model, and in section 1.3 the linearized heat and mass balances provided useful information on the behavior of the systems in the vicinity of a steady state.

The importance of linear methods reaches far beyond a preliminary study of nonlinear models. In the final analysis all the numerical methods of the following chapters reduce to a few standard linear problems: the computation of the solution to a set of linear algebraic equations and the eigenvalue/eigenvector analysis of a matrix. Intermediate stages in this process are the study of coupled first-order differential equations with constant coefficients and the solution of linear partial differential equations in the form of an eigenfunction expansion.

In the present section these two main subjects will be briefly reviewed with no attempt to present the subject in depth. A comprehensive treatment of linear algebraic problems is given by Wilkinson (1965), and the fundamental principles of linear algebra applied to chemical engineering models is covered by Amundson (1966).

1.7.1 *N*-coupled first-order differential equations with constant coefficients

A short notation for N-coupled equations with constant coefficients is

$$\frac{d\mathbf{y}}{dt} = \mathbf{A}\mathbf{y} + \mathbf{b}, \qquad \mathbf{y}(t = 0) = \mathbf{y}_0 \tag{124}$$

$\mathbf{A}$ is an $(N \times N)$ matrix with constant elements A_{ij}, the coefficient of $y_j(t)$ in the ith equation. $\mathbf{b}$ is a constant vector, and a particular solution $\mathbf{y}_f$ of (124) is given by (125) if all eigenvalues of $\mathbf{A}$ are different from 0. If $d\mathbf{y}/dt \to 0$ for $t \to \infty$, $\mathbf{y}_f$ is the final (or steady state) value of $\mathbf{y}$.

$$\mathbf{A}\mathbf{y}_f = -\mathbf{b} \tag{125}$$

The general solution of a linear differential equation $L(\mathbf{y}) = \mathbf{0}$ is the sum of the solution of the homogeneous equation—i.e., the equation with all explicit functions of t deleted—and a particular solution. The solution of the homogeneous equation is a linear combination of N linearly independent solutions, each multiplied by an arbitrary constant c_i. The arbitrary constants are eliminated from the general solution by means of the side conditions—in the case of (124) from an initial condition $\mathbf{y}(t = 0) = \mathbf{y}_0$, whereas in other situations the N side conditions may be given at different values of t.

A trial solution of the homogeneous equation $d\mathbf{y}/dt = \mathbf{A}\mathbf{y}$ is taken as $\mathbf{u} \exp(\lambda t)$:

$$\mathbf{u}\lambda \exp(\lambda t) = \mathbf{A}\mathbf{u} \exp(\lambda t)$$

or

$$\mathbf{Au} = \lambda \mathbf{u} \tag{126}$$

(126) is the algebraic eigenvalue problem for matrix $\mathbf{A}$ and it is seen that each of N different eigenvalues of $\mathbf{A}$ produces one independent solution of the homogeneous differential equation.

In all the problems considered in this text, $\mathbf{A}$ is diagonable and the solution of (124) is

$$\mathbf{y}(t) = \sum_{1}^{N} c_i \mathbf{u}_i \exp(\lambda_i t) + \mathbf{y}_f \tag{127}$$

The arbitrary constants $\mathbf{c}$ can be found as the solution of N algebraic equations:

$$\mathbf{y}_0 - \mathbf{y}_f = \mathbf{Uc} \tag{128}$$

where $\mathbf{U}$ is a matrix composed of the eigenvectors $\mathbf{u}_i$ of $\mathbf{A}$. $\mathbf{y}_f$ is the solution $-\mathbf{A}^{-1}\mathbf{b}$ of (125).

Very efficient computer codes exist for diagonalization of arbitrary real matrices $\mathbf{A}$:

$$\mathbf{A} = \mathbf{U}\mathbf{\Lambda}\mathbf{U}^{-1} \tag{129}$$

In a *QR* algorithm (described in chapter 4) the eigenvalues λ_i and the eigenvectors $\mathbf{u}_i$ of $\mathbf{A}$ are found simultaneously with the eigenvectors of $\mathbf{A}^T$ (or the eigenrows of $\mathbf{A}$). If the eigenvalues are all real, $\mathbf{\Lambda}$ is a diagonal matrix with λ_i in element Λ_{ii}. $\mathbf{U}$ is the eigenvector matrix and $\mathbf{U}^{-1}$ the eigenrow matrix.

If $\mathbf{A}$ is symmetric, all eigenvalues λ_i are real. A nonsymmetric matrix may have complex eigenvalues and if an eigenvalue, say λ_i is complex, $\mathbf{\Lambda}$ is not diagonal but contains the real and complex part of λ_i and its complex conjugate in a 2×2 block centered on the diagonal (see exercises 1.11 and 4.14).

When (129) is inserted into (124), one obtains

$$\frac{d\mathbf{y}}{dt} = \mathbf{U}\mathbf{\Lambda}\mathbf{U}^{-1}\mathbf{y} + \mathbf{b}$$

$$\frac{d(\mathbf{U}^{-1}\mathbf{y})}{dt} = \mathbf{\Lambda}(\mathbf{U}^{-1}\mathbf{y}) + \mathbf{U}^{-1}\mathbf{b}$$

$$\mathbf{Y} = \mathbf{U}^{-1}\mathbf{y} = \exp(\mathbf{\Lambda}t)\mathbf{c} + \mathbf{Y}_f$$

$$\mathbf{c} = \mathbf{Y}_0 - \mathbf{Y}_f \quad \text{and} \quad \mathbf{Y} = \exp(\mathbf{\Lambda}t)\mathbf{Y}_0 + [\mathbf{I} - \exp(\mathbf{\Lambda}t)]\mathbf{Y}_f$$

$$\begin{aligned} \mathbf{y} &= \mathbf{U}\exp(\mathbf{\Lambda}t)\mathbf{U}^{-1}\mathbf{y}_0 + \mathbf{U}[\mathbf{I} - \exp(\mathbf{\Lambda}t)]\mathbf{U}^{-1}\mathbf{y}_f \\ &= \mathbf{U}\exp(\mathbf{\Lambda}t)\mathbf{U}^{-1}(\mathbf{y}_0 + \mathbf{A}^{-1}\mathbf{b}) - \mathbf{A}^{-1}\mathbf{b} \end{aligned} \tag{130}$$

The inverse of $\mathbf{A}$ is $\mathbf{A}^{-1} = \mathbf{U}\mathbf{\Lambda}^{-1}\mathbf{U}^{-1}$. $\mathbf{y}_f$ is first constructed in three steps: multiplication of $\mathbf{U}^{-1}$ by $-\mathbf{b}$ to form $\mathbf{v}_1$, division of the ith element of $\mathbf{v}_1$ by λ_i to form $\mathbf{v}_2$, and multiplication of $\mathbf{U}$ by $\mathbf{v}_2$ to form $\mathbf{y}_f$.

$\mathbf{y}_0 - \mathbf{y}_f$ is multiplied by $\mathbf{U}^{-1}$ to form $\mathbf{v}_3$, which is multiplied onto the diagonal matrix $\exp(\mathbf{\Lambda}t)$ to give $\mathbf{v}_4$ with ith element $v_{3i}\exp(\lambda_i t)$. Finally the component y_i of the solution vector $\mathbf{y}$ is obtained as the scalar product of $\mathbf{v}_5 = \mathbf{e}_i^T\mathbf{U}$ and $\mathbf{v}_4$:

$$\begin{aligned} y_i &= U_{i1}v_{31}\exp(\lambda_1 t) + U_{i2}v_{32}\exp(\lambda_2 t) + \dots \\ &\quad + U_{iN}v_{3N}\exp(\lambda_N t) + y_{fi} \end{aligned} \tag{131}$$

An Mth-order differential equation with constant coefficients is converted into a system of M first-order equations by introduction of a new variable for each derivative $y^{(1)}, y^{(2)}, \dots, y^{(M-1)}$. In this manner N coupled second-order equations are converted into $2N$ first-order equations:

$$\frac{d^2\boldsymbol{\theta}}{dt^2} - \mathbf{Q}\frac{d\boldsymbol{\theta}}{dt} - \mathbf{B}\boldsymbol{\theta} = \mathbf{0} \tag{132}$$

or

$$\begin{aligned} \frac{d\boldsymbol{\phi}}{dt} &= \mathbf{B}\boldsymbol{\theta} + \mathbf{Q}\boldsymbol{\phi} \\ \frac{d\boldsymbol{\theta}}{dt} &= \mathbf{I}\boldsymbol{\phi} \end{aligned} \tag{133}$$

$$\frac{d\boldsymbol{\psi}}{dt} = \mathbf{M}\boldsymbol{\psi} \tag{134}$$

$$\boldsymbol{\phi} = \frac{d\boldsymbol{\theta}}{dt} \quad \text{and} \quad \boldsymbol{\psi} = (\theta_1, \theta_2, \dots, \theta_N, \phi_1, \phi_2, \dots, \phi_N)^T$$

$$\mathbf{M} = \begin{bmatrix} \mathbf{0} & \mathbf{I} \\ \mathbf{B} & \mathbf{Q} \end{bmatrix}$$

which is a matrix of order $(2N \times 2N)$. Equation (134) is solved by the method of (129) and (130).

1.7.2 A linear partial differential equation

Equation (49) is a fairly general example of a linear partial differential equation. It is typically solved by separation of variables assuming that a solution θ_n of the equation can be written as the product of a function of x alone and a function of ζ alone:

$$\theta_n = F_n(x)G_n(\zeta) \tag{135}$$

Equation (135) is inserted into (49):

$$(1 - x^2)F_nG_n^{(1)} = G_n\nabla^2F_n + \mathrm{Pe}^{-2}F_nG_n^{(2)} \tag{136}$$

$$G_n^{(1)} = \frac{dG}{d\zeta} \quad \text{and} \quad G_n^{(2)} = \frac{d^2G}{d\zeta^2}$$

Assume that $G_n = a_n \exp(\lambda_n\zeta)$. In this case F_n is obtained as the solution of (137) with side conditions (138).

$$\nabla^2F_n - [(1 - x^2)\lambda_n - \mathrm{Pe}^{-2}\lambda_n^2]F_n = 0 \tag{137}$$

The side conditions of the homogeneous second-order differential equation (137) are given at $x = 0$ and $x = 1$. The general form of these are

$$\begin{aligned} \beta_1\frac{dF_n}{dx} + \alpha_1F_n &= 0 \quad \text{at } x = 0 \\ \beta_2\frac{dF_n}{dx} + \alpha_2F_n &= 0 \quad \text{at } x = 1 \end{aligned} \tag{138}$$

The case of (51) and (52a) is obtained for $\alpha_1 = \alpha_2 = 0$, while the case of (51) and (52b) is obtained for $\alpha_1 = 0$, $\beta_2 = 0$.

A homogeneous linear equation (137) with homogeneous boundary conditions (138) is satisfied by the trivial solution $F_n = 0$. A nontrivial solution appears only for specific values of λ_n, the eigenvalues of the linear differential operator

$$L = \nabla^2 - [(1 - x^2)\lambda_n - \mathrm{Pe}^{-2}\lambda_n^2] \tag{139}$$

The nontrivial solution F_n that is obtained for each of these values of λ_n is an eigenfunction. Equation (139) has infinitely many eigenvalues and corresponding eigenfunctions. In fact each region $\zeta < 0$ and $\zeta > 0$ produces a separate set of eigenfunctions for the problem of (49) and (51) and (52).

When Pe $\to \infty$, only the region $\zeta > 0$ is of interest and $\theta = 1$ for $\zeta = 0$, $\partial\theta/\partial x = 0$ at $x = 0$, $\theta = 0$ at $x = 1$ specifies the required side conditions. The solution (F_n, λ_n) of

$$\begin{gathered} \frac{1}{x}\frac{d}{dx}\left(x\frac{dF_n}{dx}\right) - (1 - x^2)\lambda_nF_n = 0 \\ \frac{dF_n}{dx} = 0 \quad \text{at } x = 0, \qquad F_n = 0 \quad \text{at } x = 1 \end{gathered} \tag{140}$$

yields the set of eigenfunctions and eigenvalues for (49) when Pe $\to \infty$.

The complete solution of (49) is a linear combination of the linearly independent solutions $\theta_n = F_nG_n$ or

$$\theta = \sum_1^\infty b_n^*F_n(x)\exp(\lambda_n\zeta) \tag{141}$$

To determine the arbitrary constants b_n^*, the initial condition

$$\theta(\zeta = 0) = 1 = \sum_1^\infty b_n^* F_n(x) \tag{142}$$

must be used. The method of computing b_n^* from (142) with F_n given by (140) appears as one of the results of the theory of Sturm–Liouville equations. A very brief review of the major features of this theory may serve to classify many of the differential equations to be treated in the following chapters. For a satisfactory general treatment the reader is referred to either Courant and Hilbert (1953) or to Titchmarsh (1946).

The general Sturm–Liouville problem is

$$\frac{d}{dx}\left[p(x)\frac{dy}{dx}\right] - [q(x) + \lambda W(x)]y = 0 \tag{143}$$

$$\begin{aligned} \beta_1 \frac{dy}{dx} + \alpha_1 y &= 0 \quad \text{at } x = a \\ \beta_2 \frac{dy}{dx} + \alpha_2 y &= 0 \quad \text{at } x = b \end{aligned} \tag{144}$$

$p(x)$, $q(x)$, $W(x)$ are given continuous functions of x. Assume $y_i(x)$ and $y_j(x)$ to be solutions of (143) for $\lambda = \lambda_i$ and λ_j, respectively:

$$\frac{d}{dx}\left(p\frac{dy_i}{dx}\right) - (q + \lambda_i W)y_i = 0 \tag{145}$$

$$\frac{d}{dx}\left(p\frac{dy_j}{dx}\right) - (q + \lambda_j W)y_j = 0 \tag{146}$$

Multiply (145) by y_j, (146) by y_i, and subtract

$$y_j\frac{d}{dx}\left(p\frac{dy_i}{dx}\right) - y_i\frac{d}{dx}\left(p\frac{dy_j}{dx}\right) = (\lambda_i - \lambda_j)Wy_iy_j$$

This equation is integrated over the finite interval a to b—using integration by parts on the left-hand side:

$$\begin{aligned} &\int_a^b \{y_j[py_i^{(1)}]^{(1)} - y_i[py_j^{(1)}]^{(1)}\}\,dx \\ &\quad = [y_jpy_i^{(1)} - y_ipy_j^{(1)}]\big|_a^b - \int_a^b py_i^{(1)}y_j^{(1)}\,dx + \int_a^b py_j^{(1)}y_i^{(1)}\,dx \\ &\quad = (\lambda_i - \lambda_j)\int_a^b Wy_iy_j\,dx \end{aligned} \tag{147}$$

If the eigenfunctions satisfy the homogeneous boundary conditions (144), the left-hand side of (147) is zero; since λ_i was assumed different from λ_j, for $i \neq j$ one obtains

$$\int_a^b W(x)y_i(x)y_j(x)\,dx = 0 \tag{148}$$

Equation (148) is the fundamental relation for orthogonal functions. The solutions F_n of boundary value problems associated with linear partial differential equations satisfy homogeneous boundary conditions. If the boundary value problem is of the Sturm–Liouville type (143), then the eigenfunctions F_n are orthogonal over $[a, b]$ with weight function $W(x)$ and all eigenvalues are real.

Equation (140) is a Sturm–Liouville problem and the eigenfunctions are orthogonal with weight function $x(1 - x^2)$. Equation (137) is not a Sturm–Liouville problem. For such eigenvalue problems it is not possible to prove in general that the eigenvalues are real and the nonorthogonality of the eigenfunctions is a serious numerical problem when the Fourier coefficients b_i^* of (141) are to be determined.

The orthogonality property of the solutions to Sturm–Liouville problems is used as follows:

Consider a function $f(x)$ that is integrable in $[a, b]$ but otherwise arbitrary. The Fourier series expansion for $f(x)$ on a set of eigenfunctions $\{y_i(x)\}$ of a given Sturm–Liouville problem with positive weight function $W(x)$ is

$$f(x) \sim S_n = \sum_{i=1}^{n} b_i^* y_i(x) \tag{149}$$

The partial sum S_n of the Fourier series converges to $f(x)$ in the following sense:

$$\int_a^b W(x)[f(x) - S_n(x)]^2\,dx \to 0 \quad \text{for } n \to \infty \tag{150}$$

The coefficient b_j^* in the n-term Fourier expansion $f(x) \sim S_n(x)$ is obtained when (149) is multiplied by $W(x)y_j(x)$ and integrated.

$$\begin{aligned}
&\int_a^b W(x)y_j(x)\left[\sum_{i=1}^{n} b_i^* y_i(x)\right] dx \\
&\quad = \sum_{i=1}^{n} b_i^* \int_a^b W(x)y_i(x)y_j(x)\,dx \\
&\quad = b_j^* \int_a^b W(x)[y_j(x)]^2\,dx = b_j^* C_j = \int_a^b W(x)y_j(x)f(x)\,dx
\end{aligned} \tag{151}$$

Every integral in the sum of n integrals is zero except that for which $i = j$. Formula (151) allows a one-by-one determination of the coefficients b_i^* of (149).

Furthermore, if $f(x)$ is differentiable in $[a, b]$, the series (149) converges uniformly to $f(x)$ in $[a, b]$; that is, $|S_n - f(x)| \to 0$ for any x in $[a, b]$. Quite a few functions $f(x)$ of practical interest (e.g., the square wave function) lack this additional analytical property and oscillations of $S_n - f(x)$ with nondecreasing amplitude are observed also for $n \to \infty$ near points of discontinuity of $f(x)$. For most purposes the "convergence in the mean" property of (150) is, however, what is desired.

The eigenfunctions that correspond to the one-dimensional diffusion equation in the x interval [0, 1] are particularly important. The Sturm–Liouville problem is

$$\frac{d}{dx}\left(x^s\frac{dF}{dx}\right) + \lambda x^s F = 0 \tag{152}$$

For $s = 0$, the eigenfunctions are harmonic functions $\sin x\sqrt{\lambda}$ or $\cos x\sqrt{\lambda}$. The eigenvalues λ depend on the coefficients (β_i, α_i) of (144).

For $s = 1$ or 2, (152) has one solution that is finite for all $x \in [0, 1]$. This solution is $J_0(x\sqrt{\lambda})$ when $s = 1$ and $(\sin x\sqrt{\lambda})/x$ when $s = 2$. The other solution (e.g., $\cos x\sqrt{\lambda}/x$ for $s = 2$) is infinite at $x = 0$.

The constants C_j of (151) can be found analytically for all three types of eigenfunctions in (152) and if $f(x)$ is a simple function (e.g., 1, x, or x^2), the integrals on the right-hand side of (151) can also be determined analytically.

This is an exceptional situation. In most cases the solutions of the Sturm–Liouville problem are only available in numerical form—for instance, as power series. Consequently the determination of b_j^* by (151), which looks deceptively simple, may in fact require a considerable numerical effort.

The two linear partial differential equations (82) and (83) are analyzed in chapter 9. We shall see that $\hat{y}$ (or $\hat{\theta}$) can be eliminated when $\text{Bi} = \text{Bi}_M$ (or if $\hat{y} = \hat{\theta} = 0$ at $x = 1$) and when at the same time $\text{Le} = 1$. The resulting equation is a Sturm–Liouville problem. $\boldsymbol{R}_y$ and $\boldsymbol{R}_\theta$ are complicated functions of x that are only available through numerical solution of the steady state equation. Even though the Sturm–Liouville problem can only be solved numerically, it is of great theoretical significance that (for $\text{Le} = 1$, $\text{Bi} = \text{Bi}_M$) the linear model for pellet dynamics is a Sturm–Liouville problem with real eigenvalues and mutally orthogonal eigenfunctions.

In the remaining part of this section we shall define the so-called Jacobi polynomials $P_n(x)$, the polynomial solutions of (154). They are orthogonal on [0, 1] with weight function $W(x) = x^\beta(1 - x)^\alpha$,

$[(\alpha, \beta) > -1]$. P_n does not satisfy homogeneous boundary conditions (144) at $x = 1$ and 0 but (154) is still a Sturm–Liouville problem because

$$p(x) = x^{\beta+1}(1 - x)^{\alpha+1} = 0 \quad \text{at } x = 0 \text{ and } x = 1 \quad \text{for } \alpha > -1, \beta > -1 \tag{153}$$

$$[p(x)y^{(1)}]^{(1)} + \lambda x^{\beta}(1 - x)^{\alpha} y = 0 \tag{154}$$

Differentiating the first term and dividing through by $W(x)$ yields

$$x(1 - x)\frac{d^2y}{dx^2} + [\beta + 1 - (\beta + \alpha + 2)x]\frac{dy}{dx} + \lambda y = 0 \tag{155}$$

It is easily seen that (155) has a solution $y_n(x)$ equal to a polynomial of degree n in x. The eigenfunctions (here polynomials) are only determined to within an arbitrary scalar factor, and if the polynomial is taken as

$$P_n(x) = \gamma_n x^n - \gamma_{n-1}x^{n-1} + \gamma_{n-2}x^{n-2} - \ldots + (-1)^n \tag{156}$$

it is shown in chapter 3 [equation (3.9)] that

$$\gamma_k = \frac{n - k + 1}{k}\,\frac{n + \alpha + \beta + k}{\beta + k}\gamma_{k-1} \tag{157}$$

$$k = 1, 2, \ldots, n \quad \text{and} \quad \gamma_0 = 1$$

It is now easy to show that the eigenvalues of (155) are

$$\lambda_n = n(n + \alpha + \beta + 1) \tag{158}$$

As an example, take $\alpha = \beta = -\frac{1}{2}$, in which case

$$x(1 - x)y_n^{(2)} + (\tfrac{1}{2} -_{a} x)y_n^{(1)} + n^2 y_n = 0 \tag{159}$$

The three first polynomials that satisfy (159) are

$$\begin{aligned} P_0 &= y_0 = 1 && \text{for } n = 0 \\ P_1 &= 2x - 1 && \text{for } n = 1 \\ P_2 &= 8x^2 - 8x + 1 && \text{for } n = 2 \end{aligned} \tag{160}$$

The resulting family of polynomials is named after Chebyshev.

The orthogonal polynomials that are solutions to (154) are used in various circumstances throughout the remainder of this text. Specifically, for the partial differential equations that have been discussed in this section, it will be shown that an expansion in an N-term series of orthogonal polynomials of each of the N eigenfunctions F_n, $n = 1, 2, \ldots, N$, in a Fourier series (141) truncated after N terms yields just as good results as the "true Fourier" series. In this way a standardized method for solution of partial differential equations is obtained. The "true" eigenfunctions F_n that may be complicated transcendental functions never appear explicitly. A detailed description of this approximation process appears in section 4.3.

EXERCISES

1. A tubular reactor for a homogeneous reaction has the following specifications: $L = 2$ m, $R = 0.1$ m, $T_W = 600$ °K. Inlet reactant concentration is $c_0 = 0.03$ kmol/m^3 and inlet temperature, 700 °K. Heat of reaction $-\Delta H = 10^4$ kJ/kmol, $U = 70$ J/m^{2}s °K, $c_p = 1$ kJ/kg °K, and $\rho = 1.2$ kg/m^2. $v_z = 3$m/s, $k_0 = 5$ s^{-1}.
Calculate Da, β, H_W, and θ_W of (1.32)
The activation energy is assumed to be zero and the reaction is first order in the reactant.
Calculate $\theta(\zeta)$ and $y(\zeta)$ and locate the hot spot.

2. The debate on the proper choice of boundary conditions for equations (1.35) and (1.36) has not yet been concluded. Deckwer (1974)[a] presents experimental results that seem to contradict the assumption of continuity of concentration across boundary surfaces. Give a comparison of the theoretical treatment in Deckwer's paper and that of section 1.4. Is there any clue in the experimental setup (figure 6 of the reference) that may explain the author's results?

3. Give a qualitative discussion of the early papers (ref. 19 and 20) on multiple steady states for model (1.35) and (1.36) compared to the newest papers (ref. 22). What are the major developments over the 10-year period between these papers?

4. Rony (1968, 1969)[b] was the first to give a theoretical model for transport restriction in a supported liquid phase (SLP) catalyst.
Give an analysis of his single pore model and present a list of the required dimensionless numbers and boundary conditions.
Calculate the effectiveness factor by solution of his model.
Are all quantities in the model available from experiments?

5. Compare the SLP model of Rony with that of the much later paper by Livbjerg, et al. (1974)[c].
What is the major development?
Are all quantities in the model now directly measurable?

[a] Deckwer, W. D., and Maehlmann, E. A. *Advances in Chemistry*, Series 133 (1974):334–47.

[b] Rony, P. R. *Chem. Eng. Sci.* 23 (1968):1021. *J. Catalysis* 14 (1969):142.

[c] Livbjerg, H., Sørensen, B., and Villadsen, J. *Advances in Chemistry*, Series 133 (1974):242–58.

6. Yu and Douglas[d] discuss a model for oscillations in catalyst particles. Elnashaie[e] has criticized their model.
Make your own "review" of Yu's and Douglas' model, and follow an eventual continuing "letters to the editor" debate.
What is the numerical problem in the model?
Give your own suggestion of how this should be treated.

7. Another way of obtaining oscillating reactions on catalyst pellets is suggested by Elnashaie and Cresswell[f].
Give an analysis of their model and explain why simultaneous adsorption and reaction is more likely to give oscillating behavior of the pellet than a chemical reaction alone.

8. Leonard and Bay Jørgensen[g] have recently given an up-to-date review of convection and diffusion in capillary beds. They treat the bed as composed of microcirculatory units—capillary tubes surrounded by tissue. A number of models is listed in their table 1. Make a careful comparison of models 1.2 to 1.5 for the capillary and 2.2 to 2.4 for the tissue. For each model a list of necessary side conditions should be given and a numerical treatment suggested.
In their discussion of capillary penetration (p.334), one reference [to D. G. Levitt in *Am. J. Physiol.* 220 (1971):250] seems to give an interesting modification of generally accepted boundary conditions between the microcirculatory units.
Give your own summary of this discussion, and study Lightfoot's treatment of the same problem (ref. 10, p. 331ff).

9. Two papers by Davis, et al.[h] and Cooney, et al.[i] propose numerical solutions to the mass transfer problem from capillaries to tissue.
Discuss their models and their proposals for a numerical solution in light of the review cited in Exercise 8.

10. Give your own summary of Lightfoot's discussion of the blood oxygenator (ref. 10, p. 380).
What are the specific problems that have to be mathematically modeled?
Make a list of references for a detailed quantitative study of this design problem.

11. (a) Prove that

$$\mathbf{A} = \begin{pmatrix} a & b \\ -b & a \end{pmatrix} = \frac{1}{2}\begin{pmatrix} 1 & i \\ i & 1 \end{pmatrix}\begin{pmatrix} a+ib & 0 \\ 0 & a-ib \end{pmatrix}\begin{pmatrix} 1 & -i \\ -i & 1 \end{pmatrix}$$

[d] Yu, K. M., and Douglas, J. M. *Chem. Eng. Sci.* 29 (1974):163.
[e] Elnashaie, S. S. *Chem. Eng. Sci.* 29 (1974):2133.
[f] Elnashaie, S. S., and Cresswell, D. L. *Chem. Eng. Sci.* 29 (1974):753.
[g] Leonard, E. F., and Bay Jørgensen, S. *Annual Review of Biophysics and Bioengineering* 3 (1974):293–339.
[h] Davis, E. J., Cooney, D. O., and Chang, R. *Chem. Eng. Journal* 7 (1974):213.
[i] Cooney, D. O., Kim, S. S., and Davis, E. J. *Chem. Eng. Sci.* 29 (1974):1731.

(b) Next show that for an arbitrary scalar t

$$\exp(\mathbf{A}t) = \begin{pmatrix} \cos bt & \sin bt \\ -\sin bt & \cos bt \end{pmatrix} \exp(at)$$

(c) When $\mathbf{A}$ has complex eigenvalues we are not able to diagonalize $\mathbf{A}$. Assume (for simplicity) that $\mathbf{A}$ has one pair of complex eigenvalues: $\lambda_j = a + ib$ and $\lambda_{j+1} = a - ib$ while all other eigenvalues $\lambda_1, \lambda_2, \ldots, \lambda_{j-1}, \lambda_{j+2}, \ldots, \lambda_N$ are real. $\mathbf{A}$ can now be transformed to

$$A = \mathbf{UDBU}^{-1}$$

where $\mathbf{DB}$ is diagonal except for a 2×2 block at position j and $j + 1$. The columns of $\mathbf{U}$ that correspond to row j and $j + 1$ of $\mathbf{DB}$ are $\mathbf{u}_j$ and $\mathbf{u}_{j+1}$.

Show that the kth component $y_k(t)$ of the solution of

$$\frac{d\mathbf{y}}{dt} = \mathbf{Ay} \quad \text{with } \mathbf{y}(t = 0) = \mathbf{y}_0$$

is

$$y_k(t) = u_{k1}v_1 \exp(\lambda_1 t) + u_{k2}v_2 \exp(\lambda_2 t) \ldots + (q \cos bt + r \sin bt) \exp(at) + \ldots + u_{kN}v_N \exp(\lambda_N t)$$

u_{ki} is the kth component of $\mathbf{u}_i$ and v_i is the ith component of $\mathbf{U}^{-1}\mathbf{y}_0$.

$$q = u_{kj}v_j + u_{k(j+1)}v_{j+1} \quad \text{and} \quad r = u_{kj}v_{j+1} - u_{k(j+1)}v_j$$

As a trivial numerical example, you may use

$$\mathbf{A} = \begin{pmatrix} -6 & 4 & -7 \\ 0 & -2 & 3 \\ 1 & -1 & -1 \end{pmatrix}$$

where the eigenvalues are $a \pm ib = -\frac{7}{2} \pm i(\sqrt{15}/2)$ and $\lambda_3 = -2$. The solution of $d\mathbf{y}/dt = \mathbf{Ay}$ with $\mathbf{y}_0 = (1, 0, 0)$ is

$$y_1 = (0.5 \cos bt - 1.67828 \sin bt) \exp(at) + 0.5 \exp(-2t)$$

$$y_2 = (-0.5 \cos bt - 0.38729 \sin bt) \exp(at) + 0.5 \exp(-2t)$$

$$y_3 = 0.51639 \sin bt \exp(at)$$

REFERENCES

Even 17 years after its publication, *Transport Phenomena* by Bird, Stewart, and Lightfoot appears to be an outstanding reference for a general treatment of transport phenomena from a chemical engineer's point of view. Bennet and Meyer's text (1962) may be somewhat easier to digest and it attempts to give a unified treatment of "theoretical" transport phenomena and applications to unit operations. Slattery's text (1972) is very similar to *Transport Phenomena* in scope as well as in details. Development in this field runs parallel in the West and in Russia, as seen from Luikov's text (1965); however, this is considerably more mathematically oriented than any of the previous references.

Many excellent books exist on fluid flow problems. Schlichting (1968) is a classic on boundary layer theory, as Batchelor (1967) is on fluid dynamics. A good introductory text is Whitaker (1968), perhaps supplemented by, e.g., Skelland (1967) for non–Newtonian flow problems. Any serious quantitative treatment of hydrodynamics will refer to Milne–Thomson's classical book (1968).

Biological applications of transport phenomena—perhaps the most promising field of further development in the subject—are described in several recent texts. Lightfoot (1974) is easily accessible to anyone familiar with *Transport Phenomena.* The cardiovascular system is treated in Middleman (1972) and the artificial kidney in Leonard, et al. (1968), to mention two specific subjects. Sonrirajan (1970) and Lakshminarayanaiah (1969) may be used as further references to the reverse osmosis example of section 1.5.

Most of what is said in sections 1.2 to 1.4 is touched upon by Froment, whose reviews at the Reaction Engineering Symposia in 1970 and 1972 (and in several papers on the same subjects) are major contributions to a systematic treatment of various chemical reactor systems. Petersen (1965) and Aris (1969) are indispensable texts for the basic design problems of chemical reactors. Chapters 8 and 9 of Aris (1969) are especially useful for an understanding of the physical background of the reactor models.

Computations on the tubular reactor with axial dispersion was started in the early 1960s by Raymond and Amundson (1964) and further development is traced through a series of papers by Amundson and co-workers: Amundson (1965), Luss and Amundson (1967), Varma and Amundson (1972, 1973, and 1974); the 1973 papers show what astonishing computer results can be obtained from the fairly simple model (1.35) and (1.36). It is a good bet that even the detailed treatment that the problem has received in the last few years does not mark the end of this research topic. Independent results on the tubular reactor with axial dispersion are reported by Hlavaçek (1971) and McGowin and Perlmutter (1971).

Taylor diffusion is treated by Wicke (1975) in a recent review of the physical processes in reactor design. The original paper by Taylor (1953) is recommended as a masterpiece of scientific writing.

There are numerous papers on the Graetz problem of subsection 1.2.4 and on the extended Graetz problem (i.e., with axial conduction). Some of these papers are referred to in chapter 9.

The steady state and transient behavior of solid catalyst particles is the subject matter of Aris' monumental treatise (1975). Apart from the almost overwhelming but always fascinating mathematical exposure that constitutes the main body of this book, two introductory chapters give an outline of the physical problem and its historical background that with respect to clarity of presentation will be difficult to surpass. The many hundreds of references cited in Aris (1975) illustrate what possibilities for research work the contents of section 1.3 has offered over the last decade. Even though Carberry (1974) wants to declare a moratorium on steady state effectiveness factor calculations, he will probably not succeed in enforcing it—he may not even be willing to abstain from further publications in the field himself [cf. Carberry (1975)]. A short but very readable review that also includes a number of references to experimental investigations on steady state and transient effectiveness factor problems is given by Schmitz (1975).

The transient models for coupled fluid phase–solid phase systems may be studied in the text by Aris and Amundson (1973). Their object is to give a detailed treatment of first-order partial differential equations and their applications. A great number of models for adsorbers, chemical reactors with poisoning of the catalyst, and other chemical engineering systems are set up and solved.

Waede Hansen has contributed significantly to the subject of model simplification for catalyst pellets and for pellets coupled to a fluid phase. Three of his papers are listed in the references.

Reverse osmosis is treated in references 10, 13, and 14. The paper analyzed in section 1.5 (Gill and Bansal, 1973) has been followed up by an experimental investigation (Dandavati, 1975).

The paper by Zamodits and Pearson (1969) of section 1.6 refers only to a small portion of the work carried out by Pearson and co-workers in cooperation with British industry. Several reports and computer programs for design of extruders have been published as a result of this work.

Without further comment, references 38 to 51 are listed in their order of appearance in the text.

1. BIRD, R. B., STEWART, W. E., and LIGHTFOOT, E. N. *Transport Phenomena*, New York: Wiley (1960).

2. BENNET, C. O., and MEYERS, J. E. *Momentum, Heat and Mass Transfer*, New York: McGraw-Hall (1962).

3. SLATTERY, J. C. *Momentum, Energy and Mass Transfer in Continua*, New York: McGraw-Hill (1972).

4. LUIKOV, A. V., and MIKHAILOV, YU A. *Theory of Heat and Mass Transfer*. Translated from Russian by Israel Program for Scientific Translations, Jerusalem (1965).

5. SCHLICHTING, H. *Boundary Layer Theory*, New York: McGraw-Hill (1968).

6. BATCHELOR, G. K. *An Introduction to Fluid Dynamics*, Cambridge: Cambridge University Press (1967).

7. WHITAKER, S. *Introduction to Fluid Mechanics*, Englewood Cliffs, N.J.: Prentice-Hall (1968).

8. SKELLAND, A. H. P. *Non-Newtonian Flow and Heat Transfer*, New York: Wiley (1967).

9. MILNE-THOMSON, L. M. *Theoretical Hydrodynamics*, 5th ed. New York: Macmillan (1968).

10. LIGHTFOOT, E. N. *Transport Phenomena and Living Systems*, New York: Wiley (1974).

11. MIDDLEMAN, S. *Transport Phenomena in the Cardiovascular System*. Biomedical Engineering Series. New York: Wiley (1972).

12. LEONARD, E. F., ET AL., ed. "The Artificial Kidney." *Chem. Eng. Progress Symposium*, Series 84, Vol. 64 (1968).

13. SONRIRAJAN, S. *Reverse Osmosis*, London: Logos Press (1970).

14. LAKSHMINARAYANAIAH, N. *Transport Phenomena in Membranes*, New York: Academic Press (1969).

15. FROMENT, G. F. *Advances in Chemistry*, Series 109 (1972):1–34. Proceedings 1st *ISCRE*, Washington (1970).

16. FROMENT, G. F. A5-1 to A5-20 in Proceedings 2nd *ISCRE*, Amsterdam (1972). Elsevier (1972).

17. PETERSEN, E. E. *Chemical Reaction Analysis*, Englewood Cliffs, N.J.: Prentice-Hall (1965).

18. ARIS, R. *Elementary Chemical Reactor Analysis*, Englewood Cliffs, N.J.: Prentice-Hall (1969).

19. RAYMOND, L. R., and AMUNDSON, N. R. *Can. J. Chem. Eng.* 42 (1964):173.

20. AMUNDSON, N. R. *Can. J. Chem. Eng.* 42 (1965):49.

21. LUSS, D., and AMUNDSON, N. R. *Can. J. Chem. Eng.* 45 (1967):341.

22. VARMA, A., and AMUNDSON, N. R. *Can. J. Chem. Eng.* 50 (1972):470. *Ibid.*, 51 (1973):206; and 52 (1974):580.

23. HLAVAÇEK, V., HOFMANN, H., and KUBIÇEK, M. *Chem. Eng. Sci.* 26 (1971):1639.

24. MCGOWIN, C. R., and PERLMUTTER, D. D. *Chem. Eng. Journal* 2 (1971):125.

25. WICKE, E. *Advances in Chemistry*, Series 148 (1975):82.

26. TAYLOR, G. *Proc. Roy. Soc.*, London, Series A, 219 (1953):186.

27. ARIS, R. *The Mathematical Theory of Diffusion and Reaction in Permeable Catalysts*, Oxford: Clarendon Press (1975).

28. CARBERRY, J. J., and BUTT, J. B. *Catalysis Reviews—Sci. Eng.* 10 (1974):221–42.

29. SMITH, T. G., ZAHRADNIK, J., and CARBERRY, J. J. *Chem. Eng. Sci.* 30 (1975):763.

30. ARIS, R., and AMUNDSON, N. R. *Mathematical Methods in Chemical Engineering. Vol 2: First-Order Partial Differential Equations with Applications*, Englewood Cliffs, N.J.: Prentice-Hall (1973).

31. GILL, W. N., and BANSAL, B. *AIChE J.* 19 (1973):826.

32. DANDAVATI, M. S., DOSHI, M. R., and GILL, W. N. *Chem. Eng. Sci.* 30 (1975):877.

33. ZAMODITS, H., and PEARSON, J. R. A. *Trans. Soc. Rheol.* 13 (1969):357.

34. SCHMITZ, R. *Advances in Chemistry*, Series 148 (1975):156.

35. WAEDE HANSEN, K. *Chem. Eng. Sci.* 26 (1971):1555.

36. WAEDE HANSEN, K. *Chem. Eng. Sci.* 28 (1973):723.

37. WAEDE HANSEN, K., and BAY JØRGENSEN, S. *Advances in Chemistry*, Series 133 (1974):505.

38. GUNN, D. J., and ENGLAND, R. *Chem. Eng. Sci.* 26 (1971):1413.

39. BISCHOFF, K. B. *Chem. Eng. Sci.* 16 (1961):731.

40. SATTERFIELD, C. N. *Mass Transfer in Heterogeneous Catalysis*. Cambridge, Mass: MIT Press (1970).

41. LIVBJERG, H., and VILLADSEN, J. *Chem. Eng. Sci.* 27 (1972):21.

42. VILLADSEN, J. *Selected Approximation Methods for Chemical Engineering Problems*, Lyngby, Denmark: Instituttet for Kemiteknik (1970).

43. HOIBERG, J. A., LYCKE, B. C., and FOSS, A. S. *AIChE J.* 17 (1971):1434.

44. SØRENSEN, J. P. *Chem. Eng. Sci.* 31 (1976):719.

45. FROMENT, G. F. *Ind. Eng. Chem.* 59 (1967):18.

46. KJAER, J. *Computer Methods in Catalytic Reactor Calculations*, rev. ed. Vedbæk, Denmark: Haldor Topsøe (1972).

47. TROWBRIDGE, E. A., and KARRAN, J. H. *Int. J. Heat Mass Transfer* 16 (1973):1833.

48. WILKINSON, J. H. *The Algebraic Eigenvalue Problem*, Oxford: Clarendon Press (1965).

49. AMUNDSON, N. R. *Mathematical Methods in Chemical Engineering. Vol. 1: Matrices and Their Application*. Englewood Cliffs, N.J.: Prentice-Hall (1966).

50. Courant, R., and Hilbert, D. *Methods of Mathematical Physics*, New York: Interscience Publishers (1953).

51. Titchmarsh, E. C. *Eigenfunction Expansions*, Oxford: Clarendon Press (1946).

Polynomial Approximation—A First View of Construction Principles 2

Introduction

In this chapter (sections 2.2 to 2.4), the methods of weighted residuals (MWR) are introduced.

We use a fairly simple example—that of an isothermal, irreversible nth-order reaction on cylindrical catalyst pellets—to illustrate differences in quality between various approximations by polynomials of degree N. It is seen that the approximation is improved when N is increased and also that quite different accuracy is obtained with different MWR.

The example is sufficiently simple to allow an analytical solution y to be obtained when the reaction order is $n = 1$. In section 2.1 a Taylor series solution of the analytical solution is discussed. It is shown that the analytical solution is not necessary to obtain a Taylor series approximation of order N.

Differences between the Nth-order polynomial approximation y_N by Taylor series and by MWR are studied by means of the distance function $E_N = y - y_N$ and it is noted that the properties of the Taylor series approximation are less desirable than the properties of the approximations obtained by MWR.

Section 2.2 treats the first approximation by MWR; in section 2.3 a generalization to the Nth order approximation is given. The linear case $n = 1$ is used in both sections.

A nonlinear case ($n = 2$) is studied in section 2.4 and an important analogy between two MWR—one which gives very accurate results and one which is very easy to apply—is discussed.

No new approximation methods are introduced in section 2.5; however, this section contains some important material. In the previous sections approximations in ascending powers of the independent variable are always used. Here it is shown that these finite series can be reformulated either into a finite series in polynomials of ascending degree or into a sum of Nth-degree polynomials—the Lagrange interpolation polynomial for the approximated function.

Finally it is shown that one particular MWR—the Galerkin method—gives the optimal set of N approximation constants a_i when applied to the solution of the linear isothermal reaction-with-diffusion problem.

2.1 A Taylor Series Approximation

In this chapter we shall treat the model for diffusion and nth-order irreversible, isothermal reaction in the radial direction of a cylindrical catalyst pellet. The dimensionless concentration profile $y(x)$ is obtained from equation (1.69), letting $\gamma = 0$ or $\theta = 1$ for all x:

$$\frac{d^2y}{dx^2} + \frac{1}{x}\frac{dy}{dx} - \Phi^2 y^n = 0 \tag{1}$$

with boundary conditions

$$\frac{dy}{dx} = 0 \quad \text{at } x = 0 \quad \text{and } y = 1 \quad \text{at } x = 1 \tag{2}$$

In particular we wish to calculate the effectiveness factor η, which is given by (1.72)

$$\eta = \int_0^1 y^n \, dx^2 \tag{3}$$

or, by integration of (1),

$$\eta = \frac{2}{\Phi^2}\left(\frac{dy}{dx}\right)_{x=1} \tag{4}$$

An analytical solution of (1) and (2) is possible for $n = 0$ and 1. For the case $n = 1$ one obtains

$$y(x) = \frac{I_0(\Phi x)}{I_0(\Phi)} \quad \text{and} \quad \eta = \frac{2}{\Phi}\frac{I_1(\Phi)}{I_0(\Phi)} \tag{5}$$

Where I_0 and I_1 are modified Bessel functions of order zero and one.

These are expressed by the following series in ascending powers of the argument:

$$I_\nu(z) = \left(\frac{1}{2}z\right)^\nu \sum_{i=0}^{\infty} \frac{(\frac{1}{4}z^2)^i}{i!\Gamma(\nu + i + 1)} \tag{6}$$

$$\nu = 0: I_0(z) = \sum_{i=0}^{\infty} \frac{(\frac{1}{4}z^2)^i}{(i!)^2} = 1 + \frac{1}{4}z^2 + \frac{1}{4}\left(\frac{1}{4}z^2\right)^2 + \frac{1}{36}\left(\frac{1}{4}z^2\right)^3 + \dots$$

$$\nu = 1: I_1(z) = \frac{1}{2}z \sum_{i=0}^{\infty} \frac{(\frac{1}{4}z^2)^i}{i!(i+1)!} = \frac{1}{2}z\left(1 + \frac{1}{2}\left(\frac{1}{4}z^2\right) + \frac{1}{2\cdot 2\cdot 3}\left(\frac{1}{4}z^2\right)^2 + \dots\right)$$

$$= \frac{dI_0(z)}{dz}$$

$$y(x) = \frac{1 + (\Phi^2x^2/4) + \frac{1}{4}(\Phi^2x^2/4)^2 + \frac{1}{36}(\Phi^2x^2/4)^3 + \dots}{1 + (\Phi/2)^2 + \frac{1}{4}(\Phi/2)^4 + \frac{1}{36}(\Phi/2)^6 + \dots} \tag{7}$$

$$\begin{aligned}\eta &= \frac{2}{\Phi}\,\frac{\frac{1}{2}\Phi(1 + \frac{1}{2}(\Phi/2)^2 + \frac{1}{12}(\Phi/2)^4 + \frac{1}{144}(\Phi/2)^6 + \dots)}{1 + (\Phi/2)^2 + \frac{1}{4}(\Phi/2)^4 + \frac{1}{36}(\Phi/2)^6 + \dots}\\ &= \frac{1 + \frac{1}{2}(\Phi^2/4) + \frac{1}{12}(\Phi^2/4)^2 + \frac{1}{144}(\Phi^2/4)^3 + \dots}{1 + (\Phi^2/4) + \frac{1}{4}(\Phi^2/4)^2 + \frac{1}{36}(\Phi^2/4)^3 + \dots}\end{aligned} \tag{8}$$

Both series (7) and (8) are rapidly convergent for small values of z ($\Phi/2$ or $\Phi x/2$); for $z < 1$

$$y_1(x) = \frac{1 + (\Phi x/2)^2}{1 + (\Phi/2)^2} \quad \text{and} \quad \eta_1 = \frac{1 + \frac{1}{2}(\Phi/2)^2}{1 + (\Phi/2)^2} \simeq 1 - \frac{\Phi^2}{8} \tag{9}$$

are reasonably good approximations to the profile and to the effectiveness factor. Note that the same number of terms should be used in the numerator and denominator of (7) to satisfy the boundary condition at $x = 1$. It is seen from the form of (7) and (8) that approximations $y_N(x)$

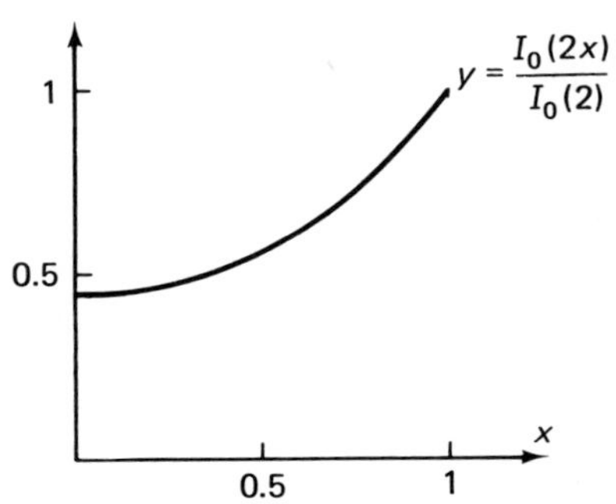

Figure 2-1. Solution of equation (2.1) for $n = 1$.

and η_N with $N + 1$ terms in the numerator and denominator are always above the exact values $y(x)$ and η for any Φ and $x < 1$.

Figure 2-1 shows $y(x)$ as computed by (7) for $\Phi = 2$. The shape of the curve indicates that the approximation (9) by a parabola is quite satisfactory. This is substantiated by table 2.1, which gives values of the so-called linear distance function:

$$E_N(x) = y(x) - y_N(x)$$

and by table 2.2, which shows the rapid convergence of η_N to η. $E_N(x)$ is of one sign throughout the interval as explained above; for $N = 1$, the error level is approximately 0.065 except close to $x = 1$. $E_N(1) = 0$ by definition of the method for obtaining $y_N(x)$.

TABLE 2.1
ERROR IN TAYLOR SERIES APPROXIMATIONS FOR $\Phi = 2$

x	$-E_1(x)\,10^2$	$-E_2(x)\,10^3$	$-E_3(x)\,10^4$	$y(x)$
0	6.1	5.7	3.3	0.439
0.2	6.4	6.0	3.5	0.456
0.4	6.8	6.7	3.9	0.512
0.6	6.9	7.4	4.6	0.611
0.8	5.2	6.7	4.5	0.768
1.0	0.0	0.0	0.0	1

TABLE 2.2
EFFECTIVENESS FACTORS FROM
TAYLOR SERIES APPROXIMATIONS FOR $\Phi = 2$

$$\eta_1 = 0.75$$
$$\eta_2 = 0.704$$
$$\eta_3 = 0.6982$$
$$\eta_{ex} = 0.6978$$

A serious objection to the procedure that leads to the polynomial approximation $y_N(x)$ by (7) truncated after the $(N + 1)$ term in the numerator via the exact solution (5) is that this procedure assumes a knowledge of the function $I_0(z)$ and published series representations of this function. Only a very limited number of differential equations have solutions that are named and discussed in literature, and in our example (1) any value of n except $n = 0$ and 1 will bring the problem outside the haven of these "known" exact solutions.

Furthermore the detour via (5) on the route from (1) to (7) is completely unnecessary since an algorithm exists for calculation of the coefficients in the Taylor series expansion of $y(x)$ for any differential equation. The differential equation (1) and its boundary conditions (2) contain exactly the same information regarding $y(x)$ as the exact solution (5). When we wish to study the Taylor series expansion of (5) truncated after $N + 1$ terms, we may just as well start from (1) and (2). The derivation of the Taylor series follows most conveniently if (1) is rewritten in terms of $u = x^2$ for $0 \le x \le 1$:

$$\frac{dy}{dx} = \frac{dy}{du} \cdot \frac{du}{dx} = \frac{dy}{du} \cdot 2\sqrt{u} \quad \text{for } u \ge 0 \tag{10}$$

$$\frac{d^2y}{dx^2} = \frac{d}{dx}\left(2\sqrt{u}\frac{dy}{du}\right) = 2\frac{d}{du}\left(\sqrt{u}\frac{dy}{du}\right)\frac{du}{dx} = 2\frac{dy}{du} + 4u\frac{d^2y}{du^2}$$

Equation (1) is reformulated to

$$u\frac{d^2y}{du^2} + \frac{dy}{du} = \frac{\Phi^2}{4}y = py \tag{11}$$

The boundary condition $dy/dx = 0$ at $x = 0$ is automatically satisfied in (11) since y is only a function of $u = x^2$. The boundary conditions of (11) are consequently $y(u = 1) = 1$ and $d^ky/du^k = y^{(k)}$ finite at $u = 0$ for $k = 0, 1, \ldots.$

Differentiating (11) k times with respect to u yields

$$uy^{(k+2)} + (k + 1)y^{(k+1)} = py^{(k)} \tag{12}$$

Introduce $y_0 = y(u = 0) = y^{(k)}$ $(k = 0, u = 0)$. Equation (12) can now be used to construct derivatives of any order at $u = 0$ by the following recurrence formula:

$$y_0^{(k+1)} = y^{(k+1)}(u = 0) = \frac{p}{k + 1}y_0^{(k)} \tag{13}$$

$$y_0^{(k+1)} = \frac{p^2}{(k + 1)k}y_0^{(k-1)} = \ldots = \frac{p^{k+1}}{(k + 1)!}y_0 \tag{14}$$

The Taylor series for $y(u)$ from $u = 0$ is defined by

$$y(u) = \sum_0^\infty \frac{1}{k!}\frac{d^{(k)}y}{du^k}\bigg|_{u=0} u^k = \sum_0^\infty b_k u^k \tag{15}$$

Equation (14) for $y_0^{(k)}$ inserted into (15) gives the following expression for b_k, $k = 0, 1, 2, \ldots.$

$$b_k = \frac{p^k}{(k!)^2}y_0$$

and y_0 is finally determined by the requirement $y = 1$ at $u = 1$:

$$y_N(u) = \sum_0^N b_k u^k = \frac{\sum_0^N [p^k/(k!)^2]u^k}{\sum_0^N [p^k/(k!)^2]} \qquad N = 1, 2, \ldots, \infty \tag{16}$$

Equation (16) is identical to (7).

Exactly the same procedure may be followed for $n \neq 1$ although the computational work is heavier since no simple recurrence formula (13) exists for the derivatives $y^{(k)}$ at $u = 0$.

The reason we do not wish to close the subject of polynomial approximation at this point is that much better polynomial approximations exist than those based on ascending powers of x (or u) multiplied by Taylor coefficients (16).

The constant negative sign of $E_N(x)$ for any Φ and N immediately signifies that a better approximation can be found.

Take approximation (9) by a second-degree polynomial y_1. E_1 is negative throughout (0, 1) and has a minimum value somewhere in (0, 1). Hence addition of $a \cdot (1 - x^2)$ with $a < 0$ to y_1 will give a new second-degree polynomial approximation to y with a smaller maximum value of $-E_1$ or in general a smaller value of the functional

$$F = \int_0^1 W(x)|y - y_1|^s \, dx \tag{17}$$

for any nonnegative function $W(x)$ and $s > 0$.

This improvement can be obtained when one of the construction principles of section 2.2 is used to find n coefficients a_i of y_N in place of the N Taylor coefficients b_k of (16). All these methods construct the coefficients a_i by operations on the residual $R_N(x)$, which is the function that appears when y_N is substituted into the differential equation in place of y. Hence it may be interesting to make a brief study of the residual $R_N(u)$ for the Taylor series method.

The manner in which the recurrence formula (13) was used to construct the Taylor coefficients b_k of (15) immediately shows that

$$R_N^{(k)}[y_N(u)] = R_N^{(k)}(u) = 0 \quad \text{at } u = 0 \quad \text{for } k = 0, 1, \ldots, N-1 \tag{18}$$

i.e., that the residual $R_N(u)$ and its first $N - 1$ derivatives with respect to u are zero at $u = 0$. Conversely the Taylor series method may be defined (somewhat artificially) as a method in which the N constants b_i of (16) are found as the solution of the N equations (18).

The methods in section 2.2 apply information on $R_N(u)$ throughout the interval $0 \leq u \leq 1$ to construct the coefficients. Intuitively such procedures would appear to give "better" approximations by y in the given u-interval (0, 1) than the extremely one-sided Taylor series approximation (16).

These remarks concerning the quality of Taylor series approximations are suggested by the peculiar fashion in which $R_N(u)$ is treated in this method. In general, such evidence is at best circumstantial: A small residual at some u-value does not always imply that $E_N = y - y_N$ is small at this point or that any other relevant measure of approximation accuracy such as $\eta - \eta_N$ in (3) is small. The constant sign of E_N that shows that a better approximation of the same order can be constructed is much stronger evidence against the Taylor series.

Even though quantitative information on important measures of the approximation accuracy [e.g., the maximum value of $|E_N(u)|$] is difficult to extract from a study of the residual $R_N(u)$, this function is of great importance in approximation theory. In the absence of an analytical solution $y(u)$, the distance function $E_N = y(u) - y_N(u)$ is not available but the residual $R_N(u)$ is a known function of u once the coefficients a_i are given. Some results on bounds for $|E_N(u)|$ based on the residual have been reported by Ferguson and Finlayson (1972) for linear problems.

It may be proved that some of the methods in section 2.2 will guarantee that $R_N(u)$ converges uniformly to zero for $N \to \infty$ in the range of the independent variable. From this it may be proved that $E_N(u)$ converges uniformly to zero for $N \to \infty$ also when $y(x)$ does not admit to a convergent Taylor series throughout the interval.

At present it is sufficient to note that Taylor series approximations are unsatisfactory from a practical viewpoint because more rapidly convergent approximations can be constructed. Also in some problems they may give completely erroneous results when the radius of convergence of the series is less than the interval length.

2.2 The Lowest-Order MWR Approximations $y_1(x)$

We shall now study different methods of obtaining the coefficient a_1 in the first approximation $y_1(x)$ to $y(x)$ of equation (1):

$$y_1(x) = 1 + a_1(1 - x^2) \tag{19}$$

Equation (19) is a slight reformulation of the expression that was used in the previous section. Here the term $a_1(1 - x^2)$ appears as a perturbation function to the known boundary value $y = 1$ at $x = 1$. Equation (19) satisfies both boundary conditions (2), and for the Taylor series one obtains

$$a_1 = -\frac{\Phi^2}{4 + \Phi^2} \tag{20}$$

by rewriting (9) in the form (19).

Inserting $y_1(x)$ from (19) into the differential equation (1) with $n = 1$, one obtains the residual $R_1(x)$ or $R_1(a_1, x)$ to accentuate that the function $R_1(x)$ depends on the choice of a_1:

$$R_1[y_1(x)] = R_1(x) = R_1(a_1, x) = -4a_1 - \Phi^2[1 + a_1(1 - x^2)] \tag{21}$$

No choice of a_1 will make $R_1(x)$ zero throughout (0, 1) since this would mean that (1) was satisfied by a second-degree polynomial. One adjustable constant is, however, enough to obtain a mean value of R_1 in (0, 1) or the mean value of a weight function multiplied by R_1 equal to zero. This looks like an entirely logical procedure when we wish to avoid giving undue preference to any particular part of the interval (0, 1).

Thus we determine a_1 by the following equation:

$$\int_0^1 R_1(a_1, x)W(x)\,dx^2 = 0 \tag{22}$$

or, in general, a_1 is determined such that the weighted residual $W \cdot R$ has a mean value of zero in the volume of the system.

Each of the various methods of weighted residuals (MWR) is characterized by a specific choice of the weight function $W(x)$; we now shall compare four choices of MWR:

1. Define $W(x)$ by

$$W(x) = \begin{cases} \infty & \text{for } x = x_1 \\ 0 & \text{for } x \neq x_1 \end{cases} \quad \text{or} \quad W(x) = \delta(x - x_1) \tag{23}$$

in which case (22) degenerates to

$$R_1(a_1, x_1) = 0 \tag{24}$$

a_1 is determined such that the residual R_1 of (21) is zero at one interior point x_1 in (0, 1).

As an example, $x_1 = \frac{1}{2}$ is chosen:

$$-4a_1 - \Phi^2[1 + a_1(1 - \tfrac{1}{4})] = 0$$

$$a_1 = \frac{-4\Phi^2}{16 + 3\Phi^2} \tag{25}$$

For $\Phi = 2$, one obtains $a_1 = -\frac{4}{7}$, $y_1 = \frac{3}{7} + \frac{4}{7}x^2$, and $\eta_1 = \frac{5}{7} = 0.714$.

2. Let the mean value of R_1 be zero; i.e., $W(x) = 1$:

$$\int_V R_1 W\,dV = \int_0^1 \{-4a_1 - \Phi^2[1 + a_1(1 - x^2)]\}\,dx^2 = 0$$

or

$$-4a_1 - \Phi^2(1 + \tfrac{1}{2}a_1) = 0$$

$$a_1 = \frac{-2\Phi^2}{8 + \Phi^2} \tag{26}$$

For $\Phi = 2$, $a_1 = -\frac{2}{3}$, $y_1 = \frac{1}{3} + \frac{2}{3}x^2$, and $\eta_1 = \frac{2}{3} = 0.667$.

3. Choose $W(x) = \partial y_1/\partial a_1 =$ the perturbation function $(1 - x^2)$ of (19).

$$\int_V R_1(1 - x^2)\,dV = \int_0^1 \{-4a_1 - \Phi^2[1 + a_1(1 - x^2)]\} \times (1 - x^2)\,dx^2 = 0$$

or

$$-2a_1 - \Phi^2(\tfrac{1}{2} + \tfrac{1}{3}a_1) = 0$$

$$a_1 = \frac{-3\Phi^2}{12 + 2\Phi^2} \tag{27}$$

For $\Phi = 2$, $a_1 = -\frac{3}{5}$, $y_1 = \frac{2}{5} + \frac{3}{5}x^2$, and $\eta_1 = \frac{7}{10} = 0.7$.

4. Choose $W(x) = \partial R_1/\partial a_1$.

$$2\int_V R_1 \cdot \frac{\partial R_1}{\partial a_1}\,dV = \frac{\partial}{\partial a_1}\int_V R_1^2\,dV = 0$$

We see that this method implies a choice of a_1 such that the integral of the square of the residual is minimum.

$$\int_0^1 \{-4a_1 - \Phi^2[1 + a_1(1 - x^2)]\}[-4 - \Phi^2(1 - x^2)]\,dx^2 = 0$$

$$a_1 = \frac{-\Phi^2[1 + \frac{1}{8}\Phi^2]}{4 + \Phi^2 + \frac{1}{12}\Phi^4} \tag{28}$$

For $\Phi = 2$, $a_1 = -\frac{9}{14}$, $y_1 = \frac{5}{14} + \frac{9}{14}x^2$, and $\eta_1 = \frac{19}{28} = 0.679$.

Table 2.3 compares the four approximations with the Taylor series approximation (9) (which we have seen in section 2.1 corresponds to method 1 with $x_1 = 0$). All values of E_1 are multiplied by 100.

Note that each MWR method 1 to 4 gives one node point in the distance function E_1, and although this should not be ascribed to the node point in R_1, which is a built-in feature of methods 1 to 4, we suspect that approximations 1 to 4 might have better properties than the Taylor series approximation for which $E_N(x)$ is negative for any x and N. The maximum error of approximations 2 and 4 is larger than the maximum error of the Taylor series approximation, but methods 1 to 4 give a smaller

TABLE 2.3
DISTANCE FUNCTION $E_1(x) = y - y_1(x)$ FOR DIFFERENT MWR
(y IS THE SOLUTION OF EQ. (1) FOR $n = 1$ AND $\phi = 2$)

	Method				
x	1	2	3	4	Taylor
0	1.0	10.6	3.9	8.2	− 6.1
0.2	0.4	9.6	3.2	7.3	− 6.4
0.4	− 0.9	7.2	1.6	5.2	− 6.8
0.6	− 2.3	3.8	− 0.5	3.7	− 6.9
0.8	− 2.7	0.8	− 1.6	− 0.1	− 5.2
0.9	− 1.8	− 0.3	− 1.3	− 0.5	− 3.2
1	0	0	0	0	0
$\eta - \eta_1$	− 0.016	0.031	− 0.002	0.019	− 0.052

error, $\int_0^1 E_1\, dx^2$ in η_1, the reason being that positive and negative errors partly cancel in the integration. In particular we note the very small error of method 3.

Throughout this text we shall restrict our attention to polynomial approximations. Approximations with other functions—especially harmonic functions in analogy with the classical Fourier series—and also with functions in which the parameters enter nonlinearly are well studied in literature [see, e.g., Rice (1964) and (1969)].

We shall briefly illustrate the use of a trigonometric approximation that is analogous to (19) and thereafter leave this subject, noting that the computational work in obtaining the coefficients a_i and the subsequent application of y_N for approximation purposes is in general heavier than for a polynomial approximation with an equal number of adjustable parameters.

$$y_1 = 1 + a_1 \cos\left(\frac{\pi}{2}x\right) \tag{29}$$

The perturbation function $\cos[(\pi/2)x]$ is again zero at $x = 1$, and y_1 satisfies both boundary conditions.

$$R_1 = -\left(\frac{\pi^2}{4}\cos\frac{\pi x}{2} + \frac{\pi}{2x}\sin\frac{\pi x}{2}\right)a_1 - \Phi^2\left(1 + a_1\cos\frac{\pi x}{2}\right)$$

We choose method 3 $[W(x) = \partial y_1/\partial a_1 = \cos \pi/2x]$:

$$\int_0^1 R_1 \cos\frac{\pi x}{2}\, dx^2 = 0$$

or, after considerable manipulation,

$$a_1\left[\left(\frac{\pi^2}{4} + \Phi^2\right)\left(\frac{1}{4} - \frac{1}{\pi^2}\right) + \frac{1}{2}\right] = \Phi^2\left(\frac{4}{\pi^2} - \frac{2}{\pi}\right) \tag{30}$$

For $\phi = 2$, $a_1 = -0.6331$ and $\eta_1 = 0.6834$.

The accuracy of approximation (29) with $a_1 = -0.6331$ is comparable with that obtained by the polynomial approximations in table 2.3, but the computational effort is considerable.

2.3 Higher-Order MWR Approximations $y_N(x)$

2.3.1 Definition of the methods

Having been acquainted with several MWR to obtain the first approximation y_1 for equation (1) with $n = 1$, we proceed to study higher-order approximations. Our general approximation y_N is defined by

$$y_N = T_0 + \sum_{i=1}^{N} a_i T_i \tag{31}$$

T_0 satisfies the boundary conditions of the given differential equation.

The trial functions (or perturbation functions) T_i are contained in the N-term sum. Each trial function satisfies homogeneous boundary conditions, i.e., for the problem (1), (2),

$$\frac{dT_i}{dx}(x = 0) = T_i(x = 1) = 0 \qquad i > 0 \tag{32}$$

The choice of trial functions T_i is in general quite free as long as functions T_i satisfy conditions (32) and are linearly independent.

We restrict our attention to polynomial trial functions and for our case study (1), (2), $T_0 = 1$ and $T_i = (1 - x^2)x^{2i-2}$, $i = 1, 2, \ldots$ satisfy conditions (32).

$$y_N = 1 + (1 - x^2)\sum_{1}^{N} a_i x^{2i-2} \tag{33}$$

It is more convenient to work with (11):

$$u\frac{d^2y}{du^2} + \frac{dy}{du} - py = 0$$

$$p = \frac{\Phi^2}{4}, \qquad y(u = 1) = 1, \qquad y^{(k)}(u = 0) \quad \text{finite for } k \geq 0$$

Here an expansion that includes all positive integer powers of the independent variable u can be used.

$$y_N(x^2) = y_N(u) = 1 + (1 - u) \sum_{i=1}^{N} a_i u^{i-1} \tag{34}$$

We shall determine the coefficients a_i of (34) and finally compute

$$\eta_N = \int_0^1 y_N \, dx^2 = \int_0^1 y_N \, du \tag{35}$$

$$= \frac{2}{\Phi^2} \frac{dy}{dx}\bigg|_{x=1} = \frac{1}{p} \frac{dy}{du}\bigg|_{u=1} \tag{36}$$

The residual $R_N(\mathbf{a}, u)$ is formed as before by substitution of y_N into the differential equation (11):

$$\begin{aligned} R_N(\mathbf{a}, u) &= u \frac{d^2 y_N}{du^2} + \frac{dy_N}{du} - p y_N \\ &= \sum_{i=1}^{N} a_i[(i-1)^2 u^{i-2} - i^2 u^{i-1}] - p\left[1 + (1-u) \sum_{i=1}^{N} a_i u^{i-1}\right] \end{aligned} \tag{37}$$

The following N relations determine the coefficients a_i:

$$\int_V R_N(\mathbf{a}, u) W_j(u) \, dV = 0, \qquad j = 1, 2, \ldots, N$$

or

$$\int_0^1 R_N(\mathbf{a}, u) W_j(u) \, du = 0, \qquad j = 1, 2, \ldots, N \tag{38}$$

Each MWR is characterized by a different choice of the sequence of N weight functions $W_j(u)$ in (38). In methods 1, 3a, 3b, and 4, we shall discuss the Nth-order analog of each of the four MWR that were introduced in section 2.2; by addition of a fifth MWR—method 2—a comprehensive catalog of the most popular MWR is established.

1. The collocation method
   Choose

   $$W_j(u) = \delta(u - u_j), \qquad j = 1, 2, \ldots, N \tag{39}$$

   where u_j are N points in (0, 1). This is equivalent to

   $$R_N(\mathbf{a}, u_j) = 0, \qquad j = 1, 2, \ldots, N$$

2. The subdomain method
   Divide the region V into N subregions V_j. Choose $W_j(u) = 1$ inside V_j and 0 outside V_j. In our case study we divide the range $0 \leq u \leq 1$ into N subintervals by the interior points u_1, u_2,

$u_3, \ldots, u_{N-1}$ ($u_0 = 0$, $u_N = 1$) and the subinterval method can be written:

$$\int_{u_{j-1}}^{u_j} R_N(\mathbf{a}, u)\, du = 0, \qquad j = 1, 2, \ldots, N \tag{40}$$

3a. The method of moments
Here $W_j(u)$ in (38) is chosen as u^{j-1}, $j = 1, 2, \ldots, N$.

$$\int_0^1 R_N(\mathbf{a}, u) u^{j-1}\, du = 0, \qquad j = 1, 2, \ldots, N \tag{41}$$

Other functions $\phi_j(u)$, $j = 1, 2, \ldots, N$ could be used instead of the monomials u^{j-1} in (40), e.g., $\cos[(j-1)\pi x/2]$ or an arbitrary polynomial of degree $(j-1)$.

The name *method of moments* is, however, cutomarily given to the method based on (41).

3b. Galerkin's method
Here $\quad W_j(u) = T_j = \partial y_N/\partial a_j \qquad$ or in our case $\hfill (42)$

$$\int_0^1 R_N(\mathbf{a}, u)(1-u)u^{j-1}\, du = 0, \qquad j = 1, 2, \ldots, N \tag{43}$$

4. The least squares method

$$W_j(u) = \frac{\partial R_N}{\partial a_j} \tag{44}$$

$$\int_0^1 R_N(\mathbf{a}, u) \frac{\partial R_N}{\partial a_j}\, du = 0, \qquad j = 1, 2, \ldots, N$$

or, in an alternative formulation, determine $\mathbf{a}$ such that

$$\int_0^1 [R_N(\mathbf{a}, u)]^2\, du \quad \text{is minimum} \tag{45}$$

The method owes its name to this formulation.

Note that $\partial R_N/\partial a_j$ is usually a much more complicated expression than $\partial y_N/\partial a_j$, the weight function in Galerkin's method (42), and that whereas W_j is always a polynomial of degree j in (43), this is not necessarily true in (44). If the differential equation is nonlinear, W_j is also a nonlinear function of u^j.

We shall now make a brief numerical study of the approximate solution y_N of (1), (2) with $n = 1$ for the following four MWR:

1. The collocation method with $u_j = j/(N+1)$ (equidistant collocation).
2. The subdomain method with $u_j = j/N$.

3a. The method of moments ($W_j = u^{j-1}$).
3b. Galerkin's method [$W_j = (1 - u)u^{j-1}$].

In each of these four cases one obtains a set of N linear algebraic equations of the form

$$(\mathbf{A} + p \cdot \mathbf{B})\mathbf{a} = p \cdot \mathbf{c} \tag{46}$$

The elements of the $(N \times N)$ matrices $\mathbf{A}$ and $\mathbf{B}$ and the N-vector $\mathbf{c}$ are

1.

$$\begin{aligned} A_{ji} &= \left[(i-1)^2 - i^2 \frac{j}{N+1}\right] \cdot \left(\frac{j}{N+1}\right)^{i-2} \\ B_{ji} &= \left(\frac{j}{N+1}\right)^{i-1} \cdot \left(\frac{j}{N+1} - 1\right) \\ c_j &= 1 \end{aligned} \tag{47}$$

2.

$$\begin{aligned} A_{ji} &= \int_{(j-1)/N}^{j/N} [(i-1)^2 u^{i-2} - i^2 u^{i-1}]\, du \\ &= (i-1)\left[\left(\frac{j}{N}\right)^{i-1} - \left(\frac{j-1}{N}\right)^{i-1}\right] - i\left[\left(\frac{j}{N}\right)^{i} - \left(\frac{j-1}{N}\right)^{i}\right] \\ B_{ji} &= \int_{(j-1)/N}^{j/N} (u^i - u^{i-1})\, du \\ &= \frac{1}{i+1}\left[\left(\frac{j}{N}\right)^{i+1} - \left(\frac{j-1}{N}\right)^{i+1}\right] - \frac{1}{i}\left[\left(\frac{j}{N}\right)^{i} - \left(\frac{j-1}{N}\right)^{i}\right] \\ c_j &= \int_{(j-1)/N}^{j/N} du = \frac{1}{N} \end{aligned} \tag{48}$$

3a.

$$\begin{aligned} A_{ji} &= \int_0^1 [(i-1)^2 u^{i-2} - i^2 u^{i-1}] u^{j-1}\, du \\ &= \begin{cases} \dfrac{(i-1)^2}{i+j-2} - \dfrac{i^2}{i+j-1} & i \neq 1 \\ \dfrac{-1}{j} & i = 1 \end{cases} \\ B_{ji} &= \int_0^1 (u^i - u^{i-1}) u^{j-1}\, du = \frac{-1}{(i+j)(i+j-1)} \\ c_j &= \int_0^1 u^{j-1}\, du = \frac{1}{j} \end{aligned} \tag{49}$$

3b.

$$A_{ji} = \int_0^1 [(i-1)^2 u^{i-2} - i^2 u^{i-1}] u^{j-1}(1-u)\,du$$

$$= \begin{cases} \dfrac{i+j-2ij}{(i+j-2)(i+j-1)(i+j)} & i \neq 1 \\[2ex] \dfrac{-1}{j(j+1)} & i = 1 \end{cases}$$

$$B_{ji} = \int_0^1 (u^i - u^{i-1})(1-u)u^{j-1}\,du \tag{50}$$

$$= \frac{-2}{(i+j-1)(i+j)(i+j+1)}$$

$$c_j = \int_0^1 (1-u)u^{j-1}\,du = \frac{1}{j(j+1)}$$

The approximate effectiveness factor may be found from

$$\eta_N = \int_0^1 y_N\,du = \int_0^1 \left[1 + (1-u)\sum_{i=1}^{N} a_i u^{i-1}\right] du = 1 + \sum_{i=1}^{N} \frac{a_i}{i(i+1)} \tag{51}$$

2.3.2 A sample calculation of $y_2(u)$

Let us consider a simple numerical example, e.g., method 3b, $N = 2$. We find that

$$\mathbf{A} = \begin{Bmatrix} -\frac{1}{2} & -\frac{1}{6} \\ -\frac{1}{6} & -\frac{1}{6} \end{Bmatrix}, \quad \mathbf{B} = \begin{Bmatrix} -\frac{1}{3} & -\frac{1}{12} \\ -\frac{1}{12} & -\frac{1}{30} \end{Bmatrix}, \quad \text{and} \quad \mathbf{c} = \begin{Bmatrix} \frac{1}{2} \\ \frac{1}{6} \end{Bmatrix}$$

For $\Phi = 2$, i.e., $p = \Phi^2/4 = 1$, we obtain

$$\begin{pmatrix} -\frac{1}{2}-\frac{1}{3} & -\frac{1}{6}-\frac{1}{12} \\ -\frac{1}{6}-\frac{1}{12} & -\frac{1}{6}-\frac{1}{30} \end{pmatrix} \begin{pmatrix} a_1 \\ a_2 \end{pmatrix} = \begin{pmatrix} \frac{1}{2} \\ \frac{1}{6} \end{pmatrix}$$

$$-\tfrac{5}{6}a_1 \quad -\tfrac{1}{4}a_2 = \tfrac{1}{2}$$

$$-\tfrac{1}{4}a_1 \quad -\tfrac{1}{5}a_2 = \tfrac{1}{6}$$

$$a_1 = -\tfrac{14}{25}, \qquad a_2 = -\tfrac{2}{15}$$

$$y_2 = 1 - \tfrac{14}{25}(1-u) - \tfrac{2}{15}(1-u)\cdot u$$

$$\eta_2 = 1 + \tfrac{1}{2}a_1 + \tfrac{1}{6}a_2 = \tfrac{157}{225} = 0.6977778$$

A general parametric representation in $p = \Phi^2/4$ may also be obtained:

$$\begin{pmatrix} -\frac{1}{2}-\frac{1}{3}p & -\frac{1}{6}-\frac{1}{12}p \\ -\frac{1}{6}-\frac{1}{12}p & -\frac{1}{6}-\frac{1}{30}p \end{pmatrix} \begin{pmatrix} a_1 \\ a_2 \end{pmatrix} = \begin{pmatrix} \frac{1}{2}p \\ \frac{1}{6}p \end{pmatrix}$$

$$a_1 = \frac{\begin{vmatrix} \frac{1}{2}p & -\frac{1}{6}-\frac{1}{12}p \\ \frac{1}{6}p & -\frac{1}{6}-\frac{1}{30}p \end{vmatrix}}{\begin{vmatrix} -\frac{1}{2}-\frac{1}{3}p & -\frac{1}{6}-\frac{1}{12}p \\ -\frac{1}{6}-\frac{1}{12}p & -\frac{1}{6}-\frac{1}{30}p \end{vmatrix}} = \frac{-2p^2-40p}{3p^2+32p+40}$$

$$a_2 = \frac{\begin{vmatrix} -\frac{1}{2}-\frac{1}{3}p & \frac{1}{2}p \\ -\frac{1}{6}-\frac{1}{12}p & \frac{1}{6}p \end{vmatrix}}{\begin{vmatrix} -\frac{1}{2}-\frac{1}{3}p & -\frac{1}{6}-\frac{1}{12}p \\ -\frac{1}{6}-\frac{1}{12}p & -\frac{1}{6}-\frac{1}{30}p \end{vmatrix}} = \frac{-10p^2}{3p^2+32p+40}$$

$$\eta_2 = 1 + \tfrac{1}{2}a_1 + \tfrac{1}{6}a_2 = \frac{\frac{1}{3}p^2+12p+40}{3p^2+32p+40}$$

2.3.3 Tabular results for $N = 2, 3, 4$

The form of equation (46) shows that the solution for the a_i and for η may be written as the ratio of two Nth-degree polynomials in the parameter p, the denominator polynomial being common for all the a_i and η. The simple dependence on the parameter p makes it possible to derive parametric expressions as demonstrated in the example above but, except when the influence of p is studied in general, it is easier to substitute the appropriate value of $p = \Phi^2/4$ in (46) before solving the equations.

In table 2.4 parametric representations of a_1 and η are given for the four methods described above and $N = 2, 3, 4$.

Next we compare the approximate solutions obtained using $N = 3$ and $\Phi = 2$ with the exact profile, $y(x) = I_0(2x)/I_0(2)$.

The following approximations are found:

1. *Equidistant Collocation*

$$\begin{aligned} y_3(u) &= \tfrac{1}{2233}(979 + 982u + 240u^2 + 32u^3) \\ -R(u)\cdot 2233 &= 32u^3 - 48u^2 + 22u - 3 \end{aligned} \tag{52}$$

2. *Subdomain Method with Equally Sized Subdomains*

$$\begin{aligned} y_3(u) &= \tfrac{1}{1254}(550 + 551u + 135u^2 + 18u^3) \\ -R(u)\cdot 1254 &= 18u^3 - 27u^2 + 11u - 1 \end{aligned} \tag{53}$$

TABLE 2.4

FIRST EXPANSION COEFFICIENT AND EFFECTIVENESS FACTOR FOR VARIOUS MWR WITH $N = 2, 3,$ AND 4

Method 1: Equidistant collocation in u

N	a_1	η
2	$\dfrac{-36p}{2p^2+27p+36}$	$\dfrac{-0.5p^2+9p+36}{2p^2+27p+36}$
3	$\dfrac{-6p^3-96p^2-1152p}{3p^3+118p^2+960p+1152}$	$\dfrac{22p^2+384p+1152}{3p^3+118p^2+960p+1152}$
4	$\dfrac{-562.5p^3-11{,}250p^2-90{,}000p}{6p^4+500p^3+12{,}125p^2+78{,}750p+90{,}000}$	$\dfrac{\frac{19}{24}p^4+62.5p^3+2750p^2+33{,}750p+90{,}000}{6p^4+500p^3+12{,}125p^2+78{,}750p+90{,}000}$

Method 2: Subdomain method with equal-sized subintervals

N	a_1	η
2	$\dfrac{-24p}{p^2+18p+24}$	$\dfrac{6p+24}{p^2+18p+24}$
3	$\dfrac{-2p^3-54p^2-648p}{p^3+65p^2+540p+648}$	$\dfrac{11p^2+216p+648}{p^3+65p^2+540p+648}$
4	$\dfrac{-520p^3-11{,}520p^2-92{,}160p}{3p^4+470p^3+12{,}360p^2+80{,}640p+92{,}160}$	$\dfrac{50p^3+2760p^2+34{,}650p+92{,}160}{3p^4+470p^3+12{,}360p^2+80{,}640p+92{,}160}$

Method 3a: Method of moments, $W_j = u^{j-1}$		
2	$\dfrac{-24p}{p^2 + 18p + 24}$	$\dfrac{6p + 24}{p^2 + 18p + 24}$
3	$\dfrac{-2p^3 - 60p^2 - 720p}{p^3 + 72p^2 + 600p + 720}$	$\dfrac{12p^2 + 240p + 720}{p^3 + 72p^2 + 600p + 720}$
4	$\dfrac{-220p^2 - 5040p^2 - 40{,}320p}{p^4 + 200p^3 + 5400p^2 + 35{,}280p + 40{,}320}$	$\dfrac{20p^3 + 1200p^2 + 15{,}120p + 40{,}320p}{p^4 + 200p^3 + 5400p^2 + 35{,}280p + 40{,}320}$
Method 3b: Galerkin's method, $W_j = u^{j-1}(1-u)$		
2	$\dfrac{-2p^2 - 40p}{3p^2 + 32p + 40}$	$\dfrac{\frac{1}{3}p^2 + 12p + 40}{3p^2 + 32p + 40}$
3	$\dfrac{-2.5p^3 - 67.5p^2 - 630p}{2p^3 + 75p^2 + 540p + 630}$	$\dfrac{\frac{1}{8}p^3 + 15p^2 + 225p + 630}{2p^3 + 75p^2 + 540p + 630}$
4	$\dfrac{-4p^4 - 504p^3 - 10{,}800p^2 - 72{,}576p}{5p^4 + 480p^3 + 10{,}584p^2 + 64{,}512p + 72{,}756}$	$\dfrac{\frac{1}{5}p^4 + 60p^3 + 2520p^2 + 28{,}224p + 72{,}576}{5p^4 + 480p^3 + 10{,}584p^2 + 64{,}512p + 72{,}756}$

3a. *Method of Moments with* $W_j = u^{j-1}$

$$y_3(u) = \tfrac{1}{1393}(611 + 612u + 150u^2 + 20u^3)$$
$$- R(u) \cdot 1393 = 20u^3 - 30u^2 + 12u - 1 \tag{54}$$

3b. *Galerkin's Method*

$$y_3(u) = \tfrac{1}{1247}(547 + 547.5u + 135u^2 + 17.5u^3)$$
$$- R(u) \cdot 2494 = 35u^3 - 45u^2 + 15u - 1 \tag{55}$$

4. *Taylor Series*

$$y_3(u) = \tfrac{1}{82}(36 + 36u + 9u^2 + u^3)$$
$$- R(u) \cdot 82 = u^3 \tag{56}$$

Figure 2-2 shows computed values of $-E_3(u) = y_3(u) - y(u)$ and of $R_3(u)$ for the four approximations (52) to (55).

From the plots of $y_3(x) - y(x)$ we notice that all these MWR approximations are superior to the corresponding Taylor series approximation. The Galerkin approximation is noteworthy for its very even error distribution (notice in particular the three node points) and a maximum error of only 5% of that of the Taylor series, followed closely by the method of moments with a maximum error of 10% of the Taylor series error. The subdomain method yields an approximation with a rather large error at $x = 0$, and the result obtained from the collocation method is rather disappointing, the error being comparable to that of the Taylor series.

As mentioned earlier, the evenly distributed errors for the method of moments and Galerkin's method indicate that very precise estimates of the effectiveness factor ($\int_0^1 y_N\,du$) can be obtained due to error cancellation in the integration.

We further notice that all the MWR approximations (1, 2, 3a, and 3b) have their maximum error at $x = 0$, this error being equal to the error on the first expansion coefficient a_1 [$y_N(0) = 1 + a_1$]. We may therefore judge the accuracy of the profile by the accuracy of a_1 {the exact value of a_1 is $-1 + [1/I_0(\Phi)]$}.

The effect of the parameter Φ on the error of η and a_1 is shown in figures 2-3 and 2-4, respectively, for $y_3(u)$ obtained by the different MWR.

It appears that for any given value of Φ the Galerkin method yields the best results, followed by the method of moments, the subdomain method, and the collocation method. We notice that the difference

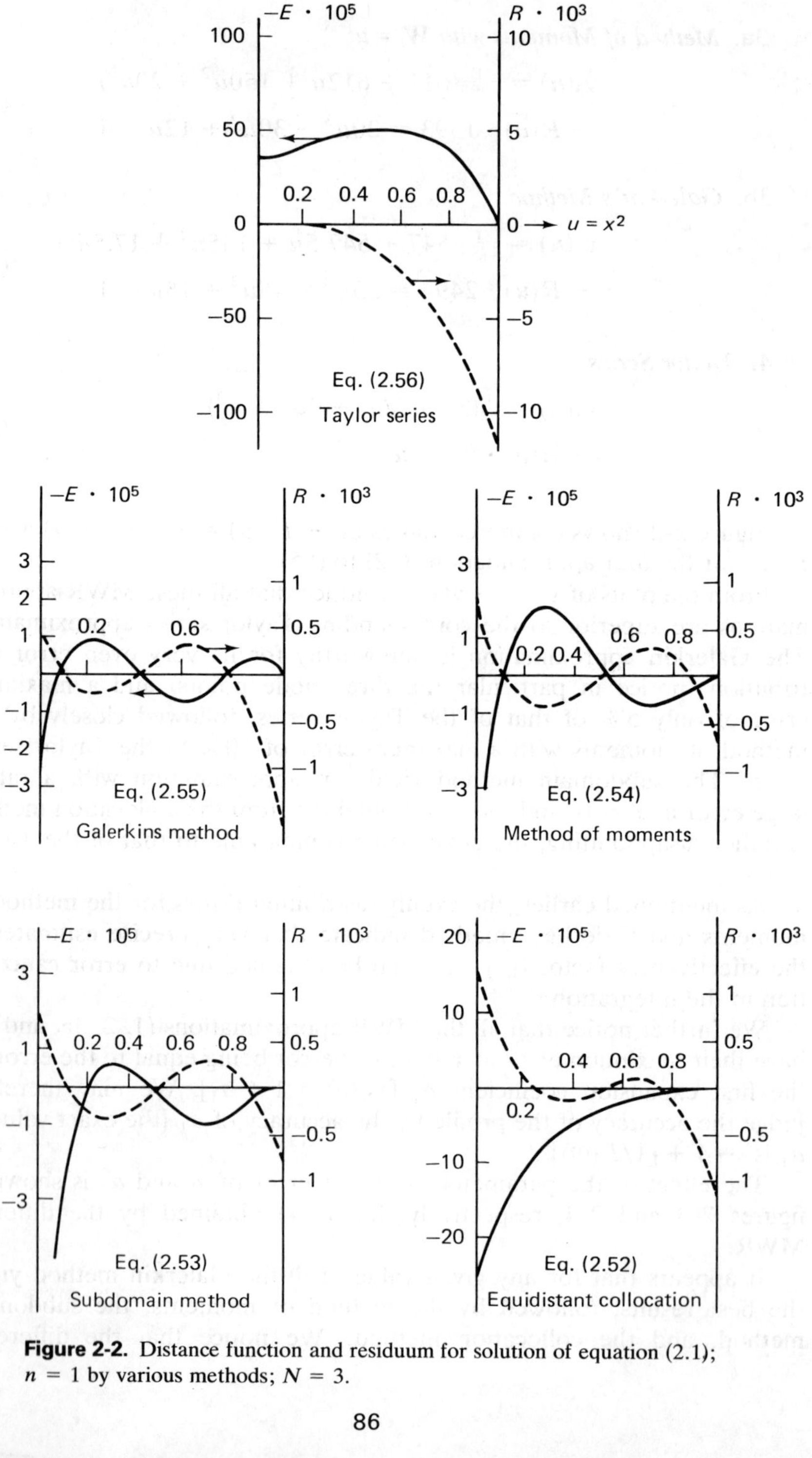

Figure 2-2. Distance function and residuum for solution of equation (2.1); $n = 1$ by various methods; $N = 3$.

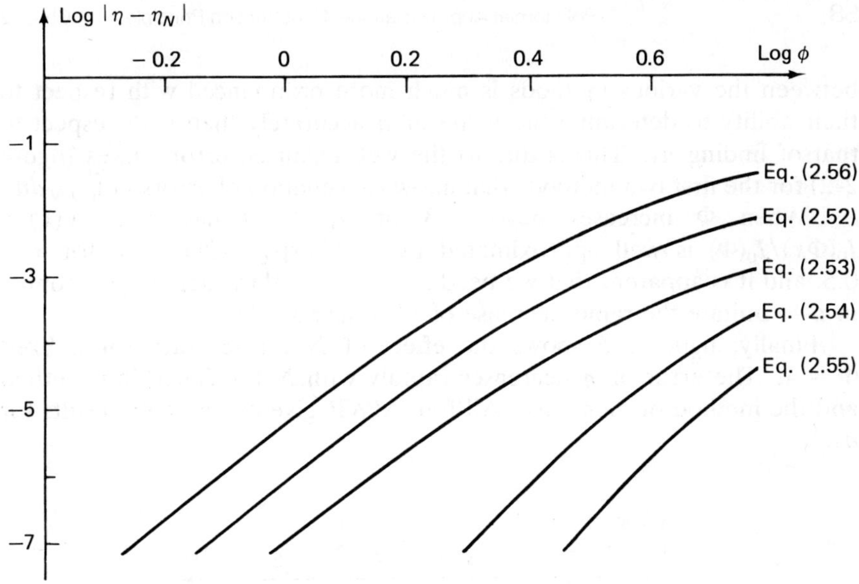

Figure 2-3. Error of $\int_0^1 y_N \, dV = \eta_N$ for various $y_N(x)$; $N = 3$.

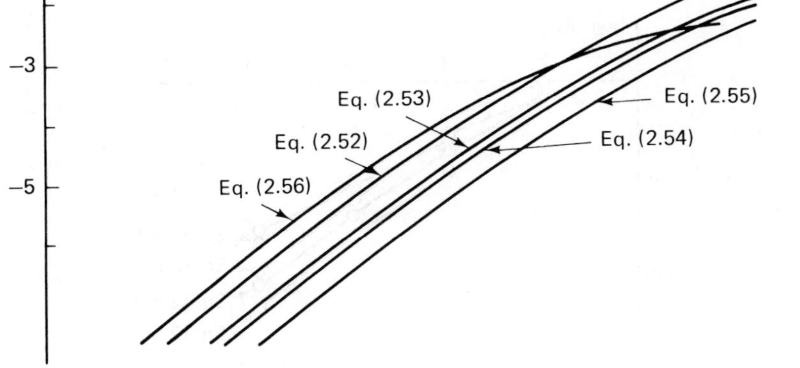

Figure 2-4. Error of first expansion coefficient; $N = 3$.

between the various methods is much more pronounced with respect to their ability to determine the value of η accurately than with respect to that of finding a_1. This is due to the well-balanced error curves (figure 2-2) for the first two methods that allow cancellation of errors in $\int_0^1 y_N\,du$.

When Φ increases beyond 3 or 4, the behavior of $y(x) = I_0(\Phi x)/I_0(\Phi)$ is well approximated by $x^{-1/2}\exp[-\Phi(1-x)]$ for $x > 0.5$, and it is apparent that we need more than a third-degree polynomial in u to imitate the rapid decrease of $y(x)$ near $x = 1$.

Finally, figure 2-5 shows the effect of N on accuracy for a fixed $\Phi = 4$. The error of η decreases rapidly with N for Galerkin's method and the method of moments. All four MWR give comparable results for a_1.

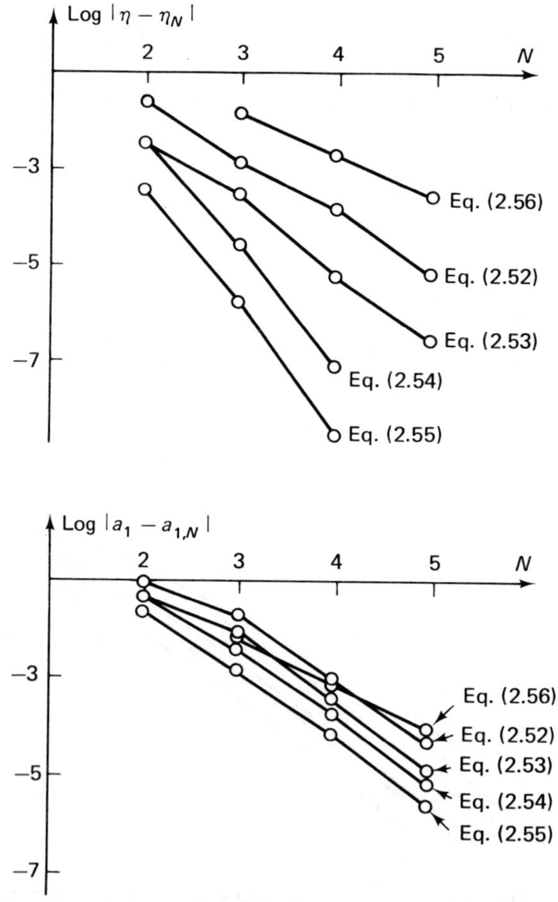

Figure 2-5. Error of η_N and $a_{1,N}$ by various methods; $\phi = 4$.

2.3.4 Expansion of η and of a_1 in powers of the parameter Φ^2

The quantities η and a_1 can be expressed by power series in the parameter Φ (or more conveniently in $p = \Phi^2/4$) by a Taylor series expansion of the exact solution or of any of the approximate solutions. We have

$$\eta(\Phi) = \frac{2}{\Phi}\left[\frac{I_1(\Phi)}{I_0(\Phi)}\right]$$
$$= 1 - \tfrac{1}{2}p + \tfrac{1}{3}p^2 - \tfrac{11}{48}p^3 + \tfrac{19}{120}p^4 - \tfrac{473}{4320}p^5 + \ldots \tag{57}$$

and

$$a_1 = \frac{1}{I_0(\Phi)} - 1 = -p + \frac{3}{4}p^2 - \frac{19}{36}p^3 + \frac{211}{576}p^4 - \ldots \tag{58}$$

The Galerkin approximations for $N = 1$ are

$$a_1 = \frac{-\Phi^2}{4 + \frac{2}{3}\Phi^2} = \frac{-3p}{3 + 2p} \quad \text{and} \quad \eta_1 = 1 + \frac{1}{2}a_1 = \frac{3 + \frac{1}{2}p}{3 + 2p}$$

and expansion of these quantities in powers of p yields

$$\eta_1 = 1 - \tfrac{1}{2}p + \tfrac{1}{3}p^2 - \tfrac{2}{9}p^3 + \ldots \tag{59}$$

$$a_1 = -p + \tfrac{2}{3}p^2 - \ldots \tag{60}$$

Similar expansions for $N = 2$ are found in the example in subsection 2.3.2:

$$\eta_2 = \frac{\frac{1}{3}p^2 + 12p + 40}{3p^2 + 32p + 40} = 1 - \frac{1}{2}p + \frac{1}{3}p^2 - \frac{11}{48}p^3 + \frac{19}{120}p^4 - \frac{1051}{9600}p^5 + \ldots \tag{61}$$

$$a_1 = \frac{-2p^2 - 40p}{3p^2 + 32p + 40} = -p + \frac{3}{4}p^2 - \frac{21}{40}p^3 + \ldots \tag{62}$$

A comparison of (59) and (61) with (57) shows that the series are identical up to and including the term p^{2N}, and comparison of the expansions for a_1 shows that the terms are identical up to the term p^N.

Table 2.5 gives the results of similar expansions for the other methods.

Addition of one extra term in methods 3a and 3b gives two extra correct terms in the expansion of η in powers of p, compared to one term (in the average) for the other methods.

TABLE 2.5
POWERS OF p CORRECTLY REPRESENTED IN TAYLOR SERIES EXPANSION OF NTH-ORDER APPROXIMATION

	a_1	η
Collocation (1)	N	N
Subdomain (2)	N	$\begin{Bmatrix} N \text{ for } N \text{ odd} \\ N+1 \text{ for } N \text{ even} \end{Bmatrix}$
Moments (3a)	N	$2N-1$
Galerkin (3b)	N	$2N$

The numerical factor of the first incorrect term is furthermore very close to its exact value as seen from the example with $N = 2$:

$$\text{Exact:}\quad -\frac{473}{4320}p^5$$

$$\text{Galerkin:}\quad -\frac{1051}{9600}p^5$$

$$\text{Difference:}\quad \frac{1}{86{,}400}p^5$$

One final point is worth noting. We remember that the effectiveness factor may also be calculated as

$$\eta = \frac{2}{\Phi^2}\left(\frac{dy}{dx}\right)_{x=1} = \frac{1}{p}\cdot\left(\frac{dy}{du}\right)_{u=1}$$

Approximate values η_N calculated from this relation with y_N instead of y may differ from those found by integration. We obtain

$$\begin{aligned}\eta_N &= \frac{1}{p}\frac{d}{du}(y_N)_{u=1} = \frac{1}{p}\frac{d}{du}\left[1 + (1-u)\sum_{i=1}^{N} a_i u^{i-1}\right]_{u=1} \\ &= \frac{1}{p}\left(-\sum_{i=1}^{N} a_i\right)\end{aligned} \tag{63}$$

For the Galerkin method and $N = 2$, the result is

$$\eta_N = \frac{40 + 12p}{40 + 32p + 3p^2} = 1 - \frac{1}{2}p + \frac{13}{40}p^2 - \ldots \tag{64}$$

that is, the Taylor series expansion is only correct for terms of powers $\leq N - 1$. The method of moments (3a) and the subdomain method, however, give identical results for effectiveness factors calculated in this manner and from the integral, the reason being that both of these methods require the relationship

$$\int_0^1 R_N \, du = 0$$

to be satisfied; i.e.,

$$\int_0^1 \left[\frac{d}{du}\left(u\frac{dy_N}{du}\right) - p \cdot y_N\right] du = 0$$

or

$$u\frac{dy_N}{du}\Big|_0^1 = \left(\frac{dy_N}{du}\right)_{u=1} = p \cdot \int_0^1 y_N \, du$$

2.3.5 Estimates of approximation accuracy

In general it is impossible to ascertain when a given approximation accuracy has been reached since the exact solution is unknown. It is, however, important to establish empirical rules for discontinuing further computation at a certain N and every loose estimate of approximation accuracy is valuable for this purpose. The simplest approach is to compare results obtained with different approximation order N. As a rule of thumb, the error in the Nth-order approximation will normally be much smaller than the difference between the Nth-order and the $(N-1\text{st})$-order approximation; e.g., $|\eta_4 - \eta_3| \gg |\eta_{ex} - \eta_4|$, cf. figure 2.5.

Some a priori guidelines may, however, be given. Returning to our original equation (1) for $n = 1$,

$$\frac{d^2y}{dx^2} + \frac{1}{x}\frac{dy}{dx} = \Phi^2 y, \qquad y(1) = 1, \qquad y'(0) = 0$$

we notice that

$$\max\left(\frac{d^2y}{dx^2} + \frac{1}{x}\frac{dy}{dx}\right) = \max\,(\Phi^2 y) = \Phi^2 \tag{65}$$

Now, if a trial solution of the type

$$y_N = 1 + (1 - x^2)\sum_{i=1}^{N} a_i x^{2i-2}$$

is used and we require (by physical arguments) that y_N is positive and has positive derivatives in (0, 1), we obtain

$$\max\left(\frac{d^2y_N}{dx^2} + \frac{1}{x}\frac{dy_N}{dx}\right) = 4N^2$$

(the steepest possible function satisfying these conditions being $y_N = x^{2N}$); i.e., the approximation order should at least satisfy

$$4N^2 > \Phi^2 \quad \text{or} \quad N > \frac{\Phi}{2}$$

Thus the results for $N = 3$ are only expected to be reliable up to $\Phi = 6$. A comparison with figure 2-4 shows that, indeed, the error on a_1 is becoming unacceptable ($\sim 10^{-2}$) for $\Phi = 6$, but the value of η is still unexpectedly accurate with an error of only about 10^{-4} for the best MWR, Galerkin's method.

2.3.6 Concluding remarks

In the given example, the "best" approximations are definitely obtained with Galerkin's method. The method of moments gives comparable but slightly less accurate results, whereas the subdomain method and the collocation method both fall rather far behind. One should, however, bear in mind that for the last two methods we have quite arbitrarily chosen equal-sized subdomains and equidistant collocation points. It might well be that better results are obtainable with different choices.

With regard to the computational effort for this linear problem, all the MWR methods require the solution of a set of N-linear algebraic equations. These equations are almost directly available in the collocation method, whereas all the other methods require integrations. This presents no problems here, but in nonlinear problems the formulation of the equations for the a_i may actually be more cumbersome than solving the equations.

2.4 Nonlinear Problems

2.4.1 The first approximation $y_1(x)$

We now consider (11) with $n \neq 1$:

$$u\frac{d^2y}{du^2} + \frac{dy}{du} = \frac{\Phi^2}{4} \cdot y^n = py^n \qquad y = 1 \quad \text{at} \quad u = 1 \tag{66}$$

$$\eta = \int_0^1 y^n \cdot du = \frac{1}{p}\left(\frac{dy}{du}\right)_{u=1} \tag{67}$$

The one-term approximation (19)

$$y_1 = 1 + a_1(1 - u)$$

is again used as a first approximation. The residual is

$$R_1(a_1, u) = -a_1 - p[1 + a_1(1 - u)]^n \tag{68}$$

The formulation of the equation for a_1 is just as simple as in the linear case when the collocation method is used:

$$R_1(a_1, u_1) = 0$$

or

$$-a_1 - p[1 + a_1(1 - u_1)]^n = 0 \tag{69}$$

The equation is now nonlinear and has to be solved numerically. In general, a parametric solution in p cannot be given.

Using Galerkin's method, we obtain

$$\int_0^1 R_1(a_1, u)(1 - u)\, du = 0$$

or

$$\int_0^1 \{-a_1 - p[1 + a_1(1 - u)]^n\}(1 - u)\, du = 0 \tag{70}$$

If n is noninteger, this integral normally has to be evaluated numerically and an explicit equation in a_1 cannot be obtained. Even for integer values of n, say $n = 2$ or $n = 3$, considerable algebraic manipulation is necessary to evaluate the integral. It appears that we have to choose between the easily applicable but in general rather inaccurate collocation method and the trustworthy but cumbersome Galerkin method.

A compromise is, however, possible. The integral in (70) may be approximated by a numerical quadrature. Quadrature formulas are treated in section 3.3 and here we shall only state some results:

$$\int_0^1 F(u)\, du \sim \sum_1^M w_j F(u_j) \tag{71}$$

or, in a more general formulation,

$$\int_0^1 W(u)F(u)\, du \sim \sum_1^M w_j F(u_j) \tag{72}$$

In case a weight function $W(u)$ can be extracted from the integrand as in (72), it is possible to build this weight function into the quadrature by means of a different choice of weights w_j and quadrature abscissas u_j in (71) and (72)—thereby improving the accuracy of the quadrature.

In our present case (70), we shall use

$$W(u) = 1 - u \quad \text{and} \quad F(u) = -a_1 - p[1 + a_1(1 - u)]^n$$

For a weight function $W(u) = u^{\beta}(1-u)^{\alpha}$ there exists an optimal choice of quadrature points in the sense that the highest possible power of u in a power series expansion of $F(u)$ is correctly integrated when these u-values are used as quadrature points.

The optimal quadrature points are zeros of $P_M^{(\alpha,\beta)}(u)$ where $P_M^{(\alpha,\beta)}(u)$ is a polynomial of degree M in u that satisfies the following relation:

$$\int_0^1 u^{\beta}(1-u)^{\alpha}u^j P_M^{(\alpha,\beta)}(u)\,du = 0 \quad \text{for } j = 0, 1, \ldots, M-1 \tag{73}$$

The quadrature is exact when $F(u)$ is a polynomial of degree $\leq 2M-1$ in u and in general the terms in a power series expansion of $F(u)$ are correctly integrated up to and including u^{2M-1}.

A discussion of the properties of orthogonal polynomials $P_M^{(\alpha,\beta)}(u)$ and the proof of the optimality of the proposed quadrature are deferred to chapter 3. Here it suffices to give the result

$$\int_0^1 R_1(u)(1-u)\,du \sim \sum_1^M w_j R_1(u_j) = \sum_1^M w_j\{-a_1 - p[1 + a_1(1-u_j)]^n\} \tag{74}$$

where $\{u_j\}$ are the M zeros of $P_M^{(1,0)}(u)$.

All zeros of $P_M^{(\alpha,\beta)}(u)$ are real and distinct and $0 < u_j < 1$.

All weights w_j are real and positive.

Using (74) instead of (70) relieves us of the task of evaluating an analytical expression for the integral without really sacrificing anything: The values computed from (74) with a given value of a_1 and increasing M form a rapidly convergent sequence. Also the original Galerkin expression (70) only provides us with a first approximation (19) to the solution of the differential equation (1), and a reasonable approximation in the determination of a_1 cannot conceivably give a final result (19), which is in principle worse than that obtained by means of (70).

If n is an integer, the residual $R_N(u)$ of equation (1) is a polynomial of degree nN in u. Exact integration of (70) and a quadrature with M quadrature points consequently give the same result when $M \geq (n+1)/2$.

An even simpler version of this method is obtained when $M = N$, i.e., when one quadrature point is used for the first approximation. Equation (74) can now be written

$$\begin{gathered} w_1 R_1(u_1) \sim \int_0^1 R_1(u)(1-u)\,du = 0 \\ -a_1 - p[1 + a_1(1-u_1)]^n = 0 \end{gathered} \tag{75}$$

u_1 is the zero of $P_1^{(1,0)}(u) = 3u - 1$ or $u_1 = \frac{1}{3}$.

The collocation method (69) contains u_1 as an adjustable parameter.

We have now shown that the specific choice $u_1 = \frac{1}{3}$ is optimal in the sense that the collocation method becomes identical to the approximate Galerkin method (74) for $M = 1$, and we have a feeling that the Galerkin approximation is in general better than other MWR.

We further notice that the linear case $n = 1$ requires $M \geq 1$ for identity between (74) and (70). Hence a collocation method (75) with $u_1 = \frac{1}{3}$ would give the same result as Galerkin's method in the example of section 2.2.

For other values of n, (75) and (70) give different results—but since the outcome of the calculations in (70) is an approximation for y, we cannot conclude that a_1 determined from (70) is always "better" than a_1 from (75).

As an example, $n = 2$, $\Phi = 3$ $(p = \frac{9}{4})$ will be chosen.

The optimal collocation method gives

$$-a_1 - \tfrac{9}{4}[1 + a_1(1 - \tfrac{1}{3})]^2 = 0$$
$$a_1 = \begin{cases} -0.6771 \\ (-3.32) \end{cases} \tag{76}$$

The solution $a_1 = -3.32$ is discarded since it leads to a physically unreasonable approximation $y_1 = 1 - 3.32(1 - u)$, which is negative for $u < \frac{2}{3}$.

The Galerkin method gives

$$\int_0^1 \{-a_1 - \tfrac{9}{4}[1 + a_1(1 - u)]^2(1 - u)\}\, du = 0$$
$$-\frac{a_1}{2} - \tfrac{9}{4}(\tfrac{1}{2} + \tfrac{2}{3}a_1 + \tfrac{1}{4}a_1^2) = 0 \tag{77}$$
$$a_1 = \begin{cases} -0.7005 \\ (-2.86) \end{cases}$$

The solution $a_1 = -2.86$ is also discarded. The effectiveness factor is found from

$$\eta_1 = \int_0^1 [1 + a_1(1 - u)]^2\, du = 1 + a_1 + \frac{a_1^2}{3}$$

that is,

Optimal collocation:	0.476
Galerkin:	0.463
(Exact result:	0.44621)

2.4.2 Optimal "Galerkin with quadrature" method

The trial solution

$$y_N = 1 + (1 - u) \sum_{i=1}^{N} a_i u^{i-1} \tag{78}$$

yields

$$R_N(\mathbf{a}, u) = \sum_{i=1}^{N} a_i[(i-1)^2 u^{i-2} - i^2 u^{i-1}] - p\left[1 + (1-u)\sum_{i=1}^{N} a_i u^{i-1}\right]^n \tag{79}$$

The Galerkin method gives the following N relations:

$$\int_0^1 \mathrm{R_N}(\mathbf{a}, u)(1-u)u^{j-1}\,du = 0, \qquad j = 1, 2, \ldots, N \tag{80}$$

or

$$\int_0^1 F_j(u)(1-u)\,du = 0, \qquad j = 1, 2, \ldots, N \tag{81}$$

where $F_j(u) = R_N(\mathbf{a}, u)u^{j-1}$.

Each of the N integrals in (81) is evaluated by quadrature:

$$\int_0^1 F_j(u)(1-u)\,du \simeq \sum_{k=1}^{M} w_k F_j(u_k) = \sum_{k=1}^{M} w_k R_N(\mathbf{a}, u_k)u_k^{j-1} \tag{82}$$

where $P_M^{(1,0)}(u_k) = 0, \qquad k = 1, 2, \ldots, M.$

We shall again choose $M = N$, and the following N equations are obtained for the N unknown coefficients a_i:

$$\sum_{k=1}^{N} w_k R_N(\mathbf{a}, u_k)u_k^{j-1} = 0, \qquad j = 1, 2, \ldots, N \tag{83}$$

The N equations can be satisfied for all Φ when

$$R_N(\mathbf{a}, u_k) = 0 \quad \text{at } u = u_1, u_2, \ldots, u_N \tag{84}$$

where u_k are the zeros of $P_N^{(1,0)}(u)$.

Again we notice that the best approximate Galerkin method is a collocation method, provided we choose as collocation points the roots of $P_N^{(1,0)}(u) = 0$.

In the case $n = 1$, F_j is a polynomial in u of degree $N + j - 1$, or at most of degree $2N - 1$, which means that for the linear problem the optimal collocation method and the Galerkin method will give identical results. In the general nonlinear case we see that the optimal collocation method can be identified with a Galerkin method where the integrals are evaluated by optimal quadrature formulas.

2.4.3 Computer results for $n = 2$

Nth-order approximations for $n = 2$ and $\Phi = 3$ are found by the optimal collocation method and by the Galerkin method (70). The accuracy of the effectiveness factor η_N [computed from (3)] and of a_1 is shown in figure 2-6 as a function of the approximation order N. In this

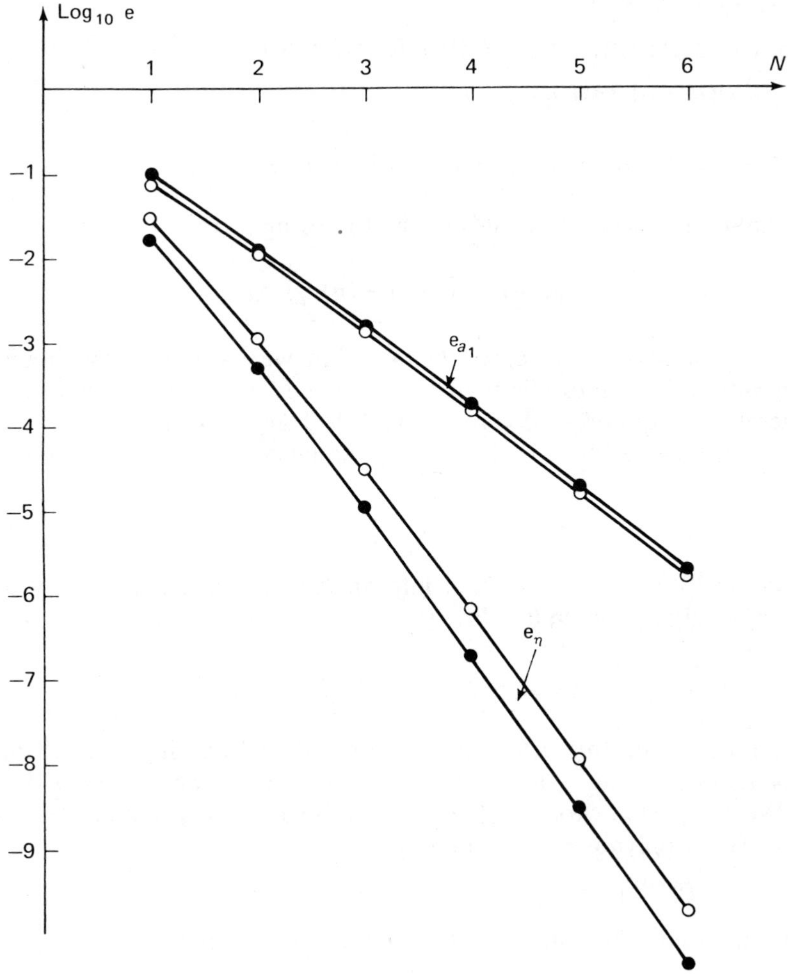

Figure 2-6. Approximation error by Galerkins method (line with closed circles) and by optimal collocation (line with open circles): $y^{(2)} + (1/x)y^{(1)} - 9y^2 = 0$.

example the optimal collocation method yields results for η that for the same N are about 2.5 times less accurate than results obtained by Galerkin's method. But more important, the same dependence of the accuracy on N is found by the two methods. Furthermore, the optimal collocation method gives a more accurate value of a_1.

In comparison with the tremendous increase in accuracy that is obtained with an increase of N, the difference between the two methods is insignificant, making the optimal collocation method our preferred choice of MWR due to the ease with which the equations (84) for **a** are set up.

2.5 Reformulations of the Nth-Order Approximation $y_N(x)$

2.5.1 An optimal Fourier-type expansion

Approximation (34) to y is of the following form:

$$y_N(u) = 1 + (1 - u) \sum_{i=1}^{N} a_i u^{i-1}$$

y_N is a polynomial of degree N in u, but we have used the boundary condition $y(1) = 1$ to eliminate one of the $N + 1$ parameters a_i that in general are required to determine an Nth-degree polynomial.

The following $(N - 1)$-degree polynomial can be extracted from (34):

$$g_{N-1}(u) = \frac{y_N - 1}{1 - u} = \sum_{i=1}^{N} a_i u^{i-1} \tag{85}$$

Equation (85) can be rewritten into an N-term expansion in any set of $(i - 1)$-degree polynomials P_{i-1} $(i = 1, 2, \ldots, N)$.

$$g_{N-1} = \sum_{i=1}^{N} a_i u^{i-1} = \sum_{i=1}^{N} b_i P_{i-1}(u) \tag{86}$$

One purpose of this reformulation may be to obtain a more rapidly convergent series, i.e., a series where b_i decreases more rapidly than a_i.

We shall often choose $P_i(u)$ as the orthogonal polynomials defined by (73). These polynomials can be written

$$P_i^{(\alpha,\beta)}(u) = \gamma_i u^i - \gamma_{i-1} u^{i-1} + \gamma_{i-2} u^{i-2} - \ldots + (-1)^i \tag{87}$$

or normalized differently such that the leading coefficient is 1:

$$\begin{aligned} p_i^{(\alpha,\beta)} &= u^i - \frac{\gamma_{i-1}}{\gamma_i} u^{i-1} + \ldots + \frac{(-1)^i}{\gamma_i} \\ &= u^i - \gamma'_{i-1} u^{i-1} + \ldots + (-1)^i \gamma'_0 \end{aligned} \tag{88}$$

The coefficients γ_i and γ'_i of (87) and (88) are all positive. If we specifically wish to stress that $(-1)^{i-j}\gamma_j$ is the coefficient of u^j in P_i, a double indexing γ_{ij} is used.

Recurrence formulas for γ_{ij} are given in chapter 3 but a few examples are shown in table 2.6.

TABLE 2.6
EXAMPLES OF ORTHOGONAL POLYNOMIALS

$P_i^{(\alpha,\beta)}$	i		
α, β	0	1	2
0, 0	1	$2u - 1$	$6u^2 - 6u + 1$
1, 0	1	$3u - 1$	$10u^2 - 8u + 1$
2, 0	1	$4u - 1$	$15u^2 - 10u + 1$

It is easily seen that the polynomials of table 2.6 satisfy the orthogonality relation:

$$\int_0^1 u^\beta(1-u)^\alpha P_i(u)P_j(u)\,du = C_i\,\delta_{ij} \tag{89}$$

To illustrate the equivalence of (85) and (86) we shall express one series in terms of the other for the specific case $P_i^{(1,0)}(u)$ and $N = 3$.

$$1 = P_0^{(1,0)}, \qquad u = \tfrac{1}{3}P_0^{(1,0)} + \tfrac{1}{3}P_1^{(1,0)}$$
$$u^2 = \tfrac{1}{6}P_0^{(1,0)} + \tfrac{4}{15}P_1^{(1,0)} + \tfrac{1}{10}P_2^{(1,0)}$$

or the following values for b_i in (86)

$$\begin{aligned} b_1 &= a_1 + \tfrac{1}{3}a_2 + \tfrac{1}{6}a_3 \\ b_2 &= \tfrac{1}{3}a_2 + \tfrac{4}{15}a_3 \\ b_3 &= \tfrac{1}{10}a_3 \end{aligned} \tag{90}$$

A systematic procedure for obtaining b_i from a_i follows from the orthogonality property (73) of the polynomials:

$$\begin{aligned} b_iC_{i-1} &= \sum_{j=1}^{N} a_j \int_0^1 W(u)P_{i-1}(u)u^{j-1}\,du \\ &= \sum_{j=i}^{N} a_j \int_0^1 W(u)P_{i-1}(u)u^{j-1}\,du \\ &= \sum_{j=i}^{N} a_j \sum_{k=1}^{N} w_kP_{i-1}(u_k)u_k^{j-1} \end{aligned} \tag{91}$$

The quadrature points in (91) are the N zeros of $P_N^{(\alpha,\beta)}(u)$. $u^{j-1} \cdot P_{i-1}(u)$ is at most a polynomial of degree $(N-1)+(N-1)=2N-2$ in u and the quadrature is exact.

Similarly the a_j can be derived from b_i of (86) by differentiating both sides of the equation and setting $u = 0$.

$$a_j = \frac{1}{(j-1)!}\sum_{i=1}^{N} b_i \frac{d^{j-1}P_{i-1}^{(\alpha,\beta)}}{du^{j-1}}\Bigg|_{u=0} = \sum_{i=j}^{N}(-1)^{i-j}\gamma_{i-1,j-1}b_i \tag{92}$$

where $(-1)^{i-j}\gamma_{i-1,j-1}$ is the coefficient of u^{j-1} in $P_{i-1}^{(\alpha,\beta)}$ as defined in (87).

In (55), a second-degree polynomial approximation for $g(u)$ is obtained using Galerkin's method.

$$g_2(u) = \frac{y_3 - 1}{1-u} = -\frac{1}{2494}(35u^2 + 305u + 1400) \tag{93}$$

i.e.,

$$\mathbf{a} = -\frac{5}{2494}(280, 61, 7)$$

Correspondingly the coefficients $\mathbf{b}$ in an expansion of $y_2(u)$ in $P_i^{(1,0)}(u)$ are determined from (90).

$$\mathbf{b} = -\frac{5}{2494}\cdot\frac{1}{10}(3015, 222, 7)$$

It is seen that the sequence of coefficients b_i is decreasing much more rapidly than the a_i coefficients, an obvious advantage for approximation purposes since a rapid convergence of the series for increasing N is expected when addition of an extra term contributes very little to the approximation.

Table 2.7 illustrates that the low-order coefficients b_i of an N-term expansion in orthogonal polynomials $P_i^{(1,0)}$ are very rapidly stabilized compared to the coefficients a_i in the expansion in monomials, which for every N is obtained by reformulation of the polynomial expansion.

TABLE 2.7
EXPANSION COEFFICIENTS OF (85) AND (86) FOR $p = 1$

N	1	2	3	4	a_i^*, b_i^*
$a_1^* - a_1$	$3.86 \cdot 10^{-2}$	$1.3 \cdot 10^{-3}$	$2.3 \cdot 10^{-5}$	$-2.6 \cdot 10^{-7}$	-0.56132
$b_1^* - b_1$	$-4.45 \cdot 10^{-3}$	$-3.8 \cdot 10^{-6}$	$-2.4 \cdot 10^{-9}$	$-3.6 \cdot 10^{-13}$	-0.60445
$a_2^* - a_2$		$1.07 \cdot 10^{-2}$	$-3.5 \cdot 10^{-4}$	$6.3 \cdot 10^{-6}$	-0.12265
$b_2^* - b_2$		$-6.2 \cdot 10^{-5}$	$-2.4 \cdot 10^{-8}$	$-3.6 \cdot 10^{-12}$	-0.044507

One further advantage of expansion (86) with $P_i = P_i^{(1,0)}$ is that the mean value of y is determined from b_1 alone.

$$y_N = 1 + (1 - u) \sum_{i=1}^{N} b_i P_{i-1}^{(1,0)}(u) \tag{94}$$

$$\eta = \int_0^1 y_N \, du = 1 + \sum_{i=1}^{N} b_i \int_0^1 (1 - u) P_{i-1}^{(1,0)}(u) \, du = 1 + b_1 \cdot C_0$$

since all higher terms in the sum are integrated to zero due to the orthogonality relation (89).

The value of

$$C_0 = \int_0^1 (1 - u) \, du = \frac{1}{2} \quad \text{and} \quad \eta = 1 + \frac{b_1}{2} = 1 - \frac{3015}{9976} = \frac{6961}{9976}$$

which is the same value that can be obtained from table 2.4, Galerkin's method, $N = 3$ for $p = 1$.

We observed in table 2.5 that terms up to and including p^{2N} in a series expansion of η in powers of p were represented exactly when the Galerkin MWR was used to determine $y_N(u)$.

It has now been shown that b_1 of expansion (94) has the same accuracy as η, i.e., that terms up to and including p^{2N} in a series expansion in powers of p of b_1^* obtained from the exact solution are correctly represented.

$$b_1^* = -\frac{1}{C_0} \int_0^1 \left[1 - \frac{I_0(2\sqrt{pu})}{I_0(2\sqrt{p})} \right] du = f(p) \tag{95}$$

by the formula for determination of Fourier series coefficients in the infinite orthogonal polynomial expansion:

$$g = \frac{y - 1}{1 - u} = \sum_{i=1}^{\infty} b_i^* P_{i-1}^{(1,0)}(u) \tag{96}$$

This high accuracy of b_1 is not exceptional and it may in fact be shown that an arbitrary coefficient in (94) is accurate up to and including the term p^{2N+1-i} in a power series representation of the corresponding coefficient b_i^* of (96). In contrast the coefficients a_i of expansion (34) in monomials are accurate up to and including p^N in a power series representation of a_i^* as obtained in the infinite Taylor series solution of $g(u)$.

Consequently the full strength of Galerkin's method as an approximation tool is first realized when the trial functions are chosen in the optimal way—in this example as $(1 - u)P_{i-1}^{(1,0)}(u)$, $i = 1, 2, \ldots, N$, or if the original expansion in monomials is converted into the optimal expansion using (91).

To prove the high accuracy of the b_i obtained by Galerkin's method with trial functions $(1 - u)P_{i-1}^{(1,0)}(u)$, we return to (11):

$$\frac{d}{du}\left(u\frac{dy}{du}\right) = py$$

$$y(1) = 1 \quad \text{and} \quad y(0) \text{ is finite}$$

We now introduce an approximation that is entirely different from the expansions in increasing powers of u:

$$y(u, p) \sim \sum_{k=0}^{M} f_k(u)p^k \tag{97}$$

This is a perturbation solution to (11). The increasing powers of $p = \Phi^2/4$ are multiplied by unknown perturbation functions $f_k(u)$. We choose $f_0(u)$ to satisfy the boundary conditions of (11) while $f_k(u)$ $(k > 0)$ satisfy homogeneous boundary conditions:

$$f_0(1) = 1 \quad \text{while } f_k(1) = 0, \quad k = 1, 2, \ldots, M$$

Equation (97) is inserted into (11) and coefficients of equal powers of p on both sides of the equation are identified:

$$\frac{d}{du}\left(u\frac{df_k}{du}\right) = \begin{cases} 0, & k = 0 \\ f_{k-1}, & k > 0 \end{cases} \tag{98}$$

The first differential equation yields

$$f_0(u) = 1$$

The second is integrated to

$$f_1(u) = u + A = u - 1$$

when $f_1(1) = 0$ is inserted. The next two perturbation functions are found in the same way and generally it is noted that $f_k(u)$ is a polynomial of degree k in u.

$$f_2(u) = \tfrac{1}{4}u^2 - u + \tfrac{3}{4}$$
$$f_3(u) = \tfrac{1}{36}u^3 - \tfrac{1}{4}u^2 + \tfrac{3}{4}u - \tfrac{19}{36}$$

Next consider expansion (94) and Galerkin's method to obtain b_i:

$$\sum_{i=1}^{N} b_i \int_0^1 \left(\frac{d}{du}\left\{u\frac{d}{du}[(1 - u)P_{i-1}]\right\} - p(1 - u)P_{i-1}\right)(1 - u)P_{j-1}\,du$$
$$= p\int_0^1 (1 - u)P_{j-1}\,du \qquad j = 1, 2, \ldots, N$$

or, in matrix notation, $(\mathbf{A} + p\mathbf{B})\mathbf{b} = p\mathbf{c}$ where $\mathbf{A}$, $\mathbf{B}$, and $\mathbf{c}$ are given by expressions similar to (50).

$$A_{ji} = \int_0^1 \left(\frac{d}{du}\left\{u\frac{d}{du}[(1-u)P_{i-1}]\right\}\right)(1-u)P_{j-1}\,du \tag{99}$$

$$B_{ji} = -\int_0^1 (1-u)^2 P_{i-1}P_{j-1}\,du \tag{100}$$

$$c_j = \int_0^1 (1-u)P_{j-1}\,du \tag{101}$$

Integration by parts in (99) gives

$$A_{ji} = u\frac{d}{du}[(1-u)P_{i-1}](1-u)P_{j-1}\Big|_0^1 - \int_0^1 u\frac{d}{du}[(1-u)P_{i-1}]\frac{d}{du}[(1-u)P_{j-1}]\,du$$

The first term is zero and the integral is symmetric in i and j.

Hence $\mathbf{A}$ is symmetric, and we can show furthermore that $\mathbf{A}$ is a diagonal matrix.

$$q_{i-1} = \frac{d}{du}\left\{u\frac{d}{du}[(1-u)P_{i-1}]\right\}$$

in (99) is a polynomial of degree $i-1$. Using the orthogonality property of the expansion polynomials $P = P^{(1,0)}(u)$, it follows that all A_{ji} with $i < j$ must be zero. Since $\mathbf{A}$ is symmetric, it must be diagonal.

Equation (100) shows that $\mathbf{B}$ is also symmetric. $(1-u)P_{i-1}$ is a polynomial of degree i and all B_{ji} with $i < j-1$ are zero. Hence $\mathbf{B}$ is a symmetric tridiagonal matrix.

Finally, $c_j = 0$ for $j > 1$, while

$$c_1 = \int_0^1 (1-u)P_0\,du = \tfrac{1}{2}$$

Our final result is

$$\left[\begin{pmatrix} A_{11} & & & & 0 \\ & A_{22} & & & \\ & & A_{23} & & \\ & & & \ddots & \\ 0 & & & & A_{NN} \end{pmatrix} + p\begin{pmatrix} B_{11} & B_{12} & & & 0 \\ B_{12} & B_{22} & B_{23} & & \\ & B_{23} & B_{33} & \ddots & \\ & & \ddots & \ddots & B_{N-1,N} \\ 0 & & & B_{N-1,N} & B_{NN} \end{pmatrix}\right]\mathbf{b} = p\begin{pmatrix} \frac{1}{2} \\ 0 \\ \vdots \\ 0 \end{pmatrix} \tag{102}$$

In a discrete analog of (97) we now expand $\mathbf{b}$ in a power series in p:

$$\mathbf{b} = \sum_{k=0}^{M} \mathbf{f}_k p^k \tag{103}$$

where $\mathbf{f}_k$ are perturbation vectors that are determined one by one when equal powers of p are collected in (102).

$$\mathbf{A}(\mathbf{f}_0 + \mathbf{f}_1 p + \mathbf{f}_2 p^2 + \ldots + \mathbf{f}_M p^M) + \mathbf{B}(\mathbf{f}_0 p + \mathbf{f}_1 p^2 + \ldots + \mathbf{f}_M p^{M+1}) = p\mathbf{c}$$

or

$$\mathbf{A}\mathbf{f}_0 = 0$$

$$\mathbf{A}\mathbf{f}_k + \mathbf{B}\mathbf{f}_{k-1} = \begin{cases} \mathbf{c} & k = 1 \\ \mathbf{0} & k > 1 \end{cases} \tag{104}$$

The solution of (104) is

$$\mathbf{f}_0 = \mathbf{0}, \qquad \mathbf{f}_1 = \mathbf{A}^{-1}\mathbf{c} = \left(\frac{1}{2A_{11}}, 0, \ldots, 0\right)$$
$$\mathbf{f}_k = -\mathbf{A}^{-1}\mathbf{B}\mathbf{f}_{k-1} \quad \text{for } k > 1 \tag{105}$$

$\mathbf{A}^{-1}\mathbf{B}$ is a tridiagonal matrix that is obtained from $\mathbf{B}$ by division of row i by A_{ii}. Hence $\mathbf{f}_k$ from (105) has k components different from 0:

$$\mathbf{f}_2 = -\frac{1}{2A_{11}}\left(\frac{B_{11}}{A_{11}}, \frac{B_{12}}{A_{22}}, 0, \ldots\right)$$

$$\mathbf{f}_3 = \frac{1}{2A_{11}}\left(\frac{B_{11}^2}{A_{11}^2} + \frac{B_{12}^2}{A_{11}A_{22}}, \frac{B_{12}B_{11}}{A_{11}A_{22}} + \frac{B_{12}B_{22}}{A_{22}^2}, \frac{B_{23}B_{12}}{A_{22}A_{33}}, 0, \ldots\right)$$

Let us mark the components of $\mathbf{f}_k$ by an index (I) that is obtained as the highest index from B_{ij} or A_{ii} that occurs in the component:

$$\mathbf{f}_1 = [(1), 0, 0, \ldots]$$

$$\mathbf{f}_2 = [(1), (2), 0, 0, \ldots]$$

$$\mathbf{f}_3 = [(2), (2), (3), 0, 0, \ldots]$$

$$\mathbf{f}_4 = [(2), (3), (3), (4), 0, 0, \ldots]$$

$$\mathbf{f}_{2k-1} = [(k), (k), (k+1), (k+1), \ldots, (2k-2), (2k-2), (2k-1), 0, 0, \ldots]$$

$$\mathbf{f}_{2k} = [(k), (k+1), (k+1), \ldots, (2k-1), (2k-1), (2k), 0, 0, \ldots] \tag{106}$$

The $\mathbf{f}_k$ determined in (105) are inserted into (94):

$$y_N = 1 + (1-u)\sum_{i=1}^{N} b_i P_{i-1} = 1 + (1-u)\sum_{i=1}^{N}\left(\sum_{k=0}^{M} f_{ik}p^k\right)P_{i-1}(u)$$
$$= 1 + \sum_{k=0}^{M} p^k\left[\sum_{i=1}^{N} f_{ik}(1-u)P_{i-1}(u)\right] \tag{107}$$

The double indexing f_{ik} is used to indicate the ith component of vector $\mathbf{f}_k$.

A comparison of (107) and (97) with the appropriate expressions for the polynomials $f_i(u)$ shows that all polynomials $f_i(u)$ of degree $i \leq N$ can be exactly represented by the linear combination of polynomials $(1 - u)P_{i-1}$, $i = 1, 2, \dots, N$ in (107). For these terms (107) is simply a reformulation of (97) and the coefficients b_i found by a Galerkin approximation of order N will all contain the correct coefficients f_{ik} in a power series expansion of the true Fourier series coefficients b_i^* up to and including the coefficient of p^N.

The low-order coefficients $b_1, b_2, \dots$ are, however, considerably more accurate. Consider a 2Nth-order Galerkin approximation similar to (107). Now all coefficients b_i are correct up to and including the coefficient f_{i2N} of p^{2N}. The schematic solution (106) for $\mathbf{f}_k$ does, however, show that only the first N rows in (102) enter into the determination of the first element f_{1k} for $k \leq 2N$, i.e., that f_{1k} are identical for $k \leq 2N$ whether they are determined by an Nth-order or a 2Nth-order Galerkin's method. In the same way (106) shows that f_{ik} are identical whether they are determined by an Nth-order or a $(2N - i + 1)$-order Galerkin approximation.

Hence for an Nth-order Galerkin approximation (94) in orthogonal polynomials $P_{i-1}^{(1,0)}(u)$, the same coefficients f_{ik} appear in a power series expansion $b_i = \sum_0^M f_{ik}p^k$ up to and including $f_{i,2N-i+1}$ as those that would have been obtained from a similar power series expansion of b_i^* in (96). This completes the proof of the optimal character of Galerkin's method using expansion (94) for the linear differential equation (11).

The result explains the accurate value of η in table 2.5 for Galerkin's method (η is given by b_1) and also the rapid stabilization of b_i in table 2.7.

The method of moments with the same set of trial functions has an accuracy of $2N - i$ ($i = 1, 2, \dots, N$) for b_i in the sense defined above.

2.5.2 Lagrange interpolation polynomials

Two formulations of the Nth-degree polynomial approximation y_N have hitherto been discussed:

$$y_N = \sum_{i=0}^{N} a_i x^i \tag{108}$$

$$y_N = \sum_{i=0}^{N} b_i P_i(x) \tag{109}$$

A third approximation operator will, however, be frequently used in this text:

$$y_N = \sum_{i=1}^{N+1} y_i l_i(x) \tag{110}$$

This is the Lagrange interpolation polynomial. The "building blocks" are now $N + 1$ polynomials $l_i(x)$ that are all of degree N:

$$l_i = \frac{p_{N+1}(x)}{(x - x_i)p_{N+1}^{(1)}(x_i)} \tag{111}$$

$p_{N+1}(x) = (x - x_1)(x - x_2)\dots(x - x_{N+1})$ is a polynomial of degree $N + 1$ with leading coefficient 1. It is called the node polynomial while $l_i(x)$ are the Lagrange polynomials. x_i are $N + 1$ interpolation points and y_i of (110) are the $N + 1$ ordinates at these points.

To verify that (110) is a polynomial of degree N passing through the $N + 1$ points $(x_1, y_1), (x_2, y_2), \dots, (x_{N+1}, y_{N+1})$, it is sufficient to note that l_i are all polynomials of degree N and that $l_i(x_j) = 0$ for $i \neq j$ and 1 for $i = j$.

Another formulation of l_i can be given:

$$l_i = \prod_{j=1,i}^{N+1} \frac{x - x_j}{x_i - x_j} \tag{112}$$

where the notation $j = 1, i$ means that the product consists of all factors $(x - x_j)/(x_i - x_j)$, $j = 1, 2, \dots, N + 1$, except $j = i$.

$$\prod_{j=1,i}^{N+1} (x - x_j) = \frac{p_{N+1}(x)}{x - x_i}$$

$$\prod_{j=1,i}^{N+1} (x_i - x_j) = \frac{dp_{N+1}(x)}{dx} \quad \text{for } x = x_i$$

and the identity of (112) and (111) is proved.

To obtain the distance function $E_N(x)$ for interpolative approximation of a given function $y(x)$, define $\varepsilon(x)$ by the following relation:

$$\varepsilon(x) = E_N(x) - Kp_{N+1}(x) \tag{113}$$

For x equal to any of the $N + 1$ interpolation points, $\varepsilon(x)$ is obviously zero whatever the value of K. For any other choice of x in the approximation interval $[a, b]$ we can choose K such that $\varepsilon(x) = 0$. Hence $\varepsilon(x)$ has at least $N + 2$ zeros $\in [a, b]$, namely $x_1, x_2, \dots, x_{N+1}$ and the current x. A repeated application of Rolles theorem shows that $\varepsilon^{(N+1)}(x)$ has at least one zero $\in [a, b]$. This (unknown) zero is at θ.

$$\varepsilon^{(N+1)}(\theta) = 0 = E_N^{(N+1)}(\theta) - K(N + 1)!$$

Now $y_N(x)$ is a polynomial of degree N in x and $E_N^{(N+1)}(\theta) = y^{(N+1)}(\theta)$; since $E_N(x) = Kp_{N+1}(x)$ at the current x, we obtain

$$E_N(x) = \frac{y^{(N+1)}(\theta)}{(N + 1)!} p_{N+1}(x) \tag{114}$$

for any given x-value $\in [a, b]$.

Equation (114) is completely analogous to the distance function for the Taylor series approximation from $x = x_0$.

$$E_N(x) = \frac{y^{(N+1)}(\theta_1)}{(N+1)!}(x - x_0)^{N+1} \tag{115}$$

The two values θ and θ_1 in (114) and (115) are, of course, different but the real difference between the two distance functions lies in the last factor, as easily seen by comparison of (114) and (115) for $y(x) = x^{N+1}$ and $x_0 = 0$:

(114):
$$E_N(x) = p_{N+1}(x) \tag{116}$$

(115):
$$E_N(x) = x^{N+1} \tag{117}$$

In both cases $E_N(x)$ is an $(N + 1)$-degree polynomial with leading coefficient 1 but the $N + 1$ zeros of $p_{N+1}(x)$ that all lie in the interpolation interval, e.g., $[0, 1]$, will render the maximum value of $|E_N(x)|$ in $[0, 1]$ in (116) much smaller than 1, the value obtained in (117). This is one of the major entries into the field of general approximation theory and a "best polynomial" approximation of degree N can be obtained by minimization of $|p_N(x)|$ in $[0, 1]$ with a suitable choice of the $N + 1$ interpolation points.

It is natural to close the discussion of construction principles for Nth-order polynomial approximations by noting that once $y_N(x)$ has been constructed in any one of the three approximation modes (108), (109), (110), it is a simple matter to compute the coefficients of the other two approximations. As an example, take the Fourier-type approximation (109) with $P_i = P_i^{(\alpha,\beta)}$ and the interpolation approximation (110):

$$y_N(x) = \sum_{i=0}^{N} b_i P_i^{(\alpha,\beta)}(x) = \sum_{i=1}^{N+1} y_i l_i(x)$$

The coefficients $\mathbf{b}$ are determined in the usual way for Fourier series:

$$b_i = \frac{1}{C_i}\int_0^1 W(x) y_N(x) P_i^{(\alpha,\beta)}(x)\, dx$$

where

$$W(x) = x^{\beta}(1 - x)^{\alpha} \quad \text{and} \quad C_i = \int_0^1 W P_i^2\, dx$$

The integrand $y_N(x)P_i^{(\alpha,\beta)}(x)$ is at most a polynomial of degree $N + N = 2N$. Hence b_i can be calculated exactly by a quadrature formula that accurately integrates a $2N$-degree polynomial using $N + 1$ quadrature points.

$$C_i b_i = \sum_{k=1}^{N+1} w_k P_i^{(\alpha,\beta)}(x_k) y_N(x_k) \tag{118}$$

In subsection 3.3.4 we derive a node polynomial $p_{N+1}(x)$ that permits exact integration of a $2N$-degree polynomial when one of the nodes is either $x = 0$ or $x = 1$ and the remaining N nodes are zeros of an orthogonal polynomial $p_N^{(\alpha,\beta)}(x)$.

Thus with $N + 1$ interpolation ordinates $y_N(x_k)$ the exact values of b_i in (109) can be found. Equation (118) supplements (91) and (92), which are other conversion formulas between identical expansions of the Nth-degree polynomial $y_N(x)$.

EXERCISES

1. Consider steady state heat conduction through a slab of thickness L. The two faces of the slab are maintained at $T(s = 0) = T_0$ and $T(s = L) = T_1$. The heat conductivity of the solid is a linear function of T:

$$k(T) = k_0 + \alpha(T - T_0)$$

Show that the heat balance takes the following form:

$$\frac{d}{dx}\left[(1 + b\theta)\frac{d\theta}{dx}\right] = 0$$

$$\theta(0) = 0 \qquad \theta(1) = 1$$

in the dimensionless variables x and θ.
What is b in terms of the given physical parameters?
What is the solution for $b = 0$?
Is the solution for $b > 0$ above or below the solution for $b = 0$?
Select a proper set of trial functions and state the form of the approximation with one parameter a_1.
Find a_1 for $b = 1$ by collocation at $x = 1$, $x = \frac{1}{2}$, $x = 0$.
For $b \ll 1$ an expansion of a_1 in powers of b is rapidly convergent.

$$a_1 = q_0 + q_1 b + q_2 b^2$$

Determine q_0 and q_1 using collocation at $x = \frac{1}{2}$.

2. Write a computer program for solution of equation (1) by equidistant collocation and trial functions given by equation (34). n and N are program input data, and $N = 1, 3, \ldots, 8$ should be used for $n = 0.5$, 1, and 2.

3. a. Expand the solution of $y^{(2)} - y^2 = 0$ with $y^{(1)}(x = 0) = 0$ in a power series from $x = 0$. Find the first seven terms (up to and including the term in x^{12}) of the series.
b. Assume that the parameter $y_0 = y(x = 0)$ of part a is 1. Solve the differential equation in closed form.
c. Prove that the radius of convergence for the power series in part a with $y_0 = 1$ is smaller than or equal to 2.9745.
d. Find y_0 in part a when $y(x = 1) = 1$.

e. Consider $y^{(2)} - \Phi^2 y^2 = 0$. What is the largest value of y_0 for which the series developed in part a is convergent for $x = 1$?
f. Find a perturbation series for the solution of $y^{(2)} - \Phi^2 y^2 = 0$, $y^{(1)}(0) = 0$, $y(1) = 1$. Terms up to and including x^8 are required.
g. Use the series in part f to find $y(x = 0)$ for $\Phi = 1$. If the result is disappointing, then use continued differencing on the sequence for y_0 that you obtain from the partial sum of series with $N = 1, 2, \ldots$ terms.
h. Reflect on the range of applicability of the perturbation series. Can it be used for the same Φ values as the Taylor series?
i. Repeat for the trivial equation $y^{(2)} - \Phi^2 y = 0$. Show that the radius of convergence for the perturbation series is $\Phi = \pi/2$.

4. The eigenvalues λ for the heat conduction problem in slab geometry and with film transport resistance are given by

$$\lambda \tan \lambda = \mathrm{Bi}$$

Show by perturbation analysis that the eigenvalue of smallest magnitude is approximately

$$\lambda_1 = \mathrm{Bi}^{1/2}\left(1 - \frac{\mathrm{Bi}}{6} + \frac{11}{360}\mathrm{Bi}^2\right)$$

for small values of Bi.

REFERENCES

Weighted residual methods have been used to solve mathematical models of physical science for the last 50 years, and naturally a vast number of references to applications exist.

A substantial part of these references is listed in Finlayson (1972). Over the years, numerous papers have appeared on weighted residual methods from research groups at the University of Minnesota (Sparrow, Scriven, Amundson, etc.), and Finlayson's review (1966) with Scriven introduced a number of chemical engineers to the subject. Finlayson's authoritative textbook from 1972 is today undoubtedly the best reference to applications of MWR in the various fields of transport phenomena.

Rice (1964, 1969) is a highly readable introduction to approximation theory and Natanson (1955) cites much of the older literature on the subject.

Morse and Feshback (1953) has a long chapter on perturbation methods but it is difficult to read and lacks newer references. The best text is probably Nayfeh (1973), which contains numerous simple examples. Other texts are Van Dyke (1964), which exclusively treats problems from fluid mechanics, and finally Cole (1968).

1. FINLAYSON, B. A. *The Method of Weighted Residuals and Variational Principles*, New York: Academic Press (1972).

2. FINLAYSON, B. A., and SCRIVEN, L. E. *Appl. Mech. Rev.* 19 (1966): 735.

3. RICE, J. R. *The Approximation of Functions*, Vols. 1–2. Reading, Mass.: Addison-Wesley (1964, 1969).

4. NATANSON, I. P. *Konstruktive Funktionentheorie*. Berlin: Akademie Verlag (1955).

5. MORSE, P. M., and FESHBACK, H. *Methods of Theoretical Physics*. Chapter 9, pp. 1001–105. New York: McGraw-Hill (1953).

6. NAYFEH, A. H. *Perturbation Methods*. New York: Wiley (1973).

7. VAN DYKE, M. *Perturbation Methods in Fluid Mechanics*, Academic Press Series in Applied Mathematics and Mechanics. New York: Academic Press (1964).

8. COLE, J. D. *Perturbation Methods in Applied Mathematics*, New York: Blaisdell (1968).

3 Some Important Properties of Orthogonal Polynomials—Formulation of a Standard Collocation Procedure

Introduction

Chapter 2 shows that a very efficient collocation method results when the collocation points are chosen as zeros of certain orthogonal polynomials, the so-called Jacobi polynomials defined by equation (2.73).

In section 3.1 methods of constructing Jacobi polynomials and of determining their zeros will be developed.

The MWR of chapter 2 are based on either power series or Fourier-type approximations of the unknown function. The object of the MWR is to determine the coefficients of these expansions. In section 2.5 the Lagrange interpolation polynomial is introduced as a reformulation of the primary expansions of y_N in increasing powers of x.

A computer-oriented collocation algorithm is much easier to set up when the N ordinates at the collocation points are used as unknowns rather than the coefficients a_i or b_i of expansions in terms of increasing powers of x or polynomials of increasing degree.

Consequently formulas for differentiation of an arbitrary Lagrange polynomial $l_i(x)$ are set up in subsection 3.3.2 and quadratures based on integration of certain Lagrange polynomials are developed in subsection 3.3.3.

The results of sections 3.1 to 3.3 are utilized in section 3.4 where computer programs for calculation of zeros of Jacobi polynomials, derivatives of y_N at the collocation points, and quadratures based on the collocation ordinates are described.

Finally in section 3.5 a rigorous formulation of our collocation procedure is given. This section is the basis on which the standard problem types of chapter 4 are treated.

3.1 The Power Series Representation of $P_N^{(\alpha,\beta)}(x)$

In section 2.5, equation (2.87), $P_N^{(\alpha,\beta)}(x)$ was written in the following form

$$P_N^{(\alpha,\beta)}(x) = \sum_{i=0}^{N} (-1)^{N-i}\gamma_i x^i \tag{1}$$

γ_0 is taken to be 1 and the remaining N coefficients can be found either directly from the orthogonality property

$$\int_0^1 x^\beta(1-x)^\alpha P_j(x)P_N(x)\,dx = 0 \qquad j = 0, 1, \ldots, N-1 \tag{2}$$

or from any of a number of other relations that are all in one way or the other derived from the fundamental relation (2). The orthogonal polynomials defined by (2) are called Jacobi polynomials. A more convenient form of (2) is

$$\int_0^1 x^\beta(1-x)^\alpha x^j P_N(x)\,dx = 0 \qquad j = 0, 1, \ldots, N-1 \tag{3}$$

x^j can be expressed as a linear combination of P_k $(k = 0, 1, \ldots, j)$—see (2.90)—and (3) is obviously equivalent to (2).

The following set of linear equations for γ_i is derived from the integrals (3).

$$\mathbf{M\gamma} = \mathbf{0}$$

$$M_{ji} = \frac{\Gamma(\beta+1+i+j)\Gamma(\alpha+1)}{\Gamma(\alpha+\beta+2+i+j)}(-1)^{N-i} \tag{4}$$

$$i = 0, 1, \ldots, N, \quad j = 0, 1, \ldots, N-1$$

since

$$\int_0^1 x^m(1-x)^n\,dx = \frac{\Gamma(m+1)\Gamma(n+1)}{\Gamma(m+n+2)} \tag{5}$$

For $\alpha = \beta = 0$, one obtains the Legendre polynomials:

$$M_{ji} = \frac{\Gamma(i+j+1)\Gamma(1)}{\Gamma(i+j+2)}(-1)^{N-i} = \frac{(-1)^{N-i}}{i+j+1}$$

For $N = 2$,

$$\begin{pmatrix} 1 & -\frac{1}{2} & \frac{1}{3} \\ \frac{1}{2} & -\frac{1}{3} & \frac{1}{4} \end{pmatrix} \begin{pmatrix} \gamma_0 \\ \gamma_1 \\ \gamma_2 \end{pmatrix} = \mathbf{0} \qquad \text{or} \qquad \begin{pmatrix} -\frac{1}{2} & \frac{1}{3} \\ -\frac{1}{3} & \frac{1}{4} \end{pmatrix} \begin{pmatrix} \gamma_1 \\ \gamma_2 \end{pmatrix} = \begin{pmatrix} -1 \\ -\frac{1}{2} \end{pmatrix}$$

since $\gamma_0 = 1$. The solution is

$$\gamma_1 = \gamma_2 = 6 \quad \text{and} \quad P_2^{(0,0)}(x) = 6x^2 - 6x + 1 \tag{6}$$

Another method is based on Rodrigues' formula, which may be derived from (2) [see, e.g., Villadsen (1970), p. 37, or Szegö (1967), p. 67]:

$$P_N^{(\alpha,\beta)}(x)(1-x)^\alpha x^\beta = \frac{(-1)^N \Gamma(\beta+1)}{\Gamma(N+\beta+1)} \frac{d^N}{dx^N}[(1-x)^{N+\alpha} x^{N+\beta}] \tag{7}$$

For $\alpha = \beta = 0$ and $N = 2$,

$$P_2^{(0,0)}(x) = \frac{1}{2} \frac{d^2}{dx^2}[(1-x)^2 x^2] = 6x^2 - 6x + 1$$

Computing the *N*th derivative in (7) and comparing coefficients of equal powers of x on the two sides of the equation, one may—after considerable algebra—arrive at the following explicit expression for γ_i [Villadsen (1970), p. 40]:

$$\gamma_i = \binom{N}{i} \frac{\Gamma(N+i+\alpha+\beta+1)\Gamma(\beta+1)}{\Gamma(N+\alpha+\beta+1)\Gamma(i+\beta+1)} \tag{8}$$

where

$$\binom{N}{i} = \frac{N!}{i!(N-i)!}$$

An even simpler recursive computation of γ_i is obtained from (8) by taking the ratio of γ_{i-1} and γ_i:

$$\gamma_i = \frac{N-i+1}{i} \frac{N+i+\alpha+\beta}{i+\beta} \gamma_{i-1} \tag{9}$$

with

$$\gamma_0 \equiv 1 \quad \text{and} \quad i = 1, 2, \ldots, N$$

For $N = 3$ and $\alpha = \beta = 0$, (9) immediately gives

$$\gamma_0 \equiv 1, \qquad \gamma_1 = 3 \cdot 4 \cdot \gamma_0 = 12$$

$$\gamma_2 = \tfrac{2}{2} \cdot \tfrac{5}{2} \cdot \gamma_1 = 30, \qquad \gamma_3 = \tfrac{1}{3} \cdot \tfrac{6}{3} \cdot \gamma_2 = 20$$

The simplicity with which γ_i can be calculated from (9) makes it the recommended formula whenever an explicit expression for $P_N^{(\alpha,\beta)}$ is desired.

Similar formulas can be derived from (7) for the coefficients δ_i in

$$x^N = \sum_{i=0}^{N} \delta_i P_i^{(\alpha,\beta)}(x) \tag{10}$$

The final result is an analog of (9):

$$\delta_{i-1} = \frac{i(2i + \alpha + \beta - 1)(N + i + \alpha + \beta + 1)}{(N - i + 1)(2i + \alpha + \beta + 1)(i + \alpha + \beta)}\delta_i \tag{11}$$

with

$$\delta_N = \frac{1}{\gamma_N} \quad \text{and} \quad i = N, N - 1, \ldots, 1$$

For $\alpha = 1$, $\beta = 0$, and $N = 2$, one obtains $\gamma_1 = 8$ and $\gamma_2 = 10$ from (9) and $\delta_2 = \frac{1}{10}$, $\delta_1 = \frac{4}{15}$, and $\delta_0 = \frac{1}{6}$ from (11) in accordance with the results of table 2.6 and (2.90).

Whereas (9) is excellently suited for a calculation of the coefficients in the explicit expression (1) for P_N, it is not at all suited for machine computation of P_N at a specific x-value. γ_{Ni} are rapidly increasing with N and since the terms in (1) occur with alternating sign, a significant loss of digits is unavoidable. As an example, the tenth-degree Legendre polynomial may be considered. Here $\gamma_{10,7} = 2{,}333{,}760$, while the maximum value of $P_{10}^{(0,0)}$ is 1.

Hence another computational scheme is used to obtain $P_N(x_k)$. The following recurrence relation is recommended:

$$P_N = (F_N x - G_N)P_{N-1} - H_N P_{N-2} \tag{12}$$

where F_N, G_N, and H_N are functions of N, α, and β.

A three-term recurrence relation (12) can be shown to hold for any family of orthogonal polynomials and the functions F_N, G_N, and H_N can be calculated by an application of the fundamental definition of the orthogonal family—in our case, equation (2) [Villadsen (1970), p. 55].

The coefficients g_N and h_N of a recurrence formula for the rescaled polynomials $p_N = P_N/\gamma_{NN}$ are given in equations (13) to (15):

$$p_N = [x - g_N(N, \alpha, \beta)]p_{N-1} - h_N(N, \alpha, \beta)p_{N-2} \tag{13}$$

$$g_1 = \frac{\beta + 1}{\alpha + \beta + 2} \qquad g_N = \frac{1}{2}\left[1 - \frac{\alpha^2 - \beta^2}{(2N + \alpha + \beta - 1)^2 - 1}\right] \tag{14}$$

for $N > 1$

$$h_1 = 0, \qquad h_2 = \frac{(\alpha + 1)(\beta + 1)}{(\alpha + \beta + 2)^2(\alpha + \beta + 3)}$$
$$h_N = \frac{(N - 1)(N + \alpha - 1)(N + \beta - 1)(N + \alpha + \beta - 1)}{(2N + \alpha + \beta - 1)(2N + \alpha + \beta - 2)^2(2N + \alpha + \beta - 3)} \tag{15}$$
$$\text{for } N > 2$$

The recursive evaluation of $p_N(x_k)$ is started with $N = 1$, $p_{-1}(x_k)$ arbitrary, and $p_0(x_k) = 1$.

The following computation on a 10-digit machine of $P_{10}^{(0,0)}(0.98) = \gamma_{NN} p_{10}^{(0,0)}(0.98)$ by (1) and by (13) with

$$g_N = \frac{1}{2} \quad (N \geq 1), \qquad h_N = \frac{(N - 1)^2}{4(2N - 1)(2N - 3)} \quad (N \geq 2)$$

$$\gamma_{10,10} = \frac{\Gamma(2N + 1)}{\Gamma(N + 1)\Gamma(N + 1)} = \frac{20!}{[(10)!]^2} = 184{,}756$$

shows that (13) is far superior to (1);

$$\begin{aligned} P_{10}^{(0,0)}(0.98) &= -0.2552433289 \quad \text{(exact)} \\ &= -0.2552433288 \quad \text{(by recurrence)} \\ &= -0.255252 \quad \text{(by power series)} \end{aligned}$$

The advantage of (13) over (1) is even more obvious when an expansion y_N in orthogonal polynomials is used to calculate $y_N(x_k)$:

$$y_N(x_k) = \sum_{i=1}^{N+1} b_i p_{i-1}(x_k) \tag{16}$$

In this case every term $p_i(x_k)$ calculated in the sequence (13) is used.

3.2 Zeros of Orthogonal Polynomials

In the example in chapter 2 it was shown that an "optimal" collocation method with accuracy comparable to—or even equal to—the accuracy of Galerkin's method was obtained when the collocation points were chosen as the zeros of an orthogonal polynomial, $P_N^{(1,0)}$.

We shall henceforth call a collocation method with collocation points equal to the zeros of any Jacobi polynomial an *orthogonal collocation method.*

In this section, various methods for an accurate computation of the zeros of $P_N^{(\alpha,\beta)}(x)$ [or $p_N^{(\alpha,\beta)}(x)$] are discussed.

It is easily proved from the fundamental relation (2) that $P_N(x)$ has N distinct, real-valued zeros x_k and that $0 < x_k < 1$:

1. $$\int_0^1 x^\beta(1-x)^\alpha P_0 P_N\,dx = \int_0^1 W(x)P_N\,dx = 0$$

   Since $W(x)$ is positive for $x \in (0, 1)$, there is at least one real-valued zero of $P_N(x)$ in the open interval (0, 1).
2. Let $x_1, x_2, \ldots, x_k$ $(k \leq N)$ be the total number of real-valued, distinct zeros in (0, 1). The remaining $N - k$ zeros are either complex, of multiplicity >1, or outside [0, 1]. $P_N(x)$ changes sign at each x_j, $j = 1, 2, \ldots, k$, but the polynomial

   $$(x - x_1)(x - x_2)\ldots(x - x_k)P_N(x) = Q_k(x)P_N(x)$$

   is of constant sign throughout [0, 1]. Consequently,

   $$I = \int_0^1 W(x)Q_k(x)P_N(x)\,dx \neq 0$$

   On the other hand if $Q_k(x)$ is a polynomial of degree $<$ N, the orthogonality relation (3) shows that $I = 0$. Hence $k = N$, and it is proved that P_N has exactly N distinct, real-valued zeros in (0, 1).

These properties of the zeros of P_N considerably facilitate their numerical determination, and they are used in all the algorithms discussed in the following:

1. Algorithms based on the explicit power series coefficients γ_i.
2. Algorithms with $p_N(x)$ and $p_N^{(1)}(x)$ computed by (13).
   a. Bisection followed by Newton's method.
   b. Newton's method with suppression of previously determined zeros.
3. *QD* algorithm on an equivalent tridiagonal matrix.

The classical methods for computation of the N distinct zeros of a polynomial $P_N(x)$ [e.g., Henrici (1964)] compute the value of $P_N(x_k^{(i)})$ where $x_k^{(i)}$ is the current estimate of the zero x_k by nested multiplication (Horner's scheme) using the coefficients of the polynomial. In our case,

$$P_N(x) = [(\gamma_N x - \gamma_{N-1})x + \gamma_{N-2}]x - \ldots \tag{17}$$

We have already seen that $P_N(x_k^{(i)})$ is inaccurately determined by this method, and serious errors in the zeros close to $x = 1$ may result for large N.

It is just as easy to compute $p_N[x_k^{(i)}]$ by (13) and the accuracy of the computation is usually of the order of the machine accuracy.

The numbers $p_0[x_k^{(i)}] \equiv 1, p_1[x_k^{(i)}], \ldots, p_N[x_k^{(i)}]$ computed from (13) form a so-called Sturm sequence and it may be shown that the number of sign changes in this sequence is equal to the number of real-valued zeros of p_N larger than $x_k^{(i)}$. Hence if K sign changes are observed for a given $x_k^{(i)}$, there are K zeros larger than $x_k^{(i)}$:

$$\underbrace{0 < x_1 < x_2 < \ldots <}_{N-K} x_{N-K} < x_k^{(i)} < \underbrace{x_{N-K+1} < \ldots < x_N < 1}_{K} \tag{18}$$

In this manner an interval δ_k in which each zero x_k is located can be determined. The final computation of the location of x_k within δ_k is made by Newton iteration since a parallel development of $p^{(1)}[x_k^{(i)}]$ and $p[x_k^{(i)}]$ is easily performed.

$$\begin{aligned} p_j &= [x_k^{(i)} - g_j]p_{j-1} - h_j p_{j-2} \\ p_j^{(1)} &= [x_k^{(i)} - g_j]p_{j-1}^{(1)} - h_j p_{j-2}^{(1)} + p_{j-1} \qquad j = 1, 2, \ldots, N \\ p_0 &= 1, \qquad p_0^{(1)} = 0, \qquad p_{-1} = p_{-1}^{(1)} \text{ arbitrary} \end{aligned} \tag{19}$$

If the Newton method fails to converge, δ_k can be reduced by bisection, but the final iteration is always done by Newton's method, which is strongly convergent locally.

The following routine avoids the initial Sturm sequence—bisection calculations to obtain intervals δ_k small enough for the Newton method to be convergent.

A Newton iteration starting from $x = 0$ will produce a sequence $x_1^{(1)}, x_1^{(2)}, \ldots$ that converges to x_1 from below—the reason being that $|P_N^{(1)}(x)|$ is monotonously decreasing in $(-\infty, x_1)$.

In the normal Newton–Horner scheme based on the coefficients of the power series representation of P_N, one automatically obtains the coefficients of $G_{N-1} = P_N/(x - x_1)$ when the smallest zero x_1 has been determined. Hence the iteration may be repeated on G_{N-1}, producing a sequence $x_2^{(1)}, x_2^{(2)}, \ldots$ that converges to x_2 from below when $x_2^{(0)} = x_1$ since $|G_{N-1}^{(1)}(x)|$ is monotonously decreasing in $(-\infty, x_2)$.

By this deflation process the zeros of P_N are determined one by one and there is no danger that the iterations will converge to one of the previously determined (smaller) zeros.

We shall use (19) with $x_1^{(0)} = 0$ to compute x_1. Even though the explicit coefficients of G_{N-k} never occur when this procedure is used to determine x_{k+1}, it is, however, still possible to suppress the previously determined zeros $x_1, x_2, \ldots, x_k$.

$$G_{N-k}(x) = p_N(x) \Big/ \prod_{i=1}^{k} (x - x_i) \tag{20}$$

$$\frac{d[\ln G_{N-k}(x)]}{dx} = \frac{G_{N-k}^{(1)}}{G_{N-k}} = \frac{p_N^{(1)}}{p_N} - \frac{\sum_{j=1}^{k}\prod_{i=1,j}^{k}(x - x_i)}{\prod_{i=1}^{k}(x - x_i)}$$

$$\frac{G_{N-k}^{(1)}}{G_{N-k}} = \frac{p_N^{(1)}}{p_N} - \sum_{i=1}^{k}\frac{1}{x - x_i} \tag{21}$$

$$\delta(x) = \frac{G_{N-k}}{G_{N-k}^{(1)}} = \frac{(p_N/p_N^{(1)})}{1 - (p_N/p_N^{(1)})\sum_{i=1}^{k}1/(x - x_i)} \tag{22}$$

The Newton iteration for x_{k+1} with $x_1, x_2, \ldots, x_k$ previously determined is

$$x_{k+1}^{(i)} = x_{k+1}^{(i-1)} - \delta[x_{k+1}^{(i-1)}] \qquad \text{for } i = 1, 2, \ldots \tag{23}$$

and

$$x_{k+1}^{(0)} = x_k + \varepsilon$$

where ε is a small positive quantity, e.g., 10^{-4}.

It is seen that the iteration for x_{k+1} proceeds on G_{N-k} just as would be the case if the coefficients of this polynomial had been known, but in the actual calculations the recursive relations (19) are all that is needed.

An entirely different method may be set up by noticing that the recursive calculation of $p_N[x_k^{(i)}]$ by (19) is exactly what is done when the determinant of a tridiagonal, symmetric matrix with off-diagonal elements $T_{j,j-1} = T_{j-1,j} = \sqrt{h_j}$ and diagonal elements $T_{jj} = x_k^{(i)} - g_j$ is calculated. If $x_k^{(i)}$ happens to be an eigenvalue of $\mathbf{T}$, the determinant $= p_N[x_k^{(i)}] = 0$, and $x_k^{(i)}$ is one of the zeros, x_k, of p_N.

Very efficient methods exist for computation of the eigenvalues of a tridiagonal matrix—e.g., the *QD* algorithm. In a sequence of operations, **T** is transformed into a diagonal matrix with the eigenvalues equal to the desired zeros of $p_N(x)$ on the diagonal.

In our experience this is the fastest method available, but the zeros are not so accurately determined as by the Newton method based on (22) and (23). The Newton algorithm has been tested for different parameters (α, β) of the Jacobi polynomial $P_N^{(\alpha,\beta)}(x)$ and for N up to 50. At most, one digit is lost in the zeros for any of the polynomials that have been investigated.

3.3 Differentiation and Integration of Lagrange Interpolation Polynomials

3.3.1 Linear operations on $y_N(x)$

In chapter 2 several approximation operators were discussed:

$$y_N(x) = \sum_{i=1}^{N+1} a_i x^{i-1} \quad \text{or} \quad \sum_{i=1}^{N+1} b_i P_{i-1}(x) \tag{24}$$

$$y_N(x) = \sum_{i=1}^{N} a_i(1-x)x^{i-1} \quad \text{or} \quad \sum_{i=1}^{N} b_i(1-x)P_{i-1}(x) \tag{25}$$

$$y_N(x) = \sum_{i=1}^{N+1} y(x_i)l_i(x) \tag{26}$$

In all three cases, y_N can be represented as the scalar product of two vectors, a coefficient vector **a**, **b**, or **y** and a vector of expansion functions $\mathbf{f}^T = (x^0, x^1, \ldots, x^N)$, $\mathbf{g}^T = (P_0, P_i, \ldots, P_N)$, or $\mathbf{h}^T = [l_1(x),\ l_2(x), \ldots, l_N(x),\ l_{N+1}(x)]$.

Linear operations are easily performed on the power series expression in (24):

$$\frac{d^k(y_N)}{dx^k} = \sum_{i=1}^{N+1} a_i\left[\frac{d^k}{dx^k}(x^{i-1})\right] = \left(\frac{d^k}{dx^k}\mathbf{f}\right)^T\mathbf{a} \tag{27}$$

$$\int_0^1 W(x)y_N\,dx = \sum_{i=1}^{N+1} a_i\left[\int_0^1 W(x)x^{i-1}\,dx\right] = \left[\int_0^1 W(x)\mathbf{f}\,dx\right]^T\mathbf{a} \tag{28}$$

Similar linear operations on the other expansions in (24) and (25) are obtained with almost equal ease but the Lagrange interpolation polynomial (26) presents some problems since $l_i(x)$ is not differentiated and integrated as easily as x^{i-1} or P_{i-1}. In the present section we shall develop methods of obtaining expressions similar to (27) and (28) for the Lagrange interpolation polynomial

$$\frac{d_k(y_N)}{dx^k} = \left[\frac{d^k}{dx^k}\mathbf{l}(x)\right]^T\mathbf{y} \tag{29}$$

$$\int_0^1 W(x)y_N\,dx = \left[\int_0^1 W(x)\mathbf{l}(x)\,dx\right]^T\mathbf{y} \tag{30}$$

3.3.2 Differentiation of a Lagrange interpolation polynomial

Differentiation of (26) yields

$$\frac{d}{dx}[y_N(x)] = \sum_{i=1}^{N+1}\frac{dl_i(x)}{dx}y(x_i) = [\mathbf{l}^{(1)}]^T\mathbf{y} \tag{31}$$

or in general

$$\frac{d^k y_N(x)}{dx^k} = [\mathbf{l}^{(k)}]^T\mathbf{y} \tag{32}$$

In particular, we are interested in obtaining the derivatives $(dy_N/dx)_{x=x_j}$ at the interpolation points x_j, $j = 1, 2, \ldots, N + 1$.

$$\left(\frac{dy}{dx}\right)_{x=x_j} = \sum_{i=1}^{N+1}\left[\frac{dl_i(x)}{dx}\right]_{x=x_j} y_i = [\mathbf{l}^{(1)}(x_j)^T]\mathbf{y} \tag{33}$$

and similarly for higher-order derivatives.

We now wish to obtain expressions for or method of calculating $l_i^{(k)}(x_j)$, $i = 1, 2, \ldots, N + 1$, $j = 1, 2, \ldots, N + 1$.

From the definition (2.111) of $l_i(x)$, one obtains

$$\frac{p_{N+1}(x)}{p_{N+1}^{(1)}(x_i)} = (x - x_i)l_i(x) \tag{34}$$

Equation (34) is differentiated:

$$\frac{p_{N+1}^{(1)}(x)}{p_{N+1}^{(1)}(x_i)} = (x - x_i)l_i^{(1)}(x) + l_i(x) \tag{35}$$

$$\frac{p_{N+1}^{(2)}(x)}{p_{N+1}^{(1)}(x_i)} = (x - x_i)l_i^{(2)}(x) + 2l_i^{(1)}(x) \tag{36}$$

$$\frac{p_{N+1}^{(k)}(x)}{p_{N+1}^{(1)}(x_i)} = (x - x_i)l_i^{(k)}(x) + kl_i^{(k-1)}(x) \tag{37}$$

Equation (37) with $x = x_i$ yields

$$l_i^{(k-1)}(x_i) = \frac{1}{k} \cdot \frac{p_{N+1}^{(k)}(x_i)}{p_{N+1}^{(1)}(x_i)} \tag{38}$$

and, for $x = x_j \neq x_i$,

$$l_i^{(k)}(x_j) = \frac{1}{x_j - x_i}\left[\frac{p_{N+1}^{(k)}(x_j)}{p_{N+1}^{(1)}(x_i)} - kl_i^{(k-1)}(x_j)\right] \tag{39}$$

Normally only $l_i^{(1)}(x)$ and $l_i^{(2)}(x)$ are of interest. For these derivatives, we obtain

$$l_i^{(1)}(x_i) = \frac{1}{2}\frac{p_{N+1}^{(2)}(x_i)}{p_{N+1}^{(1)}(x_i)} \tag{40}$$

$$l_i^{(2)}(x_i) = \frac{1}{3}\frac{p_{N+1}^{(3)}(x_i)}{p_{N+1}^{(1)}(x_i)} \tag{41}$$

$\Bigg\}\; x = x_i$ (for (40) and (41))

and, for $x = x_j \neq x_i$,

$$l_i^{(1)}(x_j) = \frac{1}{x_j - x_i}\frac{p_{N+1}^{(1)}(x_j)}{p_{N+1}^{(1)}(x_i)} \qquad [\text{noting that } l_i(x_j) = 0] \tag{42}$$

$$
\begin{aligned}
l_i^{(2)}(x_j) &= \frac{1}{x_j - x_i}\left[\frac{p_{N+1}^{(2)}(x_j)}{p_{N+1}^{(1)}(x_i)} - 2l_i^{(1)}(x_j)\right] \\
&= l_i^{(1)}(x_j)\left[\frac{p_{N+1}^{(2)}(x_j)}{p_{N+1}^{(1)}(x_j)} - 2\frac{1}{x_j - x_i}\right] \\
&= 2l_i^{(1)}(x_j)\left[l_j^{(1)}(x_j) - \frac{1}{x_j - x_i}\right]
\end{aligned}
\tag{43}
$$

All the coefficients $l_i^{(1)}(x_j)$ and $l_i^{(2)}(x_j)$, $i = 1, 2, \ldots, N + 1$ and $j = 1, 2, \ldots, N + 1$ may thus be calculated from $p_{N+1}^{(1)}(x_j)$, $p_{N+1}^{(2)}(x_j)$, and $p_{N+1}^{(3)}(x_j)$, $j = 1, 2, \ldots, N + 1$.

The vector of derivatives

$$
\mathbf{y}^{(1)} = \left[\left(\frac{dy}{dx}\right)_{x=x_1}, \left(\frac{dy}{dx}\right)_{x=x_2}, \ldots, \left(\frac{dy}{dx}\right)_{x=x_N}\right]^T \tag{44}
$$

may be expressed

$$
\frac{d}{dx}(\mathbf{y}) = \mathbf{Ay} \tag{45}
$$

where $A_{ji} = l_i^{(1)}(x_j)$ and similarly

$$
\frac{d^2}{dx^2}(\mathbf{y}) = \mathbf{By} \tag{46}
$$

where $B_{ji} = l_i^{(2)}(x_j)$.

A simple algorithm exists for numerical calculation of the derivatives $p_{N+1}^{(k)}(x_i)$ of the node polynomial $p_{N+1}(x)$ at the interpolation points x_i:

$$
p_{N+1}(x) = \prod_{j=1}^{N+1} (x - x_j)
$$

may be defined by the recurrence formula

$$
\begin{aligned}
p_0(x) &= 1 \\
p_j(x) &= (x - x_j)p_{j-1}(x) \qquad j = 1, 2, \ldots, N + 1
\end{aligned}
\tag{47}
$$

Differentiate (47) to obtain

$$
p_j^{(1)}(x) = (x - x_j)p_{j-1}^{(1)}(x) + p_{j-1} \tag{48}
$$

$$
p_j^{(2)}(x) = (x - x_j)p_{j-1}^{(2)}(x) + 2p_{j-1}^{(1)}(x) \tag{49}
$$

$$
p_j^{(3)}(x) = (x - x_j)p_{j-1}^{(3)}(x) + 3p_{j-1}^{(2)}(x) \tag{50}
$$

with

$$
p_0^{(1)}(x) = p_0^{(2)}(x) = p_0^{(3)}(x) = 0
$$

Values of $p_{N+1}^{(k)}(x_i)$ may thus be found by simultaneous evaluation of (47) to (50) with x_i inserted for x.

A further simplification is obtained when the interpolation points are reordered such that the x_i value at which $p_{N+1}^{(k)}(x)$ is to be evaluated is always placed first:

$$x_i, x_1, x_2, \ldots, x_{i-1}, x_{i+1}, \ldots, x_{N+1}$$

Starting from $p_0(x_i) = 1$, $p_0^{(k)}(x_i) = 0$, $k = 1, 2, 3$, we obtain

$$p_1(x_i) = (x_i - x_i)p_0(x_i) = 0$$
$$p_1^{(1)}(x_i) = (x_i - x_i)p_0^{(1)}(x_i) + p_0(x_i) = 1$$
$$p_1^{(2)}(x_i) = (x_i - x_i)p_0^{(2)}(x_i) + 2p_0^{(1)}(x_i) = 0$$
$$p_1^{(3)}(x_i) = (x_i - x_i)p_0^{(3)}(x_i) + 3p_0^{(2)}(x_i) = 0$$

$p_j(x_i)$ remains zero for all $j \geq 1$, and (48) to (50) become

$$p_j^{(1)}(x_i) = (x_i - x_j)p_{(j-1)}^{(1)}(x_i) \tag{51}$$
$$p_j^{(2)}(x_i) = (x_i - x_j)p_{(j-1)}^{(2)}(x_i) + 2p_{(j-1)}^{(1)}(x_i) \tag{52}$$
$$p_j^{(3)}(x_i) = (x_i - x_j)p_{(j-1)}^{(3)}(x_i) + 3p_{(j-1)}^{(2)}(x_i) \tag{53}$$

with $j = 2, 3, \ldots, i-1, i+1, \ldots, N+1$ and, starting from $p_1^{(1)}(x_i) = 1$, $p_1^{(2)}(x_i) = p_1^{(3)}(x_i) = 0$.

We finally notice that (40) to (43) and (51) to (53) are valid for any choice of distinct interpolation points x_i.

3.3.3 Gauss–Jacobi quadrature

Consider $y_{N-1}(x)$, the $(N-1)$-degree Lagrangian interpolation polynomial given by the N interpolation abscissas x_i, $i = 1, 2, \ldots, N$, and the corresponding y values $y(x_i)$.

$$y_{N-1}(x) = \sum_{i=1}^{N} l_i(x)y(x_i) = \sum_{i=1}^{N} l_i(x)y_i \tag{54}$$

where

$$l_i(x) = \frac{p_N(x)}{(x - x_i)p_N^{(1)}(x_i)}, \qquad p_N(x) = \prod_{j=1}^{N}(x - x_j) \tag{55}$$

For a given weight function $W(x)$, we obtain

$$\int_0^1 W(x)y_{N-1}(x)\,dx = \sum_{i=1}^{N} y_i \int_0^1 l_i(x)W(x)\,dx = \mathbf{w}^T\mathbf{y} \tag{56}$$

where

$$w_i = \int_0^1 l_i(x) W(x)\, dx \tag{57}$$

In subsection 3.3.2, simple expressions for derivatives of the Lagrange polynomials l_i at the interpolation points have been developed for any choice of interpolation points.

Similar expressions for the quadrature weights w_i cannot be found for an arbitrary weight function $W(x)$ and arbitrary interpolation points. We shall, however, only be interested in the specific choice $W(x) = x^\beta(1 - x)^\alpha$, $(\alpha, \beta) > -1$, and $p_N(x)$ = a Jacobi polynomial, in which case the resulting formulas are called *Gauss–Jacobi quadratures.*

In the following, simple expressions for the weights $\mathbf{w}$ of Gauss–Jacobi quadratures will be developed.

The node polynomials $p_N^{(\alpha,\beta)}$ satisfy the orthogonality relationship

$$\int_0^1 p_i^{(\alpha,\beta)}(x) p_j^{(\alpha,\beta)}(x)(1 - x)^\alpha x^\beta\, dx = c_i^{(\alpha,\beta)} \delta_{ij} \tag{58}$$

where

$$c_i^{(\alpha,\beta)} = \frac{C_i^{(\alpha,\beta)}}{(\gamma_{i,i})^2}$$

We evaluate

$$w_i = \int_0^1 l_i(x)(1 - x)^\alpha x^\beta\, dx \tag{59}$$

where

$$l_i(x) = \frac{p_N^{(\alpha,\beta)}}{(x - x_i) p_N^{(1)(\alpha,\beta)}(x_i)} \tag{60}$$

The short notation p_N or $p_N(x)$ will be used for $p_N^{(\alpha,\beta)}(x)$.

Certain properties of the expansion functions l_i are first proved:

1. Any set of Lagrange polynomials l_i of degree $N - 1$ satisfies

$$\sum_{i=1}^{N} l_i(x) = 1 \tag{61}$$

Proof: The only polynomial of degree $< N$ having the value $y(x_i) = 1$ at N distinct points x_i is the polynomial $y = 1$. Inserting y and y_i in (54), equation (61) is immediately obtained.

2. The expansion functions $l_i(x)$ defined by (60) are mutually orthogonal with weight function

$$W(x) = (1-x)^\alpha x^\beta$$

$$\begin{aligned}\int_0^1 l_i(x)l_j(x)(1-x)^\alpha x^\beta\,dx &= \int_0^1 \frac{1}{p_N^{(1)}(x_i)p_N^{(1)}(x_j)}\frac{p_N^2(x)}{(x-x_i)(x-x_j)}(1-x)^\alpha x^\beta\,dx\\ &= \frac{1}{p_N^{(1)}(x_i)p_N^{(1)}(x_j)}\int_0^1 p_N(x)\left[\prod_{k=1,i,j}^{N}(x-x_k)\right](1-x)^\alpha x^\beta\,dx = 0\end{aligned}$$

since the polynomial

$$Q_{N-2}(x) = \prod_{k=1,i,j}^{N}(x-x_k)$$

is of degree $N-2$.

With the preliminary results of (1) and (2), we next proceed to find w_i from (59):

$$\begin{aligned}w_i &= \int_0^1 l_i(x)(1-x)^\alpha x^\beta\,dx = \int_0^1 l_i(x)\left[\sum_{j=1}^{N} l_j(x)\right](1-x)^\alpha x^\beta\,dx\\ &= \int_0^1 l_i^2(x)(1-x)^\alpha x^\beta\,dx\end{aligned} \tag{62}$$

Let

$$s_i = w_i x_i(1-x_i)[p_N^{(1)}(x_i)]^2 \quad\text{and}\quad q_i(x) = \frac{p_N(x)}{(x-x_i)}$$

Then

$$s_i = \int_0^1 [q_i(x)]^2 x_i(1-x_i)(1-x)^\alpha x^\beta\,dx$$

Substitute

$$x_i = x-(x-x_i) \quad\text{and}\quad 1-x_i = (1-x)+(x-x_i)$$

Hence

$$\begin{aligned}s_i &= -\int_0^1 p_N^2(x)(1-x)^\alpha x^\beta\,dx + \int_0^1 (x-x_i)q_i^2(x)(2x-1)(1-x)^\alpha x^\beta\,dx\\ &\quad+\int_0^1 q_i^2(x)(1-x)^{\alpha+1}x^{\beta+1}\,dx\\ &= -c_N^{(\alpha,\beta)} + \int_0^1 p_N(x)[(2x-1)q_i(x)](1-x)^\alpha x^\beta\,dx\\ &\quad+\int_0^1 q_i^2(x)(1-x)^{\alpha+1}x^{\beta+1}\,dx\end{aligned}$$

The last integral is integrated by parts using $d(x - x_i) = dx$:

$$\int_0^1 q_i^2(x)(1-x)^{\alpha+1}x^{\beta+1} \cdot 1\, dx$$

$$= (x - x_i)q_i^2(x)(1-x)^{\alpha+1}x^{\beta+1}|_0^1$$

$$- \int_0^1 (x - x_i)\frac{d}{dx}[q_i^2(x)(1-x)^{\alpha+1}x^{\beta+1}]\, dx$$

$$= 0 - \int_0^1 (x - x_i)q_i(x)(1-x)^{\alpha}x^{\beta}$$

$$\times \left[x(1-x)2\frac{dq_i(x)}{dx} - (\alpha + 1)xq_i(x) + (\beta + 1)(1-x)q_i(x)\right] dx$$

$$= \int_0^1 p_N(x)(1-x)^{\alpha}x^{\beta}$$

$$\times \left[2(x^2 - x)\frac{dq_i(x)}{dx} + (\alpha + 1)xq_i(x) + (\beta + 1)(x - 1)q_i(x)\right] dx$$

This result is inserted into the expression for s_i:

$$s_i = -c_N^{(\alpha,\beta)} + \int_0^1 p_N(x)(1-x)^{\alpha}x^{\beta}$$

$$\times \left\{2(x^2 - x)\frac{dq_i(x)}{dx} + q_i(x)[2x - 1 + (\alpha + 1)x + (\beta + 1)(x - 1)]\right\} dx$$

$$= -c_N^{(\alpha,\beta)} + \int_0^1 p_N(x)(1-x)^{\alpha}x^{\beta}G_N(x)\, dx$$

$G_N(x)$ is a polynomial of degree N with leading coefficient

$$2(N-1) + 2 + (\alpha + 1) + \beta + 1 = 2N + \alpha + \beta + 2$$

We may write

$$G_N = (2N + \alpha + \beta + 2)p_N(x) + G_{N-1}$$

where G_{N-1} is some polynomial of degree $N - 1$.

$$s_i = -c_N^{(\alpha,\beta)} + (2N + \alpha + \beta + 2)c_N^{(\alpha,\beta)} = (2N + \alpha + \beta + 1)c_N^{(\alpha,\beta)}$$

or

$$w_i = \frac{(2N + \alpha + \beta + 1)c_N^{(\alpha,\beta)}}{x_i(1 - x_i)[p_N^{(1)}(x_i)]^2} \tag{63}$$

In the development of (63) we have only considered quadrature points in the interior of (0, 1)—i.e., the zeros of a Jacobi polynomial. When $y_N(x)$ is obtained as the solution of a differential equation, the side conditions contain information on y at the interval end points (e.g., $y = 1$ at $u = 1$ in the example of chapter 2). It is thus of interest to include the interval end points as additional interpolation points and to develop quadrature formulas that include the ordinates at these points.

As an example the following Nth-degree polynomial approximation is considered:

$$y_N(x) = \sum_{i=1}^{N+1} l_i(x)y(x_i) \tag{64}$$

with the node polynomial

$$p_{N+1}(x) = p_N^{(\alpha,\beta)}(x)(x - 1) \tag{65}$$

The quadrature weights are obtained as before:

$$w_i = \int_0^1 l_i(x)(1 - x)^\alpha x^\beta \, dx = \int_0^1 \frac{p_{N+1}(x)}{(x - x_i)p_{N+1}^{(1)}(x_i)}(1 - x)^\alpha x^\beta \, dx \tag{66}$$

Take $x_{N+1} = 1$ and $i \neq N + 1$:

$$p_{N+1}^{(1)}(x_i) = (x_i - 1)p_N^{(1)}(x_i)$$

$$\begin{aligned} w_i &= \int_0^1 \frac{p_N(x)(x - 1)}{(x - x_i)p_N^{(1)}(x_i)(x_i - 1)}(1 - x)^\alpha x^\beta \, dx \\ &= \int_0^1 \frac{p_N(x)(x - x_i + x_i - 1)}{(x - x_i)p_N^{(1)}(x_i)(x_i - 1)}(1 - x)^\alpha x^\beta \, dx = \int_0^1 \frac{p_N(x)(1 - x)^\alpha x^\beta \, dx}{(x - x_i)p_N^{(1)}(x_i)} \end{aligned} \tag{67}$$

The integrand is identical to that of (59), and the w_i are unchanged.

For $i = N + 1$, we obtain

$$w_{N+1} = \int_0^1 l_{N+1}(x)(1 - x)^\alpha x^\beta \, dx = \int_0^1 \frac{p_N^{(\alpha,\beta)}}{p_{N+1}^{(1)}(1)}(1 - x)^\alpha x^\beta \, dx = 0 \tag{68}$$

If an expression for the w_i containing the derivative of the actual node polynomial $p_{N+1}(x)$ is desired, we may substitute $p_{N+1}^{(1)}(x_i) = (x_i - 1)p_N^{(1)}(x_i)$ in (63):

$$w_i = \frac{(1 - x_i)(2N + \alpha + \beta + 1)c_N^{(\alpha,\beta)}}{x_i[p_{N+1}^{(1)}(x_i)]^2}, \qquad i = 1, 2, \ldots, N + 1 \tag{69}$$

Similarly, if $x_0 = 0$ (but not $x_{N+1} = 1$) is included, the result is

$$\tilde{w}_i = \frac{x_i(2N + \alpha + \beta + 1)c_N^{(\alpha,\beta)}}{(1 - x_i)[p_{N+1}^{(1)}(x_i)]^2}, \qquad i = 0, 1, 2, \ldots, N \tag{70}$$

If both $x_0 = 0$ and $x_{N+1} = 1$ are included,

$$w_i = \frac{x_i(1 - x_i)}{[p_{N+2}^{(1)}(x_i)]^2}(2N + \alpha + \beta + 1)c_N^{(\alpha,\beta)}, \qquad i = 0, 1, \ldots, N + 1 \tag{71}$$

The quadrature weights (63) were developed by integration of an $(N - 1)$-degree polynomial (54). With N free parameters w_i, any function $y_{N-1}(x)$ can be integrated correctly by (56) whatever the choice of quadrature points x_i. The remarkable property of Gauss–Jacobi quadrature formulas is that they integrate a polynomial of degree $2N - 1$ correctly. We might say that N extra parameters—the quadrature abscissas—appear beside the N weights, and this makes these formulas particularly accurate. The disarmingly simple proof follows:

Let y_{N+j} be an arbitrary polynomial of degree $N + j$. This may be written

$$y_{N+j}(x) \equiv y_{N-1}(x) + G_j(x)p_N(x) \tag{72}$$

in which the coefficients of the jth-degree polynomial G_j are chosen such that the identity (72) is satisfied.

$$\begin{aligned}\int_0^1 W(x)y_{N+j}\,dx &= \int_0^1 W(x)y_{N-1}\,dx + \int_0^1 W(x)G_j(x)p_N(x)\,dx \\ &= \int_0^1 W(x)y_{N-1}\,dx = \mathbf{w}^T\mathbf{y}\end{aligned} \tag{73}$$

when $j < N$ and $p_N(x)$ is the appropriate Jacobi polynomial.

As a corollary to this proof of optimality of Gauss–Jacobi quadrature, the "Galerkin–collocation" method of subsection 2.4.2 follows directly:

$$\int_0^1 R_N(u, \mathbf{a})W(u)u^{j-1}\,du = \sum_1^N w_k R_N(u_k, \mathbf{a})u_k^{j-1} \tag{74}$$

for $j = 1, 2, \ldots, N$, R_N a polynomial of degree N in u and $\{u_k\}$ = the N zeros of the Jacobi polynomial with weight function $W(u)$.

3.3.4 Radau and Lobatto quadrature

In chapter 2 it is shown that the functional

$$\eta = \int_0^1 y_N(u)\,du \tag{75}$$

is calculated with an exceptionally high accuracy—$2N$ in a power series expansion of η in the parameter p of the differential equation—when the

N coefficients a_i of (2.34) were determined by a collocation procedure based on the zeros of $p_N^{(1,0)}(u)$. Expansion (2.34) is identical to

$$y_N = \sum_{i=1}^{N+1} l_i(u) y_i \tag{76}$$

with

$$l_i(u) = \frac{p_{N+1}(u)}{(u - u_i) p_{N+1}^{(1)}(u_i)}$$
$$p_{N+1}(u) = (u - 1) p_N^{(1,0)}(u) \tag{77}$$

The high accuracy of η indicates that a quadrature formula based on (76) with one quadrature point at $u = 1$ has the highest possible accuracy $2N$, when the remaining N quadrature points are chosen as zeros of $p_N^{(1,0)}(u)$.

The development in (64) to (71) shows that an extra quadrature point at $x = 0$, $x = 1$, or at both locations does not at all improve the accuracy of the quadrature (59) and (60) since the quadrature weights at these points are zero. In fact (73) implies that N extra interpolation points besides the zeros of $p_N^{(\alpha,\beta)}(x)$ have no effect on the quadrature. The reason is that the N ordinates $y(x_k)$ at the zeros of $p_N^{(\alpha,\beta)}(x)$ are sufficient to integrate a polynomial of degree $2N - 1$ exactly.

We shall prove that the correct choice of N interior quadrature points in

$$\int_0^1 x^\beta (1 - x)^\alpha y_{N-1}(x)\, dx = w(1) y_{N-1}(1) + \sum_1^N w_k y_{N-1}(x_k) \tag{78}$$

are the zeros of $p_N^{(\alpha+1,\beta)}(x)$. When this has been proved, it immediately follows that the collocation ordinates $y(u_k)$ at the zeros of $p_N^{(1,0)}(u)$ are indeed the best we could wish for when (75) is evaluated by (78).

Let y_N be represented by (76) and (77) (with x in place of u).

Any polynomial of degree $2N$ can be written in a form similar to (72):

$$y_{2N} = y_N + G_{N-1}(1 - x) p_N^{(\alpha+1,\beta)}$$

Multiply by $W(x) = x^\beta (1 - x)^\alpha$ and integrate:

$$\int_0^1 y_{2N} W\, dx = \textstyle\sum_{k=1}^{N+1} w_k y_k + \displaystyle\int_0^1 G_{N-1} x^\beta (1 - x)^{\alpha+1} p_N^{(\alpha+1,\beta)}\, dx = \mathbf{w}^T \mathbf{y}$$

where $y_1, y_2, \ldots, y_N$ are the interior ordinates and y_{N+1} is the ordinate at $x = 1$.

Having proved that the best node polynomial for the quadrature (78) is $p_{N+1}(x) = (1 - x) p_N^{(\alpha+1,\beta)}(x)$, we shall next determine the weights w_k of the formula. The result is given by (83). Formulas with $x = 0$ or with both $x = 0$ and $x = 1$ as extra quadrature points are (84) and (85), which can be derived similarly to (83).

First we note without proof that (62) also holds when $l_i(x)$ are given by (77). s_i is defined by

$$s_i = x_i[p_{N+1}^{(1)}(x_i)]^2 w_i = \int_0^1 x_i q_i^2(x)(1-x)^\alpha x^\beta \, dx \tag{79}$$

First, consider $i \neq N + 1$:

$$\begin{aligned} s_i &= \int_0^1 x_i \prod_{j=1,i}^{N+1} (x - x_j)^2 (1-x)^\alpha x^\beta \, dx \\ &= \int_0^1 x_i(1-x) \prod_{j=1,i}^{N} (x - x_j)^2 (1-x)^{\alpha+1} x^\beta \, dx \\ &= \int_0^1 (1-x)[x - (x - x_i)] \prod_{j=1,i}^{N} (x - x_j)^2 (1-x)^{\alpha+1} x^\beta \, dx \\ &= \int_0^1 \prod_{j=1,i}^{N} (x - x_j)^2 (1-x)^{\alpha+2} x^{\beta+1} \, dx + \int_0^1 p_N^{(\alpha+1,\beta)} \prod_{j=1,i}^{N+1} (x - x_j) \\ &\quad \times (1-x)^{\alpha+1} x^\beta \, dx \end{aligned} \tag{80}$$

The first integral in (80) is completely similar to the third integral in the expression for s_i for Gaussian quadrature (α being changed to $\alpha + 1$), and its value is $[2(N-1) + \alpha + 2 + \beta + 1]c_N^{(\alpha+1,\beta)} = (2N + \alpha + \beta + 1)c_N^{(\alpha+1,\beta)}$.

The second integral is simply $c_N^{(\alpha+1,\beta)}$ since the product is an Nth-degree polynomial with leading coefficient 1 which may be exchanged with $p_N^{(\alpha+1,\beta)}$ in the integral.

Thus

$$s_i = (2N + \alpha + \beta + 2)c_N^{(\alpha+1,\beta)} \qquad i = 1, 2, \ldots, N \tag{81}$$

Also

$$\begin{aligned} s_{N+1} &= \int_0^1 [p_N^{(\alpha,\beta)}]^2 (1-x)^\alpha x^\beta \, dx \\ &= \int_0^1 (1 - x + x) p_N^2 (1-x)^\alpha x^\beta \, dx \\ &= \int_0^1 (1-x)^{\alpha+1} x^\beta p_N^2 \, dx + \int_0^1 (1-x)^\alpha x^{\beta+1} p_N^2 \, dx \end{aligned}$$

The first integral is $c_N^{(\alpha+1,\beta)}$. The second integral is integrated by parts:

$$
\begin{aligned}
\int_0^1 (1-x)^\alpha x^{\beta+1} p_N^2\,dx
&= \frac{-1}{\alpha+1}(1-x)^{\alpha+1}x^{\beta+1}p_N^2\Big|_0^1 \\
&\quad + \frac{1}{\alpha+1}\int_0^1 (1-x)^{\alpha+1}\frac{d}{dx}(x^{\beta+1}p_N^2)\,dx \\
&= \frac{1}{\alpha+1}\int_0^1 (1-x)^{\alpha+1}p_N x^\beta\left[2x\frac{dp_N}{dx} + (\beta+1)p_N\right]dx \\
&= \frac{1}{\alpha+1}(2N+\beta+1)c_N^{(\alpha+1,\beta)}
\end{aligned}
\tag{82}
$$

$$s_{N+1} = \frac{2N+\alpha+\beta+2}{\alpha+1}c_N^{(\alpha+1,\beta)}$$

Hence the following weights are obtained:

$$w_i = \frac{(2N+\alpha+\beta+2)c_N^{(\alpha+1,\beta)}}{x_i[p_{N+1}^{(1)}(x_i)]^2}\cdot\begin{cases} 1 & i \neq N+1 \\ \dfrac{1}{\alpha+1} & i = N+1 \end{cases} \tag{83}$$

A similar formula using the value y_0 at $x_0 = 0$ (but not y_{N+1} at $x_{N+1} = 1$) may be constructed:

$$w_i = \frac{(2N+\alpha+\beta+2)c_N^{(\alpha,\beta+1)}}{(1-x_i)[p_{N+1}^{(1)}(x_i)]^2}\cdot\begin{cases} \dfrac{1}{\beta+1} & i = 0 \\ 1 & i \neq 0 \end{cases} \tag{84}$$

Finally, using both end points $x_0 = 0$ and $x_1 = 1$,

$$w_i = \frac{(2N+\alpha+\beta+3)c_N^{(\alpha+1,\beta+1)}}{[p_{N+2}^{(1)}(x_i)]^2}\cdot\begin{cases} \dfrac{1}{\beta+1} & i = 0 \\ 1 & i \neq 0, N+1 \\ \dfrac{1}{\alpha+1} & i = N+1 \end{cases} \tag{85}$$

The weight factors w_i of (83) and (84) are called *Radau quadrature weights* and the w_i of (85) are *Lobatto quadrature weights.*

In the solution of a differential equation we may wish to include $x_0 = 0$ as an interpolation point as well as $x_{N+1} = 1$ because boundary conditions are given at both these points. Equation (83) may also be used to calculate an integral derived from y_{N+1} when $p_{N+2} = x(x-1)p_N(x)$ is used as a node polynomial. As shown in (69) to (71), the interior weights and w_{N+1} are the same as in (83), while w_0, the weight of the ordinate at $x = 0$, is zero.

A Radau formula based on N interior points and one end point will integrate any polynomial y_M exactly, provided $M \leq 2N$.

Similarly, a Lobatto formula (N interior points and two end points) will give exact results for polynomials of degree $\leq 2N + 1$.

In general the quadratures of this section may be used for exact evaluation of integrals of the type

$$\int_0^1 y_M(x)W(x)f(x)\,dx = \sum w_i f(x_i) y_i \tag{86}$$

where $M = (N - 1)$ (Gaussian), N (Radau), or $(N + 1)$ (Lobatto) for an arbitrary polynomial $f(x)$ of degree less than or equal to N.

3.4 Program Description

In Section 3.4, computer programs for obtaining zeros of Jacobi polynomials, derivatives of node polynomials, interpolation weights, differentiation weights, and quadrature weights are described. The programs are all in double precision and in FORTRAN IV. They are listed in the Appendix.

3.4.1 Zeros and derivatives of Jacobi polynomials

This subroutine calculates the zeros of $p_N^{(\alpha,\beta)}(x)$ and also the three first derivatives of the node polynomial

$$p_{NT}(x) = (x)^{N0} p_N(x)(x - 1)^{N1} \tag{87}$$

at the interpolation points. Each of the parameters N0 and N1 may be given the value 0 and 1. NT = N + N0 + N1.

The subroutine call is

```
CALL JCOBI (ND, N, N0, N1, ALFA, BETA, DIF1, DIF2, DIF3, ROOT)
```

Input parameters:

INTEGER ND: The DIMENSION of vectors DIF1, DIF2, DIF3, ROOT. In all subroutines described in this text an automatic dimensioning of vectors and matrices is made via one or two input parameters: ND for vectors and ND, NCOL for matrices. The values of these parameters are given in the DIMENSION statement of the main program. In the description of the following subroutines, ND (and NCOL) will not be commented on further.

INTEGER N:	The degree of the Jacobi polynomial, i.e., the number of interior interpolation points.
INTEGER N0:	Decides whether $x = 0$ is included as an interpolation point. N0 must be set equal to 1 (including $x = 0$) or 0 (excluding this point).
INTEGER N1:	As for N0, but for the point $x = 1$.
REAL ALFA, BETA:	The values of α and β.

The output is

REAL ARRAY ROOT:	One-dimensional vector containing on exit the N + N0 + N1 zeros of the node polynomial used in the interpolation routine.
REAL ARRAY DIF1, DIF2, DIF3:	Three one-dimensional vectors containing on exit the first, second, and third derivatives of the node polynomial at the zeros.

The vectors ROOT, DIF1, DIF2, and DIF3 may be declared in the main program with any dimension greater than or equal to the actual value of N + N0 + N1.

The program consists essentially of three parts. First the values of the coefficients g_i and h_i, $i = 1, 2, \ldots, N$ of the recurrence formula (13) are computed according to formulas (14) and (15). The 2N coefficients are temporarily stored in DIF1 and DIF2.

Next the zeros of p_N are determined as described in section 3.2 by the Newton method with root suppression. The smallest zero x_1 is stored in ROOT(1), x_2 in ROOT(2), and x_N in ROOT(N).

If $x_0 = 0$ is included as an interpolation point (N0 = 1), the N zeros are shifted one position so that ROOT(1) = 0, ROOT(2) = x_1 and ROOT(N + 1) = x_N. Finally an extra element ROOT(NT) = 1 is added to ROOT if N1 = 1. ROOT now contains the interpolation points in ascending order.

Finally the derivatives of the node polynomial are evaluated at the interpolation points using equations (51) to (53).

3.4.2 Lagrange interpolation

The value of y at any desired point $x = x_A$ may be found from

$$y(x_A) = \sum_{i=1}^{NT} l_i(x_A) y_i \tag{88}$$

where

$$l_i(x_A) = \frac{p_{NT}(x_A)}{(x_A - x_i) p_{NT}^{(1)}(x_i)} \tag{89}$$

Hence, if the interpolation points x_i (ROOT) and the first derivative of the node polynomial $p_{NT}^{(1)}(x_i)$ (DIF1) are known, the interpolation weights are easily calculated.

The subroutine call is

```
CALL INTRP (ND, NT, X, ROOT, DIF1, XINTP)
```

With input data:

NT: The total number of interpolation points (=N + N0 + N1) for which the value of the dependent variable y is known.

REAL X: The abscissa x where $y(x)$ is desired.

REAL ARRAY

ROOT, DIFI: Interpolation points and derivatives of node polynomial, derived in JCOBI.

The output data is

REAL ARRAY XINTP: The vector of interpolation weights $l_i(x)$

$y(x)$ is then found from

$$y(x) = \sum_{I=1}^{NT} \text{XINTP(I)} \cdot \text{Y(I)} \tag{90}$$

INTRP can, of course, be used for any choice of NT interpolation points, but DIF1 and the interpolation points ROOT have to be specified in other applications.

In the main program the dimension of XINTP must be equal to or larger than NT.

3.4.3 Differentiation weights and Gaussian quadrature weights

Values of $(dy/dx)_{x=x_i}$ or $(d^2y/dx^2)_{x=x_i}$ are found from

$$\left(\frac{dy}{dx}\right)_{x=x_i} = \sum_{k=1}^{NT} l_k^{(1)}(x_i)y_k \tag{91}$$

$$\left(\frac{d^2y}{dx^2}\right)_{x=x_i} = \sum_{k=1}^{NT} l_k^{(2)}(x_i)y_k \tag{92}$$

Integrals of the type

$$\int_0^1 y(x)(1-x)^\alpha x^\beta\, dx = \sum_{k=1}^{NT} w_k y_k \tag{93}$$

are determined by Gaussian quadrature.

The subroutine call is

```
CALL DFOPR (ND, N, N0, N1, I, ID, DIF1, DIF2, DIF3, ROOT, VEC)
```

Input data:

INTEGER N, N0, N1:	As in JCOBI.
INTEGER I:	The index of the node for which the weights in (91) or (92) are to be calculated.
INTEGER ID:	Indicator. ID = 1 gives the weights for dy/dx, ID = 2 for d^2y/dx^2, and ID = 3 gives the Gaussian weights. The value of I is irrelevant in this last case.
REAL ARRAY ROOT, DIF1, DIF2, DIF3:	The one-dimensional vectors computed in JCOBI.

Output data:

REAL ARRAY VEC:	The computed vector of weights $[l_k^{(1)}(x_i), l_k^{(2)}(x_i), \quad \text{or } w_k, k = 1, 2, \ldots, NT]$

This dual-purpose algorithm applies the expressions developed in section 3.3. The Gaussian weights are normalized, however, such that their sum equals 1. If the true Gaussian weights of (51) are desired, the calculated weights should be multiplied by

$$I^{(\alpha,\beta)} = \int_0^1 (1 - x)^\alpha x^\beta \, dx = \frac{\Gamma(\alpha + 1)\Gamma(\beta + 1)}{\Gamma(\alpha + \beta + 2)} \tag{94}$$

The computation of $l_k^{(1)}$ and $l_k^{(2)}$ by (91) and (92) is valid for any choice of interpolation points and may be used whenever the appropriate values of ROOT, DIF1, DIF2, and DIF3 are available. The computation of quadrature weights is restricted to node polynomials of the form (87) with $p_N(x)$ a Jacobi polynomial.

3.4.4 Radau or Lobatto quadrature

The weights of a quadrature

$$\int_0^1 y(x)(1 - x)^\alpha x^\beta \, dx = \sum_{k=1}^{NT} w_k y_k \tag{95}$$

are determined. The interior interpolation points are zeros of $p_N^{(\alpha',\beta')}$ where α' and β' are defined below.

The subroutine call is

```
CALL RADAU (ND, N, N0, N1, ID, ALFA, BETA, ROOT, DIF1, V)
```

Input data:

INTEGER N, N0, N1: As in JCOBI.

INTEGER ID: ID is an indicator. ID = 1 gives Radau quadrature weights including $x = 1$. ID = 2 gives Radau quadrature weights including $x = 0$. ID = 3 gives Lobatto weights including both.

REAL ALFA, BETA: The exponents α and β of the weight function that appears in the integral (95).

REAL ARRAY ROOT, DIF1: The node polynomial is given by

$$p_{NT}(x) = x^{N0} p_N^{(\alpha',\beta')}(x)(x-1)^{N1} \tag{96}$$

The arguments α' and β' to be used in JCOBI for calculation of ROOT and DIF1 depend on whether $x = 1$, $x = 0$, or both are used as extra quadrature points. Thus:

ID = 1: $\alpha' = \alpha + 1, \beta' = \beta$ (N1 = 1 and N0 = 0 or 1)

ID = 2: $\alpha' = \alpha,$ $\beta' = \beta + 1$ (N0 = 1 and N1 = 0 or 1)

ID = 3: $\alpha' = \alpha + 1,$ $\beta' = \beta + 1$ (N0 = N1 = 1)

ID = 1: N1 = 1 and N0 = 0 or 1

ID = 2: N0 = 1 and N1 = 0 or 1

ID = 3: N0 = 1 and N1 = 1

ROOT is the vector of zeros of $p_{NT}(x)$ in (96) and DIF1 is $p_{NT}^{(1)}$ (ROOT). Both are determined in JCOBI.

Output data:

REAL ARRAY V: The NT computed quadrature weights. These are also normalized such that

$$\sum_1^{NT} w_k = 1$$

To obtain the true weights of (95), each w_k must be multiplied by $I^{(\alpha,\beta)}$ in (94).

3.5 Discretization of Differential Equations in Terms of Ordinates

3.5.1 The dependent variable known at the boundary

An approximate solution to the problem of chapter 2

$$u\frac{d^2y}{du^2} + \frac{dy}{du} = py, \qquad y_{u=1} = 1 \tag{97}$$

was obtained in the following form:

$$y_N = 1 + (1 - u) \sum_{i=1}^{N} a_i u^{i-1} \tag{98}$$

y_N was inserted into the differential equation and in a collocation method the residual was equated to zero at N interior points u_i. The resulting N equations were solved for **a**.

y_N of (98) is a polynomial of degree N in u and it may be reformulated into an Nth-degree Lagrangian interpolation polynomial:

$$y_N = \sum_{i=1}^{N+1} l_i(u) y(u_i) \tag{99}$$

where

$$y_{N+1} = y_{u=1} = 1$$

Equation (99) contains N unknown ordinates $y(u_i)$, $i = 1, 2, \ldots, N$ instead of the N unknown coefficients **a**.

Hence, N interior collocation points $u_1, u_2, u_3, \ldots, u_N$ are selected, yielding the node polynomial

$$p_{N+1}(u) = \sum_{j=1}^{N+1} (u - u_j)$$

in which $u_{N+1} = 1$.

The residual

$$R_N = u\frac{d^2 y_N}{du^2} + \frac{dy_N}{du} - p y_N$$

is evaluated and equated to zero at the N interior u_j.

$$R_N(u_j) = u_j\left(\frac{d^2 y_N}{du^2}\right)_{u=u_j} + \left(\frac{dy_N}{du}\right)_{u=u_j} - p(y_N)_{u=u_j}$$

$$= u_j\left(\sum_{i=1}^{N+1} B_{ji} y_i\right) + \sum_{i=1}^{N+1} A_{ji} y_i - p y_j = 0, \qquad j = 1, 2, \ldots, N$$

or

$$u_j \sum_{i=1}^{N} B_{ji} y_i + \sum_{i=1}^{N} A_{ji} y_i - p y_j = -u_j B_{jN+1} - A_{jN+1} \qquad j = 1, 2, \ldots, N \tag{100}$$

The elements of **B** and **A** are evaluated by the algorithms of section 3.4 and N linear equations (101) in the unknown y_i are immediately obtained:

$$
\begin{aligned}
(\mathbf{B}^* - p\mathbf{I})\mathbf{y} &= \mathbf{c} \\
B^*_{ji} &= u_j B_{ji} + A_{ji} \qquad (101) \\
c_j &= -(u_j B_{jN+1} + A_{jN+1})
\end{aligned}
$$

3.5.2 Unknown boundary ordinates

Collocation at N interior points supplied the exact number of equations that were required to find the N unknown ordinates of (99).

In a similar problem with mass transfer resistance at the pellet surface, $y_{N+1} = y(u = 1)$ is given in terms of mass transfer coefficient h_M, pellet radius R, and pellet diffusivity D by the following relation:

$$\left(\frac{dy}{dx}\right)_{x=1} = 2\left(\frac{dy}{du}\right)_{u=1} = \mathrm{Bi}_M\,[1 - y(u = 1)] \qquad (102)$$

An expression for $(dy_N/du)_{u=1}$ may be obtained from (91) applied for $u_{\mathrm{NT}} = 1$:

$$\left(\frac{dy_N}{du}\right)_{u=1} = \sum_{i=1}^{N+1} A_{N+1,i} \cdot y_i = \frac{\mathrm{Bi}_M}{2}(1 - y_{N+1}) = K(1 - y_{N+1}) \qquad (103)$$

Equation (103) and the N collocation equations (100) supply $N + 1$ equations for the $N + 1$ unknown y-values.

For a linear boundary condition it is usually simpler to eliminate the unknown boundary ordinate from the system of equations:

$$y_{N+1}(K + A_{n+1,N+1}) = -\sum_{i=1}^{N} A_{N+1,i}y_i + K$$

or

$$y_{N+1} = \sum_{i=1}^{N} \frac{-A_{N+1,i}}{K + A_{N+1,N+1}} \cdot y_i + \frac{K}{K + A_{N+1,N+1}} \qquad (104)$$

Equation (104) is inserted into each of the N collocation equations (100) and a system of N equations for the N interior ordinates is obtained:

$$(\mathbf{B}^* - p\mathbf{I})\mathbf{y} = \mathbf{c} \qquad (105)$$

The elements of $\mathbf{B}^*$ and $\mathbf{c}$ are slightly different from the corresponding elements in (101):

$$
\begin{aligned}
B^*_{ji} &= u_j\left(B_{ji} - \frac{B_{j,N+1}A_{N+1,i}}{K + A_{N+1,N+1}}\right) + \left(A_{ji} - \frac{A_{j,N+1}A_{N+1,i}}{K + A_{N+1,N+1}}\right) \\
c_j &= -(u_j B_{j,N+1} + A_{j,N+1})\frac{K}{K + A_{N+1,N+1}}
\end{aligned}
\qquad (106)
$$

The most general type of linear boundary conditions are

$$\left(\frac{dy}{dx}\right)_{x=0} + a_1 y(x = 0) = a_2$$

$$\left(\frac{dy}{dx}\right)_{x=1} + b_1 y(x = 1) = b_2$$

We choose an interpolation polynomial $p_{NT}(x)$ with $NT = N + 2$, $x_0 = 0$, $x_1, x_2, \ldots, x_N$ = zeros of $p_N(x)$, and $x_{N+1} = 1$. In the computer programs, the index of x_i is shifted one position, but the nomenclature (x_0, y_0) for $[x = 0, y(x = 0)]$ is easier to use here and in the following chapter.

The boundary conditions are

$$\begin{aligned} \sum_{i=0}^{N+1} A_{0,i} y_i + a_1 y_0 &= a_2 \\ \sum_{i=0}^{N+1} A_{N+1,i} y_i + b_1 y_{N+1} &= b_2 \end{aligned} \tag{107}$$

$$\mathbf{M}\begin{pmatrix} y_0 \\ y_{N+1} \end{pmatrix} = \begin{matrix} (A_{00} + a_1) y_0 + A_{0,N+1} y_{N+1} = -\sum_{i=1}^{N} A_{0i} y_i + a_2 \\ A_{N+1,0} y_0 + (A_{N+1,N+1} + b_1) y_{N+1} = -\sum_{i=1}^{N} A_{N+1,i} y_i + b_2 \end{matrix}$$

$$\begin{aligned} \begin{pmatrix} y_0 \\ y_{N+1} \end{pmatrix} = \mathbf{M}^{-1}\Bigg[\begin{pmatrix} a_2 \\ b_2 \end{pmatrix} - \begin{pmatrix} A_{01} \\ A_{N+1,1} \end{pmatrix} y_1 - \begin{pmatrix} A_{02} \\ A_{N+1,2} \end{pmatrix} y_2 - \cdots \\ - \begin{pmatrix} A_{0,N} \\ A_{N+1,N} \end{pmatrix} y_N \Bigg] \end{aligned} \tag{108}$$

In each of the N collocation equations, y_0 and y_{N+1} are inserted from the solution (108) of the boundary equations and again N equations of the form (101) or (105) appear for the interior ordinates $y_1, y_2, \ldots, y_N$.

In subsection 7.2.1 the problem of eliminating a number of variables that are given through linear equations is described in a general matrix notation. In the present context the explicit elimination of (y_0, y_{N+1}) given by (108) is, however, sufficient.

EXERCISES

1. Show that

$$P_N^{(\alpha,\beta)}(1) = \frac{\Gamma(N + \alpha + 1)\Gamma(\beta + 1)}{\Gamma(N + \beta + 1)\Gamma(\alpha + 1)}$$

for the polynomials defined by equations (7) or (8).

2. Show by comparison of the power series for the two orthogonal polynomials that

$$\frac{dP_N^{(\alpha,\beta)}}{dx} = \frac{N(N+\alpha+\beta+1)}{\beta+1} P_{N-1}^{(\alpha+1,\beta+1)}$$

3. Let $y^{(1)} = f(x)$ and approximate $f(x)$ by

$$f(x) = \sum_{i=0}^{N-1} b_i P_i^{(0,0)}(x)$$

How would you find b_i from N ordinates $f(x_i)$ at the zeros of $P_N^{(0,0)}(x)$?
Write an explicit expression for $y_N(x)$ in terms of a given $y(x=0) = y_0$ and the coefficients b_i.
Show that the accuracy of $f(1)$ is determined solely by b_0.

4. In the Graetz problem a Newtonian fluid in laminar flow is being cooled from the tube wall, which is at temperature $T = 0$, in appropriate units. The fluid enters the tube at $z = 0$ with uniform temperature 1 and, during the flow through the tube $T(x)$, the radial temperature profile gradually falls to 0 for all x.
The following integral is an expression for temperature $T(x) \cdot v(x)$ averaged over the tube cross section.

$$\int_0^1 x(1-x^2)T(x)\,dx = I$$

We wish to approximate I by $I_1 = w_1T_1 = w_1T(x_1)$ with the purpose of finding the best measuring point x_1 for this average temperature, and we wish to compute the corresponding "best" weight, such that a temperature profile of the highest possible degree in x is integrated correctly by I_1.

a. The following proposals are first studied:
   i. Take $F(x) = x(1-x^2)T(x)$ and find the optimal quadrature point x_1 and the corresponding weight.
   ii. Take $F(x) = (1-x^2)T(x)$, $W(x) = x$ and repeat.
   iii. Take $F(x) = (1+x)T(x)$, $W(x) = x(1-x)$ and repeat.
   iv. Use your knowledge that $F(x) = (1-x^2)T(x) = 0$ at $x = 1$. Thus take $F(x) = (1-x^2)T(x)$, $W(x) = x$ and two quadrature points with $x_2 = 1$. [You will still obtain a quadrature of the form $I_1 = w_1F(x_1)$ since $F(x_2)$ is zero.]

   Apply (i) to (iv) to $T =$ constant (say 1) and to $T = x$. Explain your results by comparison with the exact value of I and your knowledge of the accuracy of the formulas for I_1.

b. We know that $T(x)$ is symmetric in x. Thus introduce $u = x^2$ and convert the integral to $I = \frac{1}{2}\int_0^1 (1-u)T(u)\,du$.
   i. What is now the optimal quadrature formula? Check the accuracy by integrating $T = 1$, u, and u^2.

c. We also know that $T(u = 1) = 0$. This can be applied to find an even better average temperature based on measurement of only one value of T inside the tube.
   i. Find the optimal one-interior-point formula. Check the accuracy by integrating $T = (1 - u)$ and $T = (u - u^2)$. (Now T *has* to be zero at $u = 1$, therefore the slight alteration of "trial" functions.)

d. The previous formulas can be used to find the smallest eigenvalue of

$$\frac{1}{x}\frac{d}{dx}\left(x\frac{dY}{dx}\right) - \lambda(1 - x^2)Y = 0 \tag{1}$$

which is the eigenproblem (1.140) that was discussed in section 1.7.
Write $Y_1(x)$, the first eigenfunction of equation (1), as a first-degree interpolation polynomial in $u = x^2$ with u_1 and 1 as interpolation points. Discretize the equation to obtain λ_1 as a function of u_1.
When u_1 is the optimal quadrature point of question c, one obtains $\lambda_1 = -7.11$. This is close to the true value of λ_1, which is -7.31. When u_1 is the optimal point in question b, one obtains $\lambda_1 = -9$. For larger N the best results are, however, found by the latter choice of u_1.)

5. Consider the polynomials $P_N^{(-1/2,-1/2)}(x)$.
   a. Construct $P_1(x)$, $P_2(x)$, and $P_3(x)$ using formula (3.9).
   b. Show that $P_N(x) = \cos N\theta$ where $x = (1 + \cos\theta)/2$. Determine (exact) values for the zeros of $P_3(x)$.
   c. Find the recurrence formula for $P_N(x)$ [or $p_N(x)$].
   d. Determine $\int_0^1 W(x)[P_N(x)]^2\,dx$.
   e. Determine weights in $\int_0^1 W(x)F(x)\,dx = \sum_1^N w_k F(x_k)$.
   f. Find an interpolation polynomial of degree 2 for e^x based on the zeros of $P_3(x)$.
   g. Find the expression for a_i in $y_{N-1} = \sum_1^N a_i P_{i-1}(x)$.
   h. Find $e^x \sim e_{\text{app}}(x) = \sum_1^3 a_i P_{i-1}(x)$. What is the general algorithm for finding a_i in an expansion based on $P_{i-1}^{(-1/2,-1/2)}$?
   i. Make a sketch of $e^x - e_{\text{app}}(x)$ in $[0, 1]$ and note that the maximum error is approximately evenly distributed.

6. Solve the equation

$$1000\sqrt{\frac{2}{3}}\,y_0^{1/2} = \int_1^{1/y_0} \frac{d\xi}{(\xi^3 - 1)^{1/2}}$$

for the quantity y_0.
The integrand is singular at $\xi = 1$ (which does present a problem): but since $y_0 \sim 10^{-5}$, there will be a large part of the interval where a simple Gauss formula will work well. Thus it is suggested that the interval $[1, 1/y_0]$ is split up in several parts.
In the first subinterval $[1, \varepsilon]$ a suitable weight function is separated out, and for $\xi \gg 1$ the integrand may be simplified by neglecting 1 in comparison with ξ.
Make up your own computer program following these guidelines and show that

$$y_0 = 0.8804 \cdot 10^{-5}$$

This result is applied in subsection 7.1.3.

7. Go through the statements of JCOBI and make a careful analysis of how the algorithm in equations (19) to (23) is applied.

8. Use JCOBI and RADAU to make a list of Radau formulas for integration of

$$I = \frac{1}{2}\int_0^1 u^{-1/2} f(u)\, du$$

with $N = 1(1)10$ quadrature points.
The list should contain weights and quadrature abscissas.

9. a. Compute $\exp(-x)$ at the zeros of $P_N^{(0,0)}(x)$ $(N = 1, 2, \ldots, 8)$ to machine accuracy, and use these ordinates and INTRP to compute a table of $\exp(-x)$ at $x = 0(0.05)1$.
Compare with results obtained directly at these abscissas.

b. Repeat for interpolation points chosen as zeros of $P_N^{(-1/2,-1/2)}(x)$ and note whether the maximum error for a given N is smaller here than in a.

c. Finally use collocation at the zeros of $P_N^{(1/2,1/2)}(x)$ to solve the equation

$$\frac{d^2y}{dx^2} - y = 0$$

$$y(0) = 1, \quad y(1) = 1/e$$

Note that the maximum error is comparable to that obtained in a or b.

REFERENCES

Many results in section 3.1 are derived in Villadsen (ref. 42 of chapter 1). Szegö (1967) is the standard reference to properties of orthogonal polynomials.

Zeros of polynomials are treated in all textbooks on numerical analysis, e.g., Henrici (1964). *Modern Computing Methods* (1961) contains some disturbing examples of how difficult it may be to determine the zeros. The implicit deflation process equations (20) to (23) is also described in Wilkinson (ref. 48 of chapter 1).

The method of section 3.3 to differentiate Lagrange polynomials was first proposed by Michelsen and Villadsen (1972). It is substantially better than previously published methods, e.g., Villadsen and Stewart (1967).

Quadrature formulas are treated in Kopal (1961) and in Davis and Rabinowitz (1967). Villadsen (ref. 42 of chapter 1) makes a summary of convergence properties of quadratures based on orthogonal polynomials, but far more complete presentations of this subject are given in Szegö (1967) and in Natanson (ref. 4 of chapter 2).

1. SZEGÖ, G. "Orthogonal Polynomials." American Math. Soc. Colloquium Publications 23 (1959).

2. HENRICI, P. *Elements of Numerical Analysis.* Wiley (1964).

3. *Modern Computing Methods.* Published as No. 16 in National Physical Laboratory's Series "Notes on Applied Science," H. M. Stationery Office (London) 2nd ed. (1961).

4. MICHELSEN, M. L., and VILLADSEN, J. Chem. Eng. Journal 4 (1972): 64.

5. VILLADSEN, J., and STEWART, W. E. Chem. Eng. Sci. 22 (1967): 1483.

6. KOPAL, Z. *Numerical Analysis*, 2nd ed. London: Chapman and Hall (1961).

7. DAVIS, P. J., and RABINOWITZ, P. *Numerical Integration.* New York: Blaisdell (1967).

4 Solution of Linear Differential Equations by Collocation

Introduction

The examples and programs that are contained in the present chapter are intended to form the main body of our text and a large number of linear models can be solved approximately by the methods described here.

1. Solution of two-point boundary value problems.
2. Solution of one or several coupled ordinary differential equations.
3. Solution of parabolic partial differential equations.

Two-point boundary value problems are discussed in chapters 2 and 3. In section 4.1 the collocation solution for the example of chapter 2 is briefly recapitulated on the basis of a computer program. It is shown that the problem may also be solved without using the known symmetry of the solution and different choices of collocation points are discussed.

Initial value problems are integrated using a marching technique with collocation within each step. Three efficient collocation algorithms are developed in section 4.2 for a single first-order equation, and it is shown that higher-order initial value problems can be solved by the same techniques without increasing the dimensionality of the problem.

Parabolic partial differential equations (PPDE) are solved in section 4.3 by a collocation treatment of the underlying eigenvalue problem combined with a standard method for solving linear differential equations with constant coefficients. A general program EISYS for diagonalization of a matrix is described.

In section 4.4 the collocation solution is compared with the infinite Fourier series solutions. The two solutions are very similar and both break down in the entry region. A penetration solution is developed that gives a very accurate solution for this region.

The eigenfunctions of the PDE can be found with an arbitrary accuracy by an application of the methods of section 4.3. A general algorithm for solution of boundary value problems by an initial value technique is discussed in section 4.5.

Finally a PDE with an ordinary differential equation as one of the boundary conditions is treated in section 4.6. Models of this type are shown to be of considerable importance for chemical engineering systems.

4.1 Solution of a Linear Boundary Value Problem

The linear boundary value problem

$$\frac{d^2y}{dx^2} + \frac{s}{x}\frac{dy}{dx} = \Phi^2 y = R(y)$$
$$y(1) = 1 \qquad y^{(1)}(0) = 0 \tag{1}$$

or

$$u\frac{d^2y}{du^2} + \frac{s+1}{2}\frac{dy}{du} = \frac{\Phi^2}{4}y = py$$
$$y(1) = 1 \qquad y(0) \text{ finite} \tag{2}$$

is treated in detail in chapter 2 for $s = 1$. Analogous equations for first-order, isothermal reaction in slab geometry and in spherical geometry appear when $s = 0$ and $s = 2$.

A major object of the calculation has been to find the catalyst effectiveness η as a function of the Thiele modulus Φ.

$$\eta = \frac{1}{VR(1)}\int_V R(y)\,dV$$
$$= \int_0^1 y(x)\,dx^{s+1} = \frac{s+1}{2}\int_0^1 y(u)u^{(s-1)/2}\,du \tag{3}$$

In the present section the details of a collocation solution by the procedure of section 3.5 will be described for this problem.

It is first noted that the weight function $W(u)$ in (3) is of the form $u^\beta(1-u)^\alpha$ with $\alpha = 0$ and $\beta = (s-1)/2$. Furthermore, $y(u = 1)$ is known, and if we wish to include this ordinate in a Radau quadrature representation of (3), the interior quadrature points should be chosen as the zeros of $P_N^{(\alpha+1,\beta)}(u)$—i.e., as the zeros of $P_N^{[1,(s-1)/2]}(u)$.

Our desire to compute η as accurately as possible using N interior quadrature points and the boundary point $u = 1$ thus leads us to a collocation process in which the collocation points are chosen as the zeros of $P_N^{[1,(s-1)/2]}(u)$.

The interpolation polynomial p_{N+1} is $(u - 1)p_N^{[1,(s-1)/2]}$ and the collocation equations (3.100) take the following form for equation (2):

$$u_j \sum_{i=1}^{N+1} B_{ji}y_i + \frac{s+1}{2} \sum_{i=1}^{N+1} A_{ji}y_i - py_j = 0 \tag{4}$$

with $y_{N+1} = y(u = 1) = 1$ and $j = 1, 2, \ldots, N$. The N linear equations for the interior ordinates $y_1, y_2, \ldots, y_N$ are solved by Gauss elimination and the concentration profile is next found by Lagrangian interpolation using the program INTRP.

Finally the effectiveness factor is evaluated by Radau quadrature:

$$\eta \sim \eta_N = \frac{s+1}{2} \sum_{i=1}^{N+1} w_i y_i \tag{5}$$

In section 3.4 it is noted that program RADAU determines quadrature weights w_i' with a fixed sum of 1 and that the proper weights of (5) are obtained from w_i' by

$$w_i = w_i' \int_0^1 W(u)\,du = w_i' \int_0^1 u^{(s-1)/2}\,du = \frac{2}{s+1} w_i'$$

Consequently when the weights w_i' are used, (5) should read

$$\eta_N = \frac{s+1}{2} \frac{2}{s+1} \sum_1^{N+1} w_i' y_i = \sum_1^{N+1} w_i' y_i \tag{6}$$

The complete computer formulation of the problem applies five subprograms: JCOBI, DFOPR, GAUSL, INTRP, and RADAU. With the exception of GAUSL these are all described in section 3.4. Even though a program for solution of linear equations is probably available anywhere, for the sake of completeness we shall also describe our Gauss elimination routine, GAUSL:

Let **A** be an $[N \times (N + NS)]$ matrix that contains the square matrix **B** in its first N columns and an $(N \times NS)$ matrix **C** in columns $N + 1$ to $N + NS$. The call of GAUSL will replace **C** by $\mathbf{B}^{-1}\mathbf{C}$. The **B** part of matrix **A** is also destroyed, but the determinant of **B** is obtained as $\prod_{i=1}^{N} A_{ii}$.

The program is thus seen to solve N linear algebraic equations with a fixed coefficient matrix **B** and NS different right-hand sides **C**. Specifically, GAUSL can be used to invert **B**, in which case **C** is an $(N \times N)$ unity matrix.

The subroutine call is

```
CALL GAUSL (ND, NCOL, N, NS, A)
```

The input parameters are

INTEGER ND,
NCOL: Explained in input parameter list of JCOBI.

INTEGER N: Current size of the system of equations to be solved $N \leq ND$.

INTEGER NS: Number of right-hand sides for which the system of equations is to be solved.

ARRAY A: The coefficient matrix augmented by the NS right-hand sides in columns N + 1 to N + NS.

The output is

ARRAY A: The last NS columns of **A** contain the solution of the system of equations for the NS right-hand sides. The determinant of the coefficient matrix is $\prod_{i=1}^{N} A_{ii}$.

The main program for solution of (2) and (3) using the five subroutines mentioned above is shown in the Appendix, p. A12.

The input data are N, s, and Φ (N = approximation order, s = geometry factor, Φ = Thiele modulus).

The zeros of $p_N^{[1,(s-1)/2]}(u)$ are found by JCOBI using $N0 = 0$, $N1 = 1$. The derivatives DIF1, DIF2, and DIF3 of $p_{N+1}(u) = (u-1)p_N^{[1,(s-1)/2]}(u)$ at the interpolation points ROOT ($u = u_i$ = the collocation points and $u_{N+1} = 1$) are used in DFOPR.

This routine is called N times and with ID = 1 and 2. In this manner the discretization matrices for the first and the second derivative of y with respect to u are found row by row starting with the collocation point nearest to $u = 0$.

The rows of **A** and of **B** are next combined to give the matrix of left-hand sides of equation (2). Finally the right-hand side of (2) is subtracted in the diagonal.

The last column of the $[N \times (N+1)]$ matrix **BMAT** contains the known boundary terms

$$\left(u_j B_{j,N+1} + \frac{s+1}{2} A_{j,N+1}\right) y_{N+1} = u_j B_{j,N+1} + \frac{s+1}{2} A_{j,N+1} \tag{7}$$

in the jth row of **BMAT**.

Thus when **BMAT** is used in the argument of GAUSL ($NS = 1$), the vector of N collocation ordinates is obtained as the last column of the output matrix from GAUSL after a sign change.

RADAU is called with values $\alpha = 0$, $\beta = (s-1)/2$ of the weight function in (3). The $NT = N + 1$ quadrature weights are used to compute $\eta = \mathbf{RADW}^{\mathrm{T}} \cdot \mathbf{Y}$.

Finally a profile is obtained at a fixed grid of interpolation points $x = 0, 0.1, \ldots, 0.9, 1$ by means of the interpolation program INTRP. Note that the Lagrange polynomials are

$$l_i(u) = \frac{(u-1)p_N^{[1,(s-1)/2]}(u)}{(u-u_i)p_{N+1}^{(1)}(u_i)}, \qquad i = 1, 2, \ldots, N+1 \tag{8}$$

and that $u = x^2$ should be used in the argument of INTRP.

The substitution $u = x^2$ was introduced because it was known that the Taylor series contains only even powers of x and it was assumed that for the same N a more accurate approximation could be obtained when the trial functions (here orthogonal polynomials) contain only "correct" powers of x.

In other situations it might not be immediately obvious that the solution has certain symmetry properties and the collocation procedure does, of course, also work when both even and odd powers of x are included in $y_N(x)$.

Consider (1) for $s = 1$ and use collocation at the zeros of $P_N^{(\alpha,\beta)}(x)$.

There are several attractive choices of α and β. The weight function $W(x)$ of (3) is $x^1(1-x)^0$ and this suggests a collocation procedure at the zeros of $P_N^{(1,1)}(x)$ when a Radau quadrature with $y(x=1)$ as the extra quadrature point is used to evaluate η. One might also collocate at the zeros of $P_N^{(0,1)}(x)$ and subsequently evaluate η by a Gauss quadrature. Finally the factor x of (3) might be included in the integrand $z = x \cdot y(x)$ and (3) could be evaluated either by Gauss quadrature [collocation at the zeros of $P_N^{(0,0)}(x)$] or by a Radau quadrature [collocation at the zeros of $P_N^{(1,0)}(x)$].

In all these examples, $x = 0$ is included as an interpolation point—JCOBI is called with $N0 = N1 = 1$. The boundary condition $y^{(1)} = 0$ at $x = 0$ is used to eliminate the ordinate y_0 as described in section 3.5, while the known ordinate $y(x = 1) = 1$ appears as shown in (7).

The error of η_N is given in table 4.1 for $\Phi = 2$ and for each of the choices of (α, β) mentioned above.

TABLE 4.1

$|\eta - \eta_N|$ FOR $\Phi = 2$, AND CYLINDER SYMMETRY

	α, β				Optimal polynomials
N	(1, 1)	(0, 1)	(1, 0)	(0, 0)	$P^{(1,0)}(u)$ $u = x^2$
2	$4.0 \cdot 10^{-4}$	$2.4 \cdot 10^{-3}$	$2.0 \cdot 10^{-3}$	$9.0 \cdot 10^{-3}$	$3.0 \cdot 10^{-6}$
3	$6.1 \cdot 10^{-6}$	$4.4 \cdot 10^{-5}$	$4.6 \cdot 10^{-5}$	$2.5 \cdot 10^{-4}$	$1.2 \cdot 10^{-9}$
4	$5.3 \cdot 10^{-8}$	$5.7 \cdot 10^{-7}$	$6.0 \cdot 10^{-7}$	$4.2 \cdot 10^{-6}$	$<10^{-12}$

The choice $(\alpha, \beta) = (1, 1)$, which applies the known structure of the weight function and the known boundary value in the optimal way, does give the best value of η_N, as expected. At the same time it appears that the difference in quality between the methods is less than between two successive N values. In practice all four methods are applicable and an increase of N by one corresponds to a decrease in the error of η by a factor of 50–100.

The polynomials in u are obviously better than the polynomials in x—we have seen in chapter 2 that a polynomial approximation based on these polynomials yields the best possible value of η for a given N—but they are not necessarily the best when other measures of quality are used. Table 4.2 shows that approximation by means of $P_N^{(1,1)}(x)$, which was selected as probably the best in table 4.1, does not fall far behind when the maximum error of $y_N(x)$ in $[0, 1]$ is compared.

TABLE 4.2
MAXIMUM VALUE OF $|y - y_N|$ FOR $x \in [0, 1]$ $(\Phi = 2)$

N	Collocation at zeros of $P_N^{(1,0)}(u)$	Collocation at zeros of $P_N^{(1,1)}(x)$
2	$1.3 \cdot 10^{-3}$	$3.3 \cdot 10^{-3}$
3	$2.4 \cdot 10^{-5}$	$1.3 \cdot 10^{-4}$
4	$2.6 \cdot 10^{-7}$	$1.5 \cdot 10^{-5}$

4.2 Integration of Initial Value Problems

In all examples of the previous sections MWR have been used to integrate boundary value problems. They are, however, just as applicable for integration of initial value problems. In the present section several very attractive collocation procedures are derived following the arguments of section 2.4 and especially of 2.5 where the optimality of a Galerkin procedure and a corresponding orthogonal collocation MWR was proved.

The general initial value problem for one first-order equation is

$$\text{Integrate } \frac{dy}{dx} = f(y, x) \quad \text{from } x = 0 \text{ to } X$$

$$\text{with initial value element } (0, y_0) \tag{9}$$

4.2.1 Truncation error and stability in forward integration

Usually the interval $[0, X]$ is divided into M subintervals of length h and (9) is integrated through each subinterval, using the right-hand end point ordinate of one interval as initial value in the next interval. It is desired to tabulate y at $x = n \cdot h, n \cdot 2h, \ldots, X$; as far as y is accurately determined at these x-values, it does not matter if intermediate y-values are inaccurate.

The repetitive application of the right-hand interval end point ordinate as initial value for the next interval will lead to an accumulation of error. This may not only give inaccurate y-values at the entries of the table $x = n \cdot h, \ldots$ but the whole integration process may be ruined through instability of the method.

The most common instability phenomenon appears when h is larger than a certain critical value. Any explicit method, of which the Runge–Kutta methods are the best known, exhibits this steplength–induced instability. It may be remedied by taking small integration steps h, but only at the cost of a large and unproductive computational work between the tabular points.

Stability properties of various integration methods are discussed in chapter 8. In this chapter we only study the error within each integration step—the truncation error.

The truncation error at the end of each step $x = h, 2h, \ldots$ is usually represented as some functional of y multiplied by h^k. If k is large (and h sufficiently small), this seems to indicate that the integration method is well constructed. One way of obtaining a high value of k is to use a high-order explicit method: Euler's method applies y and $y^{(1)}$ at x_0 to integrate y from x_0 to $x_0 + h$, while Runge–Kutta's fourth-order method applies the first four derivatives of y in an indirect way. The less complicated Euler method has a truncation error proportional to h^2, while Runge–Kutta's fourth-order method, which demands four evaluations of $f(y, x)$, has a truncation error proportional to h^5.

In these explicit "look ahead" methods a high-order truncation term is indicative of a small error per integration step, but the critical h value, above which instability occurs, does not increase significantly with increasing order of the integration method.

Another way of obtaining a high-order truncation term is to use previously determined ordinates $y(x_0 - h)$, $y(x_0 - 2h) \ldots$ and $f[(x_0 - h), y(x_0 - h)] \ldots$ besides $y(x_0)$ to determine $y(x_0 + h)$. Both explicit and implicit variants of these "backward interpolative" methods exist and they are widely used in practice often in a predictor–corrector scheme where the predictor formula is explicit and the corrector formula is implicit. For a purely explicit method the critical value of h decreases when several previously determined y-values are included (i.e., when the truncation error is diminished) and these formulas may be quite dangerous when used improperly.

It is characteristic for the interpolative methods mentioned above that at most one y-value is implicitly given [$y(x_0 + h)$ is found from an algebraic equation in which $f(x_0 + h)$ also appears]. The high order of the method is obtained by filling in with previously determined y- or f-values. A truly "forward interpolative" method determines $y(x_0 + h)$ and y at certain intermediate abscissas between x_0 and $x_0 + h$ by simultaneous solution of a corresponding number of algebraic equations. Only $y(x_0)$ is assumed to be known during this computation. An interpolation polynomial $y_N(x)$ of degree N can be constructed on the basis of $y(x_0)$ and N ordinates taken at x-values between x_0 and $x_0 + h$ and this polynomial is used to represent $y(x)$ in the interval.

In subsection 4.2.2, three interpolative methods are developed, and it is shown in general that optimal interpolative methods may be constructed for any number N of interpolation points. The truncation error for these methods is much smaller than for an explicit method of the same order N and this may compensate for the heavier computational work per step.

The residual $R(y_N) = y_N^{(1)} - f(y_N, x)$ can be interpolated to zero at any set of N x-values in $(x_0, x_0 + h)$ and the $N + 1$ constants of y_N can be determined from the N collocation equations and the known ordinate at x_0. Our main objective will be to choose the N points such that $y(x_0 + h)$ is especially accurate. The preferred interior interpolation—or collocation—points are again zeros of an orthogonal polynomial $p_N^{(\alpha,\beta)}(x)$. Here we shall prove this in detail for the extremely simple differential equation

$$y^{(1)} - Ky = 0 \qquad y(0) = 1 \qquad [y(x) = \exp(Kx)] \tag{10}$$

but the proof also holds for an equation of the general form (9) as shown in chapter 6 for a one-point collocation method.

4.2.2 Development of three attractive collocation methods

By a transformation of the independent variable the interval $[0, h]$ is first transformed into $[0, 1]$. As trial functions for y_N in $[0, 1]$ we use $T_0 = 1$ and

$$T_i = P_{i-1}^{(0,0)}(x) + P_i^{(0,0)}(x) \tag{11}$$

$$y_N = 1 + \sum_{i=1}^{N} b_i[P_{i-1}^{(0,0)} + P_i^{(0,0)}] \tag{12}$$

The value of $P_i^{(\alpha,\beta)}(x = 0)$ is $(-1)^i$ and all trial functions $T_i(i > 0)$ satisfy a homogeneous boundary condition $T_i(x = 0) = 0$. Equation (12) is a polynomial of degree N and it may if desired be converted into the monomial expansions of chapter 2 by means of the explicit power series for $P_i^{(0,0)}$.

The N constants $\mathbf{b}$ are determined by the method of moments.

$$\int_0^1 R(y_N)w_j(x)\,dx = 0 \qquad j = 1, 2, \ldots, N \tag{13}$$

but the weight functions w_j are not chosen as the monomials u^{j-1} used in chapter 2.

Our first choice is

$$w_j(x) = P_{j-1}^{(0,0)}(x) - P_j^{(0,0)}(x) \tag{14}$$

$$R_N = \sum_{i=1}^{N} b_i[P_{i-1}^{(1)} + P_i^{(1)}] - K - K\sum_{i=1}^{N} b_i(P_{i-1} + P_i) \tag{15}$$

R_N is inserted into (13):

$$\int_0^1 R_N(P_{j-1} - P_j)\,dx = 0 \qquad j = 1, 2, \ldots, N \tag{16}$$

in which the integrand is a polynomial of degree $N + j$. Equation (16) is written in the same way as (2.99) to (2.101):

$$(\mathbf{A} - K\mathbf{B})\mathbf{b} = K\mathbf{c} \tag{17}$$

$$A_{ji} = \int_0^1 [P_{i-1}^{(1)} + P_i^{(1)}](P_{j-1} - P_j)\,dx \tag{18}$$

$$B_{ji} = \int_0^1 (P_{i-1} + P_i)(P_{j-1} - P_j)\,dx \tag{19}$$

$$c_j = \int_0^1 (P_{j-1} - P_j)\,dx = \begin{cases} 1 & \text{for } j = 1 \\ 0 & \text{for } j > 1 \end{cases} \tag{20}$$

In (20) we have used

$$\int_0^1 P_n^{(0,0)}(x)\,dx = 0 \qquad \text{for any } n > 0$$

Next we evaluate the elements of $\mathbf{A}$ and $\mathbf{B}$:

$P_{i-1}^{(1)} + P_i^{(1)}$ is a polynomial of degree $i - 1$. If $i < j$, it may be reformulated into a sum of Legendre polynomials $P_k^{(0,0)}(x)(k < j - 1)$ and each of these is orthogonal on P_{j-1} and P_j. Hence $A_{ji} = 0$ for $i < j$.

Integration by parts yields

$$A_{ji} = (P_{j-1} - P_j)(P_{i-1} + P_i)|_0^1 - \int_0^1 (P_{i-1} + P_i)[P_{j-1}^{(1)} - P_j^{(1)}]\,dx \tag{21}$$

The first term is zero since by a result from (Exercise 3.1)

$$P_N^{(\alpha,\beta)}(x = 1) = \frac{\Gamma(N + \alpha + 1)\Gamma(\beta + 1)}{\Gamma(N + \beta + 1)\Gamma(\alpha + 1)} = 1 \quad \text{for } \alpha = \beta$$

and (22)

$$P_N^{(\alpha,\beta)}(x = 0) = -P_{N-1}^{(\alpha,\beta)}(x = 0) \quad \text{for any } (\alpha, \beta)$$

The integral in (21) is zero for $j < i$ since the last factor is a polynomial of degree $j - 1$. Consequently $\mathbf{A}$ is diagonal.

The diagonal elements are easily calculated:

$$\begin{aligned}\int_0^1 [P_{i-1}^{(1)} + P_i^{(1)}](P_{i-1} - P_i)\,dx &= \int_0^1 P_i^{(1)} P_{i-1}\,dx \\ &= P_i P_{i-1}|_0^1 - \int_0^1 P_i P_{i-1}^{(1)}\,dx \\ &= P_i P_{i-1}|_0^1 = 2\end{aligned} \tag{23}$$

$$A_{ij} = \begin{cases} 2 & \text{for } i = j \\ 0 & \text{for } i \neq j \end{cases}$$

Matrix $\mathbf{B}$ is tridiagonal since $i < j - 1$ or $j < i - 1$ both give $B_{ji} = 0$. The main diagonal of $\mathbf{B}$ is

$$\begin{aligned}B_{ii} &= \int_0^1 (P_{i-1} + P_i)(P_{i-1} - P_i)\,dx \\ &= \int_0^1 (P_{i-1}^2 - P_i^2)\,dx \\ &= C_{i-1}^{(0,0)} - C_i^{(0,0)} \\ &= \frac{1}{2(i-1)+1} - \frac{1}{2i+1} = \frac{2}{4i^2 - 1}\end{aligned} \tag{24}$$

in which the result of Villadsen (1970), p. 55, has been used for $\alpha = \beta = 0$:

$$C_N^{(\alpha,\beta)} = \frac{[\Gamma(\beta + 1)]^2 N!\,\Gamma(N + \alpha + 1)}{\Gamma(N + \beta + 1)\Gamma(N + \alpha + \beta + 1)(2N + \alpha + \beta + 1)} \tag{25}$$

The side diagonals $B_{i,i+1}$ and $B_{i+1,i}$, $i = 1, 2, \ldots, N - 1$, are found from

$$B_{i,i+1} = \int_0^1 (P_i + P_{i+1})(P_{i-1} - P_i)\,dx = -\int_0^1 P_i^2\,dx = -\frac{1}{2i+1} \tag{26}$$

$$B_{i+1,i} = \int_0^1 (P_{i-1} + P_i)(P_i - P_{i+1})\,dx = \int P_i^2\,dx = \frac{1}{2i+1} \tag{27}$$

The only difference between (17) and the corresponding equation in section 2.5 is that **B** is nonsymmetric and this is of no consequence for a word by word repetition of the proof in section 2.5 that leads to the number of correct terms in a power series expansion of b_i in powers of K:

$$b_i = \sum_{k=0}^{M} b_{ik}K^k \tag{28}$$

where b_{ik} are correct for $k \le 2N + 1 - i$. In particular, b_{ik} is accurate for $k \le 2N$. The ordinate at $x = 1$ can be obtained from

$$y_N(x = 1) = 1 + \sum_{i=1}^{N} b_i[P_{i-1}(1) + P_i(1)] = 1 + 2\sum_{1}^{N} b_i \tag{29}$$

The least accurate term in (29) is b_N, which is accurate up to and including K^{N+1}, but the first missing term b_{N+1} in (29) is itself of the order of K^{N+1} and the error term in (29) is thus $\mathcal{O}(K^{N+1})$.

A much better evaluation of $y_N(x = 1)$ is possible by an integration of (10):

$$y(1) = 1 + K\int_0^1 y(x)\,dx \tag{30}$$

The integral is evaluated by (12):

$$y_N(1) = 1 + K\int_0^1 \left[1 + \sum_{1}^{N} b_i(P_{i-1} + P_i)\right] dx = 1 + K + Kb_1 \tag{31}$$

In this formulation the accuracy of $y_N(1)$ is seen to depend only on the accuracy of b_1, which is much higher than the accuracy of the series in (29).

The total accuracy of y_N by (31) is K^{2N+1} in a power series representation of the exact solution at $x = 1$ in powers of K.

A few examples are given:

$N = 1$:

$$y_1(1) = \frac{6 + 4K + K^2}{6 - 2K} = 1 + K + \frac{1}{2}K^2 + \frac{1}{6}K^3 + \frac{1}{18}K^4 + \dots$$

$$= \sum_{j=0}^{3} \frac{1}{j!}K^j + \frac{1}{18}K^4 + \dots$$

$N = 2$:

$$y_2(1) = \frac{60 + 36K + 9K^2 + K^3}{60 - 24K + 3K^2} = \sum_{j=0}^{5} \frac{1}{j!}K^j + \frac{K^6}{800}$$

$N = 3$:

$$y_3(1) = \frac{840 + 480K + 120K^2 + 16K^3 + K^4}{840 - 360K + 60K^2 - 4K^3} = \sum_{j=0}^{7} \frac{1}{j!}K^j + \frac{K^8}{39{,}200}$$

Even the first erroneous term is quite well represented (for $N = 3$, the denominator of the ninth term in the series should be 40,320). An explicit formula for the difference Δ between the first erroneous term in the series derived by MWR and the corresponding Taylor series term can be found:

$$\Delta = 2\left[\frac{(N+1)!}{(2N+2)!}\right]^2 (-1)^N K^{2N+2} \tag{32}$$

For small values of K ($K < 1$), Δ is approximately equal to the difference $y(1) - y_N(1) = \exp(K) - y_N(1)$.

We shall now return to (13) and (14) and deduce a collocation principle that corresponds to this excellent MWR.

It is first noted that each $w_j(x)$ is zero at $x = 1$ [equation (22)] and that consequently $1 - x$ is a factor in $w_j(x)$ for $j = 1, 2, \ldots, N$. This means that

$$w_j(x) = (1 - x)Q_{j-1}(x) \tag{33}$$

where Q_{j-1} is a polynomial of degree $j - 1$ or at most of degree $N - 1$. If the N constants of the Nth-degree polynomial R_N are determined such that R_N is proportional to $P_N^{(1,0)}(x)$, all orthogonality relations (16) are automatically satisfied. It is thus seen that the procedure (13) to (16) is identical to a collocation procedure at the N zeros of $P_N^{(1,0)}(x)$ when R_N is a polynomial of degree N in x.

The very accurate value of $y_N(1)$ was obtained in (31) by an integration of $y_N(x)$, which showed that only b_1 entered into the final result. If the collocation ordinates are used in a quadrature analog of the exact integration (31) of $P_i^{(0,0)}(x)$, we must resolve the dilemma that the collocation ordinates are available at the zeros of $P_N^{(1,0)}(x)$ and not at the zeros of $P^{(0,0)}(x)$.

For (31):

$$y_N(1) = 1 + K\int_0^1 y_N(x)\,dx \tag{34}$$

The Lagrange interpolation polynomial for the Nth-degree polynomial $y_N(x)$ is

$$y_N(x) = \sum_{i=0}^{N} l_i(x) y_i \tag{35}$$

with $(x_0, y_0) = (0, 1)$ and $p_{N+1} = xp_N^{(1,0)}(x)$.

Using the $N + 1$ available quadrature points it is, of course, possible to compute the weights of a quadrature formula that exactly integrates the Nth-degree polynomial $y_N(x)$—this can be done with an arbitrary choice of $N + 1$ quadrature points.

It is, however, easier to use the Gaussian-type weights that are found in chapter 3 and for which efficient computational routines are available.

One method is to use (35) to find $y_N(x)$ at the N zeros of $P_N^{(0,0)}(x)$ and integrate (34) by a Gauss–Legendre quadrature. A more convenient method is to evaluate a preliminary value of $y_N(x = 1)$ from (35) and subsequently use this preliminary $y_N(x = 1)$ in a Radau quadrature—all interior ordinates are now at their correct positions.

$$y_N(x = 1) = 1 + K\left[\sum_{k=1}^{N} w_k y_k + w_{N+1} y_{\text{prel}}(x = 1)\right] \tag{36}$$

It is immediately clear that all three integration procedures give identical results when the integrand of (34) is a polynomial of degree N in x.

The Radau quadrature method based on extrapolation of $y_N(x)$ to $x = 1$ and using the preliminary y-value at $x = 1$ in a corrector formula (36) is, however, not only the simplest to use but it is also the most accurate whenever the integrand of (34) is a nonpolynomial function or a polynomial of degree higher than N. All ordinates $y_1, y_2, \ldots, y_{N+1}$ $[y_{\text{prel}}(x = 1)]$ are of equal accuracy N, and all terms up to and including x^{2N} in the integrand of (34) are correctly integrated. This important feature of the recommended method is discussed further in chapter 6.

At first sight the weight functions (14) appear to be polynomials of degree $1, 2, \ldots, N$. The normal method of moments of chapter 2 uses weight functions that are monomials u^{j-1}, $j = 1, 2, \ldots, N$ or polynomials P_{j-1}, $j = 1, 2, \ldots, N$. A closer analysis of (14), however, shows that $w_j(x) = (1 - x)Q_{j-1}$ with Q_{j-1} a polynomial of the usual type and $(1 - x)$ interpreted as a weight function in (16).

A strictly conventional weight function for the method of moments is

$$\begin{aligned} w_j(x) &= P_{j-1}^{(0,0)}(x) - P_j^{(0,0)}(x) \qquad j = 1, 2, \ldots, N - 1 \\ w_N(x) &= P_{N-1}^{(0,0)}(x) \end{aligned} \tag{37}$$

With this $w_j(x)$, the integrand of (13) is at most a polynomial of degree $2N - 1$ and collocation at the zeros of $P_N^{(0,0)}$ is identical to (12) and (13) with $w_j(x)$ given by (37).

Note that since R_N is orthogonal on each weight function of (37), it is also orthogonal on their sum:

$$w_0(x) = \sum_{1}^{N} w_j(x) = (P_0 - P_1) + (P_1 - P_2) + \ldots + P_{N-1} = 1$$

and $w_0(x)$ may be conceived as the "missing" zero-degree weight function of (37).

The approximation error for $y(1)$ can be derived in the same way as for collocation at the zeros of $P_N^{(1,0)}(x)$ following the development of (17) to (28). The elements of matrix **A** in (17) are the same as in (23). **B** is again tridiagonal but $B_{NN} = c_{N-1} = 1/(2N - 1)$. It is easily shown that the expansion of b_1 in powers of K is correct up to and including the term K^{2N-1}.

The dependent variable $y_N(x)$ is now available at the correct values of x for optimal integration of (34) and

$$y_N(x = 1) = 1 + K \sum_{k=1}^{N} w_k y_k \tag{38}$$

where the $\{w_k\}$ are the Gauss–Legendre quadrature weights. When the collocation points are zeros of $P_N^{(0,0)}(x)$, we can prove that the same value of $y(1)$ is obtained by quadrature and by direct extrapolation from the $N + 1$ known ordinates $y_0, y_1, \ldots, y_N$:

$$y_N(x) = \sum_{i=0}^{N} l_i(x) y_i \quad \text{and} \quad y_N(1) = \sum_{i=0}^{N} l_i(1) y_i \tag{39}$$

The residual R_N is orthogonal on $w_0(x) = 1$ and

$$\int_0^1 R_N\, dx = 0 = \int_0^1 \left(\frac{dy_N}{dx} - Ky_N\right) dx$$

$$y_N(1) = y_N(0) + K\int_0^1 y_N(x)\, dx = y_0 + K\int_0^1 y_N(x)\, dx \tag{40}$$

On the left-hand side of (40) the interpolation value (39) for $y_N(1)$ appears and the integral on the right-hand side of (40) is the same as is evaluated by (38).

It is noteworthy that the identity between the interpolated value (39) for $y_N(1)$ and the quadrature value (38) for $y_N(1)$ also holds when the original differential equation is nonlinear as in (9).

The internal consistency of the MWR based on weight functions (37) or collocation at the zeros of $P_N^{(0,0)}(x)$ is an attractive feature of this procedure in comparison with the MWR based on weight functions (14) or collocation at the zeros of $P_N^{(1,0)}(x)$. This consistency of a true method of moments is observed in chapter 2, p. 90 where the same value of the effectiveness factor was found from $dy_N/dx|_{x=1}$ and from $\int_0^1 y_N\, dx$ when y_N was determined by the method of moments [collocation at the zeros of $P_N^{(0,0)}(u)$] rather than by Galerkin's method [collocation at the zeros of $P_N^{(1,0)}(u)$].

The method based on (37) is, however, less accurate than the method based on (14), as seen by comparison of the dominant error term for small K with the corresponding formula (32) for (14).

For (37): $$\exp(K) - y_N(1) \simeq \left(\frac{N!}{2N!}\right)^2 \frac{(-1)^N}{2N + 1} K^{2N+1} \tag{41}$$

A few examples may illustrate the results obtained for $y_N(1)$:

$$y_1(1) = \frac{2+K}{2-K} = \sum_{j=0}^{2} \frac{K^j}{j!} + \frac{1}{4}K^3 + \ldots$$

$$y_2(1) = \frac{12+6K+K^2}{12-6K+K^2} = \sum_{j=0}^{4} \frac{K^j}{j!} + \frac{1}{144}K^5 + \ldots$$

$$y_3(1) = \frac{120+60K+12K^2+K^3}{120-60K+12K^2-K^3} = \sum_{j=0}^{6} \frac{K^j}{j!} + \frac{1}{4800}k^7$$

Collocation at the zeros of $P_N^{(0,0)}(x)$ is a simple, consistent method with a high integration accuracy (41) per step. Collocation at the zeros of $P_N^{(1,0)}(x)$ is even more accurate and only slightly more difficult to apply: A preliminary $y_N(1)$ must be found by extrapolation from y_0 and the internal ordinates and subsequently used in (36) to find the corrected, very accurate $y_N(x = 1)$.

If an even higher accuracy is desired, one may use the known value of y and of $(dy/dx) = f$ at the left-hand interval end point. In this way a collocation method is constructed with $y_N(1)$ accurate up to and including K^{2N+2}.

y is approximated by the following $(N + 1)$-degree polynomial:

$$y_{N+1}(x) = y_0 + f(x_0, y_0)(x - x_0) + (x - x_0)y_N^*(x) \tag{42}$$

The auxiliary polynomial $y_N^*(x)$ is given by

$$y_N^*(x) = \sum_{1}^{N+1} l_i(x)y_i^* \quad \text{with } l_i(x) = \frac{xp_N^{(1,1)}(x)}{(x - x_i)(xp_N^{(1,1)}(x))^{(1)}_{x=x_i}} \tag{43}$$

Differentiation of (42) shows that $y_N^*(x_0) = 0$ since we demand that $y_{N+1}^{(1)}(x = x_0) = f(x_0, y_0)$.

Consequently collocation might be applied to the differential equation to find N values of $y_N^*(x)$ at the collocation points $\{x_i\}$. The reason for choosing zeros of $p_N^{(1,1)}$ as interior interpolation points in (43) is that we wish to compute a high accuracy value of $y_{N+1}(1)$ by extrapolation followed by quadrature, just as in the Radau method but now working with Lobatto quadrature to include y_0 and thus obtaining an error term $\mathcal{O}(K^{2N+3})$ for the test example $y^{(1)} = Ky$.

For this equation, with $y(x_0) = y(0) = 1$, one obtains

$$y_{N+1}(x) = 1 + Kx + xy_N^*(x)$$

$$y_{N+1}(1) = y_0 + K\int_0^1 [y_0 + Kx + xy_N^*(x)]\,dx \tag{44}$$

$$= 1 + K + \frac{K^2}{2} + K\sum_{i=0}^{N+1} [x_i y_N^*(x_i)]w_i$$

The lower limit of the sum can just as well be taken as $i = 1$ since x_0 and $y^*(x_0) = 0$. $y_N^*(x_{N+1}) = y_N^*(1)$ is the preliminary value of $y_N^*(1)$ obtained by extrapolation and w_i are the Lobatto quadrature weights.

The first approximation for $y(1)$ by this Lobatto method is

$$y_1(1) = \frac{24 + 18K + 6K^2 + K^3}{24 - 6K} = \Sigma_{j=0}^{4} \frac{K^j}{j!} + \frac{K^5}{96}$$

with an error $\exp(K) - y_1(1) \sim -(K^5/480)$ for small K. In general, the dominant error term is

$$\exp(K) - y_N(1) \sim (-1)^N \frac{N!(N+2)!}{(2N+2)!(2N+3)!} K^{2N+3}$$

The approximations that we have obtained for $\exp(K)$ by (36), (38), (39), or (44) are all of the following form:

$$\exp(K) \sim y_N(x = 1) = \frac{Q_n(K)}{Q_d(K)} \tag{45}$$

Q_n and Q_d are both polynomials in K. The numerator polynomial Q_n is of the same degree N as the denominator polynomial Q_d for the Gauss method (38) or (39). In the Radau and the Lobatto methods, the degree of Q_n is, respectively, one and two higher than the degree of Q_d.

4.2.3 Collocation of right-hand interval end point

Approximations of the type (45) are called Padé approximations and they are widely used in the synthesis of process control mechanisms. The Padé approximations that we have constructed by three different collocation principles are all characterized by Q_n being of a degree higher than or equal to the degree of Q_d. Formulas with $n < d$ can be constructed if $x = 1$ is included as a collocation point besides the N interior collocation points. There are certain advantages to this procedure when solving coupled differential equations, and the end point ordinate is obtained without the extrapolation or integration step. But a serious drawback of all collocation procedures with the right-hand interval end point included as a collocation point is that the approximation error is of the same order of magnitude as that obtained with interior collocation points only.

A few Padé approximations [from Villadsen (1970), pp. 152–53] are shown to illustrate this point:

Collocation at zeros of $P_N^{(0,0)}$ and at $x = 1$:

$$y_1(1) = \frac{4 + K}{4 - 3K + K^2} = \sum_{j=0}^{2} \frac{K^j}{j!} + \frac{1}{8}K^3 + \ldots$$

which may be compared with

$$y_1(1) = \frac{2 + K}{2 - K} = \sum_{j=0}^{2} \frac{K^j}{j!} + \frac{1}{4}K^3$$

obtained by (38) or (39) on the basis of interior points only.

Collocation at zeros of $P_N^{(1,0)}$ and at $x = 1$:

$$y_1(1) = \frac{6 + 2K}{6 - 4K + K^2} = \sum_{j=0}^{3} \frac{K^j}{j!} + \frac{1}{36}K^4 + \ldots$$

A combination of $y_1(1)$ and $y_1(\frac{1}{3})$ using the quadrature formula (36) to find a "better" value of $y_N(1)$ does not succeed in improving the accuracy:

$$y_1\left(\frac{1}{3}\right) = \frac{6 - 2K}{6 - 4K + K^2}$$

and

$$[y_1(1)]_{\text{corr}} = 1 + \frac{K}{4}\left[3y_1\left(\frac{1}{3}\right) + y_1(1)\right] = 1 + K\frac{6 - K}{6 - 4K + K^2} = y_1(1)$$

It is again seen that $y_1(1)$ has the same order of accuracy as that obtained in (32) for $N = 1$, but two algebraic equations have to be solved rather than one.

Collocation methods with right-hand interval end point included as an extra collocation point are thus seen to be less attractive for integration of (10)—and in general for integration of any single differential equation—than collocation methods based on interior points only.

For integration of coupled differential equations with a large span of eigenvalues, the end point collocation methods might be preferable to collocation methods without end point collocation. The reason is that end point collocation methods have very desirable stability properties while collocation methods where the end point is determined by an extrapolation process may be unstable unless very small steps are used. This is further discussed in chapter 8.

In the present context, however, we do conclude that with regard to single-step error the Lobatto method is best followed by the Radau and the Gauss method—all based on interior points only. The local truncation error for the three methods is $\mathcal{O}(h^{2N+3})$, $\mathcal{O}(h^{2N+2})$, and $\mathcal{O}(h^{2N+1})$ for stepsize h.

4.2.4 Detailed solution of $y^{(1)}=y$ with one collocation point

The mechanical application of the three collocation methods of subsection 4.2.2 are illustrated for equation (10) with $K = 1$:

$$\frac{dy}{dx} = y \quad \text{with } y(0) = y_0 = 1$$

One collocation point ($N = 1$) is used, and the values obtained for $y_1(x = 1)$ are used to compare the accuracy of the three methods.

1. Collocation at the zero $x_1 = \frac{1}{3}$ of $P_1^{(1,0)}$.

$$p(x) = x(x - \tfrac{1}{3}) = x^2 - \tfrac{1}{3}x, \qquad p^{(1)}(x) = 2x - \tfrac{1}{3}$$

$$p^{(1)}(0) = -\tfrac{1}{3}, \qquad l_0(x) = \frac{x - \frac{1}{3}}{-\frac{1}{3}} = -3x + 1$$

$$p^{(1)}(\tfrac{1}{3}) = \tfrac{1}{3}, \qquad l_1(x) = \frac{x}{\frac{1}{3}} = 3x$$

$$y_1(x) = (-3x + 1)y_0 + (3x)y_1$$

$$\frac{dy_1(x)}{dx} = -3y_0 + 3y_1$$

$$R_1(x) = \frac{dy_1(x)}{dx} - y_1(x) = -3y_0 + 3y_1 - y_1(x)$$

$$R_1(\tfrac{1}{3}) = 0 \Rightarrow -3y_0 + 3y_1 - y_1 = 0, \qquad y_1 = \tfrac{3}{2}$$

By interpolation, a preliminary value $y_1(1)_{\text{prel}} = (-3 + 1)y_0 + 3y_1 = 2\frac{1}{2}$ is obtained.
Correction of end point value by quadrature:

$$y_1(1) - y_0 = \int_0^1 y_1(x)\,dx = w_1y_1 + w_2y_1(1)_{\text{prel}}$$

The weight factors must be determined from the node polynomial:

$$p_{N+1} = (x - 1)p_1^{(1,0)}(x) = (x - 1)(x - \tfrac{1}{3}) = x^2 - \tfrac{4}{3}x + \tfrac{1}{3}$$

$$p_{N+1}^{(1)}(x) = 2x - \tfrac{4}{3}, \qquad p_{N+1}^{(1)}(\tfrac{1}{3}) = -\tfrac{2}{3}$$

$$p_{N+1}^{(1)}(1) = \tfrac{2}{3}$$

From (3.83),

$$w_i = \frac{K'}{x_i[p_{N+1}^{(1)}(x_i)]^2}$$

$$w_1 = \tfrac{27}{4}K' \qquad w_2 = \tfrac{9}{4}K'$$

Instead of explicit evaluation of K', we may determine that the sum of the weights must be 1, yielding

$$w_1 = \tfrac{3}{4}, \qquad w_2 = \tfrac{1}{4}, \qquad \text{or} \qquad y_1(1) = 1 + \tfrac{1}{4}[3y_1 + y_1(1)_{\text{prel}}] = \mathbf{2.75}$$

Notice that the "corrected value" $y_1(1) = 2.75$ is much closer to $e = 2.718$ than the "predicted" interpolated value $y_1(1)_{\text{prel}} = 2.5$.

2. Collocation at the zero $x_1 = \frac{1}{2}$ of $P_1^{(0,0)}(x)$.

$$p(x) = x(x - \tfrac{1}{2}) = x^2 - \tfrac{1}{2}x, \qquad p^{(1)}(x) = 2x - \tfrac{1}{2}$$

$$p^{(1)}(0) = -\tfrac{1}{2}, \qquad l_0(x) = \frac{x - \frac{1}{2}}{-\frac{1}{2}} = -2x + 1$$

$$p^{(1)}(\tfrac{1}{2}) = \tfrac{1}{2}, \qquad l_1(x) = \frac{x}{\frac{1}{2}} = 2x$$

$$y_1(x) = (-2x + 1)y_0 + (2x)y_1$$

$$\frac{dy_1(x)}{dx} = -2y_0 + 2y_1$$

$$R_1(x) = \frac{dy_1(x)}{dx} - y_1(x) = -2y_0 + 2y_1 - y_1(x)$$

$$R_1(\tfrac{1}{2}) = 0 \rightarrow -2y_0 + 2y_1 - y_1 = 0, \qquad y_1 = 2$$

By interpolation:

$$y_1(1) = (-2 + 1)y_0 + 2y_1 = \mathbf{3}$$

Evaluation of $y_1(1)$ by (Gaussian) quadrature:

$$y_1(1) - y_0 = \int_0^1 y_1(x) = w_1 y_1 = y_1 = 2 \quad \text{or} \quad y_1(1) = \mathbf{3}$$

which is the same result as obtained by interpolation.

3. Collocation of modified equation at the zero $x_1 = \frac{1}{2}$ of $P_1^{(1,1)}(x)$.

Again $\quad y_1^*(x) = (-2x + 1)y_0^* + 2xy_1^* \quad$ and

$$\frac{dy_1^*(x)}{dx} = -2y_0^* + 2y_1^*$$

From (43),

$$y_1(x) = 1 + x + xy_1^*(x) \qquad \text{with } y^*(x = 0) = 0$$

$$\frac{dy_1(x)}{dx} = 1 + y_1^*(x) + x\frac{dy_1^*(x)}{dx}$$

$$R_1(x) = \frac{dy_1(x)}{dx} - y_1(x) = \left[1 + y_1^*(x) + x\frac{dy_1^*(x)}{dx}\right] - [1 + x + xy_1^*(x)]$$

and, for $x = \frac{1}{2}$,

$$R_1(\tfrac{1}{2}) = 0 = [1 + y_1^* + \tfrac{1}{2}(-2y_0^* + 2y_1^*)] - (1 + \tfrac{1}{2} + \tfrac{1}{2}y_1^*)$$

$$y_1^* = \tfrac{1}{3}$$

By interpolation,

$$y_1^*(1)_{\text{prel}} = -y_0^* + 2y_1^* = \tfrac{2}{3}$$

and, by (42),

$$y_0 = 1, \qquad y_1 = 1 + \tfrac{1}{2} + \tfrac{1}{2} \cdot \tfrac{1}{3} = \tfrac{5}{3}, \qquad y_1(1)_{\text{prel}} = \tfrac{8}{3}$$

$$y_1(1) - y_0 = \int_0^1 y_1(x) = w_0y_0 + w_1y_1 + w_2y_1(1)_{\text{prel}}$$

The Lobatto weights are

$$w_0 = \tfrac{1}{6}, \qquad w_1 = \tfrac{4}{6}, \qquad w_2 = \tfrac{1}{6}$$

$$y_1(1) = 1 + \tfrac{1}{6} \cdot 1 + \tfrac{4}{6} \cdot \tfrac{5}{3} + \tfrac{1}{6} \cdot \tfrac{8}{3} = \tfrac{49}{18} = \mathbf{2.7222}$$

There is a remarkable difference between the accuracy of $y_1(1)_{\text{prel}}$ and the corrected value $y_1(1)$.

4.2.5 Solution of coupled and higher-order initial value problems

In section 1.7 a set of M coupled first-order differential equations with constant coefficients

$$\frac{d\mathbf{y}}{dx} = \mathbf{Q}\mathbf{y} \qquad \mathbf{y} = \mathbf{y}_0 \quad \text{at } x = 0 \tag{46}$$

was solved by diagonalization of $\mathbf{Q}$:

$$\frac{d(\mathbf{U}^{-1}\mathbf{y})}{dx} = \mathbf{\Lambda}(\mathbf{U}^{-1}\mathbf{y}) \qquad \mathbf{Q} = \mathbf{U}\mathbf{\Lambda}\mathbf{U}^{-1} \text{ or } \frac{d\mathbf{Y}}{dx} = \mathbf{\Lambda}\mathbf{Y} \tag{47}$$

Each of the M decoupled equations in (47) is of the type (10), which has been treated in this section. If collocation is used to solve (47), the dominant error term of the approximate y value for each component at the end of an integration interval is $\mathcal{O}(|h\lambda_i|^q)$ where q is $2N + 3$, $2N + 2$, or $2N + 1$, depending on the choice of the N collocation points. Consequently the largest eigenvalue $|h\lambda_{\max}|$ of $\mathbf{Q}$ is a measure of the overall accuracy of the integration of (46) from x_0 to $x_0 + h$.

A system of equations with constant coefficient matrix would, of course, be solved by the methods of section 1.7, but the approximate methods of this section are equally well applied in less trivial circumstances.

The application of the Lobatto method for solution of two coupled equations is shown below in detail, while the corresponding calculations for the Radau and Gauss methods are referred to as an exercise.

Integrate

$$\begin{aligned} \frac{dy_1}{dx} &= y_1 + y_2 = f_1(x, y_1, y_2) \\ \frac{dy_2}{dx} &= y_1 + (1 - x)y_2 = f_2(x, y_1, y_2) \end{aligned} \tag{48}$$

one step $\Delta x = 1$ forward from $x = 0$ with initial element $y_1(0) = 0$ and $y_2(0) = 1$.

The auxiliary functions (42) that are used in the Lobatto method are defined by

$$\begin{aligned} y_1(x) &= y_{10} + f_1(0, y_{10}, y_{20})x + xy_1^*(x) \\ y_2(x) &= y_{20} + f_2(0, y_{10}, y_{20})x + xy_2^*(x) \end{aligned} \tag{49}$$

Index 1 or 2 is used to designate component $y_1(x)$ or $y_2(x)$ of the solution vector. y_{10} and y_{20} are the values of $y_1(x)$ and $y_2(x)$ at $x = 0$. We have dropped the index $N + 1$ or N that designates approximation order. With $N = 1$, $y_1(x)$ and $y_2(x)$ are both second-degree polynomials, while $y_1^*(x)$ and $y_2^*(x)$ are first-degree polynomials. With the initial conditions of (48), we obtain

$$\begin{aligned} y_1(x) &= x + xy_1^*(x) \\ y_2(x) &= 1 + x + xy_2^*(x) \end{aligned} \tag{50}$$

$$\frac{dy_i(x)}{dx} = 1 + y_i^*(x) + \frac{dy_i^*(x)}{dx}x \tag{51}$$

$y_i^*(x)$ is interpolated at the zero $x_1 = \frac{1}{2}$ of $p_1^{(1,1)}(x)$:

$$y_i^*(x) = (-2x + 1)y_{i0}^* + 2xy_{i1}^* \tag{52}$$

The derivative of $y_i^*(x)$ is

$$\frac{dy_i^*(x)}{dx} = -2y_{i0}^* + 2y_{i1}^* = 2y_{i1}^* \tag{53}$$

since $y_{i0}^* = 0$ by the definition of $y_i^*(x)$.

Equations (53) and (51) are inserted on the left-hand side of (48) and (50) is inserted on the right-hand side of (48) with $x = \frac{1}{2}$:

$$\begin{aligned} 1 + y_{11}^* + \tfrac{1}{2} \cdot 2y_{11}^* &= \tfrac{1}{2} + \tfrac{1}{2}y_{11}^* + 1 + \tfrac{1}{2} + \tfrac{1}{2}y_{21}^* \\ 1 + y_{21}^* + \tfrac{1}{2} \cdot 2y_{21}^* &= \tfrac{1}{2} + \tfrac{1}{2}y_{11}^* + \tfrac{1}{2} + \tfrac{1}{4} + \tfrac{1}{4}y_{21}^* \end{aligned} \tag{54}$$

The solution of (54) is

$$(y_{11}^*, y_{21}^*) = (\tfrac{15}{19}, \tfrac{7}{19}) \tag{55}$$

Preliminary values of $y_i^*(x)$ at $x = 1$ are found from equation (52):

$$(y_{12}^*, y_{22}^*) = (2y_{11}^*, 2y_{21}^*) = (\tfrac{30}{19}, \tfrac{14}{19}) \tag{56}$$

y_{ik}^* is transformed to corresponding values y_{ik} by (50):

i/k	0	1	2
1	0	$\frac{17}{19}$	$\frac{49}{19}$
2	1	$\frac{32}{19}$	$\frac{52}{19}$

Corrected values of y_{i2} are finally obtained by an integration of (48) with a Lobatto quadrature:

$$\begin{aligned} y_{12} &= y_{10} + \tfrac{1}{6}(y_{10} + 4y_{11} + y_{12}) + \tfrac{1}{6}(y_{20} + 4y_{21} + y_{22}) = \tfrac{158}{57} \\ &= 2.77 \end{aligned}$$

$$\begin{aligned} y_{22} &= y_{20} + \tfrac{1}{6}(y_{10} + 4y_{11} + y_{12}) + \tfrac{1}{6}(y_{20} + 4 \cdot \tfrac{1}{2}y_{21} + 0 \cdot y_{22}) \\ &= \tfrac{157}{57} = 2.75 \end{aligned}$$

These values are quite good considering the large stepsize $\Delta x = 1$. The true solution of (48) is $y_1(x) = xe^x$ and $y_2(x) = e^x$.

The Radau method gives $(y_{12}, y_{22}) = (3.18, 3.04)$, while the Gauss method is very poor $[(y_{12}, y_{22}) = (8, 7)]$. The step size is obviously too large for this comparatively low-order method.

A Runge–Kutta fourth-order method has a truncation error proportional to h^5 just as the Lobatto one-point method. It yields $(y_{12}, y_{22}) = (2.61, 2.625)$, a result of about the same accuracy as that obtained with the Lobatto method.

A linear nth-order differential equation can be interpreted as n-coupled, linear first-order equations. Normally collocation at N points for each of the equations would lead to a system of nN algebraic equations for the collocation ordinates. It is of great practical interest that the particular system of collocation equations that result from an nth-order equation can be solved with only slightly more work than is necessary to solve the N collocation equations from a single first-order equation. We have occasion to solve several complicated higher-order equations, e.g., in sections 4.5 and 5.5. Here a simple second-order equation with constant coefficients and a numerical example suffice to explain how the reduction in dimensionality is achieved.

Consider a second-order differential equation

$$\frac{d^2y}{dx^2} + k_1\frac{dy}{dx} + k_2 y = 0 \tag{57}$$

with given values $[y_0, y_0^{(1)}]$ of $(y, dy/dx)$ at $x = 0$. Let $y_1(x) = y(x)$ and $y_2(x) = dy(x)/dx$.

$$\begin{aligned}\frac{dy_1(x)}{dx} &= 0 \cdot y_1(x) + y_2(x)\\ \frac{dy_2(x)}{dx} &= -k_2 y_1(x) - k_1 y_2(x)\end{aligned} \tag{58}$$

The discrete version of (58) is

$$\mathbf{y}_1^{(1)} = \mathbf{A}\mathbf{y}_1 + \mathbf{A}_0 y_{10} = \mathbf{y}_2 \tag{59}$$

$$\mathbf{y}_2^{(1)} = \mathbf{A}\mathbf{y}_2 + \mathbf{A}_0 y_{20} = -k_2\mathbf{y}_1 - k_1\mathbf{y}_2 \tag{60}$$

Equation (59) is substituted into (60):

$$\mathbf{A}(\mathbf{A}\mathbf{y}_1 + \mathbf{A}_0 y_{10}) + \mathbf{A}_0 y_{20} = -k_2\mathbf{y}_1 - k_1(\mathbf{A}\mathbf{y}_1 + \mathbf{A}_0 y_{10})$$

or

$$(\mathbf{A}^2 + k_1\mathbf{A} + k_2\mathbf{I})\mathbf{y}_1 = -(\mathbf{A}\mathbf{A}_0)y_{10} - \mathbf{A}_0 y_{20} - k_1\mathbf{A}_0 y_{10} \tag{61}$$

In (59) to (61) $\mathbf{A}$ is the $(N \times N)$ matrix that results when the first row and column of the discretization matrix for the first derivative are erased. $\mathbf{A}_0$ contains the weights of the left-hand end point ordinate in the N collocation equations. Collocation at the zeros of $P_N^{(0,0)}$, $P_N^{(1,0)}$, or $P_N^{(1,1)}$ via the auxiliary functions gives N linear equations (61) for $\mathbf{y}_1$. The solution is inserted into (59) to give $\mathbf{y}_2$. Preliminary and corrected values for $y_1(1)$ and $y_2(1)$ in the Radau and Lobatto methods are obtained just as in the case of one first-order equation.

As an example, take

$$\begin{aligned}\frac{dy_1(x)}{dx} &= y_2(x)\\ \frac{dy_2(x)}{dx} &= -\frac{1}{2}y_1(x) - \frac{3}{2}y_2(x)\end{aligned} \tag{62}$$

with initial element $[y_1(0), y_2(0)] = (2, -\frac{3}{2})$.

We use one collocation point at the zero $x = \frac{1}{3}$ of $P_1^{(1,0)}(x)$.

$$\begin{aligned} y_{i1}(x) &= (-3x + 1)y_{i0} + 3xy_{i1} \\ y_{i1}^{(1)}(x) &= -3y_{i0} + 3y_{i1} \end{aligned} \tag{63}$$

Vectors $\mathbf{y}_i^{(1)}$ of (59) and (60) represent values of the first derivative at internal points only. Only the last N rows of the full $\mathbf{A}$ matrix (63) are needed. Also the influence of y_{i0} has been separated out in (59) and (60) through the term $\mathbf{A}_0 y_{i0}$. Consequently the matrix $\mathbf{A}$ of (59) and (60) is obtained from the full $\mathbf{A}$ matrix (63) by deleting its first row and column. In our example with only one unknown ordinate (y_{11}, y_{21}) we obtain

$$(3^2 + \tfrac{3}{2} \cdot 3 + \tfrac{1}{2} \cdot 1)y_{11} = -[3 \cdot (-3)] \cdot 2 - (-3) \cdot (-\tfrac{3}{2}) - \tfrac{3}{2} \cdot (-3) \cdot 2$$

with solution $y_{11} = \frac{45}{28}$; from (59),

$$y_{21} = 3 \cdot y_{11} + (-3) \cdot y_{10} = 3 \cdot \tfrac{45}{28} - 6 = -\tfrac{33}{28}$$

y_{12} and y_{22} are found by extrapolation to $x = 1$:

$$\begin{aligned} y_{12} &= -2y_{10} + 3y_{11} = -4 + \tfrac{135}{28} = \tfrac{23}{28} \\ y_{22} &= -2y_{20} + 3y_{21} = 3 - \tfrac{99}{28} = -\tfrac{15}{28} \end{aligned} \tag{64}$$

The preliminary end point values (64) are finally corrected through an integration of (62):

$$y_{12} = 2 + \int_0^1 y_{21}(x)\,dx = 2 + \tfrac{1}{4}[3(-\tfrac{33}{28}) + (-\tfrac{15}{28})] = \tfrac{55}{56} = 0.982$$

$$\begin{aligned} y_{22} &= -\tfrac{3}{2} - \tfrac{1}{2}\int_0^1 y_{11}(x)\,dx - \tfrac{3}{2}\int_0^1 y_{21}(x)\,dx \\ &= -\tfrac{3}{2} - \tfrac{1}{8}(\tfrac{135}{28} + \tfrac{23}{28} - \tfrac{297}{28} - \tfrac{45}{28}) = -\tfrac{38}{56} = -0.679 \end{aligned}$$

The result may be compared with the exact solution $y(x) = y_1(x) = e^{-x} + e^{-1/2x}$ of (62). For $x = 1$ one obtains

$$(y_{12}, y_{22}) = (0.9744, -0.6711)$$

4.3 Linear Parabolic Differential Equations

The Graetz problem of section 1.2 is discussed further in section 1.7 as a typical nontrivial representative of the linear parabolic partial differential equations that turn up every so often as models of important chemical engineering systems.

In the present section we shall develop methods of attacking this model by MWR in much the same way as a two-point boundary value problem is treated in chapter 2. We find N-term analogs of the truncated Fourier series (1.141) of the solution. The coefficients (b_i) of the MWR series are not identical to the true Fourier series coefficients (b_i^*), and the approximation functions are close approximations to the true eigenfunctions, but not identical to them. This is exactly the situation that was encountered in subsection 2.5.1: The combination of N approximate eigenfunctions will be shown to represent the solution of the model very well not only for large values of the unrestricted variable z but also for quite small values of z where the Fourier series is slowly convergent. By a specific choice of approximation function, certain functionals derived from the solution of the partial differential equation can be particularly well determined.

In section 4.4 the solution for the initial flow (or entry flow) problem is treated separately. Neither a true Fourier series nor a polynomial approximation to this can be used to calculate the heat (or mass) transfer for $z \to 0$ since the derivative of the profile at the solid wall increases to infinity at this z-value.

Finally in section 4.5 the initial value methods of section 4.2 are used to calculate the true eigenvalues and eigenfunctions of the differential equation one by one with an arbitrary accuracy. As indicated above, this imitation of the true Fourier series is not the most efficient method of solving the partial differential equation, but the principle of the method of section 4.5 can be used in many other circumstances.

Cylinder geometry with eigenfunctions that are even functions of x has been treated in most of the preceding examples. In order that the application of MWR in circumstances with less obvious symmetries shall also be sufficiently well exposed, a close analog of (1.140) is solved rather than the Graetz problem in cylinder geometry.

$$(1 - x^2)\frac{\partial y}{\partial z} = \frac{\partial^2 y}{\partial x^2} \tag{65}$$

with boundary conditions.

$$\begin{gathered} y = 0 \quad \text{at } x = 0, z > 0 \qquad \frac{dy}{dx} = 0 \quad \text{at } x = 1, z > 0, \\ y = 1 \quad \text{at } z = 0 \end{gathered} \tag{66}$$

Equations (65) and (66) are the mathematical model for desorption from a liquid film that flows in the z-direction. The flow is laminar with a fully developed parabolic velocity profile also at $z = 0$. The free surface is at $x = 0$ and there is no gas film transport resistance. Transport in the x-direction occurs by liquid diffusion with zero diffusion gradient at the solid wall $x = 1$.

This model has been discussed in Bird (1960), p. 537. Several simplified versions of the model are solved in the reference. Finlayson (1972) has used Galerkin's method to solve the problem for large z and a penetration depth concept combined with the method of moments to give a first approximation to the solution for small z.

We are interested in the solution $y(x, z)$ and also in certain derived properties:

1. The bulk mean concentration defined by

$$\bar{y}(z) \int_A v_z \, dA \equiv \int_A y v_z \, dA$$

or for a laminar velocity profile and $dA = dx$:

$$\bar{y}(z) = \tfrac{3}{2} \int_0^1 (1 - x^2) y(x, z) \, dx \tag{67}$$

2. The local Sherwood number defined similarly to (1.53) without the factor 2, which enters in (1.53) through the use of $2R$ as diffusion length scale and with a plus sign due to reversal of the direction of x

$$\text{Sh} = \frac{(\partial y/\partial x)_{x=0}}{\bar{y}} \tag{68}$$

Integration of (65) yields an expression similar to (1.55):

$$\text{Sh} = -\frac{2}{3} \frac{(d\bar{y}/dz)}{\bar{y}} \tag{69}$$

4.3.1 Trial functions for MWR solution

The Nth approximation $y_N(x, z)$ is chosen in the same way as with ordinary differential equations:

$$y_N(x, z) = T_0(x, z) + \sum_{i=1}^{N} a_i(z) T_i(x) \tag{70}$$

The trial functions $T_i(x)$, $i > 0$, are functions of x alone and we intend to determine the z-dependent coefficients $a_i(z)$ by MWR applied to the residual $R_N(x, z)$. The function $T_0(x, z)$ must satisfy the boundary conditions on x [the upper equations of (66) in the present example], and it is convenient if $T_0(x, z)$ also satisfies the differential equation (in which case it disappears from the residual). In the case of (65) and (66) one obtains $T_0(x, z) = 0$.

$T_i(x)$, $i > 0$, must satisfy homogeneous boundary conditions [here $y(0) = 0$ and $y^{(1)}(1) = 0$]. A polynomial trial function $T_i(x)$ may be

$$T_i(x) = (i + 1)x - x^{i+1}$$

The coefficients $a_i(z)$ of (70) are determined through the solution of N coupled ordinary differential equations, and the problem of finding suitable initial values $a_i(0)$ arises.

It is immediately clear that no set of N constants $a_i(0)$ can be chosen to satisfy the initial condition $y = 1$ at $z = 0$ exactly:

$$1 \neq \sum_1^N a_i(0)[(i + 1)x - x^{i+1}]$$

For each MWR however, it is possible to make y_N approximately equal to the initial condition $y = 1$ at $z = 0$. Using a collocation method, one would require the residual at $z = 0$ to be zero at the collocation points:

$$y(x_j, 0) = 1 = \sum_1^N a_i(0)[(i + 1)x_j - x_j^{i+1}]$$

In the method of moments one would require that the initial residual

$$1 - \sum_1^N a_i(0)[(i + 1)x - x^{i+1}]$$

is orthogonal on N weight functions $w_j(x)$, which may be x^{j-1} or the N first trial functions $T_j(x)$, $j = 1, 2, \ldots, N$ in close analogy with the method of moments and Galerkin's method as defined in section 2.3. We use $(1 - x^2)T_j(x)$ as weight functions in a discrete analog of (1.151). The arbitrary constants b_i^* were determined by a Fourier expansion of the initial condition function $y(x) = 1$ in the eigenfunctions of the differential equation. At each z the true eigenfunctions are here represented by linear combination of $T_i(x)$, $i = 1, 2, \ldots, N$ and $1 - x^2$ is the weight function $W(x)$ of the Sturm–Liouville problem that is obtained from (65).

The choice of weight function is, however, not critical. In any case, a more or less accurate approximation of $y(x, z = 0)$ is obtained, but when z is increased, the differences between the various approximate initial functions have a progressively diminishing influence on the profile.

4.3.2 Solution by Galerkin's method

The residual R_N of the differential equation (65) is

$$\begin{aligned} R_N(\mathbf{a}, x, z) &= (1 - x^2)\frac{\partial y_N}{\partial z} - \frac{\partial^2 y_N}{\partial x^2} \\ &= (1 - x^2)\sum_1^N \frac{da_i(z)}{dz} T_i(x) - \sum_1^N a_i(z)\frac{d^2 T_i}{dx^2} \qquad (71) \\ &= (1 - x^2)\sum_1^N \frac{da_i}{dz}[(i + 1)x - x^{i+1}] - \sum_1^N a_i[-(i + 1)ix^{i-1}] \end{aligned}$$

R_N is made orthogonal on the N trial functions $T_j(x)$:

$$\int_0^1 R_N(\mathbf{a}, x, z) T_j(x)\, dx = 0, \qquad j = 1, 2, \ldots, N \tag{72}$$

Inserting (71) in (72), one obtains

$$\mathbf{A}\frac{d\mathbf{a}}{dz} = \mathbf{Ba} \tag{73}$$

$$A_{ji} = \int_0^1 [(i+1)x - x^{i+1}][(j+1)x - x^{j+1}](1 - x^2)\, dx \tag{74}$$

$$B_{ji} = -i(i+1)\int_0^1 x^{i-1}[(j+1)x - x^{j+1}]\, dx \tag{75}$$

A is clearly symmetric and may also be shown to be positive definite.

$$\begin{aligned} B_{ji} &= \int_0^1 T_i^{(2)}(x) T_j(x)\, dx = T_i^{(1)}(x) T_j(x)\big|_0^1 - \int_0^1 T_i^{(1)}(x) T_j^{(1)}(x)\, dx \\ &= -\int_0^1 T_i^{(1)}(x) T_j^{(1)}(x)\, dx \end{aligned} \tag{76}$$

Consequently **B** is also symmetric. We note that these properties of **A** and **B** are also found in chapter 2 for Galerkin's method.

The coefficients $a_i(z)$ are found as the solution of (73):

$$\frac{d\mathbf{a}}{dz} = \mathbf{A}^{-1}\mathbf{Ba} = \mathbf{Ca} \tag{77}$$

with solution (cf. 1.130):

$$\mathbf{a}(z) = \mathbf{U} \exp(\mathbf{\Lambda} z) \mathbf{U}^{-1} \mathbf{a}_0 \tag{78}$$

Wilkinson (1965), p. 35, shows that the eigenvalues of the product **C** of the inverse of a positive definite matrix **A** and a symmetric matrix **B** is diagonable with real eigenvalues (which may or may not be distinct). The eigenvectors $\mathbf{u}_i$ can therefore be used as a basis for N-space.

The solution of (65) by Galerkin's method consequently involves the following steps:

1. $\mathbf{a}_0 = \mathbf{a}(z = 0)$ is determined as described in the previous subsection:

$$\int_0^1 \left\{1 - \sum_1^N a_{i0}[(i+1)x - x^{i+1}]\right\}(1 - x^2)[(j+1)x - x^{j+1}]\, dx = 0$$

or $\mathbf{Aa} = \mathbf{c}$ with **A** given by (74) and

$$c_j = \frac{1}{4}(j+1) - \left(\frac{1}{j+2} - \frac{1}{j+4}\right) \tag{79}$$

2. The eigenvalues $\mathbf{\Lambda}$, eigenvectors $\mathbf{U}$, and eigenrows $\mathbf{U}^{-1}$ of $\mathbf{A}^{-1}\mathbf{B}$ are determined. This can be done directly from $\mathbf{A}$ and $\mathbf{B}$ without calculating the product matrix $\mathbf{A}^{-1}\mathbf{B}$ [Wilkinson (1965), p. 54].
3. $\mathbf{U}^{-1}\mathbf{a}_0 = \mathbf{b}$ is calculated.
4. $a_i(z)$ is now given by

$$a_i(z) = \sum_{j=1}^{N} U_{ij} \exp(\lambda_j z) b_j \tag{80}$$

5. $y_N(x, z)$ is finally determined from (70).

4.3.3 Solution by collocation

The Nth approximation for the dependent variable $y_N(x, z)$ is now represented by N interior ordinates $y(x_i, z)$ at the collocation points x_i and by the boundary ordinates $y(0, z) = y_0 = 0$ and $y(1, z) = y_{N+1}$:

$$\begin{aligned} y_N(x, z) &= \sum_{0}^{N+1} l_i(x) y(x_i, z) \\ p_{N+2}(x) &= x(x-1)p_N^{(\alpha,\beta)}(x) \end{aligned} \tag{81}$$

The trial functions of (70) were chosen such that the boundary conditions $y(0, z) = 0$ and $y^{(1)}(1, z) = 0$ were automatically satisfied. In (81) these boundary conditions provide two extra equations for the $N + 2$ ordinates. It is known from chapter 2 that (81) and (70) are equivalent representations of $y_N(x, z)$.

The initial values $y(x_i, 0)$ are immediately available:

$$y(x_i, 0) = 1 \quad \text{for } i = 1, 2, \ldots, N \tag{82}$$

In the collocation procedure the residual is equated to zero at $x_1, x_2, \ldots, x_N$ and a set of N ordinary differential equations for the interior ordinates y_j are obtained:

$$(1 - x_j^2)\frac{dy_j}{dz} = \sum_{i=0}^{N+1} B_{ji} y_i \tag{83}$$

B_{ji} is an element of the discretization matrix for the second derivative. It is found by the methods of chapter 3 ($N0 = N1 = 1$).

The boundary conditions yield

$$\begin{aligned} y(0, z) = y_0 &= 0 \\ \sum_{0}^{N+1} A_{N+1,i} y_i &= 0 \end{aligned} \tag{84}$$

If the differential equation contains an explicit function $f(x, z)$, it is convenient to subtract a particular solution from y and collocate on the resulting homogeneous equation. Also if the boundary conditions are inhomogeneous, it is advantageous to subtract a solution of the differential equation that satisfies this particular inhomogeneous boundary condition.

The unknown boundary ordinate $y_{N+1}(z)$ is eliminated from (83) by means of (84) and $y_0(z) = 0$:

$$(1 - x_j^2)\frac{dy_j}{dz} = \sum_1^N B_{ji}^* y_i$$
$$B_{ji}^* = B_{ji} - \frac{B_{j,N+1}A_{N+1,i}}{A_{N+1,N+1}} \tag{85}$$

We finally obtain a set of N homogeneous first-order equations with constant coefficients when row j of $\mathbf{B}^*$ is divided by $(1 - x_j^2)$:

$$\frac{d\mathbf{y}}{dz} = \mathbf{Cy}$$
$$C_{ji} = \frac{B_{ji}^*}{1 - x_j^2} \tag{86}$$

The solution of (86) is

$$\mathbf{y} = \mathbf{U}\exp(\mathbf{\Lambda} z)\mathbf{U}^{-1}\mathbf{y}_0 \quad \text{with } \mathbf{y}_0^T = (1, 1, \ldots, 1)$$

Matrix $\mathbf{C}$ is not symmetric and the eigenvalues of $\mathbf{C}$ in general are not real for an arbitrary choice of collocation points x_i even though the original problem is known to have real eigenvalues only.

In all examples that we have studied, an orthogonal collocation scheme with any (α, β) reasonably close to zero (e.g., α and β both less than 2) does, however, lead to a matrix $\mathbf{C}$ with real eigenvalues when the differential operator eigenvalues are real.

4.3.4 Computation of derived quantities

It is desired to develop expressions for $\bar{y}$ and $d\bar{y}/dz$ from the approximate solutions $y_N(x, z)$ in order to compute Sh (z) by (69). This is easily done from either the Galerkin or the collocation solution of the partial differential equation.

The trial functions of the Galerkin approximation are integrated one by one:

$$\bar{y} = \frac{3}{2}\sum_1^N a_i(z)\int_0^1 [(i + 1)x - x^{i+1}](1 - x^2)\,dx$$
$$= \frac{3}{2}\sum_1^N a_i(z)\left[\frac{i + 1}{4} - \frac{2}{(i + 2)(i + 4)}\right] \tag{87}$$

$$\frac{d\bar{y}}{dz} = \frac{3}{2}\sum_{1}^{N}\left[\frac{i+1}{4} - \frac{2}{(i+2)(i+4)}\right]\frac{da_i}{dz}$$
$$= \frac{3}{2}\mathbf{v}^T\mathbf{u}\boldsymbol{\Lambda}\exp(\boldsymbol{\Lambda}z)\mathbf{u}^{-1}\mathbf{a}_0 \tag{88}$$

where the components of $\mathbf{v}$ are given in the square bracket of equation (88).

In the collocation method the integral in (67) may be found by a Radau quadrature based on the internal ordinates and the known ordinate at $x = 0$. $(1 - x^2)y_N(x, z)$ can be used as integrand and in this case the interior ordinates should be at the zeros of $P_N^{(0,1)}(x)$. Another possibility is to interpret $(1 - x)$ as a weight function; in this case the integrand is $(1 + x)y_N(x, z)$ and the interior quadrature ordinates should be at the zeros of $P_N^{(1,1)}(x)$.

In all circumstances the desire to calculate an accurate value of the integral dictates the choice of collocation abscissas so that the appropriate ordinates for the selected quadrature formula are available.

4.3.5 Computer program for solution of linear PDE

We choose collocation at the zeros of $P_N^{(0,1)}(x)$ and

$$\bar{y}(z) = \frac{3}{2}\sum_{i=0}^{N+1} w_i(1 - x_i^2)y_i(z) \tag{89}$$

where $y_0(z) = 0$.

The interpolation points are obtained from JCOBI using $N0 = N1 = 1$ to include the two interval end points. The RADAU program is used to generate the weight coefficients w_i of (89). These are multiplied by $\frac{3}{2}(1 - x_i^2)$ and stored in vector **V2** (see the complete computer program in Appendix, A18). Note the argument list of RADAU and compare with the description of the program in subsection 3.4.4: $ID = 2$ (since we know $y(0)$ for all z) and consequently $\alpha' = \alpha = 0$ and $\beta' = \beta + 1 = 1$. The value of $N0$ is 1 since $ID = 2$, and the value of $N1 = 1$ since we use $x = 1$ as an interpolation point.

Next vector $\mathbf{A}_{N+1}$ and the matrix **B** for the second derivative are found from DFOPR. These are combined to form $\mathbf{B}^*$ of (85), and finally **C** of (86) is constructed.

Note that the argument I in DFOPR refers to the Ith interpolation point, that is to collocation point $I - 1$. Therefore the Ith row of matrix **C** is determined from

CALL DFOPR (ND, N, 1, 1, I + 1, 2, DIF1, DIF2, DIF3, ROOT, Z2)

and the coefficient of collocation point J is the $(J + 1)$th element of **Z2**.

It is seen from the computer program that the setting up of the collocation problem is an almost trivial matter. The major computer work is done in subroutine EISYS in which the diagonalization of **C** is performed

$$\mathbf{C} = \mathbf{U}\mathbf{\Lambda}\mathbf{U}^{-1} \tag{90}$$

The subroutine call is

```
CALL EISYS (ND, NCOL, N, INDEX, EPS, NC, A).
```

The input parameters are

INTEGER ND, NCOL: Row and column dimensions of **A** (see description of GAUSL and JCOBI).

INTEGER N: Current size of **A** = number of collocation equations.

INTEGER INDEX: This parameter can have the values − 1, 0, or 1.

- INDEX = − 1: Only the eigenvalues of **A** are found. At exit, **A** is transformed into a diagonal matrix (or in case of complex eigenvalues to a block diagonal matrix) $\mathbf{\Lambda}$.
- INDEX = 0: The eigenvector matrix **U** is also found and stored in rows $N + 1$ to $2N$ of **A**.
- INDEX = 1: Eigenrows $\mathbf{U}^{-1}$ and eigenvectors **U** are found besides eigenvalues. The eigenrows are stored in row $2N + 1$ to $3N$ of **A**. Eigenvalues and eigenvectors are stored as for INDEX = 0.

REAL EPS: Desired accuracy of eigenvalues. Suggested value for EPS (in double precision) is 10^{-12}.

The output is

INTEGER ARRAY NC: Information regarding the eigenvalues is stored in this output vector. If $NC(I) = 0$, the Ith eigenvalue of **A** is real. If $NC(I) = 1$ and $NC(I + 1) = 2$, the Ith and the $(I + 1)$st eigenvalues form a complex pair.

ARRAY A: The computed eigenvalues and eigenvectors of **A** and $\mathbf{A}^T$ are stored in **A** as described under INDEX. The row dimension ND of **A** must be $\geq N$ for INDEX = − 1, $\geq 2N$ for INDEX = 0 and $\geq 3N$ for INDEX = 1.

On exit the value of INDEX is the number of iterations in the QR algorithm. Consequently the subroutine must be called with INDEX as a variable, e.g.,

```
INDEX = 1
CALL EISYS (ND, NCOL, N, INDEX, 1.D-12, NC,A)
```

The remainder of the computer program for collocation solution of the partial differential equations consists of manipulations on $\mathbf{U}$, $\mathbf{U}^{-1}$, $\exp(\mathbf{\Lambda}z)$, and the initial vector $\mathbf{y}_0 = 1$.

The differential equation is

$$\frac{d}{dz}(\mathbf{U}^{-1}\mathbf{y}) = \mathbf{\Lambda}(\mathbf{U}^{-1}\mathbf{y}) \tag{91}$$

with $\mathbf{U}^{-1}\mathbf{y} = \mathbf{U}^{-1} \cdot \mathbf{1}$ for $z = 0$. The solution of (91) is

$$\mathbf{U}^{-1}\mathbf{y} = \exp(\mathbf{\Lambda}z)(\mathbf{U}^{-1}\mathbf{y}_0) = \exp(\mathbf{\Lambda}z)\mathbf{Z1} \tag{92}$$

We specifically wish to determine $\bar{y}$ of (89):

$$\begin{aligned}\bar{y} &= \sum_1^N [\tfrac{3}{2}w_j(1 - x_j^2)]y_j = \mathbf{V2}^T\mathbf{y} \\ &= (\mathbf{V2}^T\mathbf{U})\exp(\mathbf{\Lambda}z)\mathbf{Z1} = \mathbf{Z2}^T\exp(\mathbf{\Lambda}z)\mathbf{Z1} \\ &= \sum_1^N Z2(I) \cdot Z1(I) \cdot \exp(\lambda_i z) = \sum_1^N Z3(I)\exp(\lambda_i z)\end{aligned} \tag{93}$$

It is easily seen that the coefficient $Z3(I)$ of the approximate eigenfunction $\exp(\lambda_i z)$ is the product of the Ith element of $\mathbf{Z2}^T$ and $\mathbf{Z1}$. These two vectors are matrix-vector products $\mathbf{V2}^T\mathbf{U}$ and $\mathbf{U}^{-1} \cdot \mathbf{1}$ where $\mathbf{U}$ and $\mathbf{U}^{-1}$ are found in row $N + 1$ to $2N$ and $2N + 1$ to $3N$, respectively, of $\mathbf{A}$, the output matrix from EISYS.

The final part of the computer program describes the calculation of Sh(z) at selected z-values. The derivative is

$$\frac{d\bar{y}_N}{dz} = \sum_1^N \lambda_i Z3(I)\exp(\lambda_i z) \tag{94}$$

where $\mathbf{\Lambda}$ is taken from the diagonal A_{jj} ($j = 1, 2, \ldots, N$).

4.4 Collocation Solution of a Linear PDE Compared to Exact Solution

It appears from the sample output in the Appendix that even for small N the collocation solution yields a stable value of the low-order eigenvalues and expansion coefficients. Sh(z) for $z > 0.1$ to 0.2 also seems to have attained a stable value when $N = 6$. For small z a very significant difference in Sh(z) for $N = 4$ and 6 is noticed.

The collocation solution (70) is of the same form as the infinite Fourier series solution of the PDE and the properties of the collocation solution can be explained by a comparison with this exact representation of y. Specifically it can be shown that a truncated Fourier series will also fail to represent Sh (z) for small z with a satisfactory accuracy.

In this section we make this comparison of the collocation solution of (65) and (66) with the Fourier series and with a penetration solution that is applicable for small z only.

4.4.1 Properties of the Fourier series

The exact value of $\bar{y}(z)$ for any $z \geq 0$ is given by the following infinite series:

$$\begin{aligned}\bar{y}(z) &= \int_0^1 \tfrac{3}{2}(1 - x^2)y(x, z)\,dx \\ &= \sum_{i=1}^{\infty} \exp(\lambda_i^* z) b_i^* \int_0^1 \tfrac{3}{2}(1 - x^2)F_i(x)\,dx \\ &= \sum_1^{\infty} c_i^* \exp(\lambda_i^* z)\end{aligned} \tag{95}$$

$$c_i^* = \frac{3}{2}\frac{[\int_0^1 F_i(x)(1 - x^2)\,dx]^2}{\int_0^1 [F_i(x)]^2(1 - x^2)\,dx} \tag{96}$$

$[F_i(x), \lambda_i^*]$ are eigenfunctions and eigenvalues of

$$\begin{gathered}\frac{d^2F}{dx^2} - \lambda(1 - x^2)F = 0 \\ \frac{dF}{dx} = 0 \quad \text{at } x = 1 \quad \text{and} \quad F = 0 \quad \text{at } x = 0\end{gathered} \tag{97}$$

Values of c_i^* and λ_i^* are given for $i = 1, 2, \ldots, 6$ in table 4.3 with 12-digit accuracy. These quantities have been found by the method of section 4.5, but variants of the Graetz problem have been treated in many numerical investigations and results of comparable accuracy can no doubt be found in the literature.

TABLE 4.3
EXACT FOURIER SERIES COEFFICIENTS AND EIGENVALUES

i	c_i^*	λ_i^*
1	0.789702616221	− 5.121669307365
2	0.097255111374	− 39.66083891418
3	0.036093616457	− 106.2492321836
4	0.018686373595	− 204.8560604864
5	0.011401759677	− 335.4731969788
6	0.007675970198	− 498.0970836812

For large values of i, one may obtain the eigenvalues λ_i^* and the Fourier coefficients c_i^* from the following relations:

$$(-\lambda_{i+1}^*)^{1/2} - (-\lambda_i^*)^{1/2} \to 4 \tag{98}$$

$$\lambda_i^* c_i^* \to -3.82\ldots \tag{99}$$

Both relations are derived by the method described in Courant and Hilbert (1968), p. 250—the first relation follows almost immediately after performing a variable transformation

$$f = (1 - x^2)^{1/4} F \quad \text{and} \quad t = \int_0^x \sqrt{1 - \xi^2}\, d\xi$$

while (99) is obtained by numerical integration of a definite integral.

For large values of z, only the first term of (95) is of importance:

$$\bar{y}(z) = c_1^* \exp(\lambda_1^* z) \quad \text{for } z \to \infty$$

and the asymptotic Sherwood number

$$\lim_{z\to\infty} \text{Sh}(z) = -\frac{2}{3} \frac{d/dz\,[c_1^* \exp(\lambda_1^* z)]}{c_1^* \exp(\lambda_1^* z)} = -\frac{2}{3}\lambda_1^* = 3.414446204 \tag{100}$$

For small values of z in the infinite series, representation of $\bar{y}$ is inconvenient since a large number of terms is required to give satisfactory accuracy.

For $z = 0$,

$$1 = \sum_1^\infty c_i^* \tag{101}$$

The sum of the first six c_i^* values of table 4.3 is only 0.961. The derivative at $z = 0$ is undetermined since

$$\frac{d\bar{y}}{dz} = \lim_{N\to\infty} \left(\sum_{i=1}^N \lambda_i^* c_i^* \right) = -\infty \quad \text{at } z = 0$$

by the result of (99).

The sum of the first six $\lambda_i^* c_i^*$ of table 4.3 is 23.212 and one would obtain the (erroneous) result

$$\text{Sh}(0) = \frac{2}{3} \frac{23.212}{0.961} = 16.10$$

from the Fourier series truncated after six terms. This result is only interesting insofar as the six-term collocation solution yields 16.94 for Sh $(z = 0.0001)$. This means that the collocation solution is at least no worse than the truncated Fourier series for small z. A different approach is obviously required if a solution for small z is desired.

4.4.2 A coordinate transformation and the penetration solution

A physical argument suggests that the two independent variables x and z of (65) may be combined into a single variable η for small z: Close to $z = 0$, y is different from 1 only in a thin layer Δx close to $x = 0$. For increasing z, this layer penetrates farther into the liquid film and one might expect to obtain the same y along a curve $\eta = x/z^q$ $(q > 0)$ in the (x, z)-plane.

If $\eta = f(x, z)$ is the correct "characteristic" for the PDE, a variable transformation

$$\eta = f(x, z) \qquad z_1 = z \tag{102}$$

should not only make x and $z = z_1$ disappear from the differential equation, which is now transformed into an ordinary second-order differential equation in η, but the three side conditions (66) should collapse into two boundary conditions in η.

We shall try $\eta = f(x, z) = x/z^q$ with $q > 0$ since this combination of the independent variables has a qualitatively reasonable form. The partial derivatives are recalculated in terms of η and z_1:

$$\frac{\partial y}{\partial z} = \frac{\partial y}{\partial \eta}\frac{\partial \eta}{\partial z} + \frac{\partial y}{\partial z_1}\frac{\partial z_1}{\partial z} = -\frac{q\eta}{z_1}\frac{\partial y}{\partial \eta} + \frac{\partial y}{\partial z_1} \tag{103}$$

$$\frac{\partial y}{\partial x} = \frac{\partial y}{\partial \eta}\frac{\partial \eta}{\partial x} + \frac{\partial y}{\partial z_1}\frac{\partial z_1}{\partial x} = \frac{1}{z_1^q}\frac{\partial y}{\partial \eta} \tag{104}$$

$$\frac{\partial^2 y}{\partial x^2} = \frac{1}{z_1^{2q}}\frac{\partial^2 y}{\partial \eta^2} \tag{105}$$

Substitution into the differential equation yields

$$(1 - \eta^2 z_1^{2q})\left(-q\eta z_1^{2q-1}\frac{\partial y}{\partial \eta} + z_1^{2q}\frac{\partial y}{\partial z_1}\right) = \frac{\partial^2 y}{\partial \eta^2} \tag{106}$$

It is apparent that we shall not succeed in removing z_1 from (106) for any positive q-value. The combination of variables $\eta = xz^{-q}$ (i.e., any product of powers of x and z) does not reduce (65) to an ordinary differential equation.

Let us choose $q = \frac{1}{2}$, which at least takes care of one of the z_1 dependent terms of (106): $\eta = \frac{1}{2}xz^{-1/2}$ where the numerical factor $\frac{1}{2}$ is used to give a slightly more convenient form of the resulting differential equation.

$$(1 - 4\eta^2 z_1)\left(-2\eta\frac{\partial y}{\partial \eta} + 4z_1\frac{\partial y}{\partial z_1}\right) = \frac{\partial^2 y}{\partial \eta^2} \tag{107}$$

The side conditions (66) are transformed as follows:

$$y = 0 \quad \text{at } x = 0 \quad \rightarrow \quad y = 0 \quad \text{at } \eta = 0 \tag{108}$$

$$\frac{\partial y}{\partial x} = 0 \quad \text{at } x = 1 \quad \rightarrow \quad \frac{\partial y}{\partial \eta} = 0 \quad \text{at } \eta = \frac{1}{2\sqrt{z}} \tag{109}$$

$$y = 1 \quad \text{at } z = 0 \quad \rightarrow \quad y = 1 \quad \text{for } \eta \rightarrow \infty \tag{110}$$

If the liquid film had been of infinite depth, the boundary condition (109) would have been equivalent to $y = 1$ at $x \rightarrow \infty$ (or $\eta \rightarrow \infty$) for any finite z. In this case (66) would have coalesced into two boundary conditions: one at $\eta = 0$ and one at $\eta \rightarrow \infty$. For a finite liquid film depth, $y < 1$ at $x = 1$ for any nonzero z-value and the required coalescence of the side conditions also fails to materialize.

We solve (107) for small values of $z = z_1$ using only boundary conditions (108) and (110), and we separate the two variables η and z by expanding y in the following perturbation series:

$$y[\eta(x, z), z] \sim y_M(\eta, z) = \sum_{k=0}^{M} f_k(\eta) z^k \tag{111}$$

The same procedure was used to separate u and p in (2.97). For each M, a check should be made to see whether $y_M(\eta, z)$ fits the neglected boundary condition (109) to a satisfactory degree of accuracy.

Substitution of (111) into (107) (with $z = z_1$) and equating the coefficients of equal powers of z yields the following differential equations for $f_k(\eta)$:

$$\frac{d^2 f_0}{d\eta^2} + 2\eta \frac{df_0}{d\eta} = 0 \tag{112}$$

$$\frac{d^2 f_1}{d\eta^2} + 2\eta \frac{df_1}{d\eta} - 4f_1 = 8\eta^3 \frac{df_0}{d\eta} \tag{113}$$

$$\frac{d^2 f_2}{d\eta^2} + 2\eta \frac{df_2}{d\eta} - 8f_2 = 8\eta^3 \frac{df_1}{d\eta} - 16\eta^2 f_1 \tag{114}$$

and, in general,

$$\frac{d^2 f_k}{d\eta^2} + 2\eta \frac{df_k}{d\eta} - 4kf_k = 8\eta^3 \frac{df_{k-1}}{d\eta} - 16\eta^2 (k - 1) f_{k-1} \tag{115}$$

Equation (112) is solved with boundary conditions (108) and (110), while the following equations are solved with homogeneous boundary conditions:

$$f_k(0) = f_k(\eta \rightarrow \infty) = 0 \quad \text{for } k > 0$$

The solution of (112) is

$$f_0(\eta) = C_1 \int_0^\eta \exp(-\xi^2)\, d\xi + C_2$$

with

$$C_2 = 0 \quad \text{and} \quad C_1 = \left[\int_0^\infty \exp(-\xi^2)\, d\xi\right]^{-1} = \frac{2}{\sqrt{\pi}}$$

$$f_0(\eta) = \frac{2}{\sqrt{\pi}} \int_0^\eta \exp(-\xi^2)\, d\xi = \text{erf}(\eta) \tag{116}$$

The gradient at $x = 0$ is determined from (104):

$$\left(\frac{\partial y}{\partial x}\right)_{x=0} \sim \left[\frac{\partial y_0(\eta, z)}{\partial x}\right]_{x=0} = \frac{1}{2\sqrt{z}}\left(\frac{df_0}{d\eta}\right)_{\eta=0} = (\pi z)^{-1/2} \tag{117}$$

In (69), we used the following relation (118) between

$$\bar{y} = \frac{3}{2}\int_0^1 (1 - x^2) y(x, z)\, dx \quad \text{and} \quad \left(\frac{\partial y}{\partial x}\right)_{x=0}$$

$$\frac{d\bar{y}}{dz} = -\frac{3}{2}\left(\frac{\partial y}{\partial x}\right)_{x=0} \tag{118}$$

Equation (118) is integrated from 0 to z with $y_0(\eta, z) = f_0(\eta)$ instead of y:

$$\bar{y}(z) \sim 1 - \int_0^z \frac{3}{2}(\pi s)^{-1/2}\, ds = 1 - 3\sqrt{\frac{z}{\pi}} \tag{119}$$

Finally an approximation to Sh (z) is obtained from (117) and (119):

$$\text{Sh}(z) = \frac{(\partial y/\partial x)_{x=0}}{\bar{y}} \sim \frac{1}{\sqrt{\pi z} - 3z} = \text{Sh}_0(z) \tag{120}$$

$f_0(\eta)$ is inserted into (113):

$$\frac{d^2 f_1}{d\eta^2} + 2\eta \frac{df_1}{d\eta} - 4f_1 = 8\eta^3 \frac{df_0}{d\eta} = 16\pi^{-1/2}\eta^3 \exp(-\eta^2) \tag{121}$$

A particular solution to (121)

$$f_1(\eta) = -\pi^{-1/2}(\tfrac{4}{3}\eta^3 + \eta)\exp(-\eta^2) \tag{122}$$

is found by standard methods and since $f_1(\eta)$ satisfies the boundary conditions $f_1(0) = f_1(\eta \to \infty) = 0$, it is also the full solution of (121).

$$\left.\frac{df_1(\eta)}{d\eta}\right|_{\eta=0} = -\pi^{-1/2}$$

$$\left[\frac{\partial y_1(\eta, z)}{\partial x}\right]_{x=0} = \frac{1}{2\sqrt{z}}\left(\left.\frac{df_0}{d\eta}\right|_{\eta=0} + z\left.\frac{df_1}{d\eta}\right|_{\eta=0}\right) = (\pi z)^{-1/2}\left(1 - \frac{z}{2}\right) \tag{123}$$

$$\bar{y}(z) \sim 1 - \frac{3}{2}\int_0^z (\pi s)^{-1/2}\left(1 - \frac{s}{2}\right) ds = 1 - 3\sqrt{\frac{z}{\pi}} + \tfrac{1}{2}z\sqrt{\frac{z}{\pi}}$$

$$\text{Sh}\,(z) \sim \frac{1 - (z/2)}{\sqrt{z\pi} - 3z + \frac{1}{2}z^2} = \text{Sh}_1\,(z)$$

$f_2(\eta)$, $f_3(\eta)$, etc. are found similarly—for every k, the particular solution of (115) is also the full solution.

For $\eta = 0$, we obtain

$$\left.\frac{df_2(\eta)}{d\eta}\right|_{\eta=0} = -\frac{19}{12}\pi^{-1/2} \quad \text{and} \quad \left.\frac{df_3(\eta)}{d\eta}\right|_{\eta=0} = -\frac{631}{120}\pi^{-1/2}$$

$$\left[\frac{\partial y_3(\eta, z)}{\partial x}\right]_{x=0} = \frac{1}{2\sqrt{z}}\sum_0^3 z^k \frac{df_k(\eta)}{d\eta}$$

$$= (\pi z)^{-1/2}\left(1 - \frac{z}{2} - \frac{19}{24}z^2 - \frac{631}{240}z^3\right)$$

$$\text{Sh}\,(z) \sim \frac{1 - (z/2) - \frac{19}{24}z^2 - \frac{631}{240}z^3}{\sqrt{\pi z} - 3z + \frac{1}{2}z^2 + \frac{19}{40}z^3 + \frac{631}{560}z^4} = \text{Sh}_3\,(z) \tag{124}$$

$\text{Sh}_2\,(z)$ is obtained by deleting the last term in the numerator and denominator of (124).

The sequence of functions $\text{Sh}_k\,(z)$ are extremely accurate approximations for $\text{Sh}\,(z)$ for small values of z. At $z = 0.05$, the relative error of $\text{Sh}_k\,(z)$ is given by

k	0	1	2	3
Error%	4	0.3	0.05	0.02

The improvement for $k = 4$ and 5 is still noticeable but, for $z = 0.1$, $\text{Sh}_2\,(0.1)$ and $\text{Sh}_3\,(0.1)$ both have an error of 1%.

The reason why the high-order approximations fail to give an increased accuracy of $\mathrm{Sh}_M(z)$ beyond a certain M is that the approximations $y_M(\eta, z)$ of (111) violate the neglected boundary condition $\partial y_M/\partial\eta = 0$ at $\eta = 1/2\sqrt{z}$ to an alarming degree when M is increased. When z is 0.05, the level of attainable accuracy is reached for a larger M than when z is 0.1 This behavior of the approximate method is somewhat similar to that which is observed in an asymptotic series approximation of a given function, but the approximations $\mathrm{Sh}_M(z)$ do not appear to diverge for increasing M.

A perfectly satisfactory solution of (65) and (66) is obtained, however, if the approximations of this subsection are combined with the Fourier series solution of subsection 4.4.1. A truncated Fourier series with three terms yields an approximation for $\mathrm{Sh}(z)$ with a maximum error of only 0.004% for $z \geq 0.05$.

4.4.3 Properties of the collocation solution

The Nth-order collocation method leads to a representation (93) for $\bar{y}(z)$ with exactly N terms. The value of $\mathrm{Sh}(z)$ for $z \to \infty$ is $-\frac{2}{3}\lambda_1$ where λ_1 is the eigenvalue of smallest magnitude. Consequently the accuracy of the asymptotic Sherwood number is solely determined by the accuracy of λ_1. The average value of $\bar{y}$ is furthermore influenced by the first approximate Fourier coefficient c_1. Values of the first two (λ_i, c_i) are collected in table 4.4 for collocation at the zeros of $P_N^{(0,1)}(x)$:

TABLE 4.4
FIRST- AND SECOND-COLLOCATION EIGENVALUE AND FOURIER CONSTANT

N	$-\lambda_1$	$-\lambda_2$	c_1	c_2
2	5.1358	34.6	0.807	0.0257
4	5.1215	40.19	0.78967	0.09451
6	5.1216691	39.674	0.78970249	0.09737
8	5.1216693	39.661	0.789702616	0.09726

The values for $N = 8$ are accurate for all digits shown in the table.

The higher eigenvalues will also converge to their true values λ_i^* with increasing N but at a slower rate. A general observation for problems of this type is that the first $N/2$ collocation eigenvalues are reasonably accurate approximations to the true eigenvalues. The high-order ($i > N/2$) collocation eigenvalues become much larger in magnitude than the true eigenvalues, as seen by comparison of the results of table 4.3 and the output in A19 and A20.

$Sh_N(z)$ determined by sixth-order orthogonal collocation and by the first six terms of the Fourier series are compared with the exact function $Sh(z)$ in table 4.5.

TABLE 4.5
COMPARISON OF COLLOCATION AND TRUNCATED FOURIER SERIES

z	Collocation $N = 6$	Fourier series six terms	Exact
0.001	14.9	13.7	18.865
0.002	13.06	11.99	13.634
0.005	9.48	8.911	9.039
0.010	6.83	6.7513	6.7545
0.020	5.160	5.185078	5.185081
0.050	3.92795	3.927106	3.927106
0.10	3.503935	3.503894	3.503894

The collocation solution follows the exact solution reasonably well to $z = 0.002$, while the six-term Fourier series representation seems to break down for a somewhat higher z-value. For $z > 0.05$, the truncated Fourier series is accurate to seven digits and the asymptotic value of $Sh(z)$ is, of course, found exactly by only one term in the Fourier series.

The first collocation eigenvalue is very close to λ_1^* and there is an insignificant difference between the asymptotic Sherwood number found by one or the other of the two approximate methods.

4.5 Construction of Eigenfunctions by Forward Integration

In section 4.4 we have established a frame of reference for evaluating the accuracy of collocation solutions to a particular linear partial differential equation: For $z > 0.05$, the Fourier series solution of (65) and (66) may be truncated after the fourth term with a negligible (0.004%) error; for $z < 0.05$, a penetration solution has been constructed that becomes increasingly accurate as z tends to zero.

Considering the superiority for large z of the truncated Fourier series over any other approximate solution, it becomes a matter of great interest to compute the true eigenfunctions and eigenvalues to a high degree of accuracy. We only need the first few eigenfunctions since it is intended to use this representation of y for large enough z-values to make the series rapidly convergent.

The eigenfunctions $F_i(x)$ and eigenvalues λ_i are given by the solution of

$$\begin{gathered}\frac{d^2F_i(x)}{dx^2} - \lambda_i(1 - x^2)F_i(x) = 0 \\ F_i(0) = F_i^{(1)}(1) = 0\end{gathered} \tag{125}$$

or introducing $u = 1 - x$ by

$$\begin{gathered}\frac{d^2F_i(u)}{du^2} - \lambda_i u(2 - u)F_i(u) = 0 \\ F_i^{(1)}(0) = F(1) = 0\end{gathered} \tag{126}$$

Equation (126) defines the eigenfunctions except for a scalar factor. We choose $F_i(0) = 1$. The solution of (126) can now be found as follows:

1. Insert a trial value λ for λ_i [all eigenvalues of (126) are negative and $-\lambda_1 < -\lambda_2 < \ldots$].
2. Integrate (126) with $F_i(0) = 1$ and $F_i^{(1)}(0) = 0$ from $u = 0$ to 1.
3. If $F_i(1, \lambda) = 0$ and $F_i(u, \lambda)$ has $i - 1$ sign changes in the open interval $0 < u < 1$, we have succeeded in computing the ith eigenfunction $F_i(u, \lambda_i)$ and the trial value λ is equal to λ_i.
4. If on the other hand $F_i(1, \lambda) \neq 0$ and $F_i(u, \lambda)$ has i sign changes $\in (0, 1)$, then $-\lambda_i < -\lambda < -\lambda_{i+1}$.

By this application of Sturm's equioscillation theorem, one may isolate each eigenvalue in a certain λ-interval. Within these intervals the individual eigenvalues can be determined by a search technique: bisection, inverse interpolation on the values obtained for $F_i(1, \lambda)$, or (most profitably) by Newton's method, which is excellently suited for this type of problem.

Define $G(u, \lambda)$ by

$$G(u, \lambda) \equiv \frac{\partial}{\partial \lambda}[F(u, \lambda)] \tag{127}$$

The relation between the sensitivity function F and G is the same as that existing between $\Phi(\lambda)$ and $d\Phi/d\lambda$ when solving an ordinary algebraic equation $\Phi(\lambda) = 0$ for λ. In the present case, $\Phi(\lambda)$ is the value of $F_i(1, \lambda)$ and it is given implicitly by the solution of (126) with $F(0, \lambda) = 1$, $dF/du|_{u=0} = 0$.

Inserting (127) into (126) yields the following equation for G:

$$\frac{d^2G}{du^2} - \lambda(2 - u)uG - (2 - u)uF = 0 \tag{128}$$

with initial conditions

$$G(u = 0) = 0 \qquad [F(0, \lambda) = 1 \quad \text{for all } \lambda] \tag{129}$$

$$\left.\frac{dG}{du}\right|_{u=0} = 0 \qquad \left[\left.\frac{\partial F(u, \lambda)}{\partial u}\right|_{u=0} = 0 \quad \text{for all } \lambda\right] \tag{130}$$

The two coupled second-order initial value problems (126) and (128) are solved by integration from $u = 0$ to 1. Values of $F(1, \lambda)$ and $G(1, \lambda)$ determine the correction to λ:

$$\Delta\lambda = -\frac{F(1, \lambda)}{G(1, \lambda)} \tag{131}$$

The u-interval $[0, 1]$ is divided into M subintervals of equal length $h = 1/M$, and the four first-order equations for $f_1 = F$, $f_2 = dF/du$, $g_1 = G$, $g_2 = dG/du$ that result from (126) and (128) are solved by one of the collocation methods of section 4.2 using N-collocation points in each subinterval $u_0 = (K - 1)h < u = u_0 + hv < u_0 + h$.

$$\begin{aligned}
\frac{df_1}{dv} &= hf_2 \\
\frac{df_2}{dv} &= h\lambda(2 - u)uf_1 \\
\frac{dg_1}{dv} &= hg_2 \\
\frac{dg_2}{dv} &= h\lambda(2 - u)ug_1 + h(2 - u)uf_1
\end{aligned} \tag{132}$$

The first u_0-value at $K = 1$ is zero and

$$f_1 = 1 \quad \text{while } f_2 = g_1 = g_2 = 0 \tag{133}$$

Values of the four dependent variables at $u = u_0 = (K - 1)h$ are $f_1(u_0) = f_{10}$, $f_2(u_0) = f_{20}$, etc., and the discretization of (132) yields expressions similar to (59) and (60) for $f_1(u)\ldots$ at the collocation abscissas $\mathbf{v}$:

$$\begin{aligned}
\mathbf{Af}_1 + f_{10}\mathbf{A}_0 &= h\mathbf{f}_2 \\
\mathbf{Af}_2 + f_{20}\mathbf{A}_0 &= h\lambda\,\mathbf{Uf}_1 \\
\mathbf{Ag}_1 + g_{10}\mathbf{A}_0 &= h\mathbf{g}_2 \\
\mathbf{Ag}_2 + g_{20}\mathbf{A}_0 &= h\lambda\,\mathbf{Ug}_1 + h\,\mathbf{Uf}_1
\end{aligned} \tag{134}$$

$\mathbf{U}$ is a diagonal matrix with

$$U_{ii} = (2 - u_0 - v_i h)(u_0 + v_i h) \tag{135}$$

The first and third equations are solved for $\mathbf{f}_2$ and $\mathbf{g}_2$, respectively. These are substituted into the second and fourth equations and one obtains

$$(\mathbf{A}^2 - h^2\lambda\mathbf{U})\mathbf{f}_1 = -f_{10}\mathbf{A}\mathbf{A}_0 - hf_{20}\mathbf{A}_0 \tag{136}$$

$$(\mathbf{A}^2 - h^2\lambda\mathbf{U})\mathbf{g}_1 = -g_{10}\mathbf{A}\mathbf{A}_0 - hg_{20}\mathbf{A}_0 + h^2\mathbf{U}\mathbf{f}_1 \tag{137}$$

Equation (136) is first solved for $\mathbf{f}_1$ and (137) is thereafter solved for $\mathbf{g}_1$. Altogether two $(N \times N)$ systems of linear algebraic equations must be solved to obtain the values of F and G at the collocation points of the Kth subinterval.

The coefficient matrix of (136) and (137) is the same and the reduction of this matrix to an upper triangular form needs to be done only once per subinterval.

Matrix $\mathbf{A}^2$ and vectors $(\mathbf{A}\mathbf{A}_0, \mathbf{A}_0)$ are calculated only once, at the start of the integration. The diagonal matrix $\mathbf{U}$ depends on the value of the independent variable within the current subinterval and it has to be evaluated at each of the M steps from $u = 0$ to $u = 1$.

At each K, the right-hand side of (136) is recalculated using the right-hand interval end point ordinates from the previous step to give the scalars f_{10} and f_{20}. When (136) has been solved for $\mathbf{f}_1$, the right-hand side of (137) can also be updated.

A slightly different formulation is required if the Lobatto method [with auxiliary functions $f_1^*(v) \ldots$ that are all zero at $v = 0$—i.e., $u = u_0$] is used, but the system of equations has the same properties as (136) and (137).

Once the eigenfunction $F_i(u, \lambda_i)$ has been determined with sufficient accuracy, the Fourier coefficients c_i^* of table 4.3 can also be found. In each interval, $u_0 < u < u_0 + h$, $F_i(u, \lambda_i)$ is given at the quadrature points of an optimal quadrature formula and the integrals that enter into b_i^* and c_i^* are easily evaluated.

$$\begin{aligned} b_i^* &= \frac{\int_0^1 F_i(x)(1 - x^2)\,dx}{\int_0^1 F_i^2(x)(1 - x^2)\,dx} \\ c_i^* &= \tfrac{3}{2}b_i^* \int_0^1 F_i(x)(1 - x^2)\,dx \end{aligned} \tag{138}$$

$$\begin{aligned} y(x, z) &= \sum_1^\infty b_i^* F_i(x) \exp(\lambda_i z) \\ \bar{y}(z) &= \sum_1^\infty c_i^* \exp(\lambda_i z) \end{aligned} \tag{139}$$

The Lobatto method was applied to compute λ_i^* and c_i^* of table 4.3.

If this method should succeed to determine λ_i with a high accuracy, it is imperative that the method of integration of (126) to (128) produces reliable results for $f_1(u = 1)$ and $g_1(u = 1)$. Otherwise the correction (131) will not lead to a better λ-value when $\Delta\lambda$ is in the range of uncertainty of $f_1(1)$ and $g_1(1)$.

With N interior points in each subinterval the error per step is of the order of h^{2N+3} with the Lobatto method. The total number of steps is proportional to h^{-1} and the overall error of integration is expected to be $\mathcal{O}(h^{2N+2})$.

This is verified in figure 4-1, which shows the error of the first eigenvalue as a function of M for $N = 1, 2, 3, 4$. The straight lines on the log-log plot all have a slope $\alpha = -2N - 2$ for $M > 2$ to 3.

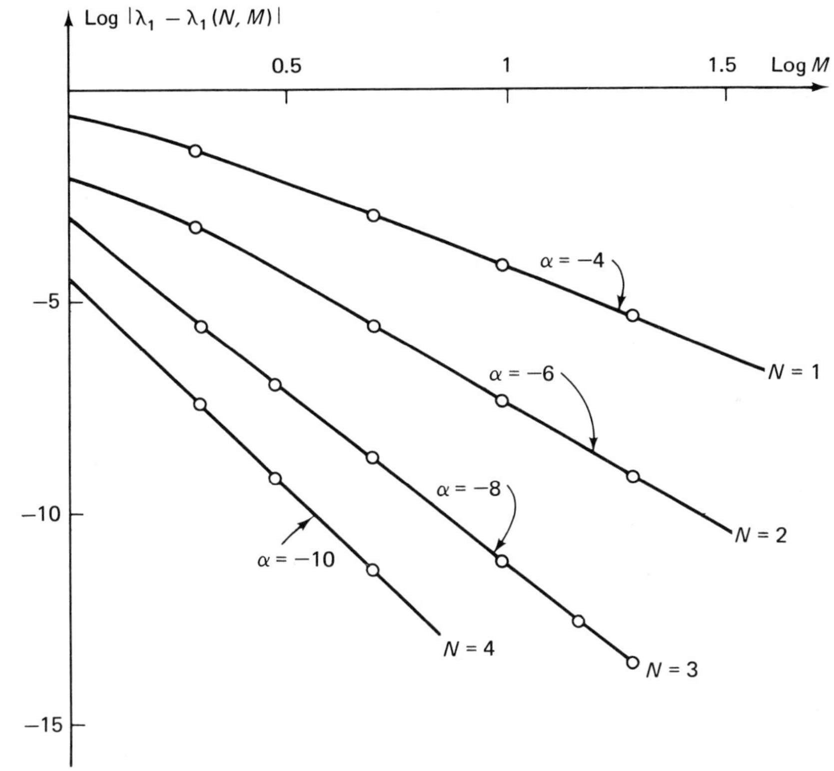

Figure 4-1. Error of first eigenvalue of equation (4.125) by an initial value integration technique using different number of subintervals M and different number of collocation points N in each subinterval.

The same dependence of accuracy on N is found for the higher eigenvalues and for the expansion coefficients c_i^*.

4.6 An Ordinary Differential Equation at the Boundary of a PDE

In the desorption problem considered in sections 4.3 to 4.5, the concentration of the volatile component in the gaseous phase $x \leq 0$ is assumed to be 0 and gas-phase mass transfer resistance is not taken into account. These assumptions are likely to be valid for desorption (or absorption) of very insoluble gases.

If a more soluble gas is desorbed, mass transfer resistance in the gas phase should be taken into account. Furthermore the concentration of the desorbing component in the gas phase also changes in the z-direction provided the flow rate of gas relative to that of the liquid is not extremely high.

The liquid-phase concentration profile would still be described by (65):

$$(1 - x^2)\frac{\partial y}{\partial z} = \frac{\partial^2 y}{\partial x^2}$$

and two of the boundary conditions remain unchanged:

$$y = 1 \quad \text{at } z = 0 \qquad \frac{\partial y}{\partial x} = 0 \quad \text{at } x = 1$$

The third boundary condition is, however, changed to

$$\frac{\partial y}{\partial x} = \mathrm{Bi}_M(y_{x=0} - y_g) \quad \text{at } x = 0 \tag{140}$$

where Bi_M (see section 1.3) is a modified gas-phase mass transfer coefficient and y_g is the liquid concentration that would correspond to equilibrium with the partial pressure of the desorbing component in the gas phase.

A total mass balance for the two streams yields

$$\frac{d\bar{y}}{dz} + q\frac{dy_g}{dz} = 0 \tag{141}$$

where q is a capacity ratio for the gas and the liquid stream. For countercurrent flow, q is negative; for cocurrent flow, q is positive.

The concentration of solute in the inlet gas stream is normally specified. This additional boundary condition is different for countercurrent and cocurrent flow:

$$\begin{aligned} &\text{Countercurrent:} \qquad y_g = y_{g,in} \quad \text{at } z = h \\ &\text{Cocurrent:} \qquad y_g = y_{g,in} \quad \text{at } z = 0 \end{aligned} \tag{142}$$

where h is a dimensionless column height.

The total mass balance (141) is integrated from $z = 0$:

$$\bar{y} + q \cdot y_g = \bar{y}_{z=0} + q \cdot (y_g)_{z=0} = 1 + q \cdot y_{g0} \tag{143}$$

For cocurrent flow, y_{g0} is known, $y_{g0} = y_{g,in}$. For countercurrent flow, however, y_{g0} is unknown.

Equation (143) may be applied to express y_g by $\bar{y}$ and the conditions $y_g = y_{g0}$ and $\bar{y} = 1$ at $z = 0$:

$$y_g = y_{g0} + \frac{1}{q}(1 - \bar{y}) \tag{144}$$

Discretization of the liquid-phase equation at N interior points x_i and at the interval end points $x_0 = 0$, $x_{N+1} = 1$, and subsequent collocation at the interior points leads to the collocation equation (83):

$$(1 - x_i^2)\frac{dy_i}{dz} = \sum_{j=0}^{N+1} B_{ij}y_j \tag{145}$$

and the boundary relations

$$\begin{aligned} x = 0: \quad & \sum_{j=0}^{N+1} A_{0,j}y_j = \mathrm{Bi}_M(y_0 - y_g) \\ x = 1: \quad & \sum_{j=0}^{N+1} A_{N+1,j}y_j = 0 \end{aligned} \tag{146}$$

From (144),

$$y_g = y_{g0} + \frac{1}{q}(1 - \bar{y}) = y_{g0} - \frac{1}{q}\sum_{j=1}^{N} w_j y_j + \frac{1}{q}$$

where w_j are the Gaussian weights for integration of $\frac{3}{2}(1 - x^2) \cdot y$. Radau quadrature is not used in this case, where the dependent variable is given implicitly at the end points.

The boundary relations may be written

$$(A_{0,0} - \mathrm{Bi}_M)y_0 + A_{0,N+1}y_{N+1} = \sum_{j=1}^{N}\left(-A_{0,j} + \frac{\mathrm{Bi}_M}{q}w_j\right)y_j - \mathrm{Bi}_M\left(y_{g0} + \frac{1}{q}\right)$$

$$A_{N+1,0}y_0 + A_{N+1,N+1}y_{N+1} = \sum_{j=1}^{N} - A_{N+1,j}y_j$$

These relations are used to express y_0 and y_{N+1} by a linear combination of interior ordinates y_i and the term $\mathrm{Bi}_M[y_{g0} + (1/q)]$. The resulting expression is substituted into (145) to give a set of equations:

$$\frac{d}{dz}\mathbf{y} = \mathbf{C} \cdot \mathbf{y} + \mathbf{d} \cdot [y_{g0} + (1/q)]$$

with initial conditions $\mathbf{y} = \mathbf{1}$ at $z = 0$.

The solution is

$$\begin{aligned}\mathbf{y} &= \exp(\mathbf{C}z)\cdot\mathbf{1} + \exp(\mathbf{C}z)\int_0^z \exp(-\mathbf{C}x)\cdot\mathbf{d}\cdot[y_{g0} + (1/q)]\,dx \\ &= \exp(\mathbf{C}z)\cdot\mathbf{1} + [\exp(\mathbf{C}z) - \mathbf{I}]\cdot\mathbf{C}^{-1}\cdot\mathbf{d}\cdot[y_{g0} + (1/q)]\end{aligned} \tag{147}$$

(provided $\mathbf{C}$ is nonsingular).

For concurrent flow we simply substitute $y_{g0} = y_{g,in}$, but for countercurrent flow the situation is more complicated. y_{g0} must be determined from the side condition.

At $z = h$:

$$y_g = y_{g,in} \quad \text{and} \quad y_{g,in} = y_{g0} + \frac{1}{q}(1 - \bar{y}_{z=h})$$

or

$$\bar{y}_{z=h} = 1 + q(y_{g0} - y_{g,in})$$

Letting $z = h$ in (147), this yields

$$\begin{aligned}1 + q(y_{g0} - y_{g,in}) &= \mathbf{w}^T\exp(\mathbf{C}h)\mathbf{1} + \mathbf{w}^T[\exp(\mathbf{C}h) - \mathbf{I}]\mathbf{C}^{-1}\mathbf{d}\left(y_{g0} + \frac{1}{q}\right) \\ &= \mathbf{w}^T\left\{\exp(\mathbf{C}h) + [\exp(\mathbf{C}h) - \mathbf{I}]\mathbf{C}^{-1}\mathbf{d}\left(1 + \frac{1}{q}\right)\right\} \\ &\quad + \mathbf{w}^T[\exp(\mathbf{C}h) - \mathbf{I}]\mathbf{C}^{-1}\mathbf{d}(y_{g0} - 1)\end{aligned} \tag{148}$$

The first term on the right-hand side of (148) is the solution that would apply if y_{g0} was 1. In this case, all driving forces would be zero, and the solution would be $\mathbf{y} = \mathbf{1}$ for all $\mathbf{z}$ or $\bar{y} = 1$ for all z.

We thus obtain

$$q(y_{g0} - y_{g,in}) = \mathbf{w}^T[\exp(\mathbf{C}h) - \mathbf{I}]\mathbf{C}^{-1}\mathbf{d}(y_{g0} - 1) = e(y_{g0} - 1)$$

or

$$y_{g0} = \frac{q\cdot y_{g,in} - e}{q - e}$$

where

$$e = \mathbf{w}^T[\exp(\mathbf{C}h) - \mathbf{I}]\mathbf{C}^{-1}\mathbf{d}$$

The problem treated in this section as a relatively trivial extension of the standard problem type of the previous three sections is quite commonly encountered in practice.

A partial differential equation with one boundary condition given as an ordinary differential equation has been discussed several times in chapter 1. Typical examples are

1. Adsorption into a solid phase or extraction from a solid phase with a limited outer volume.

2. Double-pipe ordinary heat exchangers or chemical reactors with radial and axial temperature gradients and cocurrent or countercurrent flow in the outer tube.
3. Breakthrough curves for fixed bed adsorption under constant pattern conditions (see exercise 9.7).

EXERCISES

1. In this exercise we wish to demonstrate that for $dy/dx = Ky$ the coefficients $b_{ik}^{(N)}$ of equation (4.28) are correct when $k \leq 2N + 1 - i$, as stated in the text.

a. Calculate exact expressions for the first three coefficients b_i^* of the infinite series

$$y = \exp(Kx) = 1 + \sum_1^\infty b_i^*[P_i^{(0,0)} + P_{i-1}^{(0,0)}]$$

b. Solve equation (4.17) for $N = 1$, 2, and 3 to obtain $b_i^{(j)}$ ($i = 1, 2, 3$; $j = i, i+1, \ldots, N$) where $b_i^{(j)}$ is the $(j - i + 1)$ approximation to b_i^*.

c. Expand $b_i^{(j)}$ and b_i^* in powers of K. Confirm that the following terms of $b_i^{(j)}$ are identical to the corresponding terms of b_i^*:

$$b_i^{(j)} = K^i(n_1 + n_2K + n_3K^2 + \ldots + n_{2j-2i}K^{2j-2i+1})$$

plus incorrect terms starting with K^{2j-i+2}.

2. The choice of trial function and the choice of the method of moments with a rather unfamiliar weight function in derivation of the Radau method (4.14) to (4.28) does involve a large portion of hindsight: The collocation method at the zeros of $P_N^{(1,0)}(x)$ was found to give excellent results for $y_N(1)$ when applied as shown in the text, (4.35) and (4.36), and an identical "classical" MWR was derived afterward. Whereas there can be no doubt that the highest possible accuracy is found with the proposed procedure [the accuracy corresponds exactly to the number of free parameters in the resulting Padé approximation for exp (K)], one might well ask why the method of moments was used rather than the Galerkin method, which was found to give very accurate results in chapter 2 for an apparently similar problem.
The reason why a Galerkin MWR is less suitable with our present goal—to approximate $y(1)$ well—is given in this exercise.

a. Prove—by manipulations on γ_i of formula (3.8) or more simply by using the orthogonality properties of the polynomials—that

$$P_N^{(\alpha,\beta)} + P_{N-1}^{(\alpha,\beta)} = (\gamma_{N,1} - \gamma_{N-1,1})xP_{N-1}^{(\alpha,\beta+1)} = q_NxP_{N-1}^{(\alpha,\beta+1)}$$

b. Consider the residual $R_N(\mathbf{a}, x)$ of

$$\frac{dy}{dx} - Ky = 0 \qquad y(0) = 1$$

when

$$y_N = 1 + x\sum_1^N a_iP_{i-1}^{(0,1)} = 1 + \sum_1^N \frac{a_i}{q_i}[P_i^{(0,0)} + P_{i-1}^{(0,0)}]$$

Show that collocation of the zeros of $P_N^{(0,1)}(x)$ will give the same value of a_i as Galerkin's method.

c. If the collocation ordinates are used in a Radau quadrature similar to (36) but now including the known value $y(0)$ as the extra quadrature ordinate, one might think that $y(1)$ is obtained as accurately as in the Radau method of the text. What is the result for $N = 1$?

d. In the present method $y_N(1)$ is obtained similarly to (4.31):

$$y_N(1) = 1 + K + \frac{K}{2}a_1$$

The disappointing result of part c is explained by consideration of the accuracy of a_1.
Find the structure of **A** and **B**, the matrices of (4.17), for the present method. What is the accuracy of a_1 with an N-term approximation?
The result for $N = 1$ and $N = 2$ is

$$a_1^{(1)} = \frac{K}{1 - \frac{2}{3}K}, \qquad a_1^{(2)} = \frac{K(1 - \frac{4}{15}K)}{1 - \frac{3}{5}K + \frac{3}{20}K^2}$$

3. Calculate the solution of equation (48) for $x = 1$:

$$\frac{dy_1}{dx} = y_1 + y_2$$
$$y_1(0) = 0, \qquad y_2(0) = 1$$
$$\frac{dy_2}{dx} = y_1 + (1 - x)y_2$$

by the Gauss and the Radau methods using $\Delta x = 1$.
Make a computer program that employs the Lobatto, the Radau, or the Gauss method to solve (48) with an arbitrary Δx.
Find an approximate relation between the error at $x = 1$ and Δx.

4. Consider the end point collocation method with internal collocation point x_1. Write the two collocation equations in terms of $y_1 = y(x_1)$ and $y_2 = y(1)$ for a general $f(x, y)$.
Show that

$$y_2 - y_0 = \frac{1}{2(1 - x_1)}h[f(x_0 + x_1h, y_1) + (1 - 2x_1)f(x_0 + h, y_2)]$$

where $y_0 = y(x_0)$.
It is desired to eliminate y_1 from the expression. We wish to avoid using f taken at $y = y_1$ and $y = y_2$. The elimination follows in two steps: (1) Collocation at x_1 but with an interpolation polynomial in x_0 and x_1 gives one equation for $f(x_1, y_1)$. (2) The first term of a Taylor series from (x_0, y_0) gives equations for $f(x_0 + x_1h, y_1)$ and $f(x_0 + h, y_2)$.
Derive the following expression for y_2 and show that the approximations have not resulted in a decrease of the order of accuracy for the method.

$$\left[1 - \frac{1 - 2x_1}{2(1 - x_1)}hf_y(x_0, y_0)\right](y_2 - y_0)$$
$$= \frac{h}{2(1 - x_1)}[f(x_0 + hx_1, y_1) + (1 - 2x_1)f(x_0 + h, y_0)]$$

where y_1 is given explicitly by

$$[1 - x_1 h f_y(x_0, y_0)](y_1 - y_0) = x_1 h f(x_0 + h x_1, y_0)$$

Show that $x_1 = 1 - \frac{1}{2}\sqrt{2}$ is a particularly suitable collocation point and obtain the following general $\mathcal{O}(h^3)$ method:

$$[\mathbf{I} - (1 - \tfrac{1}{2}\sqrt{2})h\mathbf{A}](\mathbf{y}_1 - \mathbf{y}_0) = (1 - \tfrac{1}{2}\sqrt{2})h\mathbf{f}(x_0 + hx_1, \mathbf{y}_0)$$

$$[\mathbf{I} - (1 - \tfrac{1}{2}\sqrt{2})h\mathbf{A}](\mathbf{y}_2 - \mathbf{y}_0)$$

$$= \frac{h}{\sqrt{2}}\left\{\mathbf{f}\left[x_0 + \left(1 - \frac{\sqrt{2}}{2}\right)h, \mathbf{y}_1\right] + (\sqrt{2} - 1)\mathbf{f}(x_0 + h, \mathbf{y}_0)\right\}$$

where $\mathbf{f} = \mathbf{f}(x_0, \mathbf{y}_0)$ and $\mathbf{A} = (\partial f_i/\partial y_j)_{x_0,\mathbf{y}_0}$.
This is Rosenbrock's method, one of the semi-implicit Runge–Kutta methods discussed in section 8.2. What are the computational operations that are involved in each integration step?

5. Solve $\partial y/\partial t = \nabla^2 y - \Phi^2 y$ with the following side conditions:

$$y(0, x) = 1, \qquad y^{(1)}(t, 0) = 0 \quad \text{and} \quad y^{(1)}(t, 1) + \mathrm{Bi}_M\, y(t, 1) = 0$$

and in the three geometries.
The solution should be computed at $x = 0(0.1)1$ and $t = 0.1(0.1)3$.
Case a. $\mathrm{Bi}_M \to \infty$, $\phi = 0$.
Case b. Bi_M a given value $< \infty$ (e.g., 2, 5, 10, 50, 200).
In each case the average value of $y = \bar{y}$ should be compared to the curves in Crank (1957).
Case c. $\Phi \neq 0$ (e.g., 1 and 2) and otherwise as for cases a and b.

6. Determine the first two eigenvalues and eigenfunctions for the model of Exercise 5 using the forward integration technique of section 4.5. Case a and examples of cases b and c should be studied. Compare the approximate eigenfunctions with their known analytical expressions for plane parallel geometry.

7. The usual procedure for calculating a penetration solution is to assume in advance that the two independent variables can be compounded into a single variable η. This is done in Mickley, Sherwood, and Reed (*Applied Mathematics in Chemical Engineering*) and in Jenson and Jeffreys (*Mathematical Methods in Chemical Engineering*), two standard texts for chemical engineering students. In fact this procedure might obscure the more general technique that is discussed in subsection 4.4.2.
Consider diffusion from a semi-infinite medium

$$\frac{\partial^2 y}{\partial x^2} = \frac{\partial y}{\partial z} \qquad y(x, 0) = 1 = y(\infty, z) \quad \text{and} \quad y(0, z) = 0$$

which has the *exact* solution $y = \operatorname{erf}(x/2\sqrt{z})$.

Proceed as in subsection 4.4.2 ($\eta = x/z^q$, $z_1 = z$) and show that the first perturbation function f_0 is erf $(x/2\sqrt{z})$, while all higher perturbation functions $f_1, f_2, \ldots$ are identically zero.

8. The differential equations (115) for the perturbation functions f_k were solved analytically in the text. Expansions similar to (111) may be obtained for other entry length problems where an analytical solution of the perturbation equation is not feasible and a numerical solution of (115) is therefore of interest.
Compute a numerical solution for f_k of (115), $k = 0, 1, 2, 3, 4$ by either of the following techniques.
a. The boundary condition at $\eta \to \infty$ is replaced by a boundary condition at $\eta_{\max}$ and the differential equation is solved by collocation on the resulting finite interval $[0, \eta_{\max}]$.
Discuss the effect of the choice of $\eta_{\max}$.
b. The semi-infinite interval $[0, \infty]$ for η is transformed to a finite interval $[0, 1]$ by a suitable coordinate transformation, e.g., $x = \exp(-a\eta)$, and the transformed equation is solved by collocation.
Discuss the effect of the parameter a.

9. The Graetz problem in cylinder geometry

$$(1 - x^2)\frac{\partial \theta}{\partial z} = \frac{1}{x}\frac{\partial}{\partial x}\left(x\frac{\partial \theta}{\partial x}\right)$$

$$\theta(1, z) = 0 \quad \text{and} \quad \frac{\partial \theta}{\partial x}(0, z) = 0 \quad \text{for } z > 0$$

$$\theta(x, 0) = 1$$

can be solved by the method of subsection 4.4.2 combined with the collocation technique of Exercise 8 for the higher perturbation functions.
Compute the first four terms of the Levèque series for $J = -4\int_0^z (\partial\theta/\partial x)|_{x=1}\,dz$ and compare with Newman's result (1969), which is given in chapter 9 [(equation (9.19))].
Hint: A reasonable η is $(1 - x)/z^q$. It appears that the transformed equation in variables (η, z_1) is simplified best when $q = \frac{1}{3}$. Next $v = z_1^{1/3}$ is introduced as a new variable and finally the perturbation functions appear from the following equation:

$$\sum_0^\infty v^i \frac{d^2 f_i}{d\eta^2} = \left[-\frac{1}{3}\eta^2(2 - \eta v) + v\sum_0^\infty (\eta v)^i\right]\sum_0^\infty v^i\frac{df_i}{d\eta} + \frac{1}{3}\eta(2 - \eta v)\sum_0^\infty i v^i f_i$$

10. Mashelkar, et al. (1973) have made a calculation similar to that of section 4.3 but for a power law fluid (see section 1.1). In the nomenclature of the reference one obtains

$$(1 - x^{n+1})\frac{\partial \theta}{\partial z} = \frac{\partial^2 \theta}{\partial x^2}$$

$$\theta(0, z) = 0 = \frac{\partial \theta}{\partial x}(1, z) \quad \text{and} \quad \theta(x, 0) = 1$$

Compute $G = 1 - \bar{\theta}$ where

$$\bar{\theta} = \frac{n+2}{n+1}\int_0^1 \theta(1 - x^{n+1})\,dx$$

and compare with the values obtained by Mashelkar for $n = 0, 0.5, 1$ (a Newton fluid), 2.5, 10 at $z = [0, (0.02), 0.3]$.

11. Davis, et al. (1974) have solved a model for capillary-tissue mass transfer. Their final model for the blood phase (capillary) is the same as used in Exercise 9, but a flux balance at the boundary capillary-tissue should be used to tie the two phases together. Only radial diffusion needs to be considered. Compare your computed results with figure 2-4 of the reference.
Hint: The problem is solved in two parts. First the ordinary differential equation for the tissue phase is solved with an arbitrary value of the interphase concentration. The result is the flux at the interphase except for a scalar factor. The collocation equations for the capillary phase can now be set up, and the unknown interphase concentration eliminated using the flux condition at the interphase.

12. The following model for laminar Newtonian flow with first-order isothermal chemical reaction in an empty tube is presented in equation (1.34) of chapter 1:

$$2(1 - x^2)\frac{\partial y}{\partial \zeta} = \frac{D_r L}{R^2 v_{av}}\frac{1}{x}\frac{\partial}{\partial x}\left(x\frac{\partial y}{\partial x}\right) - \text{Da}\, y \tag{1}$$

Integrate this equation by the method of section 4.3 to obtain the mean value

$$\bar{y}(\zeta) = 4\int_0^1 x(1 - x^2)y(x, \zeta)\,d\zeta = \sum A_i \exp(\lambda_i \zeta) \tag{2}$$

Verify the results in table 1.4 and also verify that

$$\begin{aligned}
\lambda_1 &= -\text{Da}\left(1 - \text{Da}\frac{R^2 v_{av}}{48 D_r L}\right) + \text{Da}\,\mathcal{O}\left(\text{Da}\frac{R^2 v_{av}}{48 D_r L}\right)^2 \\
A_1 &= 1 + \mathcal{O}\left(\text{Da}\frac{R^2 v_{av}}{48 D_r L}\right)^2 \\
A_i &= \mathcal{O}\left(\text{Da}\frac{R^2 v_{av}}{48 D_r L}\right)^2 \quad \text{for } i > 1
\end{aligned} \tag{3}$$

13. Modify the boundary condition at $x = 1$ of Exercise 5 to

$$y^{(1)}(t, 1) + \text{Bi}_M[y(t, 1)] = \text{Bi}_M\, y_b(t)$$

$$q y_b + \bar{y} = 1 \quad \text{and} \quad y_b(t = 0) = 0$$

The model can now be taken to represent leaching of a solute from a solid into a solvent of finite volume q per unit volume solid. y_b is the solute concentration in the solvent outside the solid and $\bar{y}$ is the volume averaged concentration on the solid.

a. Compute $\bar{y}(t)$, $y_b(t)$, and the extent of the leaching process:

$$E = \frac{1 - \bar{y}(t)}{1 - \bar{y}(t \to \infty)} = \frac{q + 1}{q}[1 - \bar{y}(t)]$$

To check the program, insert $q = 1$, plane parallel geometry, $\mathrm{Bi}_M \to \infty$, and $\Phi = 0$. Now E can be compared with figure 4.6 in Crank (1957).

b. Modify the program to design an isothermal cocurrent desorption process from a liquid film in laminar downward flow into a gas phase of volume q per unit volume liquid. As a result the distance to achieve 99% desorption may be plotted as a function of G.
Gas film transfer resistance may be neglected and for convenience the solute may be assumed to be equally soluble in the gas and liquid phases.

14. (i) Use the material of subsection 1.2.3 to derive the following transient model for a tubular reactor. Radial gradients are neglected but axial dispersion is taken into account.
The reaction is first order and isothermal

$$\frac{\partial y}{\partial \tau} + \frac{\partial y}{\partial z} - \frac{1}{Pe_M}\frac{\partial^2 y}{\partial z^2} + \mathrm{Da}\, y = 0$$

$$z = 0: \quad y - \frac{1}{Pe_M}\frac{\partial y}{\partial z} = y_{in}(\tau)$$

$$z = 1: \quad \frac{\partial y}{\partial z} = 0$$

$$\tau = 0: \quad y = y_0(z)$$

For simplicity let $y_{in}(\tau) = 1$ and $y_0(z) = 0$.

(ii) Find $y(1, \tau)$ for $\mathrm{Da} = 1$ and $\mathrm{Pe}_M = 2$, 5, 20, and 50.
Hint: Proceed as in section 4.3 with collocation at the zeros of $P_N^{(0,0)}(z)$. For certain combinations of N and Pe_M complex eigenvalues may be obtained and you should consult exercise 1.11.

(iii) Derive an analytical expression for the jth eigenvalue and compare with the numerical results.
Find $\lim_{\mathrm{Pe}_M \to 0} \lambda_j$ and $\lim_{\mathrm{Pe}_M \to \infty} \lambda_j$ and use these results to explain the behavior of the approximate solution for large and small Pe_M.

(iv) Derive an analytical solution for the transient in the two cases $\mathrm{Pe}_M \to 0$ and $\mathrm{Pe}_M \to \infty$.

REFERENCES

The Gauss elimination program of section 4.1 is similar to many other programs that use row and column pivotation. It may be slightly better to use a LU decomposition program, which is also standard at most computing centers.

The program EISYS of section 4.3 for diagonalization of a nonsymmetric matrix by the QR method is made following the recommendations of Wilkinson (1965). Program packages of this type are available by now almost everywhere.

The methods of section 4.2 for solving initial value problems are related to well-known implicit or semi-implicit methods. Some references are given in chapter 8. The end point collocation methods of subsection 4.2.3 were probably first proposed by Villadsen (1968). They are further discussed in Villadsen (1970).

Bird (1960), section 17.5, and Finlayson (1972), section 3.4 have both used the example of section 4.3. The asymptotic behavior of the large eigenvalues in subsection 4.4.1 can be derived following a method discussed in Courant and Hilbert (1968). The presentation of the penetration solution in subsection 4.4.2 closely follows Newman's derivation (1969) of a similar series for the Graetz problem in cylinder geometry that we have included as Exercise 9. The model appears to have a considerable interest in the design of trickling filters and related units for biological water treatment. An application is given in Exercise 8.13.

1. WILKINSON, J. *The Algebraic Eigenvalue Problem.* Oxford: Clarendon Press (1965).

2. VILLADSEN, J. *Transactions Norddata* 68, pp. 138–75, Helsinki (June, 1968).

3. VILLADSEN, J. *Selected Approximation Methods for Chemical Engineering Problems.* Lyngby, Denmark: Instituttet for Kemiteknik (1970).

4. BIRD, R. B., STEWART, W. E., and LIGHTFOOT, E. N. *Transport Phenomena.* Wiley (1960).

5. FINLAYSON, B. A. *The Method of Weighted Residuals and Variational Principles.* New York: Academic Press (1972).

6. COURANT, R., and HILBERT, D. *Methods of Mathematical Physics.* New York: Interscience Publishers (1953).

7. NEWMAN, J. *J. Heat Transfer* 91 (1969): 177.

8. CRANK, J. *The Mathematics of Diffusion.* Oxford: Clarendon Press (1957).

9. MASHELKAR, R. A., CHAVAN, V. V., and KARANTH, N. G. *Chemical Engr. Journal* 6 (1973): 75.

10. DAVIS, E. J., COONEY, D. O., and CHANG, R. *Chemical Engr. Journal* 7 (1974): 213.

Nonlinear Ordinary Differential Equations 5

Introduction

The examples of previous chapters have generally been linear, the only exception being the simple nonlinear problem used in section 2.4 to introduce orthogonal collocation by comparison with an approximate Galerkin method.

A linear differential equation is characterized by having constants or explicit functions of the independent variable as coefficients to the dependent variable and its derivatives. In a nonlinear differential equation at least one of these coefficients is a function of the dependent variable.

Nonlinearities do not in principle introduce new aspects of the numerical methods discussed in this text. The examples of sections 5.4 and 5.5 illustrate that an equation in which the derivatives occur linearly is treated by an almost trivial modification of the algorithms of section 4.1, but much more complicated nonlinear differential equations have been solved by MWR, especially by collocation.

The layout of the iterative calculations to solve the nonlinear algebraic equations that result from discretization of a nonlinear differential equation does, however, vary significantly from problem to problem. In the numerical treatment of the Weisz–Hicks problem in section 5.5, it becomes apparent that different methods of approach to the numerical calculations may lead to computational schemes of widely different efficiency. The major part of the computational work is, however, always made in subroutines that have been described in chapters 3 and 4.

The outstanding difference between nonlinear and linear differential equations is not of numerical nature but is of a much more fundamental character. The qualitatively different response of nonlinear and linear differential equations to a change in side conditions is well known from elementary textbooks, but the frequent occurrence of linearly independent multiple solutions to nonlinear differential equations in certain parameter intervals is an intriguing feature of these equations that is not at all well studied in theory or by computer experiments.

The occurrence of strange phenomena in physical systems is a major incentive to use nonlinear models in an attempt to obtain bounds on the regions where these phenomena, e.g., unstable flow in a heat exchanger or flickering of a catalyst between several activity levels, occur. Only nonlinear models can give any hope of a mathematical explanation of most of these phenomena.

It would be highly desirable if without actually solving the equation one could decide whether any given nonlinear boundary value problem may have multiple solutions in some domain of its parameter space. This could help the investigator to avoid wasting time by computer solution of a model that either cannot explain a certain phenomena at all or can only do so within a certain parameter range that may go undetected in the numerical experiments.

The present state of development of functional analysis and related topics of applied mathematics does not give much encouragement, however, for this rational approach to the model analysis. A large number of investigations have recently appeared in chemical engineering literature where topics from functional analysis such as fixed-point theory have been applied to estimate the size of parameter regions where multiple solutions may exist. We have chosen to illustrate this type of preliminary model analysis by means of comparison differential equations since these are extremely simple to apply and they may lead to approximate solutions that give much insight into the actual solution of the given nonlinear model. Two theorems from the theory of differential inequalities are stated without proof in section 5.2 and the few examples that illustrate the application of these theorems may whet the appetite of those who want to study this promising field of functional analysis more closely.

The paucity of the results of general nature concerning nonlinear boundary value problems does, however, emphasize the usefulness of efficient numerical methods for solving the actual problem. In the forseeable future we must expect nonlinear differential equations to be discussed on the basis of individual examples solved on a computer and with only a preliminary guidance from theoretical mathematics.

For this reason the various numerical tricks that are mentioned in connection with the Weisz–Hicks problem in section 5.5 may be of value also beyond this example since they represent the semianalytical approach to numerical work that is necessary in order to reach the often almost inaccessible regions of parameter space where the most interesting features of a mathematical model are displayed.

5.1 Multiple Solutions of a Trivial Boundary Value Problem

Consider the following problem:

$$\frac{d^2y}{dx^2} + f(y) = 0, \qquad f(y) = \begin{cases} -\Phi^2 K_1^2 y & \text{for } y_s \le y \le 1 \\ -\Phi^2 K_2^2 y & \text{for } 0 \le y \le y_s \end{cases} \tag{1}$$

$$y(1) = 1 \quad \text{and} \quad \frac{dy}{dx} = y^{(1)} = 0 \quad \text{for } x = 0$$

K_1, K_2, Φ, and y_s are given constants, and we are interested only in a solution that is contained in the closed interval $0 \le y(x) \le 1$.

The differential equation (1) is nonlinear since $f(y) = f_1(y)y$ where the coefficient $f_1(y)$ to y is a function of the solution y and not an explicit function of x. The second derivative is discontinuous at y_s, but y and $y^{(1)}$ are both continuous functions of x.

A similar but linear differential equation appears when $f(y)$ is given by

$$f(y) = \begin{cases} -\Phi^2 K_1^2 y & \text{for } x_s \le x \le 1 \\ -\Phi^2 K_2^2 y & \text{for } 0 \le x \le x_s \end{cases} \tag{2}$$

It will be shown that (1) has either one or three nonnegative solutions for $x \in [0, 1]$ while the linear equation with f given by (2) has one and only one solution $y(x)$.

The differential equation is solved in the same way whether $f(y)$ is given by (1) or by (2).

A value y_0 is chosen for $y(x = 0)$ and the corresponding Φ value is calculated. For $y_0 < y_s$ the solution of (1) is:

$$y = y_0 \cosh(\Phi K_2 x) \quad \text{for } y_0 \le y \le y_s \tag{3}$$

$$y = A \exp(\Phi K_1 x) + B \exp(-\Phi K_1 x) \quad \text{for } y_s \le y \le 1 \tag{4}$$

K_1, K_2, and y_s are given besides y_0, and the four constants Φ, x_s, A, and B of (3) and (4) can be determined from

$$y_s = y_0 \cosh(\Phi K_2 x_s) \tag{5}$$

$$y_s = A \exp(\Phi K_1 x_s) + B \exp(-\Phi K_1 x_s) \tag{6}$$

$$y_0 \frac{K_2}{K_1} \sinh(\Phi K_2 x_s) = A \exp(\Phi K_1 x_s) - B \exp(-\Phi K_1 x_s) \tag{7}$$

$$1 = A \exp(\Phi K_1) + B \exp(-\Phi K_1) \tag{8}$$

Since $y_s/y_0 > 1$, a unique (and positive) solution for Φx_s is determined from (5). This value is inserted into (6) and (7), and the (unique) solution (A, B) of the two linear equations is inserted into (8). The solution of (8) is

$$\exp(\Phi K_1) = \frac{1 \overset{+}{\underset{(-)}{}} (1 - 4AB)^{1/2}}{2A} \tag{9}$$

where $A > 0$ and $4AB$ is either negative or less than 1. The plus sign must always be chosen and a unique (positive) value of Φ is determined from (9).

The numerical results $\Phi(y_0)$ are shown in figure 5-1 for $K_1 = 4$, $K_2 = 20$, and $y_s = 0.6$.

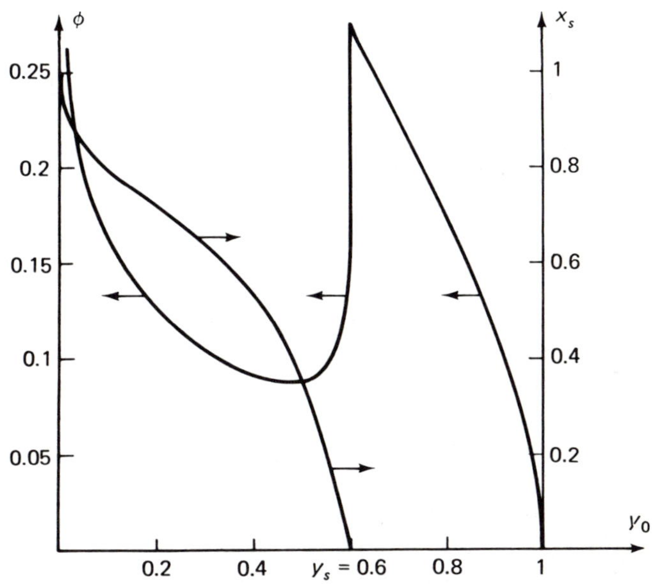

Figure 5-1. Solutions (Φ, x_s) of equations (5.3) and (5.4); $(K_1, K_2) = (4, 20)$.

For $y_0 > y_s$, $f(y)$ of (1) is $-\Phi^2 K_1^2 y$ for all $x \in [0, 1]$ and the solution is simply

$$y = y_0 \cosh(\Phi K_1 x) \quad \text{with } \Phi = \frac{1}{K_1} \operatorname{arc\,cosh}\left(\frac{1}{y_0}\right) \tag{10}$$

For any $y_0 < y_s$, the method of solution described in (5) to (9) must be used. When $y_0 \to 0$, Φ increases to infinity while $x_s \to 1$.

The interesting feature of figure 5-1 is that y_0 is a multivalued function of Φ, while x_s as well as Φ are single-valued functions of y_0. x_s increases from 0 to 1 when y_0 decreases from y_s to 0.

The following conclusions are drawn from the calculations:

1. The initial value problem (1) given by y_0, $y^{(1)}(0)$, $y(1)$, K_1, and K_2 has a unique solution $[\Phi, y(x)]$ that passes through $(x, y) = (1, 1)$.
2. The boundary value problem (1) given by $y^{(1)}(0)$, $y(1) = 1$, K_1, K_2, and Φ has three solutions $[y_0, y(x)]$ in a certain Φ range:

$$\Phi_1 < \Phi < \Phi_2$$

The upper limit of multiple solutions for $(K_1, y_s) = (4, 0.6)$ is easily found:

$$\Phi_2 = \frac{1}{K_1} \operatorname{arc\,cosh}\left(\frac{1}{y_s}\right) = 0.27465 \tag{11a}$$

The lower limit of multiple solutions can only be found iteratively:

$$\Phi_1 = 0.0884 \quad \text{at } y_0 = 0.45 \tag{11b}$$

3. The uniqueness of the linear boundary value problem (2) follows from the uniqueness of the initial value problem (1) and the monotonicity of $x_s(y_0)$: For a given parameter set x_s, Φ, K_1, K_2, only one y_0 value is found.

The very simplicity of the present example allows some important aspects of a nonlinear problem and its numerical solution to be studied without an undue burden of numerical work.

We note the uniqueness of the solution of the corresponding initial value problem: Numerical integration of a series of initial value problems with implicit determination of one of the problem parameters (here Φ) permits a tracing of all possible solutions to the problem.

Also the occurrence of multiple solutions in a finite interval of a parameter (here the Thiele modulus), which is quite typical for a nonlinear boundary value problem, is found even in this simple model.

In the present example as well as in more complicated problems (Φ_1, Φ_2) can be determined with any desired degree of accuracy by numerical solution of the differential equation. Here (Φ_1, Φ_2) was found by a relatively uncomplicated method that to a certain degree (e.g., iteration on y_0 to determine Φ_1 or Φ_2) can be applied also when no analytical solution of the differential equation is available.

The solution of the trivial model (1) gives some insight into the much more complicated model (1.69)–(1.70) that is discussed in detail in section 5.5. For a first-order irreversible reaction, plane parallel geometry, and both Bi_M and $\mathrm{Bi}_M \to 1$:

$$\frac{d^2y}{dx^2} - \Phi^2 y \exp\left[\gamma\beta \frac{1 - y}{1 + \beta(1 - y)}\right] = 0 \tag{12}$$

$$y^{(1)}(0) = 0 \quad \text{and} \quad y(1) = 1$$

The Thiele modulus Φ, dimensionless activation energy γ, and adiabatic temperature rise β have all been defined in chapter 1. Figure 5-2 shows the function $F(y)$ *for* $\gamma = 20$ and $\beta = 0.8$.

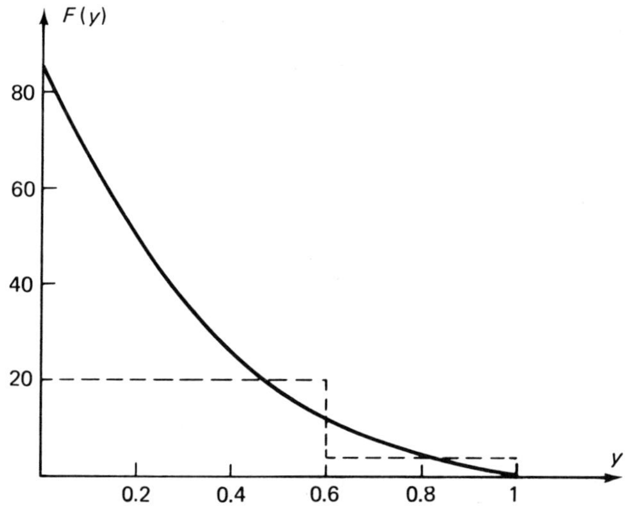

Figure 5-2. The function $F(y)$ of equation (5.13) with $(\gamma, \beta) = (20, 0.8)$ and its approximate representation [equation (5.14)].

$$F(y) = \left\{\exp\left[\gamma\beta\frac{1 - y}{1 + \beta(1 - y)}\right]\right\}^{1/2} \tag{13}$$

An approximate representation of $F(y)$ by

$$F(y) = \begin{cases} 4 & 0.6 < y \leq 1 \\ 20 & 0 \leq y < 0.6 \end{cases} \tag{14}$$

is also shown on the figure.

Equation (12) with the approximation (14) for $F(y)$ is exactly (1). Equation (12) admits to three solutions for

$$0.0691 < \Phi < 0.250 \tag{15}$$

and it is seen that the region of multiple solutions obtained from (1),

$$0.0884 < \Phi < 0.275 \tag{16}$$

is quite close to the region obtained from the exact solution. Changing K_1 from 20 to 30 and y_s from 0.6 to 0.56 yields almost the same region of multiple solutions for the two equations, but no manipulation on the approximate model can, however, give more than a qualitative impression of the solution to the original nonlinear problem or its region of unicity.

True upper and lower bounds Φ_1 and Φ_2 for Φ may be obtained by the methods of the next section, but the reader should note that the actual values of Φ_1 and Φ_2 are found with many digits accuracy by the methods of section 5.5 without excessive numerical work.

5.2 Existence and Uniqueness Theorems

We consider a second-order differential equation of the following general form:

$$\frac{d^2y}{dx^2} + f[x, y, y^{(1)}] = 0 \tag{17}$$

It is assumed that f is continuous and that it has bounded partial derivatives with respect to y and $y^{(1)}$ for all x in the interior of the x-interval $[a, b]$ and for such y and $y^{(1)}$ that are of interest in the given problem (e.g., positive y when y is a dimensionless concentration in a catalyst pellet). With these assumptions the function f is said to be Lipschizian. The following four constants K_1, K_2, L_1, L_2 limit the values of $\partial f/\partial y$ and $\partial f/\partial y^{(1)}$ within the open x-interval (a, b) and the $(y, y^{(1)})$ interval considered:

$$K_1 < \frac{\partial f}{\partial y} < K_2, \qquad L_1 < \frac{\partial f}{\partial y^{(1)}} < L_2 \tag{18}$$

The boundary conditions of (17) at the two interval end points $x = a$ and $x = b$ are

$$\begin{aligned} y^{(1)}(a) + Ay(a) &= C_1 \\ y^{(1)}(b) + By(b) &= C_2 \end{aligned} \tag{19}$$

We compare the solution y of (17) and (19) with the solutions $u_1(x)$ and $u_2(x)$ of the following comparison differential equations:

$$\frac{d^2u_1}{dx^2} + h_1[x, u_1, u_1^{(1)}] = 0 \tag{20}$$

$$\frac{d^2u_2}{dx^2} + h_2[x, u_2, u_2^{(1)}] = 0 \tag{21}$$

The side conditions of (20) and (21) are to be the same (19) as those applying to the given differential equation (17) and the two functions h_1 and h_2 will satisfy the following inequalities:

$$h_1[x, y, y^{(1)}] \le f[x, y, y^{(1)}] \le h_2[x, y, y^{(1)}] \tag{22}$$

throughout the $[x, y, y^{(1)}]$ interval to be investigated in (17).

It is assumed that the solution u_1 and u_2 of (20), (21) with the boundary conditions of (17) is unique in the x-interval $[a, b]$ or in any smaller part of this interval.

Under these conditions on f and on (u_1, u_2) the following theorem may be proved:

Theorem 1: There exists at least one solution of (17) in the given $[x, y, y^{(1)}]$ region, and any solution of (17) has the property that

$$u_1(x) < y(x) < u_2(x) \tag{23}$$

The choice of comparison differential equations is completely unrestricted as long as (22) is satisfied. Practical considerations do, however, limit the choice: In order that the uniqueness of the solutions u_1 and u_2 of the comparison equations can be easily determined, a rather simple form of h_1 and h_2 is preferable. It is shown in section 5.3 that (23) can be used to construct an approximation to $y(x)$, often in an iterative procedure. This also calls for a simple, linear form of h_1 and h_2. As a final consideration, h_1 and h_2 should be as close approximations to f as possible without sacrificing the simplicity of (20) and (21). In this case $u_1(x)$ and $u_2(x)$ may give close bounds on $y(x)$.

A specific choice of $h_i[x, y, y^{(1)}]$ $(i = 1, 2)$ of (20) and (21) is

$$h_i[x, y, y^{(1)}] = G_i[y, y^{(1)}] + f(x, 0, 0) \tag{24}$$

where $G_i[y, y^{(1)}]$ is given in table 5.1 by means of the constants in (18):

TABLE 5.1

FUNCTIONS G_1 AND G_2 OF (24) IN DIFFERENT REGIONS $[y, y^{(1)}]$

y	$y^{(1)}$	$G_1[y, y^{(1)}]$	$G_2[y, y^{(1)}]$
≥ 0	≥ 0	$K_1y + L_1y^{(1)}$	$K_2y + L_2y^{(1)}$
≥ 0	≤ 0	$K_1y + L_2y^{(1)}$	$K_2y + L_1y^{(1)}$
≤ 0	≥ 0	$K_2y + L_1y^{(1)}$	$K_1y + L_2y^{(1)}$
≤ 0	≤ 0	$K_2y + L_2y^{(1)}$	$K_1y + L_1y^{(1)}$

A theorem on uniqueness of the solution to (17) is based on the function G_2 of table 5.1:

Theorem 2: Provided the conditions of Theorem 1 are satisfied, the solution of (17) is unique if the linear differential equation

$$\frac{d^2u}{dx^2} + G_2[x, u, u^{(1)}] = 0 \tag{25}$$

with boundary conditions

$$u^{(1)}(a_1) + Au(a_1) = 0$$

$$u^{(1)}(b_1) + Bu(b_1) = 0$$

has no solution except $u = 0$ for any $(a_1, b_1) \in (a, b)$.

Theorems 1 and 2 are the major results of the theory of differential inequalities. Detailed proofs of the theorems are given by Bailey, et al. (1968) in chapters 5 to 7 for the case of $(A, B) \to \infty$ in (18) (y known at both interval end points) and for either $(A, B) = (\infty, 0)$ or $(A, B) = (0, \infty)$ [y known at one interval end point and $y^{(1)}$ at the other end point]. An extension to the more general boundary value problem with finite $(A. B)$ is, however, obvious from the proof of the theorems that is given in Bailey.

The second theorem concerning uniqueness of the solution to (17) is hinged on the solution of a linear eigenvalue problem (25). In a number of important problems this equation is of the form

$$\nabla^2 u + Ku = 0 \tag{26}$$

$$u^{(1)} + Au = 0 \quad \text{at } x = a$$

$$u^{(1)} + Bu = 0 \quad \text{at } x = b$$

with $\nabla^2 = (d^2y/dx^2) + (s/x)(du/dx)$ and $s = 0, 1,$ or 2 for plane parallel, cylindrical, and spherical geometry of the system.

The largest value of K in (26) that can be permitted in order that (25) shall have only the trivial solution $u = 0$ in $[a, b]$ or in any subinterval of $[a, b]$ is determined by the first eigenvalue of the operator ∇^2:

$$K < -\lambda_1$$

The value of $-\lambda_1$ depends on the geometry factor s and on the type of the boundary conditions (19).

The interval $[a, b]$ can be translated to $[0, 1]$ without changing the eigenvalues by more than a scalar factor. We briefly discuss the influence of s and B on the eigenvalues using $a = 0, b = 1$, and $A = 0$.

For $B \to \infty$ (y known at $x = 1$ and $y^{(1)} = 0$ at $x = 0$) $-\lambda_1$ is, respectively, $(\pi/2)^2$, 2.4048^2, and π^2 for $s = 0, 1,$ and 2.

These are the largest values that can be obtained from (26) in the three geometries. If the constant B has a finite value, a smaller value of $-\lambda_1$ is obtained. One interpretation of the effect of a decreasing $|\lambda_1|$ is that uniqueness of the solution to (17) is guaranteed only in an x-interval smaller than 1. More relevant to our problems in which $[a, b]$ is usually fixed is an equivalent formulation: If in two cases the first eigenvalue of (26) is $\lambda_1^{(1)}$ and $\lambda_1^{(2)}$ with $-\lambda_1^{(1)} < -\lambda_1^{(2)}$, respectively, then the parent problem (17) has a unique solution for a larger K in the case $\lambda_1 = \lambda_1^{(2)}$ than in the case $\lambda_1 = \lambda_1^{(1)}$.

5.3 Application of Comparison Differential Equations

The model for diffusion and chemical reaction with heat evolution in catalyst pellets is well suited for a demonstration of the theorems of section 5.2. Several results concerning existence and uniqueness of the solution can be obtained with practically no computation. On the other hand it is apparent that only very conservative bounds on the uniqueness region can be obtained even for this relatively simple example of a single nonlinear differential equation with linear derivatives. It will also be shown that Theorem 1 can be used for an approximate construction of a specific solution—but again it must be concluded that the power of the semi-analytical methods of this section is extremely limited compared to the collocation solutions of section 5.4, at least for quantitative studies.

For spherical catalyst pellets and a first-order reaction, one obtains the following equation for the pellet temperature θ when the surface temperature is $\theta(x = 1) = 1$:

$$\frac{d^2\theta}{dx^2} + \frac{2}{x}\frac{d\theta}{dx} + \Phi^2(1 - \theta + \beta)\exp\left(\gamma - \frac{\gamma}{\theta}\right) = 0 \tag{27}$$

In section 5.4 the same differential equation is analyzed numerically in terms of the dimensionless concentration $y = 1 - (1/\beta)(\theta - 1)$. This has been a tradition in numerical studies of the Weisz–Hicks problem ever since the first paper on this model appeared [Weisz and Hicks (1962)]—y varies between 1 and 0, while θ varies between 1 and the less well-defined upper limit $1 + \beta$. On the other hand, all theoretical papers on uniqueness of the solution to the Weisz–Hicks problem have been cast in terms of θ. A comparison of results obtained by the theorems of the last section and equivalent results from the literature is easier when exactly the same equation is discussed here.

Figure 5-3 shows

$$R(\theta) = (1 - \theta + \beta)\exp\left(\gamma - \frac{\gamma}{\theta}\right) \quad \text{for } \gamma = 20 \text{ and } \beta = 0.8 \tag{28}$$

which were the original choice of parameters in the Weisz–Hicks paper.

$$\frac{\partial R}{\partial \theta} = \frac{-\theta^2 - \gamma(\theta - 1) + \gamma\beta}{\theta^2}\exp\left(\gamma - \frac{\gamma}{\theta}\right) \tag{29}$$

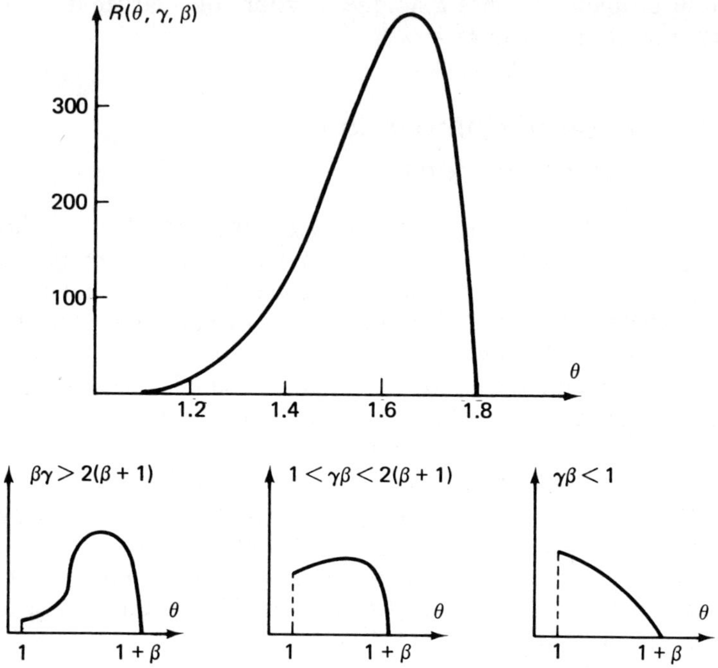

Figure 5-3. $R = (1 - \theta + \beta)\exp[\gamma - (\gamma/\theta)]$ for $(\gamma, \beta) = (20, 0.8)$ and for three more general cases.

$R(\theta)$ has a maximum at

$$\theta = \theta_{\max} = \frac{-\gamma + [\gamma^2 + 4\gamma(\beta + 1)]^{1/2}}{2} \tag{30}$$

$R(\theta)$ has an inflexion point at

$$\theta = \theta_{\text{infl}} = \frac{1 + \beta}{1 + [2(\beta + 1)/\gamma]} \tag{31}$$

At $\theta = \theta_{\text{infl}}$,

$$\frac{\partial R}{\partial \theta} = \left(\frac{\partial R}{\partial \theta}\right)_{\max} = \left[1 + \frac{4(1 + \beta)}{\gamma}\right] \exp\left[\frac{\gamma\beta - 2(\beta + 1)}{1 + \beta}\right] \tag{32}$$

Three distinct cases appear according to the position of $\theta = 1$ relative to $\theta_{\max}$ and θ_{infl}.

Case 1: $\beta\gamma > 2(\beta + 1) \rightarrow 1 < \theta_{\text{infl}}$ (33)

Case 2: $1 < \gamma\beta < 2(\beta + 1) \rightarrow \theta_{\text{infl}} < 1 < \theta_{\max}$ (34)

Case 3: $\gamma\beta < 1 \rightarrow \theta_{\max} < 1$ (35)

Each of these cases is illustrated by the small figures in figure 5-3.

The tangent at $\theta = 1 + \beta$ and the chord from $\theta = 1$ to $\theta = 1 + \beta$ are given by $h_2(\theta)$ and $h_1(\theta)$, respectively:

$$h_2(\theta) = -\exp\left(\frac{\gamma\beta}{1+\beta}\right)(\theta - 1 - \beta) \tag{36}$$

$$h_1(\theta) = -(\theta - 1) + \beta \tag{37}$$

$$h_1(\theta) \le R(\theta) \le h_2(\theta) \quad \text{for } 1 \le \theta \le 1 + \beta$$

The two comparison differential equations

$$\frac{d^2u_2}{dx^2} + \frac{2}{x}\frac{du_2}{dx} - \Phi^2 q^2 u_2 = -(1+\beta)q^2, \qquad q = \exp\frac{\gamma\beta/2}{1+\beta} \tag{38}$$

$$\frac{d^2u_1}{dx^2} + \frac{2}{x}\frac{du_1}{dx} - \Phi^2 u_1 = -\Phi^2(\beta + 1) \tag{39}$$

have unique solutions for any value of Φ. Consequently, by Theorem 1, any solution $y(x)$ of (27) lies between the solutions of (38) and (39) with $u_i(x) = 1$ at $x = 1$ and $du_i/dx = 0$ at $x = 0$:

$$u_2(x) = 1 + \beta - \beta\frac{\sinh(\Phi q x)}{x \sinh(\Phi q)} \tag{40}$$

$$u_1(x) = 1 + \beta - \beta\frac{\sin h(\Phi x)}{x \sin h(\Phi)} \tag{41}$$

$$u_1(x) < \theta(x) < u_2(x) \quad \text{for } 1 < \theta < 1 + \beta$$

The function $G_2(u)$ of Theorem 2 is different in the three cases of figure 5-3.

Case 1: $G_2(u) = \Phi^2\left[1 + \dfrac{4(1+\beta)}{\gamma}\right] u \exp\left[\dfrac{\gamma\beta - 2(\beta+1)}{1+\beta}\right]$

Case 2: $G_2(u) = (\gamma\beta - 1)\Phi^2 u \qquad (\gamma\beta > 1)$

Case 3: $G_2(u) = (\gamma\beta - 1)\Phi^2 u \qquad (\gamma\beta < 1)$

Theorem 2 guarantees that the steady state is unique for all values of Φ in case 3 since $G_2(u)$ is negative for all positive u. In case 2 or case 1, multiple solutions may be found for some Φ-values—as in case 2, when $\Phi > \pi/(\gamma\beta - 1)^{1/2}$. This result is very poor since it is shown in chapter 6 that uniqueness is guaranteed when $\gamma\beta/(1 + \beta) < 4$, which covers all situations corresponding to case 2 and even part of the case 1 region.

It may furthermore be proved [Luss (1969)] that there exists an upper bound Φ^* of the region of multiple steady states

$$(\Phi^*)^2 = \frac{\pi^2}{(\partial R/\partial\theta)_{\theta=1}} = \frac{\pi^2}{(R_\theta)_{\theta=1}} \tag{42}$$

When $\Phi > \Phi^*$, uniqueness of the steady state is guaranteed.

For $\gamma = 20$ and $\beta = 0.8$, multiple steady states may occur since $\gamma\beta/(1+\beta) > 4$. The value of $G_2(u)$ *is* $\Phi^2 \cdot 1.36 \cdot u \cdot \exp(6.8889) = 1334\Phi^2 u$, while $(\partial R/\partial\theta)_{\theta=1} = 15\Phi^2$. Thus uniqueness is guaranteed for $\Phi < \Phi_* = \pi/(1334)^{1/2} = 0.086$ and by (42) for $\Phi > \Phi^* = \pi/(15)^{1/2} = 0.811$.

Even though the comparison differential equation concept in its simplest formulation fails to predict the correct interval of γ and β for which multiple steady states can be found and cannot directly give an upper bound (42) for the region of multiple steady states, it is nevertheless quite helpful when used with proper ingenuity. Two examples follow.

First an upper bound Φ^* similar to (42) can be derived. We observe that R_θ is negative when $\theta_{\max} < \theta < 1 + \beta$ where $\theta_{\max}$ is defined in (30). For very large Φ the major part of the temperature profile $\theta(x)$ may have θ values greater than $\theta_{\max}$. If the x-value for which $\theta = \theta_{\max}$ is called x_m, we need only be concerned with the x-interval $x_m < x < 1$ where R_θ is positive. If this x-interval is small enough to give uniqueness by Theorem 2, the whole solution $\theta(x)$ is unique. x_m is, of course, unknown but an estimate for x_m may be found by means of the comparison functions u_1 and u_2.

The construction of functions $u_2(x)$ and $u_1(x)$ that by Theorem 1 are, respectively, above and below the true solution is characteristic of a proper use of the comparison differential equation concept. Our second example shows how this can be done.

Define θ_t as the abscissa of the R versus θ curve at which the tangent passes through $(\theta, R) = (1, \beta)$ and let θ^* be a value of θ in the interval $(1, \theta_t)$. For any such θ value the chord from $(1, \beta)$ to $[\theta^*, R(\theta^*)]$ is always above $R(\theta)$ and a suitable upper bounding function $u_2(x)$ is defined by

$$\frac{d^2u_2}{dx^2} + \frac{2}{x}\frac{du_2}{dx} + \Phi^2 a^2 u_2 = \Phi^2 a^2 - \Phi^2\beta$$

where

$$a^2 = \frac{R(\theta^*) - \beta}{\theta^* - 1} \quad \text{and} \quad u_2(1) = 1, \qquad u_2^{(1)}(0) = 0$$

or

$$u_2 = 1 + \frac{\beta}{a^2}\left(\frac{\sin \Phi a x}{x \sin \Phi a} - 1\right) \tag{43}$$

The maximum value of u_2 is found at $x = 0$. This value $u_2(0)$ must be smaller than θ^*; otherwise we fail to meet the condition that the solution constructed from the chord is an upper solution. The smaller we choose our θ^*, the more likely is $u_2(0)$ to be larger than θ^*; but we are obviously interested in choosing θ^* as close to 1 as possible in order that the chord shall be close to $R(\theta)$ for the θ values that correspond to the given Φ.

Thus the smallest admissible θ^* is that for which $u_2(0) = \theta^*$ or

$$u_2(0) = \theta^* = 1 + \frac{\beta}{a^2}\left(\frac{\Phi a}{\sin \Phi a} - 1\right) \tag{44}$$

which is an algebraic equation in θ^*. Its solution is inserted into (43) to give the best possible upper solution $u_2(x)$.

A tangent to $R(\theta)$ taken at any point $[\theta_*, R(\theta_*)]$ where θ_* is between 1 and θ_{infl} of (31) is certainly below $R(\theta)$ for all $\theta \in (1, \theta_{\text{infl}})$. Hence a lower bound for θ is found by solution of

$$\frac{d^2u_1}{dx^2} + \frac{2}{x}\frac{du_1}{dx} + \Phi^2 R_{\theta_*} u_1 = \Phi^2[R_{\theta_*}\theta_* - R(\theta_*)] \tag{45}$$

$$u_1^{(1)}(0) = 0 \qquad u_1(1) = 1$$

Here R_{θ_*} is $\partial R/\partial\theta$ taken at $\theta = \theta_*$. We wish by an appropriate choice of θ_* to obtain $u_1(x)$ as close to $\theta(x)$ as possible. From the solution of (44) we calculate a θ^* that is certainly above θ for any x and the given Φ. Thus θ_* of (45) should be smaller than θ^*, and a value of $\theta_* = \frac{1}{2}(1 + \theta^*)$ may give a suitable "even" distribution of errors in our approximation of $R(\theta)$ by $R_{\theta_*}u_1 - R_{\theta_*}\theta_* + R(\theta_*)$.

Table 5.2 shows that quite satisfactory results are obtained when the upper solution is constructed by (43) and (44) and the lower solution by (45) either with $\theta_* = 1$ or with $\theta_* = \frac{1}{2}(1 + \theta^*)$.

TABLE 5.2

UPPER AND LOWER LIMIT ON CENTER TEMPERATURE
SPHERICAL GEOMETRY AND $(\gamma, \beta) = (20, 0.8)$

Φ	$u_2(0) = \theta^*$ [solution of (44)]	$u_1(0)$: $\theta_* = 1$	$u_1(0)$: $\theta_* = \frac{1}{2}(\theta^* + 1)$	Exact
0.1	1.001358	1.001356	1.001357	1.001357
0.3	1.01451	1.01420	1.01444	1.01445
0.4	1.0331	1.0293	1.03153	1.03178
0.4525	1.0654	1.0417	1.0480	1.0512

For $(\gamma, \beta) = (20, 0.8)$, equation (44) has no real solution in $1 < \theta^* < 1 + \beta$ when Φ is slightly above 0.453. Consequently for $\Phi > 0.453$, we cannot construct an upper solution $u_2(x)$ by the chord process with the property that $u_2(x) \leq \theta^*$ for all x. The limiting Φ value obtained by this process has no direct connection with the upper and lower bounds for the region of multiple solutions. Rather an approximation is found for the Φ range in which the "quenched" solution to the Weisz–Hicks problem exists for spherical geometry and $(\gamma, \beta) = (20, 0.8)$. By the constructive process outlined above we conclude that the quenched solution is obtained at least in the Φ interval $(0,\ 0.453)$. This result is surprisingly good: Solution of the nonlinear differential equation shows that the quenched solution exists for $0 \leq \Phi < 0.48$.

5.4 Global Collocation Solution of a Nonlinear Differential Equation

Nonlinear differential equations encountered in chemical engineering problems are frequently linear in the derivatives:

$$a_1(x)\frac{d^2y}{dx^2} + a_2(x)\frac{dy}{dx} + f(x, y) = 0 \tag{46}$$

with boundary conditions

$$y + b_1\frac{dy}{dx} = c_1 \quad \text{at } x = 0$$

$$y + b_2\frac{dy}{dx} = c_2 \quad \text{at } x = 1$$

A typical example is the nonlinear effectiveness factor problem discussed in chapter 2 for the geometry factor $s = 1$:

$$\frac{d^2y}{dx^2} + \frac{s}{x}\frac{dy}{dx} - \Phi^2y^2 = 0 \tag{47}$$

$$\frac{dy}{dx} = 0 \quad \text{at } x = 0$$

and at $x = 1$, $y = 1$.

For solution of equations of the general form (46), orthogonal collocation using the ordinates y_i as expansion coefficients is a very convenient method, due to the simple structure of the system of algebraic equations that result after the collocation process.

The N zeros x_i of a suitable Jacobi polynomial and the interval end points $x_0 = 0$ and $x_{N+1} = 1$ are chosen as interpolation points. One of the boundary conditions can often be eliminated by a suitable change of independent variable, and collocation equations are set up for the interior points:

$$a_1(x_i)\sum_{j=0}^{N+1} B_{ij}y_j + a_2(x_i)\sum_{j=0}^{N+1} A_{ij}y_j + f(x_i, y_i) = 0 \tag{48}$$

The discretized boundary conditions

$$y_0 + b_1 \sum_{j=0}^{N+1} A_{0,j}y_j = c_1$$

and

$$y_{N+1} + b_2 \sum_{j=0}^{N+1} A_{N+1,j}y_j = c_2$$

are used to eliminate y_0 and y_{N+1} from (48) and a set of N equations in the interior ordinates is obtained:

$$F_i = \sum C_{ij}y_j + f(x_i, y_i) + d_i = 0, \qquad i = 1, \ldots, N \tag{49}$$

or

$$\mathbf{F} = \mathbf{C}\mathbf{y} + \mathbf{f}(\mathbf{x}, \mathbf{y}) + \mathbf{d} = \mathbf{0} \tag{50}$$

The matrix $\mathbf{C}$ and the vector $\mathbf{d}$ have constant elements, and the nonlinearity enters only in $\mathbf{f}$.

The solution is conveniently determined by the Newton–Raphson method. Given an estimate $\mathbf{y}^k$ of the solution vector, an improved estimate is obtained from

$$\mathbf{y}^{k+1} = \mathbf{y}^k - [\mathbf{J}^k]^{-1} \cdot \mathbf{F}^k \tag{51}$$

where the Jacobian matrix $\mathbf{J}$ is given by $J_{ij} = (\partial F_i/\partial y_j)$.

The elements of the Jacobian are easily derived:

$$\frac{\partial F_i}{\partial y_j} = C_{ij} + \frac{\partial f_i}{\partial y_j} = \begin{cases} C_{ij} & j \neq i \\ C_{ii} + \dfrac{\partial f_i}{\partial y_i} & j = i \end{cases}$$

or

$$\mathbf{J} = \mathbf{C} + \mathbf{f}^{(1)} \tag{52}$$

where $\mathbf{f}^{(1)}$ is diagonal with $f_{ii}^{(1)} = \partial f_i/\partial y_i$. Except for the diagonal elements $\mathbf{J}$ does not change from iteration to iteration.

The substitution $u = x^2$ is introduced in (47), and (49) takes the following form for $s = 1$:

$$F_i = \sum_{j=1}^{N} (4u_iB_{ij} + 4A_{ij})y_j - \Phi^2 y_i^2 + (4u_iB_{i,N+1} + 4A_{i,N+1}) = 0 \tag{53}$$

that is,

$$C_{ij} = 4u_iB_{ij} + 4A_{ij} \tag{54}$$

$$f_i = -\Phi^2 y_i^2 \tag{55}$$

$$d_i = 4u_iB_{i,N+1} + 4A_{i,N+1} \tag{56}$$

and

$$f_{ii}^{(1)} = -2\Phi^2 y_i \tag{57}$$

A computer program for solution of (47) is shown in the Appendix, (A15). The differential equation in u is

$$4u\frac{d^2y}{du^2} + 2(s+1)\frac{dy}{du} - \Phi^2 y^2 = 0 \quad \text{with } y(1) = y_{N+1} = 1 \tag{58}$$

and the effectiveness factor η is

$$\eta = \frac{s+1}{2}\int_0^1 y^2(u)u^{(s-1)/2}\,du \tag{59}$$

Following the arguments of sections 2.4 and 2.5, collocation at the zeros of $P_N^{[1,(s-1)/2]}(u)$ is used to give a high accuracy of η.

To start the iterations, all interior ordinates are chosen equal to the boundary value $y = 1$, but any other positive initial y-estimate may be used since the solution is unique for $y > 0$ by the results of Theorem 2 in section 5.3.

$$G_2(u) = \max_{u\geq 0}(-2\Phi^2 u)u = 0 \tag{60}$$

The iterations are continued until the sum of squared corrections to y_i is less than 10^{-16}. This ensures that the correction to each y_i in the last iteration is less than $\sim 10^{-8}$ and the machine accuracy is presumably reached since the Newton algorithm is quadratically convergent. Four to five iterations are sufficient to terminate the iteration. Generally the correction Δy_i in a Newton process has about the same number of correct digits as the previous estimate y_i, which indicates that iteration until each $\Delta y_i < 10^{-9}$ is sufficient on a 16-digit machine.

Clearly the example of this section is extremely simple and the nonlinearity could still be coped with by the other MWR of chapter 2. But even trivial nonlinearities such as $f = y^n$ with n noninteger or $f = \exp(y)$ require numerical integration for evaluation of the algebraic equations in the expansion coefficients and a further numerical integration to obtain the Jacobian.

Any differential equation of the form (46) can be solved by collocation using exactly the same procedure and it is apparent that the doubtful gain in accuracy that may be obtained by, e.g., Galerkin's method, is unable to compensate for the drastically increased computing effort.

5.5 Concentration Profiles for Nonisothermal Reactions

5.5.1 The purpose of the computations

Let us again consider the steady state model for a nonisothermal first-order irreversible reaction occurring inside a catalyst particle. The mass balance equation is

$$\nabla^2 y = \Phi^2 \cdot R(y) = \Phi^2 \cdot y \cdot \exp\left[\frac{\gamma\beta(1-y)}{1+\beta(1-y)}\right] \tag{61}$$

The dimensionless activation energy γ is normally positive. β is a dimensionless heat of reaction. If β is large, the temperature in the center of the pellet is much higher than the surface temperature with a corresponding increase in the rate of reaction. This temperature effect may exceed that of the lower reactant concentration; for high enough $\beta \cdot \gamma$, as indicated in section 5.1 and further discussed in section 5.3, multiple solutions are possible inside a certain Φ-interval.

For an endothermic reaction β is negative, and the reaction rate in the interior of the pellet is decreased due to both the lower temperature and the lower reactant concentration. Under these circumstances it is immediately clear that only one solution exists.

Numerical computation shows that multiple solutions are possible for certain Φ-values provided

$$\frac{\gamma\beta}{1+\beta} > \text{about } 4 \tag{62}$$

For a set of parameter values outside the uniqueness region the numerical solution [proceeding as in section 5.4, but substituting the rate term (61)] will depend on the initial estimate of the solution vector. If three solutions are possible, two of these will normally be physically stable (depending somewhat on the value of a capacitance parameter, the Lewis number Le discussed in section 1.6) while the third solution is unstable.

The two stable solutions are the "quenched" solution, with values of y and θ close to the surface value throughout the pellet, and the "ignited" solution, with $y \sim 0$ (and $\theta = 1 + \beta$) in the center region. The third unstable solution is the "intermediate" solution, which may be quite difficult to obtain by a global method since it normally requires an initial estimate of the profile that is close to the correct profile.

For a given set of parameter values β and γ, we are often interested in obtaining a plot of the effectiveness factor curve, that is, a plot of η as a function of Φ. The effectiveness factor integral is again evaluated from the collocation solution by quadrature:

$$\eta = (s + 1)\int_0^1 R(y)x^s\,dx \tag{63}$$

A set of parameter values (γ, β) permitting three solutions for certain Φ-values yields a typical S-shaped curve, as in figure 5-4 for spherical geometry with $(\gamma, \beta) = (20, 0.3)$.

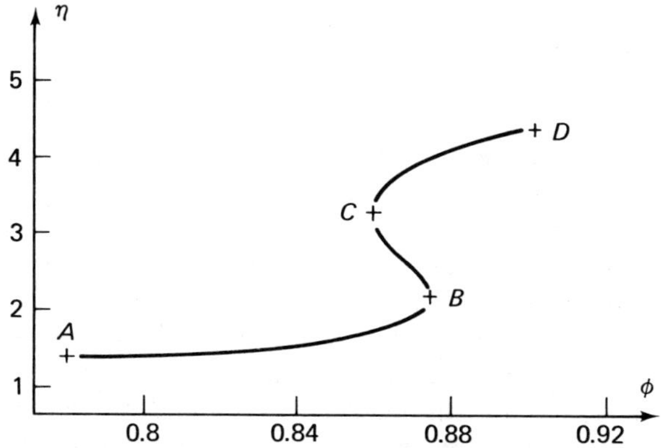

Figure 5-4. $\eta(\phi)$ for spherical pellets; $(\gamma, \beta) = (20, 0.3)$.

To obtain this curve one would start with a fairly small value of Φ yielding a solution where the concentration (and the temperature) does not differ significantly from the surface value. By gradually increasing Φ points on the branch, A–B may be obtained.

5.5.2 Sensitivity function with respect to Φ

For each new Φ-value an initial estimate of the solution vector is needed. For very small values of Φ, the initial estimate $y_i = 1$ at all interior points is satisfactory. For larger Φ-values, however, previously obtained concentration profiles may be of considerable help.

Let the solution $y(\Phi_0, x)$ for $\Phi = \Phi_0$ be known. Differentiate (61) with respect to Φ^2 and define $z(x) \equiv \partial y(\Phi, x)/\partial\Phi^2$. This yields

$$\nabla^2 z - \Phi^2\left(\frac{\partial R}{\partial y}\right)z - R(y) = 0 \tag{64}$$

with boundary conditions

$$z^{(1)}(0) = z(1) = 0 \tag{65}$$

Equation (64) is *linear* in z and is easily solved by collocation when y has been determined. Notice that the matrix of coefficients for z in the collocation equations is identical to the final Jacobian matrix in the solution of (61) so that $z(x)$ is obtained at little extra work.

A good initial estimate of $y(x)$ at $\Phi = \Phi_1$, where Φ_1 is close to Φ_0, is obtained from

$$y(\Phi_1, x) \simeq y(\Phi_0, x) + (\Phi_1^2 - \Phi_0^2)z(\Phi_0, x) \tag{66}$$

The main part of the branch A–B of the $\eta(\Phi)$-curve is easily traced in this manner, but as point B is approached, the "sensitivity" function $z(x)$ attains very large values, the reason being that the Jacobian is singular at this point [Φ cannot be increased beyond $\Phi(B)$ if we wish to remain on this branch of the curve]. It is now necessary to decrease the value of Φ and guess a new initial estimate of the y-vector to obtain a solution on branch B–C of the curve. When one point (or rather, one profile) is known, the unstable branch from B to C can be traced. Finally a point on the branch C–D must be determined to start the marching process for the "ignited" steady state.

5.5.3 Using $y(x = 0)$ as a free parameter

The solutions obtained following the $\eta(\Phi)$-curve (figure 5-4) in the direction $A \to B \to C \to D$ have the property that $y(x = 0)$ is monotonically decreasing along the curve. This may be utilized in an alternative technique where the problems of passing from branch to branch are completely avoided.

In section 5.1 it is mentioned that the solution of (61) is unique in $y(x = 0)$: The model becomes an initial value problem with y and $y^{(1)}$ given at $x = 0$. The $\eta(\Phi)$-curve may consequently be computed by specifying a sequence of values for $y_{x=0}$ and determining the corresponding values of Φ and η.

If the substitution $u = x^2$ is applied, $y_0 = y(x = 0)$ does not enter into the collocation equations. A fairly large number of collocation points must often be used, however, and the first interior collocation abscissa u_1 is very close to zero. Hence we would expect that specifying the value $y_1 = y(u = u_1)$ will also uniquely determine the value of the dependent variable at the remaining collocation points $(y_2 \ldots y_N)$ and the Thiele modulus Φ.

The following procedure is suggested. The collocation equations are set up in the usual manner, initial estimates of $y_2 \ldots y_N$ and Φ^2 being given:

$$F_i = 4u_i \sum_{j=1}^{N+1} B_{ij}y_j + 2(s + 1) \sum_{j=1}^{N+1} A_{ij}y_j - \Phi^2 R(y_i) = 0 \tag{67}$$

where now y_1 (and of course $y_{N+1} = 1$) is known, while the N unknowns are $\Phi^2, y_2, y_3, \ldots, y_N$.

If we let Φ^2 be the first element in the solution vector, the Jacobian matrix is only changed in its first column, where now

$$J_{i,1} = \frac{\partial F_i}{\partial \Phi^2} = -R(y_i) \tag{68}$$

No difficulties are encountered near points B and C in figure 5-4. Furthermore, it is possible to use a tracing technique similar to the one described above by differentiation of the equations in (67) with respect to y_1.

It is even better to specify the center concentration $y_0 = y(x = 0)$ rather than the ordinate at the somewhat arbitrarily chosen collocation point u_1. There are now $N + 1$ unknowns, $y_1, y_2, \ldots, y_N$ and Φ^2. The necessary extra equation may be furnished by the Lagrange interpolation formula where the specified y_0 is expressed as a linear combination of the interior ordinates. Another possibility is to include $u = 0$ as an extra collocation point.

If collocation in x rather than in u is preferred (and this may be advantageous when we wish to have more collocation points close to the center), the boundary condition at $x = 0$ provides the $(N + 1)$ equation. Sensitivity functions ($\partial y_i/\partial y_0$ and $d\Phi^2/dy_0$) are easily evaluated in both cases.

Collocation in x may be preferable to collocation in $u = x^2$ in a certain y_0 interval where the concentration is approximately 1 near the pellet surface and the accumulated heat effect is large enough to ignite the reaction in the center region of the pellet causing y to decrease rapidly. The substitution $u = x^2$ shifts the collocation points out of this region where a larger number of points is necessary to obtain a proper representation of the steep profile.

For moderate values of γ and β, branches A–B, B–C, and the first part of branch C–D in figure 5-4 are easily determined by the step forward process in y_0. For very large values of Φ—i.e., much below the maximum on the C–D branch—the concentration is zero throughout the pellet except in a narrow interval close to $x = 1$. No global polynomial approximation of reasonable order (e.g., $N < 20$) can effectively cope with a concentration profile of this shape and other techniques discussed in chapter 7 must be used. This "ignited" region is, however, of limited interest from a numerical point of view. It is already well known that η is inversely proportional to Φ when Φ increases to infinity, and it is possible to determine a sufficiently large portion of the C–D branch by the present methods to carry the solution far into the region where an asymptotic solution can be used.

5.5.4 More than three solutions—stretching the *x*-coordinate

Here we show how global collocation can be modified to cope with an unusually severe numerical problem. Copelowitz and Aris (1970) and Michelsen and Villadsen (1972) have studied the Weisz–Hicks problem for spherical geometry and large values of γ and β. More than three steady states are obtained and it appears that an unlimited (uneven) number of steady states may be found when γ tends to infinity. For slab and cylinder geomety, no more than three steady states have ever been found.

Figure 5-5 shows the effectiveness factor versus the Thiele modulus curve for spherical geometry and $(\gamma, \beta) = (28, 1)$.

For $0.018 < \Phi < 0.1908$ and for $0.1922 < \Phi < 0.359$, three steady state solutions are found; in the region $0.1908 < \Phi < 0.1922$, however, five steady states exist.

The $\eta(\Phi)$ curve is now composed of five branches:

A–B: The quenched steady state. Φ increases from 0 to 0.359.
B–C: An intermediate region with Φ decreasing to 0.1908.
C–D: An intermediate region with Φ increasing to 0.1922.
D–E: An intermediate region with Φ decreasing to 0.018.
E–F: The ignited steady state with Φ increasing from 0.018 to infinity.

The value of y_0 at the branching points is seen from table 5.3.

TABLE 5.3

CENTER CONCENTRATION AT BRANCH POINTS $(\gamma, \beta) = (28, 1)$

Point	$y_0 = y(x = 0)$
B	0.9345
C	0.445
D	0.33
E	0.003

The branch *C–D* is by far the most difficult to obtain numerically. Close to $x = 1$, a concentration profile from this branch is almost overlapping a concentration profile from branch *A–B*, and the ignition takes place in a very small volume close to the center of the sphere causing y to decrease steeply close to $x = 0$.

The concentration profile for $y_0 = 0.40$ obtained with $N = 17$ collocation points is shown in figure 5-6(a). Table 5.4 shows that not even this high value of N is quite satisfactory.

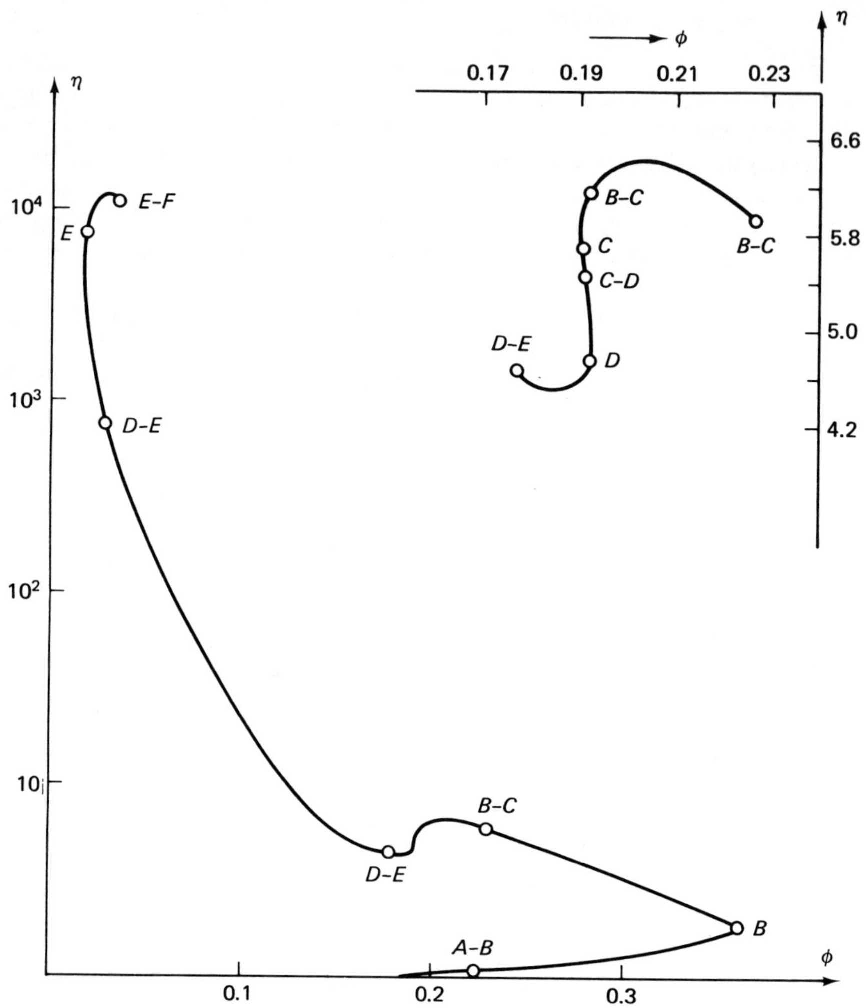

Figure 5-5. $\eta(\phi)$ for spherical pellets; $(\gamma, \beta) = (28, 1)$.

The table shows that Φ increases when y_0 decreases from 0.45 to 0.40—i.e., that $y_0 = 0.40$ belongs to branch C–D, but neither the quantitative values of Φ nor the erratic behavior of η is quite satisfactory from a numerical point of view.

The shape of the concentration profile suggests that it might be easier to represent the profile if the x-coordinate is stretched close to $x = 0$ and compressed close to $x = 1$. This may be accomplished by a simple

TABLE 5.4

CONVERGENCE OF COLLOCATION SOLUTION TO ONE OF THE INTERMEDIATE SOLUTIONS

	Φ			η		
y_0	$N = 16$	$N = 17$	Exact	$N = 16$	$N = 17$	Exact
0.50	0.1915	0.1916	0.1918	6.135	6.124	6.116
0.45	0.1911	0.1905	0.1908	5.717	5.751	5.724
0.40	0.1946	0.1922	0.1912	5.14	5.27	5.30

coordinate transformation of the type

$$u = \frac{(a + 1)x}{1 + ax} \tag{69}$$

where $a > 0$.

The differential equation is transformed into

$$\frac{d^2y}{dx^2} + \frac{2}{x}\frac{dy}{dx} = \frac{[1 + a(1 - u)]^4}{(1 + a)^2}\left(\frac{d^2y}{du^2} + \frac{2}{u}\frac{dy}{du}\right) = \Phi^2 \cdot R(y) \tag{70}$$

or

$$\frac{d^2y}{du^2} + \frac{2}{u}\frac{dy}{du} = \frac{(1 + a)^2}{[1 + a(1 - u)]^4} \cdot \Phi^2 \cdot R(y) \tag{71}$$

The factor to $\Phi^2 R(y)$ is $(1 + a)^{-2}$ at $u = 0$ and $(1 + a)^2$ at $u = 1$, and by a suitable choice of the constant a it is possible to "balance" the rate term such that the nonlinear group on the right-hand side of (71) shows a less extreme variation between $y = 1$ and $y = y_0$.

For example, when a satisfies

$$(1 + a)^4 = \frac{R(y_0)}{R(1)} = R(y_0) \tag{72}$$

the right-hand side of (71) has the same value at $u = 0$ and at $u = 1$. For $y_{x=0} = 0.4$ this criterion yields

$$(1 + a)^4 = R(0.4) = 14{,}526 \quad \text{or} \quad a = 10$$

The "transformation factor" $(1 + a)^2/[1 + a(1 - u)]^4$ increases sharply when u is close to zero and in practice the optimal value of a is slightly smaller than the one derived from the simple criterion (72).

The concentration profile for $y_{x=0} = 0.4$ is replotted as a function of u for $a = 9$ in figure 5-6(b).

Clearly, a smoother profile is obtained, and accurate values of Φ and η are found for a much smaller approximation order N.

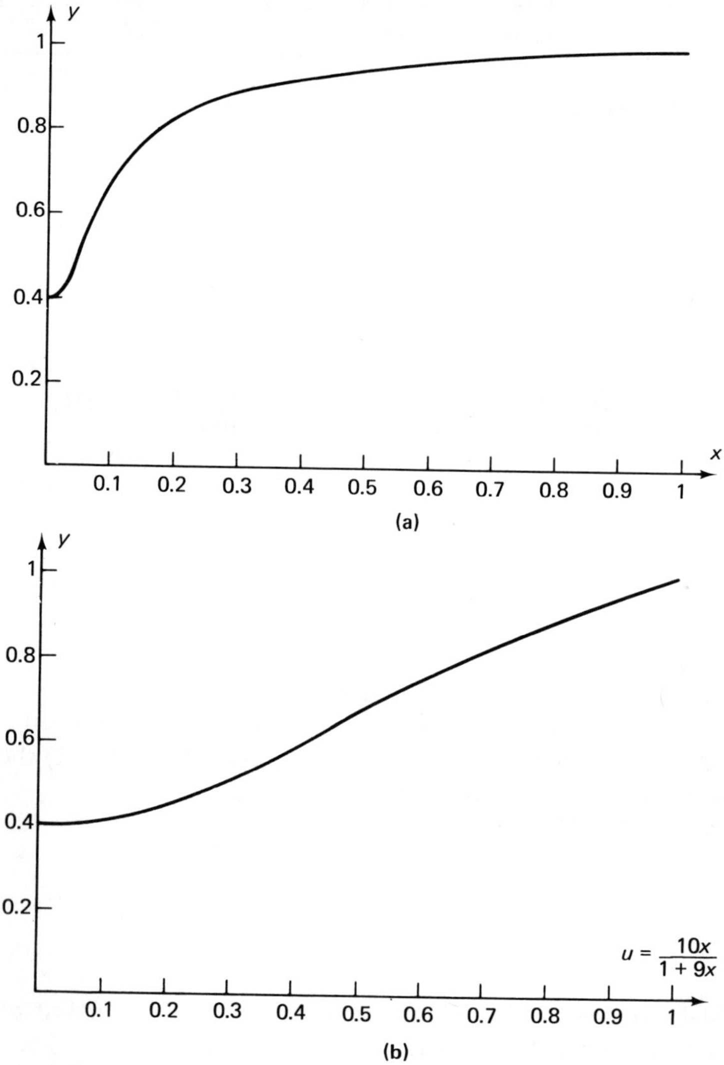

Figure 5-6. Solution of $(d^2y/dx^2) + (2/x)(dy/dx) - \phi^2 R = 0$; $y(x = 0) = 0.4$; $R = \exp\{\gamma\beta(1 - y)/[1 + \beta(1 - y)]\}y$; $(\gamma, \beta) = (28, 1)$.

TABLE 5.5
IMPROVED CONVERGENCE FOR $(y_0, \gamma, \beta) = (0.4, 28, 1)$ $a = 9$ AND $u = 10x/(1 + 9x)$

N	Φ	η
8	0.19107	5.312
10	0.191238	5.3005
12	0.1912234	5.301200
14	0.19122340	5.3012320
16	0.19122340	5.3012320

The solution for $N = 8$ with the "stretched" radial coordinate is more accurate than the solution for $N = 17$ with x as the independent variable.

5.5.5 Solution by an initial value technique for large γ and β

The transformation of subsection 5.5.4 makes it possible to obtain accurate results for the intermediate solutions when (γ, β) are large enough to admit a maximum of five steady state solutions. For larger values of (γ, β) the multiplicity increases further [21 solutions for $(\gamma, \beta) = (80, 2.5)$] and profiles of a weird shape are obtained for the intermediate solutions. In the case $(\gamma, \beta) = (80, 2.5)$ one intermediate profile increases from 0.179 to 0.9996 when x increases from $1.14 \cdot 10^{-10}$ to $6.47 \cdot 10^{-6}$ and from this point the profile is indistinguishable from the profile obtained with $y_0 = 0.9996$.

In this situation no transformation based on global approximation by polynomials is feasible.

The particular structure of equation (61) does, however, make it possible to obtain an accurate solution even for these pathologic profiles.

Introduction of a coordinate transformation $u = \Phi x$ leads to

$$\frac{d^2y}{du^2} + \frac{2}{u}\frac{dy}{du} = R(y) \tag{73}$$

with boundary conditions $dy/du = 0$ at $u = 0$ and $y = 1$ at $u = \Phi$. Again we specify y_0 and treat Φ as an unknown. The solution $y(u)$ to this initial value problem is unique, and (73) can be integrated, e.g., by the very accurate methods of chapter 4, until we reach the value of y where $y(u) = 1$. This terminal u-value is then the Thiele modulus that corresponds to the specified y_0.

Proceeding as in subsection 4.2.5, we obtain (74) for integration from u_0 to $u_0 + h$:

$$\frac{dy_1}{du} = y_2 \qquad u\frac{dy_2}{du} = -2y_2 + u \cdot R(y_1) \tag{74}$$

Let $y(u_0) = y_{10}$ and $(dy/du)_{u_0} = y_{20}$.

Introducing $u = u_0 + v \cdot h$, where h is the current step length, we obtain

$$\frac{dy_1}{dv} = hy_2 \qquad (u_0 + vh)\frac{dy_2}{dv} = -2hy_2 + h(u_0 + vh)R(y_1) \tag{75}$$

Discretization by the Gauss or Radau method yields

$$\begin{aligned} \mathbf{A} \cdot \mathbf{y}_1 + \mathbf{A}_0 \cdot y_{10} &= h \cdot \mathbf{y}_2 \\ (\mathbf{U}_0 + h\mathbf{V})(\mathbf{A}\mathbf{y}_2 + \mathbf{A}_0 \cdot y_{20}) &= -2h\mathbf{y}_2 + h(\mathbf{U}_0 + \mathbf{V} \cdot h) \cdot \mathbf{R}(\mathbf{y}_1) \end{aligned} \tag{76}$$

where $\mathbf{y}_1$ is the concentration at the N collocation points in the current integration interval, and $\mathbf{y}_2$ the value of the derivative at these points.

From the first equation of (76),

$$\mathbf{y}_2 = \frac{1}{h}(\mathbf{A}\mathbf{y}_1 + \mathbf{A}_0 \cdot y_{10}) \tag{77}$$

which is substituted into the second equation to give N nonlinear equations in the N interior y_{1i} ordinates $(y_{11}, y_{12}, \ldots, y_{1N})$.

A variable step length should be used with small steps in a region where $R(y)$ is large. The shortest computing time for a given accuracy is obtained with fairly small steps and only two or three interior points in each subinterval.

Parameter values up to $(\gamma, \beta) = (80, 2.5)$, for which 21 steady state solutions exist in a narrow range of the Thiele modulus, have been used without any numerical difficulties.

The sensitivity function $z(u) = \partial y(u, y_0)/\partial y_0$ can be found at hardly any extra cost. z is given by

$$\frac{d^2z}{du^2} + \frac{2}{u}\frac{dz}{du} = \left(\frac{\partial R}{\partial y}\right)_u z = R_y z \tag{78}$$

with initial conditions

$$z(u = 0) = \left.\frac{\partial y}{\partial y_0}\right|_{u=0} = 1 \quad \text{and} \quad \left.\frac{dz}{du}\right|_{u=0} = 0 \tag{79}$$

After each integration step, y and consequently also R_y are known at the collocation points. The coefficient matrix for collocation solution of the linear differential equation (78) from u_0 to $u_0 + h$ is simply the Jacobi matrix that was used in the final iteration on (76) and z is found at the collocation points of the current u interval by one additional back substitution.

For a given y_0 the value of u when $y = 1$ determines Φ while (in a very detailed notation)

$$\eta = \frac{3}{\Phi^2}\left(\frac{\partial y}{\partial x}\right)_{y_0}\Bigg|_{x=1} = \frac{3}{\Phi^2}\left(\frac{\partial y}{\partial u}\right)_{y_0}\Bigg|_{u=\Phi} \cdot \frac{\partial u}{\partial x} = \frac{3}{\Phi}\frac{\partial y}{\partial u}\Bigg|_{u=\Phi}$$

The parametric representation of η and Φ with y_0 as parameter is immediately available: y_0 decreases from 1 toward 0 along the curve $\eta(\Phi)$ in the direction A–D in figure 5.4 or A–F in figure 5.5.

Michelsen and Villadsen (1972) pursued the parametric representation even further and obtained the first derivative of Φ and η with respect to y_0. Their result is stated here without proof:

$$\frac{d\Phi}{dy_0} = -\frac{3z(x = 1)}{\Phi\eta} \tag{80}$$

$$\frac{d\eta}{dy_0} = \frac{3}{\Phi^2}\left[\frac{dz}{dx}\Bigg|_{x=1} + 3z(x = 1)\frac{\eta - 1}{\eta}\right] \tag{81}$$

The solutions of (78) and (79) and the two derivatives (80) and (81) are useful in subsection 5.5.6.

5.5.6 Stability of steady state solutions

Solution of the nonlinear transient equations (1.62) and (1.63) corresponding to the steady state equations of this chapter will not be attempted in this text [except as an exercise in chapter 9, (exercise 9.5)] since the mass of detail would confuse the numerical principles we wanted to demonstrate. Collocation in the spatial coordinate combined with one of the forward integration methods of chapter 8 does, however, appear to be a very favorable numerical scheme.

The linearized equations (1.82) are solved in chapter 9 for an arbitrary value of the capacitance parameter Le when $\theta = y = 1$ at $x = 1$, that is, for the case when a linear relationship $\theta_s = 1 + \beta(1 - y_s)$ holds between temperature and concentration for every point on the steady state profile.

If Le $= 1$ and $\theta = y = 1$ at $x = 1$, this linear relation also holds between θ and y, the solutions to (1.62) and (1.63). θ can be eliminated from (1.62) before the linearization that now leads to a single, linear partial differential equation in the deviation variable $\hat{y}(x, \tau)$ from a steady state profile $y_s(x)$.

Separation of variables gives

$$\hat{y}(x, \tau) = \sum_{1}^{\infty} A_k v_k(x) \exp(\lambda_k \tau)$$

where the eigenfunctions $v_k(x)$ are solutions of

$$\lambda v = \nabla^2 v - \Phi^2 \left(\frac{\partial R}{\partial y}\right)_{y_s} v \tag{82}$$

$$v(1) = \left.\frac{dv}{dx}\right|_{x=0} = 0 \tag{83}$$

and

$$\left(\frac{\partial R}{\partial y}\right)_{y_s} = \left(\left\{1 - \frac{\gamma\beta y}{[1 + \beta(1 - y)]^2}\right\} \exp\left[\frac{\gamma\beta(1 - y)}{1 + \beta(1 - y)}\right]\right)_{y = y_s} \tag{84}$$

Since $\partial R/\partial y|_{y_s}$ is a known function of x when the steady state solution $y_s(x)$ has been found, (82) is a linear eigenvalue problem with distinct, real eigenvalues λ_k. Furthermore, if all λ_k are negative, the investigated steady state $y_s(x)$ is asymptotically stable ($\hat{y} \to 0$ for $\tau \to \infty$) for sufficiently small initial deviation $\hat{y}(x, 0)$ from the steady state.

Evaluation of the eigenvalues of (82) is treated in chapter 9 as a specific case of the solution of equation (1.82) but the qualitative aspects of the solution of (82) are of interest in the present context giving further insight into the nature of the steady state solutions. As an example, consider spherical geometry, and let the value of the eigenfunctions be 1 at $x = 0$. The trivial case $\Phi = 0$ gives $\lambda_k = -k^2\pi^2$ and $v_k = (\sin k\pi x)/k\pi x$.

Let us first consider a rate expression $R(y)$ with $\gamma\beta < 1$. Here $\partial R/\partial y|_{y_s} > 0$ for any x on any profile $y_s(x)$. The eigenvalues of (82) are smaller (i.e., more negative) than the corresponding eigenvalues $-\pi^2k^2$ of the pure diffusion problem $\nabla^2 v = \lambda v$ by at least $\Phi^2 \min_x \partial R/\partial y|_{y_s}$. Thus, for any Φ in the open interval $(0, \infty)$, endothermic, isothermal, or slightly exothermic reactions are characterized by steady states with eigenvalues that are even more negative than the diffusion eigenvalues. This means that $\hat{y}(x, \tau)$ will decrease faster to zero than if no reaction occurred.

Next consider the case $\gamma\beta > 1$. Now $(\partial R/\partial y)|_{y_s}$ is negative for all x when $y_0 = y(x = 0) > y_{\max}$, the value of y at the maximum of $R(y)$. Such steady states must all have eigenvalues larger than $-k^2\pi^2$.

If on the other hand y_0 is so small that $y_s(x) < y_{\max}$ in the major part of the x-interval $(0, 1)$, then each eigenvalue λ_k will again become smaller than the corresponding diffusion eigenvalue $-k^2\pi^2$. This certainly happens when Φ is so large (or y_0 so small) that the reaction occurs almost entirely on the pellet surface.

The maximum value of $-\partial R/\partial y$ determines whether the eigenvalue λ_1 strays past 0 to become positive for steady states characterized by center concentrations in a certain y_0-interval $y_{02} < y_0 < y_{01}$. The larger the value of γ and β, the more probable is the occurrence of nonstable steady states. It may even happen that the second (λ_2) or more of the small-magnitude eigenvalues become positive in y_0 intervals inside $[y_{02}, y_{01}]$ but the large-magnitude eigenvalues will always remain negative and close to $-k^2\pi^2$ since $-\partial R/\partial y$ is finite for finite γ and β.

As a numerical method of determining $v_k(x)$ and λ_k for any given Φ, the forward integration technique of section 4.5 is used. First a trial value $\lambda = 0$ is inserted on the left-hand side of (82) and

$$\nabla^2 v - \Phi^2 \frac{\partial R}{\partial y}\bigg|_{y_s} v = 0 \tag{85}$$

is integrated from $x = 0$ ($v = 1$ and $dv/dx = 0$) to $x = 1$. If $v(\lambda = 0, x) > 0$ for all $x \in [0, 1]$, then by the arguments of section 4.5 all eigenvalues λ_k of (82) are negative. If on the other hand $v(x)$ has k sign changes in the open interval $x \in (0, 1)$, then k eigenvalues are positive.

A comparison of (85) and (83) (for spherical geometry) and (78) and (79) shows that the two problems are identical except for the scalar factor Φ^2 that has been incorporated into the independent variable of (78).

Thus if the sensitivity function $z(u)$ corresponding to a given steady state $y_s(x)$ has k sign changes in $(0, \Phi)$, then the eigenvalue problem (82) has k positive eigenvalues for that particular $y_s(x)$.

We note from (80) that each point on the $\Phi(y_0)$ curve where $d\Phi/dy_0$ is zero—the so-called bifurcation points—is characterized by $z(1) = 0$ and consequently that one of the eigenvalues λ_k of (82) is zero at these points.

It now becomes an easy task to trace the sign of the small eigenvalues along the $\eta(\Phi)$ curve in the direction of decreasing y_0.

First, if $d\Phi/dy_0$ remains negative for all $y_0 \in [0, 1]$, then all steady states are stable since λ_1 never reaches 0 before it turns back.

At point B in figures 5-4 and 5-5, $d\Phi/dy_0 = 0$ and $\lambda_1 = 0$. If η continues to increase with decreasing y_0 as in figure 5-4, then at point C in that figure $d\Phi/dy_0$ is again zero and $dz/dx|_{x=1}$ is negative. During the passage from B to C the point of intersection of $z(x)$ with $z = 0$ has retracted from $x = 1$ and then advanced again to $x = 1$ and λ_1 is positive on this branch of $\eta(\Phi)$. On branch C–D, λ_1 again becomes negative and we conclude that branch A–B and C–D are stable.

At point C, in figure 5-5, $d\Phi/dy_0 = 0$, while $d\eta/dy_0 > 0$. $z(1) = 0$ at point C, and by equation (81), $(dz/dx)|_{x=1} > 0$ at this value of y_0. $z(x)$, which is positive at $x = 0$, must have one interior zero besides the zero at $x = 1$. We conclude that it is the second eigenvalue λ_2 that is zero at point C. On branch C–D in figure 5-5, λ_1 and λ_2 are both positive.

At point D, $d\eta/dy_0$ is still positive and we conclude that λ_2 has returned to zero at this point. Finally, at point E, λ_1 retracts past 0 since $d\eta/dy_0$ is positive at this point. No bifurcation points are encountered when y_0 decreases from its value at point E, and the upper branch E–F corresponds to stable solutions.

Our final conclusion is that one eigenvalue is zero at each of the bifurcation points B, C, D, and E. At any bifurcation point where $d\eta/dy_0 < 0$, an eigenvalue with uneven index passes zero and at all points where $d\eta/dy_0 > 0$, an eigenvalue with even index passes zero. If $\eta(\Phi)$ consists of $2k - 1$ branches connecting bifurcation points with alternating sign of $d\eta/dy_0$, then k eigenvalues become positive and there are $2k + 1$ steady state solutions for certain values of Φ. Only two steady states, the "quenched" and the "burnt out" steady states, are stable irrespective of the number of branches of $\eta(\Phi)$.

The question of stability of the steady state solutions of the Weisz–Hicks problem can be finally settled for Le = 1. In chapter 9, it is seen that very few general results can be derived analytically for the stability of a steady state when Le $\neq$ 1. The reason is that a nonself-adjoint eigenvalue problem appears instead of the easily handled Sturm–Liouville problem (82).

EXERCISES

1. Solve the problem of isothermal reaction and diffusion on spherical catalyst pellets using

$$R(y) = \frac{K_1 y}{1 + K_2 y^2}$$

Derive the boundary of multiple steady states in the (K_1, K_2) plane $(K_2 \geq 0)$.

2. The global collocation procedure works well for the example in the text and for Exercise 1 above.

The following examples will present problems:

a. $\dfrac{d^2y}{dx^2} - 10^6 y^2 = 0, \qquad y(1) = 1, \qquad y^{(1)}(0) = 0$

b. $\dfrac{d^2y}{dx^2} - \Phi^2 y^{1/5} = 0, \qquad y(1) = 1, \qquad y^{(1)}(0) = 0$

Attempt to compute the solution of parts a and b. What should be the solution to part b for large enough Φ?

Better procedures for these problems are discussed in chapter 7.

3. Consider the Weisz–Hicks problem for spherical particles and $\gamma = 20$, $\beta = 0.05$.

a. Draw a sketch of $R(\theta)$ for $1 \leq \theta \leq 1 + \beta$.

b. Construct a lower bound $u_1(x)$ for the temperature profile $\theta(x)$ when $\Phi = 1$. The iterative chord process of equation (44) is used to compute θ^*.
c. Construct an upper bound $u_2(x)$ by a tangent process using either $\theta_* = 1, \theta^*$, or $\frac{1}{2}(1 + \theta^*)$ as the point of tangency.
d. Compute $\theta(x)$, $u_1(x)$, and all three cases of $u_2(x)$; compare the solutions at $x = 0, 0.1, 0.2, \ldots, 0.9$. At least seven digits must be correct to show that $u_1(x) < \theta(x) < u_2(x)$ for $x < 1$.
e. Finally, compute η from u_1, u_2, and θ; investigate whether $\eta(u_1) < \eta(\theta) < \eta(u_2)$.

4. In equation (73), y is a function of u and of y_0. Write the total derivative of y. u is again a function of x and of Φ. At $u = \Phi$, $y = 1$ irrespective of the value of y_0 and, consequently, $dy = 0$ at this point.
a. Derive formula (80) using this information.
b. dy/du is also a function of u and of y_0. Write the total derivative of this function and use (80) to derive (81).
Each step should be carefully explained to obtain the full benefit of this exercise.

5. The separation curve $G_{3,5}$ between three and five steady state solutions to the Weisz–Hicks problem is a function of γ and β as shown in Michelsen and Villadsen (1972).
In this exercise we wish to introduce yet another parameter, s = the geometry factor, into $G_{3,5}$. For $s = 2$ and $\beta = 1$, the bifurcation point is at a γ-value slightly below 28.
If $G_{3,5}(s, \gamma, \beta)$ passes through $(1, \gamma, \beta)$ for some finite values of γ and β, it has been proved that more than three steady states can occur in cylinder geometry. To investigate this the following procedure is proposed:
a. Write a computer program following the method of subsection 5.5.5 and compute the $\eta(\Phi)$ curve in figure 5-5.
b. When the program has been checked in part a, it is next used to find the γ value corresponding to $G_{3,5}(2, \gamma, 1)$. This bifurcation γ value is γ_2.
c. The program is now applied for $s = 1.95$ to find $\gamma_{1.95}$.
d. Extrapolation from γ_2 and $\gamma_{1.95}$ gives a starting point for determination of $\gamma_{1.9}$. When $\gamma_{1.9}$ has been found, parabolic extrapolation gives a starting value for $\gamma_{1.85}$, etc.
e. Finally a curve $G_{3,5}$ $(s, \gamma, 1)$ appears and if $\gamma_s \to \infty$ for some $s = s_{ass} < 1$, no more than three steady states exist for $s < s_{ass}$ and $\beta = 1$. β can now be increased from 1 to β_1 to investigate whether $G_{3,5}$ (s, γ, β_1) exists below s_{ass}.
f. If a global asymptote $s^*_{ass} > 1$ appears for large β, the hypothesis of the introduction must be rejected.

6. A nonlinear problem totally different from those studied in the text was solved by orthogonal collocation in Christiansen and Fredenslund (1975). Discuss their method for a thermodynamic consistency test. Is the problem a reformulated boundary value problem? If the answer is affirmative, eventual peculiarities of the boundary conditions should be discussed.
Does the collocation approach seem satisfactory or should an alternative procedure be suggested?

7. Discuss the GPM method applied to nonisothermal reaction on catalyst pellets as described in Kubiçek and Hlavaček (1972a). Compare the method to the methods discussed in the text.

REFERENCES

In section 5.2 we have only just touched upon a subject that has been given enormous attention during the last few decades, both from the engineering point of view and from a more formal mathematical standpoint.

Most of the chemical engineering work on the topic is centered on multiplicity of steady states in various reacting systems; the first to compute the effectiveness factor versus Thiele modulus for large enough $\beta \cdot \gamma$ to give three steady states are Weisz and Hicks in their now famous 1962 paper. The list of references adjoined to Aris' (1975, chapter 6) review of existence and uniqueness of the steady state is probably complete up to 1973–1974, and he also includes many fundamental studies on topology, e.g., Cronin (1964). Another classic on the functional analysis approach to ordinary differential equations is Hille (1969), based on a lifetime of research. Davis (1962) contains many examples from nonlinear mathematics and is in general very readable. He traces the Nth-order chemical reaction with diffusion problem back to Emden's differential equation that originated in astrophysics. Newer collections of nonlinear differential equations are found in the proceedings of a Battelle Summer Institute (1972). The biological examples of this reference may be of special interest to engineers who are accustomed to studying nonlinearity for catalyst pellet problems.

Comparison differential equations were explained in rather simple terms in a tract by Bailey, Shampine, and Waltman (1968). Their examples are often trivial, however, and the subject can certainly be further explored. The most exhaustive treatment of the subject is probably Lakshmikantham and Leela (1969), which is certainly not easy reading.

A forward integration technique for iterative solution of the Weisz–Hicks problem was in fact used in the original (1962) paper, where Weisz and Hicks integrated the equations by the Runge–Kutta method. Parameter sensitivity functions were used by Kubiçek and Hlavaček (1972a) for the Weisz–Hicks problem. In Exercise 7, their GPM (general parameter mapping) method is compared to the method in subsections 5.5.2 and 5.5.3.

Further references to the GPM method are Kubiçek and Hlavaček (1972b) and (1973), (1974).

Copelowitz and Aris (1970) first discovered the many steady states of a spherical catalyst pellet when $\gamma\beta$ is extremely large. Michelsen and Villadsen (1972) studied the phenomenon in more detail and they discussed the stability of the middle steady states. The (γ, β) of subsection 5.5.4 are those used in the latter reference.

1. ARIS, R. *The Mathematical Theory of Diffusion and Reaction in Permeable Catalysts.* Oxford: Clarendon Press (1975).
2. CRONIN, J. "Fixed Points and Topological Degree in Mathematical Analysis," AMS, Providence (1964).
3. HILLE, E. *Lectures on Ordinary Differential Equations.* Reading, Mass.: Addison Wesley (1969).
4. DAVIS, H. T. *Introduction to Non-linear Differential and Integral Equations.* New York: Dover (1962).
5. "Nonlinear Problems in the Physical Sciences and Biology," Proceedings of a Battelle Summer Institute (Seattle, July, 1972), Springer Verlag (1973), as *Lecture Notes in Mathematics* no. 322.
6. BAILEY, P. B., SHAMPINE, L. F., and WALTMAN, P. E. *Nonlinear Two Point Boundary Value Problems.* New York: Academic Press (1968).
7. LAKSHMIKANTHAM, V., and LEELA, S. *Differential and Integral Inequalities.* New York: Academic Press (1969).
8. WEISZ, P. B. and HICKS, J. S. *Chem. Eng. Sci.* 17 (1962):265.
9. LUSS, D. *Chem. Eng. Sci.* 24 (1969):879.
10. KUBIÇEK, M. and HLAVAČEK, V. *Chem. Eng. Sci.* 27 (1972):743.
11. KUBIÇEK, M. and HLAVAČEK, V. *Chem. Eng. Sci.* 27 (1972):2095.
12. KUBIÇEK, M. and HLAVAČEK, V. *J. Inst. Maths. Applics.* 12 (1973):287.
13. KUBIÇEK, M., HLAVAČEK, V., and JELINEK, J. *Chem. Eng. Sci.* 29 (1974):435.
14. COPELOWITZ, I. and ARIS, R. *Chem. Eng. Sci.* 25 (1970):906.
15. MICHELSEN, M. L. and VILLADSEN, J. *Chem. Eng. Sci.* 27 (1972):751.
16. CHRISTIANSEN, L. J. and FREDENSLUND, Aa. *AIChE Journal* 21 (1975):49.

One-Point Collocation 6

Introduction

A number of disparate subjects is treated in this chapter. In sections 6.1 and 6.2 a somewhat arbitrary distinction has been chosen between one-point collocation results for ordinary and partial differential equations. In section 6.3 it is proved that the one-point predictor-corrector collocation method based on the zero of $P_1^{(1,0)}(x)$ has an error $\mathcal{O}(h^4)$ when applied to the solution of a general nonlinear initial value problem.

It is perhaps more appropriate to emphasize that sections 6.1 and 6.2 treat different methods of model simplification (or "lumping of parameters"). In section 6.1 the boundary value problem that describes the steady states in a catalyst pellet is "lumped" by different methods. Very convenient and quite accurate methods result and it is even possible to solve the catalyst pellet model graphically. In this respect the one-point collocation method may appeal more directly to the engineer than the computer-oriented methods of chapters 8 and 9.

Simplification of the partial differential equation models of section 6.2 may lead to either a first-order differential equation (the parabolic equations of subsection 6.2.2) or to sets of algebraic equations (elliptic equations of subsection 6.2.1). In subsection 6.2.2 it becomes apparent that the one-point collocation method is only one way of obtaining a simplified model. More general methods based on perturbation techniques are developed to illustrate just what is being obtained in terms of accuracy by means of one-point collocation. In subsection 6.2.2 the subject matter of the present chapter is tied to the discussion of approximate models in chapter 1 and to the perturbation techniques that are introduced in chapter 2 and are used throughout chapter 9.

6.1 Application of One-Point Collocation to Ordinary Differential Equations

6.1.1 One-point collocation constants

First-degree orthogonal polynomials with arbitrary α and β are constructed by means of equation (3.9):

$$P_1(u) = \frac{\alpha + \beta + 2}{\beta + 1} u - 1 \tag{1}$$

$$P_1(u) = 0 \quad \text{for } u = u_1 = \frac{\beta + 1}{\alpha + \beta + 2} \tag{2}$$

A collocation method based upon the zeros of $P_N^{1,(s-1)/2}(u)$, where s (=0, 1, or 2) is the geometry factor, has been recommended in chapters 2 to 5 when the purpose of the calculations is to obtain an accurate value of the effectiveness factor

$$\eta = \frac{s+1}{2} \int_0^1 R(y) u^{(s-1)/2}\, du \tag{3}$$

and $y(u)$ is determined from

$$4u \frac{d^2 y}{du^2} + 2(s+1) \frac{dy}{du} - \Phi^2 R(y) = 0 \tag{4}$$

with $y(1) = 1$ and $u = x^2$. For $N = 1$, the collocation point u_1, the weights of a quadrature formula

$$\eta = w_1 R[y(u_1)] + w_2 R[y(1)], \tag{5}$$

weights in an interpolation formula for $\nabla^2 y$ at u_1, and the Fourier coefficient c_1 in

$$y_1(u) = 1 + c_1(1 - u) \tag{6}$$

are calculated from (2), (3.83), and (2.91).

Collocation constants for the three geometries [equation (4) with $s = 0$, 1, and 2] are collected in table 6.1.

TABLE 6.1
COLLOCATION CONSTANTS FOR $N = 1$ AND SYMMETRIC PROBLEMS

s	$u_1 = x_1^2$	w_1	w_2	$-C_{11} = C_{12}$	$c_1 = -(1 - y_1)/(1 - u_1)$
0	$\frac{1}{5}$	$\frac{5}{6}$	$\frac{1}{6}$	$\frac{5}{2}$	$-\frac{5}{4}(1 - y_1)$
1	$\frac{1}{3}$	$\frac{3}{8}$	$\frac{1}{8}$	6	$-\frac{3}{2}(1 - y_1)$
2	$\frac{3}{7}$	$\frac{7}{30}$	$\frac{1}{10}$	$\frac{21}{2}$	$-\frac{7}{4}(1 - y_1)$

6.1.2 One-point collocation for effectiveness factor problems

The one-point collocation version of (4) is

$$\frac{-C_{11}}{\Phi^2}(-y_1 + 1) = R(y_1) \tag{7}$$

where $-C_{11}$ is $2(s + 1)/(1 - u_1)$ or 2.5, 6, and 10.5 in the three geometries.

Stewart and Villadsen (1969) plotted the straight line (7) and the nonlinear function $R(y)$ as functions of y. The intersection of the two curves, i.e., the point where the diffusion rate is equal to the reaction rate, determines y_1. Now η, as well as $y_1(u)$, is easily determined from (5) and (6).

A typical result for spherical geometry and a first order irreversible reaction of the Arrhenius type ($\gamma = 30$, $\beta = 0.1$) is shown in figure 6.1.

Multiple solutions are possible for larger γ and β as shown in the construction on figures 6.2a. ($\gamma = 30$, $\beta = 0.2$), where limiting Φ-values

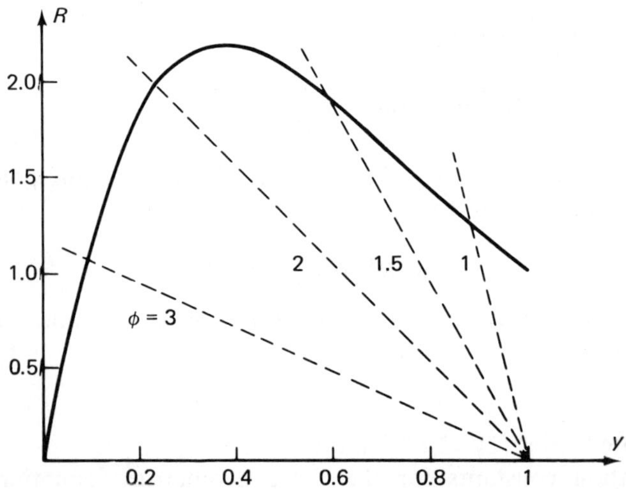

Figure 6-1. Graphical solution of equation (6.7) for spherical symmetry and $(\gamma, \beta) = (30, 0.1)$.

(bifurcation solutions) can be found from the slope of the two tangents to R that pass through $[y, R(y)] = (1, 0)$. A reformulation of (7) is, how-

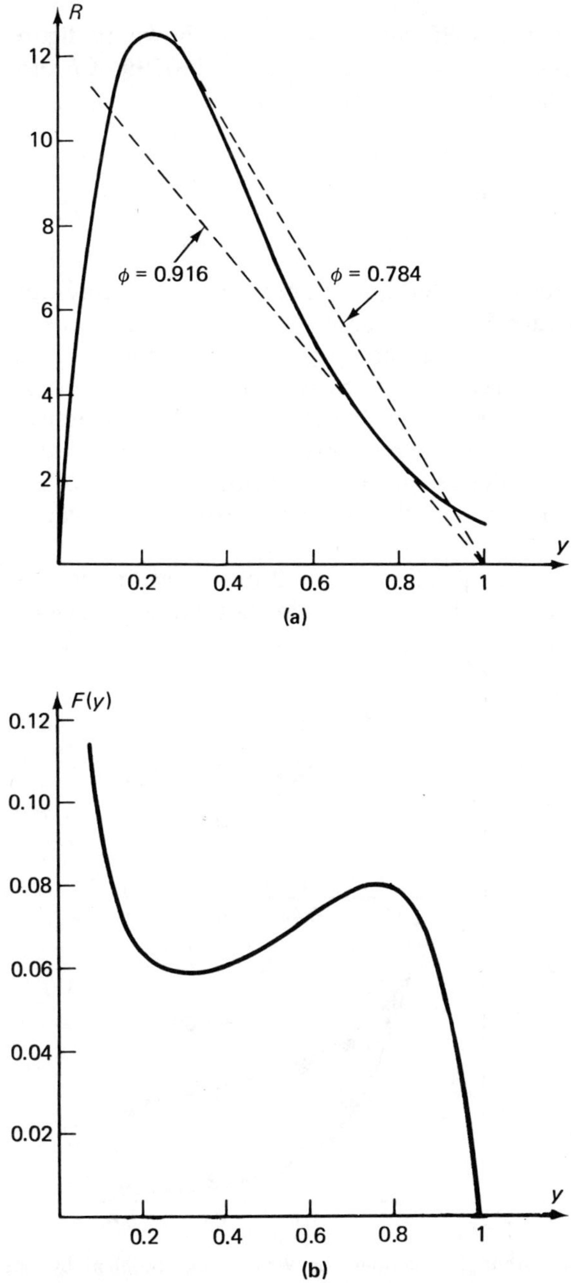

Figure 6-2. Graphical solution of equation (6.7) and the function $F(y)$ of equation (6.8); $(\gamma, \beta) = (30, 0.2)$.

ever, more convenient for manual calculations.

$$\frac{\Phi^2}{-C_{11}} = \frac{1 - y_1}{R(y_1)} = F(y_1) \tag{8}$$

A plot of F for $\gamma = 30$ and $\beta = 0.2$ is shown in figure 6.2b. The extremum values $F_{\min} = 0.0586$ and $F_{\max} = 0.0799$ of F directly give the Φ-interval in which multiple solutions exist

$$-C_{11}F_{\min} < \Phi^2 < -C_{11}F_{\max}$$

or, for spherical geometry,

$$0.784 < \Phi < 0.916$$

The predicted region of multiple solutions is remarkably well determined as seen from figure 6.3. The large Φ behavior of the $\eta(\Phi)$ curve cannot possibly be correctly represented by (7) and (5) since $\eta \rightarrow w_2$ for $y_1 \rightarrow 0$ in (5), while the true effectiveness factor decreases to zero when Φ increases to infinity and $y \rightarrow 0$ for all $u < 1$. This qualitative discrepancy at large Φ between the approximate η and its true value is of small consequence since the asymptotic behavior of $\eta(\Phi)$ can be predicted by the method of Bischoff (1967) for any particle shape and reaction rate expression.

Returning to the question of multiplicity we noticed that multiple solutions are obtained in a certain Φ-interval provided F has an

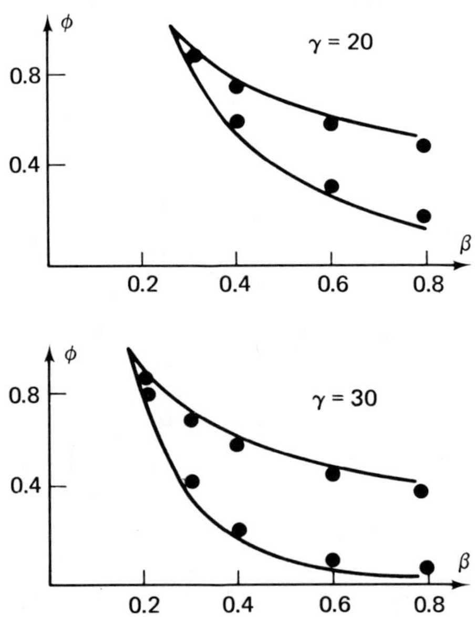

Figure 6-3. Multiple solutions for Weisz–Hicks problem by one-point collocation. Exact region: closed circles.

extremum for $y \in (0, 1)$. This observation can be put into quantitative form. Differentiate F with respect to y_1:

$$\frac{dF}{dy_1} = -\frac{1}{[R(y_1)]^2}\left[R(y_1) + (1 - y_1)\frac{dR(y_1)}{dy_1}\right] \tag{9}$$

For

$$R(y) = y \exp\left[\gamma\frac{\beta(1 - y)}{1 + \beta(1 - y)}\right]$$

(9) is reduced to

$$\frac{dF}{dy_1} = 0 \quad \text{when } [1 + \beta(1 - y_1)]^2 - \gamma\beta y_1(1 - y_1) = 0$$

or

$$\frac{dF_1}{dy_1} = 0 \text{ at the zeros of a quadratic}$$

$$P_2(y_1) = \beta(\gamma + \beta)y_1^2 - \beta[\gamma + 2(1 + \beta)]y_1 + (1 + \beta)^2$$

The discriminant of $P_2(y_1)$ is negative if

$$\beta^2[\gamma + 2(1 + \beta)]^2 - 4\beta(\gamma + \beta)(1 + \beta)^2 < 0$$

or

$$\frac{\gamma\beta}{1 + \beta} < 4 \tag{10}$$

A unique steady state is consequently found for all values of Φ if (10) is satisfied. We note that the criterion (10) depends neither on the geometry factor s nor on the choice of collocation point.

If, on the other hand, (10) is not satisfied, P_2 has two zeros, y_{11} and y_{12}, which are easily shown to be located in (0, 1). When these two y_1-values are substituted into (8) approximate Φ-limits for multiple steady states are obtained. These limits do indeed depend on the geometry factor s and on the choice of collocation point. If the collocation point is properly chosen (e.g., as shown in table 6.1 for $s = 0, 1,$ and 2) the approximate Φ-interval for multiple solutions is close to that which would be found by a high order approximation method as described in chapter 5.

Luss (1968) has derived equation (10) by an elegant analytical method. The main points of his proof will be presented here partly as an application of the sensitivity function z of subsection 5.5.5, but also to show how closely his result is connected with the one point collocation result.

The steady state equations for the profile $y(x)$ and for the sensitivity function $z(x) = (\partial y/\partial y_0)_x$ are given by

$$\nabla^2 y - \Phi^2 R(y) = 0 \tag{11}$$

$$\nabla^2 z - \Phi^2 R_y z = 0 \tag{12}$$

with side conditions $y(0) = y_0$, $y'(0) = 0$, $y(1) = 1$, $z(0) = 1$, $z'(0) = 0$.

Again Φ is given implicitly as a function of y_0 as described in subsection 5.5.5.

Consider the first bifurcation point, i.e., the largest value y_0^* of y_0 where $z(1) = 0$. We saw in subsection 5.5.6 that if no y_0^* exists for $1 > y_0 > 0$, then the steady state solution is unique for all Φ, but here we assume that y_0^* exists, and we know that $z(x) > 0$ for all x when $y_0 > y_0^*$ since y_0^* is the first bifurcation point.

The solution of (10) and (11) with $y_0 = y_0^*$ is $[y^*(x), z^*(x)]$. Multiply (11) by z^* and (12) by $(1 - y^*)$ and add (11) to (12). Integrate over the pellet volume to obtain

$$I_1 = \int_0^1 [z^*\nabla^2 y^* + (1 - y^*)\nabla^2 z^*]x^s\,dx$$

$$= \int_0^1 z^* x^s [R(y^*) + (1 - y^*)R_y(y^*)]\Phi^2\,dx = I_2$$

Partial integration of I_1 yields

$$I_1 = \left[z^*\frac{dy^*}{dx} + (1 - y^*)\frac{dz^*}{dx}\right]x^s\Bigg|_0^1 - \int_0^1 x^s\left(\frac{dz^*}{dx}\frac{dy^*}{dx} - \frac{dy^*}{dx}\frac{dz^*}{dx}\right)dx$$

I_1 is zero since $z^*(1) = 0$, $y^*(1) = 1$, and both derivatives are zero at $x = 0$.

Thus at the first bifurcation point

$$I_2 = \int_0^1 z^*(x)x^s\Phi^2\{R(y^*) + [1 - y^*(x)]R_y(y^*)\}\,dx = 0 \tag{13}$$

When (9) (with y^* instead of y_1) is inserted in I_2 one obtains

$$I_2 = \int_0^1 z^*(x)x^s\Phi^2\{R[y^*(x)]\}^2\frac{dF}{dy}\,dx = 0 \tag{14}$$

The only factor in the integrand of (14) that can change sign is dF/dy and uniqueness is consequently guaranteed for all Φ provided F has no extremum in $(0, 1)$. This, for a first order irreversible reaction, leads precisely to the inequality (10) that was obtained by one point collocation.

Aris (1975), chapter 6, uses approximately the same lumping procedure to compare a pellet and a back-mix reactor. Equation (10) is a sufficient condition for uniqueness of the solution of a back-mix reactor model. We have seen that it also provides a sufficient condition for uniqueness in the one-point collocation approximation of a pellet.

The exact boundary $G(\gamma, \beta)$ between one and three steady states is much more complicated than (10), but numerical calculations by Michelsen and Villadsen (1972) do show that (10) with the constant 4 changed to 4.3 very nearly describes $G(\gamma, \beta)$ over a large range of γ and β. In the extreme case $\beta \to \infty$, the vertical asymptote of $G(\gamma, \beta)$ is $\gamma = 4.187$, while (10) gives $\gamma = 4$.

Any lumping of a distributed system such as the one-point collocation approximation erases the fine structure of the system. Thus no more than three steady states can ever be found by one-point collocation applied to the Weisz–Hicks problem, although as seen in chapter 5 the problem admits to many more solutions for certain values of γ, β. The fine structure can only be detected by higher-order approximation.

6.1.3 The concept of a "burnt-out" and a "reaction" zone

Several attempts have been made to extend the applicability of the one-point collocation method for equation (4) to large values of Φ. Paterson and Cresswell (1971) proposed the following model simplification:

$$\begin{aligned} &y \text{ given by (4)} \quad \text{for } x \geq x_p \text{ (the ``reaction zone'')} \\ &y = 0 \quad \text{and} \quad \frac{dy}{dx} = 0 \quad \text{for } x \leq x_p \end{aligned} \tag{15}$$

They applied this concept of an "exterior reaction zone" to a case where heat and mass transfer resistance outside the pellet was included in the model, but the principle of their method can just as well be analyzed using the simpler boundary condition $\theta = y = 1$ at $x = 1$. A new variable

$$\xi = \frac{x - x_p}{1 - x_p}$$

is introduced in the mass balance:

$$\frac{d}{dx}\left(x^s \frac{dy}{dx}\right) - \Phi^2 x^s R(y) = 0$$

or

$$\frac{1}{(1 - x_p)^2} \frac{d^2 y}{d\xi^2} + \frac{s}{x_p + (1 - x_p)\xi} \frac{dy}{(1 - x_p)\, d\xi} = \Phi^2 R(y) \tag{16}$$

The collocation point is chosen at $\xi = \xi_1$. The form of the parabolic approximation for y is immediately settled by the imposed boundary condition (15) at $\xi = 0$ and the boundary condition $y(1) = 1$:

$$\begin{aligned} y &= 0 \quad \text{for } \xi \leq 0 \\ y &= \xi^2 \quad \text{for } 0 \leq \xi \leq 1 \end{aligned}$$

and the collocation ordinate is $y_1 = \xi_1^2$, independent of the rate expression $R(y)$.

The collocation equation (16) is consequently used to determine x_p. Inserting $\xi = \xi_1, y_1 = \xi_1^2, dy/d\xi|_{\xi_1} = 2\xi_1$, and $d^2y/d\xi^2 = 2$ yields the following algebraic equation for x_p:

$$\frac{2}{(1-x_p)^2} + \frac{s}{x_p + (1-x_p)\xi_1}\frac{2\xi_1}{1-x_p} = \Phi^2 R(y_1) \tag{17}$$

Finally the effectiveness factor is found in the usual way:

$$\begin{aligned}\eta &= \frac{s+1}{R(1)}\int_0^1 x^s R(y)\,dx = \frac{s+1}{\Phi^2 R(1)}\frac{dy}{dx}\bigg|_1 = \frac{(s+1)}{\Phi^2 R(1)(1-x_p)}\frac{dy}{d\xi}\bigg|_1 \\ &= \frac{2(s+1)}{\Phi^2 R(1)(1-x_p)}\end{aligned} \tag{18}$$

When x_p has been determined from (17), η is immediately available from (18).

The accuracy of the result obviously depends on the choice of collocation point and on the rate expression $R(y)$. For a *first*-order isothermal reaction, $R(y_1) = y_1 = \xi_1^2$ and with $s = 0$ one obtains

$$\frac{1}{1-x_p} = \frac{\Phi\xi_1}{\sqrt{2}} \quad \text{and} \quad \eta = \frac{\sqrt{2}\xi_1}{\Phi} \tag{19}$$

which immediately shows that $\xi_1^2 = \frac{1}{2}$ (not an unnatural choice) gives the correct asymptotic result $\eta = 1/\Phi$. Cylindrical and spherical geometry are treated analogously, and $\xi_1^2 = \frac{1}{2}$ again yields the correct asymptotic result for η when $\Phi \to \infty$.

For a finite Φ and $s \neq 0$, equation (17) is a third-degree polynomial equation for, e.g., $1/(1 - x_p)$, which must be solved numerically. For a *zero*-order isothermal reaction and $s = 0$, one of the two following solutions is obtained depending on the value of ϕ:

$$\text{a. } \Phi = L\sqrt{\frac{k}{Dc_0}} \le \sqrt{2}: \quad y = 1 - \frac{\Phi^2}{2}(1-x^2) \quad \text{and} \quad \eta = 1$$

$$\text{b. } \Phi \ge \sqrt{2}: \quad y = \xi^2 = \left(\frac{x-x_p}{1-x_p}\right)^2 \quad \text{for } x \ge x_p \tag{20}$$

and

$$y = 0 \quad \text{for } x \le x_p$$

$$\frac{1}{1-x_p} = \frac{\Phi}{\sqrt{2}} \quad \text{and} \quad \eta = \frac{\sqrt{2}}{\Phi}$$

This is exactly the solution that is obtained by Paterson and Cresswell's method irrespective of the choice of collocation point. The result is hardly surprising since the concentration profile is a parabola for $\Phi > \sqrt{2}$ as well as for $\Phi < \sqrt{2}$.

The results for a zero-order reaction are purely coincidental, however. Any other reaction order n requires an individual choice of collocation point to give the correct asymptotic dependence $\eta(\Phi)$:

$$\eta = \frac{\sqrt{2/(n+1)}}{\Phi} \quad \text{where } \Phi^2 = L^2 \frac{k}{D} c_0^{n-1} \tag{21}$$

while the result from (19) is

$$\eta = \frac{\sqrt{2}}{\Phi} \xi_1^n$$

which means that ξ_1 should be chosen equal to $(n+1)^{-1/2n}$—or increasing toward one when n increases from zero to infinity—to obtain the correct asymptotic form $\eta(\Phi)$.

Paterson and Cresswell's method is thus seen to contain a significant element of prescience that makes it difficult to apply to full advantage without a fairly good knowledge of the form of the solution. The principle of the method is, however, perfectly sound. Utilization of the intuitively reasonable concept of a reaction zone and a nonactive inner core has allowed Paterson and Cresswell and later also Van den Bosch and Padmanabhan (1974) to reproduce the results of Hatfield and Aris (1969) to a surprising degree of accuracy. In chapter 7 it is shown that the method represents the first step of a process in which collocation is applied to a number of subintervals connected by continuity of y and $y^{(1)}$ at the subinterval end points.

6.1.4 Transformation of the dependent variable and the use of nonpolynomial trial functions

Quite frequently linearization of the model leads to a differential equation that can easily be solved by analytical methods. The upper and lower bounding solutions $u_2(x)$ and $u_1(x)$ of section 5.3 are typical examples. Incorporating the solution $y_1(x)$ of the linearized problem into the solution of the full problem presumably gives some numerical advantages. In particular this might be useful when the solution $y(x)$ is poorly represented by a single polynomial.

We shall discuss two procedures that take advantage of the solution $y_1(x)$ to a linearized version of the model.

In the first method, y is represented as the sum of y_1 and a perturbation v: $y = y_1(x) + v(x)$. Inserting this sum into the given differential equation leads to a new differential equation in v, and since v is assumed to be small for all x, quite accurate results are expected even for one collocation point.

In the second method, an expansion of y in $y_1(x)$ and functions related to $y_1(x)$ are used rather than the conventional expansion of y in polynomials. This approach has been more or less discarded in section 2.2 as being too unwieldy but it might still be justifiable if so much information on the solution can be collected in the first trial function that even the one-point collocation result has a satisfactory accuracy.

The two methods are discussed on the basis of equation (22). This example is certainly not difficult to solve by standard collocation but the main ideas of the methods are best illustrated when the algebra is minimal.

$$\frac{d^2y}{dx^2} - \Phi^2 y^2 = 0$$
$$y(1) = 1, \qquad y^{(1)}(0) = 0 \tag{22}$$

Linearization of y^2 around a point $[y^*, (y^*)^2]$ yields the following linear equation in $y_1(x)$.

$$\frac{d^2y_1}{dx^2} - 2y^*\Phi^2 y_1 + \Phi^2(y^*)^2 = 0$$
$$y_1(1) = 1, \qquad y_1^{(1)}(0) = 0 \tag{23}$$

or

$$y_1(x) = \frac{[1 - (y^*/2)]\cosh x\Phi\sqrt{2y^*}}{\cosh \Phi\sqrt{2y^*}} + \frac{y^*}{2} \tag{24}$$

$y_1(x)$ is an upper bounding solution since the tangent to $-\Phi^2 R(y)$ is always above the function when $R(y) = y^2$.

By the first modified collocation method, $y = y_1(x) + v$ is inserted in (22) to give

$$\frac{d^2v}{dx^2} - (2y_1 v + v^2)\Phi^2 = \Phi^2 y_1^2 - \frac{d^2y_1}{dx^2}$$
$$v(1) = v^{(1)}(0) = 0 \tag{25}$$

The boundary conditions of (25) are homogeneous since $y_1(x)$ satisfies the boundary conditions of (22).

We choose rather arbitrarily to represent v by a parabola:

$$v(x) = \frac{x^2 - 1}{x_1^2 - 1}v(x_1) \quad \text{and} \quad y = y_1 + \frac{v(x_1)}{1 - x_1^2}(1 - x^2) \tag{26}$$

x_1 is the collocation point abscissa for solution of (25) by one-point collocation. Strictly for convenience we choose $x_1^2 = \frac{1}{5}$ and the collocation equation can be written from the entries of table 6.1:

$$\Phi^2 v^2(x_1) + [2.5 + 2\Phi^2 y_1(x_1)]v(x_1) + \Phi^2 y_1^2(x_1) - \left.\frac{d^2 y_1}{dx^2}\right|_{x_1} = 0 \quad (27)$$

The function $y_1(x)$ depends on the point of linearization. y is known at $x = 1$ and this point is consequently directly available.

$$y^* = 1: y_1(x) = \frac{1}{2}\left(\frac{\cosh x\Phi\sqrt{2}}{\cosh \Phi\sqrt{2}} + 1\right) \quad (28)$$

By (28) $y_1(0) > \frac{1}{2}$, which is undesirable for large Φ. The smallest $y_1(0)$ obtainable for a given Φ is found by differentiation of (24) with respect to y^*, and since the tangent solution is above $y(x)$ it seems reasonable to use a y^* determined in this way. For Φ between 1 and 3, $y^* = 0.5$ yields a $y_1(0)$ close to the minimum obtainable by tangent approximation.

Columns 4 and 5 of table 6.2 compare results obtained by the modified collocation method with standard one- and two-point collocation. A value of 8 has been chosen for Φ^2 and y^* is 1 and 0.5, respectively. In the first case, $v(x_1) = -0.15500$; in the second case, $v(x_1) = -0.005736$. The very small value of $v(x_1)$ for $y^* = 0.5$ means that the perturbation v to $y_1(x)$ (which by its construction is above y for all x) is insufficient to make $y_1(x) + v$ coincide with the exact solution at any x-value. It is indeed possible to have both positive and negative values of $v(x_1)$ depending on the choice of x_1. Thus for $y^* = 0.5$ a collocation point $x_1 = 0.57$ yields

$$\Phi^2 y_1^2(x_1) - \left.\frac{d^2 y_1}{dx_1}\right|_{x_1} = 0$$

and $y = y_1$ for all values of x.

Comparison of the modified collocation method with the two first columns of the table shows that a one-point modified method is significantly better than the standard one-point collocation method [which of course means that y is poorly represented by $1 + a(1 - x^2)$] but not as accurate as a two-point standard collocation method.

The last row of the table shows the effectiveness factor $\eta = \int_0^1 y^2\,dx$. The integral has been calculated by Simpson's rule with 10 subdivisions in the case of the modified collocation method. Using the quadrature weights of table 6.1 would give a less accurate result since an unnecessary error is introduced in integration of the known function $y_1^2(x)$.

Again we note that the modified method gives an η-value that is more accurate than obtained by one-point standard collocation.

$x_1^2 = 0.2$ is by no means the best collocation point in (25)—the equation has not at all the same form as the equations of chapter 2 that are used to derive "optimal" collocation points for a linear model. On the

Table 6.2

Solution of $(d^2y/dx^2) - \Phi^2 y^2 = 0$ for $\Phi^2 = 8$ by Different Collocation Methods

	y							
x	1. Point standard c	2. Point standard c	Modified 1.p.c. $y^* = 1$	Modified 1.p.c. $y^* = 0.5$	Hyperbolic c $y^* = 1$	Hyperbolic c $y^* = 0.6724$	Hyperbolic c $y^* = 0.5$	Exact
0	0.2802	0.3225	0.3245	0.3312	0.3789	0.3519	0.3354	0.3169
0.1	0.2874	0.3256	0.3279	0.3348	0.3809	0.3548	0.3390	0.3210
0.2	0.3090	0.3356	0.3385	0.3460	0.3869	0.3636	0.3500	0.3333
0.3	0.3450	0.3537	0.3569	0.3656	0.3981	0.3795	0.3693	0.3545
0.4	0.3954	0.3820	0.3844	0.3952	0.4162	0.4041	0.3985	0.3859
0.5	0.4602	0.4237	0.4236	0.4371	0.4441	0.4401	0.4400	0.4292
0.6	0.5394	0.4826	0.4778	0.4946	0.4865	0.4913	0.4970	0.4874
0.7	0.6329	0.5636	0.5522	0.5724	0.5502	0.5635	0.5741	0.5648
0.8	0.7409	0.6723	0.6552	0.6725	0.6454	0.6643	0.6777	0.6680
0.9	0.8632	0.8153	0.7985	0.8154	0.7876	0.8047	0.8159	0.8075
1	1	1	1	1	1	1	1	1
η	0.3166	0.2854	0.2790	0.2924	0.2929	0.2931	0.2949	0.2840

other hand, v is sufficiently smooth to make a search for a better collocation point a matter of small practical interest.

If, however, y is represented not as a sum of a known function $y_1(x)$ and a perturbation $v(x)$ but as an expansion in functions related to $y_1(x)$ as in the second of our two proposed modified methods, it becomes necessary to determine a completely new set of collocation constants.

Several expansions that incorporate our knowledge of the solution to the linearized model can be proposed; e.g.,

$$y = 1 + a_1\left(\frac{\cosh \Phi_1 x}{\cosh \Phi_1} - 1\right) + a_2\left(\frac{x \sinh \Phi_1 x}{\sinh \Phi_1} - 1\right) + a_3\left(\frac{x^2 \cosh \Phi_1 x}{\cosh \Phi_1} - 1\right) + \cdots \tag{29}$$

$$y = 1 + \left(\frac{\cosh \Phi_1 x}{\cosh \Phi_1} - 1\right)(a_1 + a_2 \cosh x + a_3 \cosh 2x + \cdots) \tag{30}$$

In both cases the trial functions T_i $(i > 0)$ satisfy homogeneous boundary conditions, just as in section 2.3. The expansion (29) is derived by differentiation of the first trial function $\cosh \Phi_1 x$ with respect to Φ_1, while in the second expansion (30) the higher-order trial functions $T_2, T_3, \ldots$ are simply formed as products of T_1 and low-order hyperbolic functions.

The first term of either series is taken from the solution of the linearized problem. Thus $\Phi_1 = \Phi\sqrt{2y^*}$ for equation (22) and different expansions appear with different y^*. We shall only consider the approximation $y \sim y_1(x) = 1 + a_1 T_1$ and derive a corresponding one-point collocation method. The residual $R_1(a_1, x)$ of (22) with y replaced by y_1 is

$$R_1 = \frac{a_1 \Phi_1^2 \cosh \Phi_1 x}{\cosh \Phi_1} - \Phi^2\left[1 + a_1\left(\frac{\cosh \Phi_1 x}{\cosh \Phi_1} - 1\right)\right]^2$$

We wish to determine a_1 by Galerkin's method:

$$\int_0^1 R_1(a_1, x) \cdot \left(\frac{\cosh \Phi_1 x}{\cosh \Phi_1} - 1\right) dx = 0 \tag{31}$$

Now, for equation (22), $R_1(a_1, x)$ consists of three types of functions: a constant, $\cosh \Phi_1 x$, and $\cosh^2 \Phi_1 x$. No choice of a single collocation point gives the same result as an exact determination of a_1 from (31)—just as for the similar nonlinear example in section 2.4. But if we restrict ourselves to the linear terms in R_1, it is indeed possible to integrate these correctly, namely, by demanding that R_1 is proportional to $b + \cosh \Phi_1 x$ where b is determined by

$$\int_0^1 (b + \cosh \Phi_1 x)(\cosh \Phi_1 x - \cosh \Phi_1)\, dx = 0$$

or

$$b = \frac{1 - \cosh \Phi_1(\sinh \Phi_1/\Phi_1)}{2[\cosh \Phi_1 - (\sinh \Phi_1/\Phi_1)]} \tag{32}$$

The collocation point is chosen as the zero of $\cosh \Phi_1 x + b = 0$.

With this choice of collocation point x_1 we have the added advantage that for the specific example (22) all terms of $[y_1(x)]^2$ are correctly integrated by a one-point Radau formula based on x_1 and 1 as quadrature points. Consequently we can obtain the effectiveness factor directly from $y_1(x_1)$.

The remaining collocation constants are

1. The weights C_{11} and C_{12} of the discretization matrix for $\nabla^2 = d^2/dx^2$ in (22).
2. The quadrature weights w_1 and w_2.

These are determined by

$$\begin{aligned} C_{11} &= \frac{\Phi_1^2 \cosh \Phi_1 x}{\cosh \Phi_1 x_1 - \cosh \Phi_1} = \frac{\Phi_1^2 b}{b - \cosh \Phi_1} = -C_{12} \\ w_1 &= \frac{(\sinh \Phi_1/\Phi_1) - \cosh \Phi_1}{b - \cosh \Phi_1} = 1 - w_2 \end{aligned} \tag{33}$$

The second derivative at $x = x_1$ of any function that is a linear combination of a constant and $\cosh \Phi_1 x$ is correctly represented by $C_{11}y(x_1) + C_{12}y(1)$. The weights w_1 and w_2 are the correct weights in a quadrature (5) when the function to be integrated is a linear combination of a constant, $\cosh \Phi_1 x$, and $\cosh^2 \Phi_1 x$.

Collocation constants for four different choices of y^* are given in table 6.3. $y^* = 0.672$ is near the middle of the y-range for $\Phi^2 = 8$. The results for three of the methods used on (22) are shown in columns 6, 7, and 8 of table 6.2.

There seems to be little difference between the present modified method and that in which a rather arbitrary collocation method is applied to $v = y - y_1$. This is intuitively reasonable since in the present method the collocation method must be well chosen in order to represent the large perturbation $a_1 T_1$ to the boundary value 1.

TABLE 6.3
COLLOCATION CONSTANTS FOR HYPERBOLIC COLLOCATION

y^*	Φ_1	b	x_1	w_1	C_{11}	
1	4	−4.523	0.5475	0.8767	−2.808	$\Phi_1 = \sqrt{16y^*}$
0.6724	3.28	−2.852	0.5210	0.8859	−2.935	
0.5	$\sqrt{8}$	−2.206	0.5050	0.8990	−3.176	$w_2 = 1 - w_1$
0.4389	2.65	−2.009	0.4989	0.8730	−2.764	$C_{12} = -C_{11}$

Figure 6-4 shows results from both modified methods. In the upper part of the figure are the exact solution and the exact functions $v(x)$ for $y^* = 1$ and $y^* = 0.5$. In the last case $v(x)$ is much less smooth than in

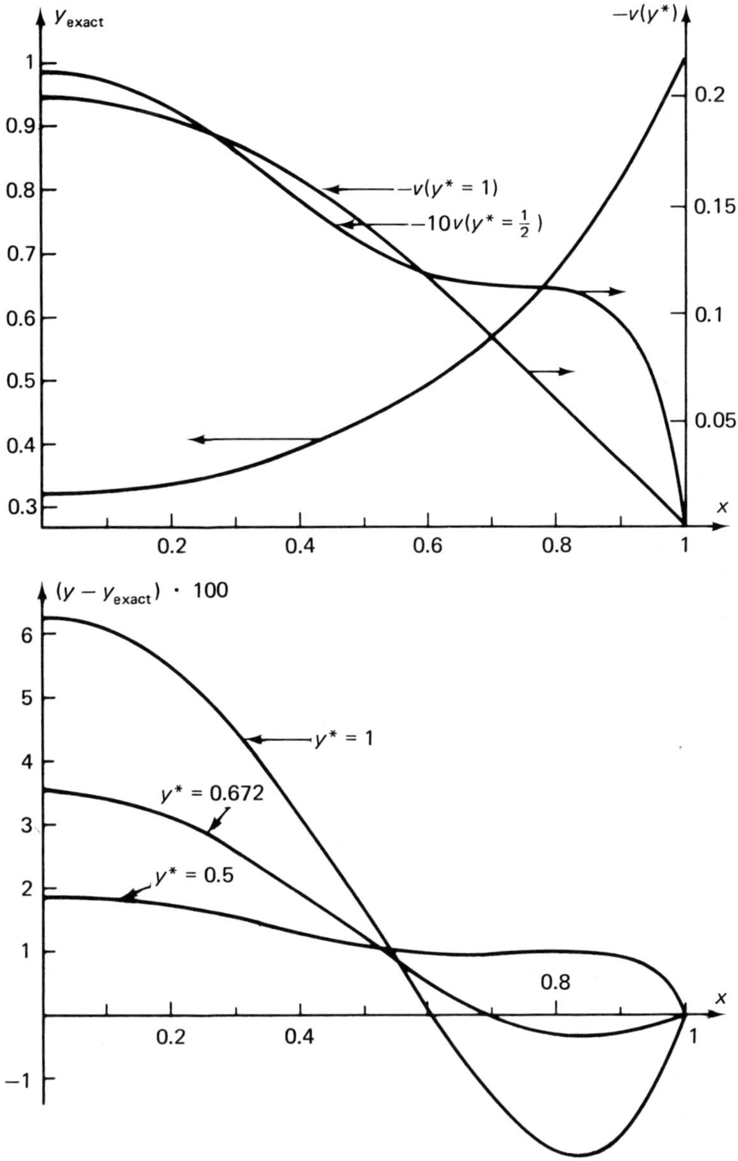

Figure 6-4. Deviation function v and one-point hyperbolic collocation for $(d^2y/dx^2) - 8y^2 = 0$.

the first case, which explains why one-point collocation on v does not significantly improve the first approximation $y_1(x)$, which is already quite close to $y(x)$. The lower part of the figure shows deviations between approximations obtained by one-point hyperbolic collocation approximation and the exact solution. Again $y^* = 0.5$ gives a one-sided error, while $y^* = 0.6724$ gives almost zero error for $x > 0.6$.

It is interesting that Guertin, Sørensen, and Stewart (1975) have constructed an Nth-order collocation method similar to our "hyperbolic collocation method." The method is used for integration of coupled first-order equations rather than for boundary value problems, and their very satisfactory results for kinetic equations with a large span of eigenvalues do prove that methods derived specifically for the given problem can be justified.

In one of their examples—a one-dimensional, isothermal reactor model with four independent reactions—they find the eigenvalues of the Jacobian of **R**, to be $\lambda_i = -0.27, -6, -20$, and -140 at the reactor inlet.

For approximation order $N = 1$ to 4, they include successively all four eigenfunctions $\exp(\lambda_i z)$ in their approximation basis, starting with $\exp(-0.27z)$ for $N = 1$. When $N = 5$ to 8, the same functions are included but are now multiplied by x [as in $T_2(x)$ in (29)]. For $N = 9$ to 12, they are multiplied by x^2, etc.

Even though the authors report that their results are considerably better than straightforward collocation (applied as in section 4.2), it is not obvious that a judicious application of spline collocation as in chapter 7 or a quite different integration method as in section 8.2 would not give equally good results. From a mathematical point of view it is, however, quite interesting that the principles of chapter 2 can be used to construct "tailor-made" expansions of solutions to differential equations.

6.1.5 Optimal collocation for a general linear boundary condition at $x = 1$

In the preceding subsections we have treated the effectiveness factor problem for large values of the Thiele modules Φ. Special techniques were developed that took advantage of the qualitative form of the profile. This was fitted either by the solution of two different differential equations as in subsection 6.1.3 or by the sum of the solution to a suitable linearized problem and a "deviation" function as in subsection 6.1.4.

Here we return to the linear problem

$$\nabla^2 y - \Phi^2 y = 0 \tag{34}$$

but now the boundary condition at $x = 1$ is

$$\frac{dy}{dx} + Ay = A \tag{35}$$

while at $x = 0$ we still use $y^{(1)} = 0$.

For $A \to \infty$, boundary condition (35) simplifies to $y = 1$ at $x = 1$ and here an optimal collocation method at the zeros of $P_N^{[1,(s-1)/2]}(u = x^2)$ was derived in chapter 2 for the three geometries $s = 0, 1$, and 2. Optimality of the method was interpreted in the sense that a series expansion of η in powers of Φ^2 was correct up to and including $(\Phi^2)^{2N}$ where N is that number of collocation points (subsection 2.3.4).

For $A < \infty$, this collocation scheme is no better than any other choice of collocation points. The boundary point is determined with an accuracy of N and the Radau quadrature that is used to determine η from the collocation ordinates and $y(u = 1)$ has no higher accuracy.

In chapter 2 we also showed that a collocation method based on the zeros of $P_N^{[0,(s-1)/2]}(u)$ was identical to the method of moments when applied to (34). With this method an accuracy $2N - 1$ is obtained for the collocation ordinates also with boundary condition (35). Since these interior ordinates are the only ones used in a Gauss–Jacobi quadrature (subsection 3.3.3) for computing η, an accuracy $2N - 1$ is obtained for the effectiveness factor when collocation based on the zeros of $P_N^{[0,(s-1)/2]}(u)$ is used to solve (34) and (35).

Thus, with respect to accuracy—in the sense of subsection 2.3.4—collocation of the zeros of $P_N^{[1,(s-1)/2]}(u)$ or $P_N^{[0,(s-1)/2]}(u)$ are distinguished from any other collocation method. The first collocation method can only be used when $A \to \infty$ in (35), while the other can be used for any A-value. When $N > 2$ or 3, there is no significant difference between the accuracy of the two methods—one increasing as $2N$, the other as $2N - 1$.

Here we derive a Galerkin-type collocation method that can be applied also for boundary condition (35). The possible increase of accuracy is only from $2N - 1$ to $2N$ and the resulting method that requires the use of A-dependent collocation points is only significantly better than a "standard" method based upon zeros of $P_N^{[0,(s-1)/2]}(u)$ when N is 1 or possibly 2. Consequently, the method is most appropriately discussed in this chapter even though it appears as a modification of the Galerkin-type collocation method of chapter 2.

It should immediately be noted that the proof of the optimality of the modified collocation method requires (as in chapter 2) that the residuum for the Nth-order method is a polynomial of degree $\leq N$ in $u = x^2$. Also no method better than collocation at the zeros of $P_N^{[0,(s-1)/2]}(u)$ can be proposed for coupled differential equations of the type (34) when the boundary condition (35) has different A-values for the different components of the solution vector.

We use an approximation of the general form (2.31):

$$y_N = T_0 + \sum_{i=1}^{N} a_i T_i \tag{36}$$

T_0 is chosen equal to the function 1, which satisfies (35) as well as $T_0^{(1)} = 0$ at $x = 0$. The following trial functions T_i must satisfy homogeneous boundary conditions at $x = 0$ and at $x = 1$. Our choice is

$$T_1 = 1 + \tfrac{1}{2}(1 - u)A \tag{37}$$

$$T_i = (1 - u)^i \qquad (i > 1) \tag{38}$$

Equation (37) as well as (38) satisfies

$$2\frac{dT_i}{du} + AT_i = 0 \quad \text{at } u = 1$$

Galerkin's method applied to (34) with y_N inserted from (36) yields the following equations for the coefficients a_i:

$$\int R_N(\mathbf{a}, u)T_i(u)u^{(s-1)/2}\,du = 0 \qquad i = 1, 2, \ldots, N, \quad y_N = 1 + \sum_1^N a_i T_i \tag{39}$$

The residuum R_N for the differential equation (34) is obviously a polynomial of degree N in u. Thus if T_i had been exclusively taken from the set (38), it would not be difficult to derive a collocation method that is identical to Galerkin's method. By the same arguments as in subsection 2.4.1, one would arrive at a collocation method based on the zeros of $P_N^{[1,(s-1)/2]}(u)$ since

$$\int_0^1 R_N(\mathbf{a}, u)(1 - u)^i u^{s-1/2}\,du = \int_0^1 (1 - u)u^{s-1/2}R_N(\mathbf{a}, u)(1 - u)^{i-1}\,du$$
$$= 0$$

for $i = 1, 2, \ldots, N$ when R_N is proportional to $P_N^{[1,(s-1)/2]}(u)$.

The first trial function is given by (37), however, and (39) is not satisfied for $i = 1$ if R_N is proportional to $P_N^{[1,(s-1)/2]}(u)$. To satisfy (39) for all $i = 1, 2, \ldots, N$, one must represent R_N by a combination of two orthogonal polynomials. Presently we derive a linear combination of orthogonal polynomials that satisfies (39) for $i \leq N$.

If two polynomials from the same family are used—this will make the calculations easier—one obtains

$$R_N(\mathbf{a}, u) = \text{constant} \cdot [qP_N^{(\alpha,\beta)}(u) + rP_{N-1}^{(\alpha,\beta)}(u)] \tag{40}$$

We first observe that two successive polynomials P_N and P_{N-1} appear in (40). If the polynomials are defined as in section 3.1, their value at $u = 0$ is $(-1)^N$. We also demand that the linear combination of the two polynomials is $(-1)^N$ at $u = 0$ and consequently $r = q - 1$. The collocation points are chosen as zeros of $P_N(u)$:

$$P_N(u) = qP_N^{(\alpha,\beta)}(u) + (q - 1)P_{N-1}^{(\alpha,\beta)}(u) \tag{41}$$

$P_N(u)$ is not itself a Jacobi polynomial but it is the sum of two Jacobi polynomials given by the same α and β for all N. Next we determine α and β.

For $N = 1$, we demand that

$$\int_0^1 [qP_1^{(\alpha,\beta)} + (q - 1)P_0^{(\alpha,\beta)}][1 + \tfrac{1}{2}A(1 - u)]u^{s-1/2}\,du = 0 \tag{42}$$

For $N = 2, 3, \ldots$ and $i = 2, 3, \ldots, N$, we demand that

$$\int_0^1 [qP_N^{(\alpha,\beta)} + (q - 1)P_{N-1}^{(\alpha,\beta)}](1 - u)^2 u^{s-1/2}(1 - u)^{i-2}\,du = 0 \tag{43}$$

in addition to

$$\int_0^1 [qP_N^{(\alpha,\beta)} + (q - 1)P_{N-1}^{(\alpha,\beta)}][1 + \tfrac{1}{2}A(1 - u)]u^{s-1/2}\,du = 0 \tag{44}$$

All equations (43) are automatically satisfied if $\alpha = 2$ and $\beta = (s - 1)/2$ since $P_N^{[2,(s-1)/2]}(u)$ and $P_{N-1}^{[2,(s-1)/2]}(u)$ are orthogonal with weight function $(1 - u)^2 u^{s-1/2}$ on any polynomial of degree less than $N - 1$ in u.

With α and β determined, we turn to (42) to find the constant q and thus the sample point polynomial for $N = 1$.

$$P_1^{[2,(s-1)/2]}(u) = \frac{s + 7}{s + 1}u - 1 \quad \text{and} \quad P_0^{[2,(s-1)/2]}(u) = 1$$

by formula (2.9). Now q is obtained from

$$\int \left[q\left(\frac{s + 7}{s + 1}u - 1\right) + (q - 1)\right]\left[1 + \frac{1}{2}A(1 - u)\right]u^{s-1/2}\,du = 0$$

Evaluation of the integral yields

$$q = \frac{(s + 5)(A + s + 3)}{(s + 7)(A + s + 5)} \quad \text{and} \quad P_1(u) = q \cdot \frac{s + 7}{s + 1}u - 1 \tag{45}$$

The collocation point is

$$u_1 = \frac{s + 1}{s + 5}\,\frac{A + s + 5}{A + s + 3}$$

For $A \to \infty$, $u_1 = (s+1)/(s+5)$, the previously used collocation point; for $A = 0$, $u_1 = (s+1)/(s+3)$. These two u_1 values represent the limit of variation for the collocation point.

The one-point collocation approximation for y is

$$y_1(u) = \frac{u-1}{u_1-1}y_1(u_1) + \frac{u-u_1}{1-u_1}y_1(1) \tag{46}$$

where

$$u_1 = \frac{(s+1)(A+s+5)}{(s+5)(A+s+3)} \tag{47}$$

Now the remaining entries of a table similar to 6.1 can be calculated.

1. Discretization of Laplacian at $u = u_1$:

$$C_{11} = -C_{12} = -\frac{2}{1-u_1} \tag{48}$$

2. The weights of a one-point Radau quadrature formula (3.83):

$$\frac{w_1}{w_2} = \frac{1 \cdot \{d/dx[(u-1)(u-u_1)]_{u=1}\}^2}{u_i\{d/dx[(u-1)(u-u_1)]_{u=u_1}\}^2} \cdot \frac{0+1}{1} = \frac{1}{u_1}$$

$$w_1 + w_2 = \frac{1}{2}\int_0^1 u^{(s-1)/2}\,du = \frac{1}{s+1}$$

or

$$w_1 = \frac{1}{(s+1)(u_1+1)} \quad \text{and} \quad w_2 = \frac{u_1}{(s+1)(u_1+1)} \tag{49}$$

$$\frac{1}{2}\int_0^1 F(y)u^{(s-1)/2}\,du = w_1F[y(u_1)] + w_2F[y(1)] \tag{50}$$

Thc resulting quadrature formula is correct whenever F is a polynomial of degree ≤ 2 in u.

Higher-order collocation formulas can be derived in a similar fashion. The key point is calculation of $q(N)$ from formula (44). This can be done without too much difficulty for $N = 2$ and 3 by manual evaluation of the integral. Since the modified collocation method is not much better than standard collocation for N larger than 3, this procedure is certainly applicable.

It is of some interest, however, that a general formula for q can be derived. The result is

$$q = \frac{N + [(s+1)/2] + 1}{2N + [(s+1)/2] + 1}\,\frac{2N + s + 1 + (A/N)}{2N + s + 3 + (A/N)} \tag{51}$$

The derivation of (51) follows by a considerable amount of manipulation on the polynomials. Formulas (3.7) to (3.10), the result of Exercises 3.1 and 3.2, and the relation (see Exercise 4.2)

$$q[P_N^{(\alpha,\beta)} + P_{N-1}^{(\alpha,\beta)}] = q \cdot (\gamma_1^N - \gamma_1^{N-1})P_{N-1}^{(\alpha,\beta+1)} \tag{52}$$

where γ_1^N and γ_1^{N-1} are the coefficients of u in $P_N^{(\alpha,\beta)}$ and $P_{N-1}^{(\alpha,\beta)}$, respectively, are all used. We refrain from giving the details of the proof and state only the result for $N = 2$:

$$q = \frac{s+7}{s+11}\,\frac{10+2s+A}{14+2s+A} \tag{53}$$

$$P_2 = q \cdot \left[\frac{(s+9)(s+11)}{(s+1)(s+3)}u^2 - \frac{2(s+9)}{s+1}u + 1\right] + (q-1)\left(\frac{s+7}{s+1}u - 1\right) \tag{54}$$

Stewart and Sørensen (1972) have used the same orthogonal polynomials $P_N^{[2,(s-1)/2]}$ and $P_{N-1}^{[2,(s-1)/2]}$ as we have to construct P_N of (41). Their results are shown as tables of collocation points $x_i = u_i^{1/2}$ for $N = 1, 2, 3$ and $s = 1, 2$.

In concluding the discussion of optimal collocation methods for boundary condition (35) it should be noticed that the lack of accuracy of the standard collocation scheme applied to (34) and (35) is due to the elimination of $y(u = 1)$ by discretization of (35) in the manner of subsection 3.5.2. If (35) is integrated over the system volume and $y(1)$ eliminated from the resulting equation, one may obtain the same accuracy $2N$ (although not exactly the same answer) by a standard collocation method at the zeros of $P_N^{[1,(s-1)/2]}(u)$. This "improved Galerkin-type collocation" is discussed in Exercise 6.3.

6.2 Application of One-Point Collocation to Partial Differential Equations

6.2.1 Equations of elliptic type

By far the largest number of applications of the collocation method in partial differential equations has been to equations of parabolic type where a boundary value problem in one space coordinate is discretized by collocation while the integration in the direction of the remaining coordinate is done by some other method.

Among the few published results for elliptic equations are two examples where the strength of one-point collocation for a preliminary study of the model is clearly demonstrated.

One of these examples is given in the first paper on orthogonal collocation [Villadsen and Stewart (1967)]: The axial velocity $v(x, y)$ of a constant density and viscosity Newtonian fluid in stationary laminar flow through a square duct is given by equation (1.3) with $v_y = v_x = \partial v_z/\partial z = \partial v_z/\partial t = 0$.

$$\mu\left(\frac{\partial^2 v_z}{\partial x^2} + \frac{\partial^2 v_z}{\partial y^2}\right) = \frac{\partial p}{\partial z} = \text{constant } K \quad v_z = 0 \text{ on the wall of the duct} \tag{55}$$

In suitable dimensionless variables one obtains

$$\frac{\partial^2 v}{\partial x^2} + \frac{\partial^2 v}{\partial y^2} = -1 \quad \text{with } v = 0 \text{ at } x = \pm 1 \text{ and } y = \pm 1 \tag{56}$$

The dimensionless flow rate

$$Q = \int_{-1}^{1}\int_{-1}^{1} v\,dx\,dy = 4\int_{0}^{1}\int_{0}^{1} v\,dx\,dy \tag{57}$$

is the quantity that is usually desired in problems of this type.

An analytical solution to (56) is available in terms of a double trigonometric series, but the rate of convergence of the series is so poor that it is highly impractical to evaluate Q by means of the analytical solution.

The symmetry of the solution $v(x, y)$ in each octant of the (x, y)-plane suggests the following polynomial approximation:

$$v_N(x, y) = (1 - x^2)(1 - y^2)\sum_1^N\sum_1^N a_{ij}P_{i-1}(x^2)P_{j-1}(y^2) \quad \text{with } a_{ij} = a_{ji} \tag{58}$$

A collocation process at the zeros of $P_N^{(1,-1/2)}(u = x^2 \text{ or } y^2)$ is clearly optimal in view of our desire to evaluate Q as accurately as possible.

With one collocation point the residual $R_N(v_N)$ is equated to zero at $x_1^2 = y_1^2 = \frac{1}{5}$ and the discrete version of (56) is

$$\begin{aligned}-2.5v_1(x_1, y_1) &+ 2.5v_1(1, y_1) - 2.5v_1(x_1, y_1) + 2.5v_1(x_1, 1)\\ &= -5v_1(x_1, y_1) = -1\end{aligned} \tag{59}$$

since $v_1(x_1, 1) = v_1(1, y_1) = 0$. The weights of the double quadrature

$$Q \sim 4\sum_1^2\sum_1^2 w_iw_jv_1(x_i, y_j)$$

are taken from table (6.1) with $s = 0$:

$$Q \simeq 4\left(\tfrac{5}{6}\cdot\tfrac{5}{6}\cdot\tfrac{1}{5} + \tfrac{5}{6}\cdot\tfrac{1}{6}\cdot 0 + \tfrac{1}{6}\cdot\tfrac{5}{6}\cdot 0 + \tfrac{1}{6}\cdot\tfrac{1}{6}\cdot 0\right) = \tfrac{5}{9}$$

This result is already in very good agreement with the value of Q obtained from the analytical solution ($Q = 0.5623$) and the collocation process may if desired be extended to more than one collocation point, as shown in table 6.4.

TABLE 6.4
CHOICE OF COLLOCATION POINTS FOR SQUARE DUCT PROBLEM

$N = 2$	(x_1, x_1) (x_1, x_2) (x_2, x_2) where $u_i = x_i^2$ are zeros of $P_2^{(1,-1/2)}(u)$
$N = 3$	(x_1, x_1) (x_1, x_2) (x_1, x_3) (x_2, x_2) (x_2, x_3) (x_3, x_3) where $u_i = x_i^2$ are zeros of $P_3^{(1,-1/2)}(u)$

The number of collocation points increases as $\frac{1}{2}N(N + 1)$ even when the symmetry of $v(x, y)$ is used to full advantage as in table 6.4, but the double quadrature formula for Q is rapidly convergent. With $N = 3$ one obtains five- or six-digit accuracy and further computation is unnecessary.

Sørensen, Guertin, and Stewart (1973) have used exactly the same technique to compute the effectiveness factor for cylindrical catalyst pellets with a finite length/diameter ratio L/d. Their steady state model for a first-order isothermal reaction is

$$\frac{1}{x}\frac{\partial}{\partial x}\left(x\frac{\partial y}{\partial x}\right) + \frac{d^2}{L^2}\frac{\partial^2 y}{\partial z^2} - \left(2 + \frac{d}{L}\right)^2 \Lambda^2 y = 0$$

$$y = 1 \quad \text{at} \quad z = 1 \quad \text{and} \quad x = 1 \tag{60}$$

$$\frac{\partial y}{\partial x} = 0 \quad \text{at } x = 0 \quad \text{and} \quad \frac{\partial y}{\partial z} = 0 \quad \text{at } z = 0$$

$$\Lambda = \frac{(d/2)\sqrt{k/D}}{[2 + (d/L)]}$$

equals the generalized Thiele modulus with a spatial length parameter equal to the volume to surface area ratio of the pellet:

$$\frac{V}{S} = \frac{\frac{1}{4}\pi d^2 \cdot L}{\pi d L + \frac{1}{2}\pi d^2} = \frac{d/2}{2 + (d/L)}$$

Replacing the derivatives in (60) by one-point collocation expressions from table 6.1, one obtains the following expression for the collocation ordinate y_{11} at $(x_1, z_1) = (\sqrt{\frac{1}{3}}, \sqrt{\frac{1}{5}})$:

$$\left(6 + 2.5\frac{d^2}{L^2}\right)(1 - y_{11}) = \left(2 + \frac{d}{L}\right)^2 \Lambda^2 y_{11}$$

$$y_{11} = \frac{6 + 2.5(d^2/L^2)}{6 + 2.5(d^2/L^2) + [2 + (d/L)]^2 \Lambda^2} \tag{61}$$

For $d/L \to \infty$ (a very flat cylinder with diffusion almost exclusively in the z-direction except for $x \sim 1$), (61) degenerates to the one-point collocation formula for flat plates. For $d/L \to 0$ (a long slender cylinder), (61) also correctly degenerates to the one-point collocation expression for an infinite cylinder. The accuracy of the limiting expressions is better than 0.5% for $\Lambda < 2$ and for these Λ values formula (61) is probably of the same order of accuracy for all values of (d/L).

The effectiveness factor

$$\begin{aligned}\eta &= 2\int_0^1 x\left(\int_0^1 y\,dz\right)dx = 2\sum_{i=1}^{N_x+1}\sum_{j=1}^{N_z+1} w_{ix}w_{jz}y_{ij}\\ &= 2\left(\tfrac{3}{8}\cdot\tfrac{5}{6}\cdot y_{11} + \tfrac{3}{8}\cdot\tfrac{1}{6}\cdot y_{12} + \tfrac{1}{8}\cdot\tfrac{5}{6}\cdot y_{21} + \tfrac{1}{8}\cdot\tfrac{1}{6}\cdot y_{22}\right)\\ &= \tfrac{1}{8}(5y_{11} + 3)\end{aligned} \tag{62}$$

$y_{12} = y_{21} = y_{22} = 1$ as shown in figure 6-5(a).

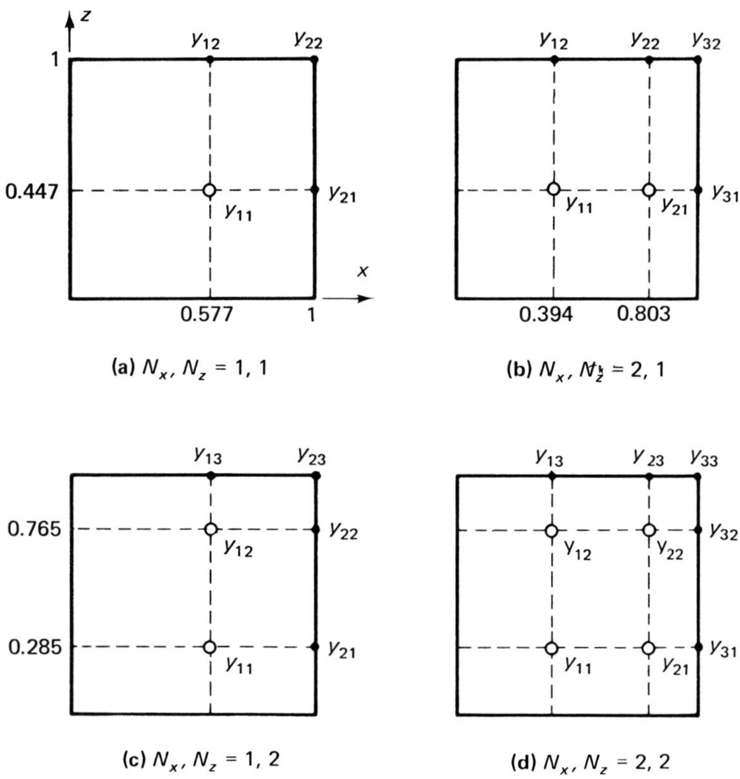

Figure 6-5. Location of collocation points in the finite cylinder effectiveness problem. Collocation points: open circles.

While (61) gives an accurate value for y_{11} at all d/L ratios, the same cannot be true for the d/L independent formula (62). For $d/L \to \infty$, the weight coefficients of y_{11} and 1 should have been $\frac{5}{6}$ and $\frac{1}{6}$; for $d/L \to 0$, one should have used $\frac{3}{4}$ and $\frac{1}{4}$.

For $d/L = 1$ and $\Lambda = 1$ the weights of (62) give approximately the same result as a quadrature formula based on the equivalent sphere model. For $\Lambda \ll 1$ and $d/L = 1$ (62) predicts a more accurate value of η than the equivalent sphere model.

Numerical results for $\Lambda = 1$ are shown in table 6.5 for various (N_x, N_z) and collocation points in figure 6-5(a)–(d). The accuracy for $N_x = N_z = 1$ is approximately 4%, while the analytical result for the equivalent sphere predicts the effectiveness factor of the cylinder with an error of about 3% in accordance with the well-known empirical result of Aris (1957). "Exact results" have been taken from Sørensen et al. (1973), who integrated (60) (for a nonisothermal reaction) using the technique described in subsection 5.5.3.

TABLE 6.5
EFFECTIVENESS FACTOR FOR FINITE CYLINDERS $d/L = 1, \Lambda = 1$

N_x, N_z		1, 1	2, 1	1, 2	2, 2	Exact	Equivalent sphere
y_{x_i, z_i} or y_{ij}	1, 1	0.4857	0.39124	0.44148	0.336713		
	2, 1	—	0.67243	—	0.648185		
	1, 2	—	—	0.64327	0.585285		
	2, 2	—	—	—	0.758938		
η		0.6786	0.66915	0.66631	0.65559	0.6550	0.6716

6.2.2 Parabolic equations

Collocation applied to the solution of a parabolic partial differential equation leads to a set of first-order ordinary differential equations. If the original model is linear, an algebraic eigenvalue problem results by what is in effect an averaging process over the system volume.

The averaging property of the collocation method, which (as we have seen already in chapter 2) ties the method to other MWR, is nowhere illustrated more clearly than in the treatment of parabolic partial differential equations. In one-point collocation, relations between parameters are derived without too much algebra; in this subsection, which serves as a supplement to the computer treatment of parabolic partial differential equations in chapters 4 and 8, we use two examples to illustrate this application of one-point collocation.

In the first example the boundary between stable and unstable solutions to the asymptotic stability problem of subsection 1.3.3 is described by an inequality relation between the Lewis number and the parameters of the steady state models Φ^2, β, and γ.

The second example is used to derive the "effective" transport coefficients that were repeatedly used in chapter 1 to simplify reactor models.

Model simplification is so common in chemical engineering literature that its numerical aspects deserve more than a passing reference. Consequently we digress somewhat from the main subject of this chapter to give a more general treatment of the subject before one-point collocation is shown as a means of obtaining typical relations such as (1.27) for the "effective" wall heat transfer coefficient.

Hellinckx, Grootjans, and van den Bosch (1972) investigated the asymptotic stability of the steady states for a first-order irreversible reaction on catalyst pellets. Their analysis starts with the linearized model (1.82) and they assume that both Biot numbers are large such that $\hat{y} = \hat{\theta} = 0$ at $x = 1$.

Replacing the Laplacian by its one-point collocation approximation gives

$$\frac{d\hat{y}}{d\tau} = (C_{11} - R_y)\hat{y} - R_\theta\hat{\theta}$$

$$\frac{d\hat{\theta}}{d\tau} = \frac{\beta R_y}{\mathrm{Le}}\hat{y} + \left(\frac{C_{11}}{\mathrm{Le}} + \frac{\beta R_\theta}{\mathrm{Le}}\right)\hat{\theta} \tag{63}$$

In (63), $(\hat{y}, \hat{\theta})$ is the value of the concentration deviation variable $(y - y_{ss})$ and temperature deviation variable $(\theta - \theta_{ss})$ at the collocation abscissa $x^2 = u = u_1$. R_θ and R_y are the values of the partial derivatives of the rate expression (1.80) and (1.81) taken at u_1 and the steady state solution (y_{ss}, θ_{ss}). The boundary ordinates $\hat{y}$ and $\hat{\theta}$ at $u = 1$ are zero.

The solution of (63) tends to zero for $\tau \to \infty$ if the eigenvalues of

$$\mathbf{M} = \begin{pmatrix} C_{11} - R_y & -R_\theta \\ \dfrac{\beta R_y}{\mathrm{Le}} & \dfrac{C_{11}}{\mathrm{Le}} + \dfrac{\beta R_\theta}{\mathrm{Le}} \end{pmatrix} \tag{64}$$

have negative real part.

For a first-order irreversible reaction, the one-point collocation expressions for R_y and R_θ are

$$R_y = \Phi^2 \exp\left(\gamma\frac{\theta_{ss} - 1}{\theta_{ss}}\right) = C_{11}\left(\frac{y - 1}{y}\right)_{ss}$$

$$R_\theta = \gamma\Phi^2\frac{y_{ss}}{\theta_{ss}^2}\exp\left(\gamma\frac{\theta_{ss} - 1}{\theta_{ss}}\right) = C_{11}\gamma\left(\frac{y - 1}{\theta^2}\right)_{ss}$$

These expressions are inserted into **M**:

$$\mathbf{M} = C_{11}\begin{pmatrix} \dfrac{1}{y} & -\gamma\dfrac{y-1}{\theta^2} \\ \dfrac{\beta(y-1)}{\mathrm{Le}\, y} & \dfrac{1}{\mathrm{Le}} + \dfrac{\beta\gamma(y-1)}{\mathrm{Le}\,\theta^2} \end{pmatrix}_{ss,u=u_1} = C_{11}\begin{pmatrix} a_1 & a_2 \\ a_3 & a_4 \end{pmatrix} \tag{65}$$

The eigenvalues of $\mathbf{M}/C_{11}$ are zeros of

$$\lambda^2 - (a_1 + a_4)\lambda + (a_1 a_4 - a_2 a_3) = 0$$

Now C_{11} is a negative constant (−2.5, −6, or −10.5) and the eigenvalues of M have negative real part if

$$(a_1 + a_4) > 0 \tag{66}$$

$$a_1 a_4 - a_2 a_3 > 0 \tag{67}$$

The inequality (66) yields

$$\mathrm{Le} > -\frac{\beta + 1 - \theta_{ss}}{\beta} + \gamma\frac{(\theta_{ss} - 1)(\beta + 1 - \theta_{ss})}{\beta\theta_{ss}^2} \tag{68}$$

while from (67)

$$1 > \frac{\gamma(\theta_{ss} - 1)(\beta - \theta_{ss} + 1)}{\beta\theta_{ss}^2} \tag{69}$$

The maximum value of the right-hand side of (69) occurs for

$$\theta_{ss} = \theta_{ss}^* = \frac{2(1+\beta)}{2+\beta}$$

and its value at the maximum is $\gamma\beta/4(1+\beta)$. Consequently, inequality (69) leads to

$$\frac{\gamma\beta}{1+\beta} < 4$$

This is the inequality that we have encountered several times in this chapter. If the steady state θ_{ss} is unique, inequality (69) is automatically satisfied, while (68) gives the desired relation between Le and the steady state parameters γ, β, and Φ^2 [where Φ^2 is hidden in $\theta_{ss}(u = u_1)$].

The steady state collocation ordinate $\theta_{ss}(u = u_1)$ is given as the solution of

$$C_{11}\theta - C_{11} + \Phi^2(\beta + 1 - \theta)\exp\left(\gamma\frac{\theta - 1}{\theta}\right) = 0 \tag{70}$$

The solution of (70) is inserted into (68) to give the smallest value of Le where the steady state is asymptotically stable—within the accuracy of the one-point collocation approximation.

The approximate stability limit that is calculated by this almost trivial method may give some valuable insight into the solution of the non-Sturm–Liouville problem (1.82), although the numerical results are, as seen in Exercise 4, not too reliable. Equation (1.82) can be solved, however, without any extreme numerical effort to a very high degree of accuracy by the methods of chapter 9.

The reactor model simplifications of chapter 1 offer a much more fertile ground for numerical studies since these simplifications are often necessary in order to limit what would otherwise be an extreme computational effort. The key to these simplifications is an accurate—or at least a well-argued—modification of transfer coefficients at the reactor wall or at the pellet surface and calculation of an effective axial dispersion coefficient as in subsection 1.2.3.

A discussion of the general principle of this type of calculation follows, and it shows that the same modifications often result from a judicious choice of one collocation point.

Consider, for example, the one-dimensional heat conduction problem

$$\frac{\partial \theta}{\partial \tau} = \frac{\rho c_p R^2}{k}\frac{\partial \theta}{\partial t} = \frac{1}{x^s}\frac{\partial}{\partial x}\left(x^s \frac{\partial \theta}{\partial x}\right) \tag{71}$$

with boundary conditions

$$\begin{aligned} \frac{\partial \theta}{\partial x} &= 0 \quad \text{at } x = 0,\ \tau \geq 0 \\ \frac{\partial \theta}{\partial x} &= -\text{Bi}\,\theta \quad \text{at } x = 1,\ \tau > 0 \end{aligned} \tag{72}$$

The solution of (71) and (72) for spherical geometry is easily obtained by the methods of chapter 1:

$$\theta = \sum_{j=1}^{\infty} a_j \frac{\sin \lambda_j x}{x} \exp(-\lambda_j^2 \tau) \tag{73}$$

$$\bar{\theta} = \int_0^1 \theta \, dx^{s+1} = \sum_{j=1}^{\infty} A_j \exp(-\lambda_j^2 \tau) \tag{74}$$

where

$$A_j = \frac{6\text{Bi}^2}{\lambda_j^2[\lambda_j^2 + \text{Bi}(\text{Bi} - 1)]} = 3\frac{\text{Bi} \sin \lambda_j}{\lambda_j^2} a_j \tag{75}$$

and

$$\lambda_j \cot \lambda_j = 1 - \text{Bi} \tag{76}$$

For any value of Bi, there is one eigenvalue in each interval $[0, \pi], [\pi, 2\pi], \ldots$. For $\text{Bi} \to \infty$, the eigenvalues are exactly at $\pi, 2\pi, \ldots$, while in the other extreme case (Bi=0) the eigenvalues are at $0, \pi, 2\pi, \ldots$.

When Bi decreases from infinity to zero, the ratio λ_2/λ_1 increases from 2 to infinity; for very small Bi, the series (73) and (74) are completely dominated by their first term.

For small values of Bi, all higher-order terms in (74) have coefficients that tend to zero as Bi^2 and we now develop an expression for A_1 that is correct up to and including Bi^1.

Expanding the $\cot\lambda_1$ in (76) in increasing powers of λ_1 yields

$$\lambda_1\left(\frac{1}{\lambda_1} - \frac{\lambda_1}{3} - \frac{\lambda_1^2}{45} - \cdots\right) = 1 - \frac{\lambda_1^2}{3} - \frac{\lambda_1^4}{45} - \cdots = 1 - \text{Bi}$$

Since λ_1 is small, we may insert $\lambda_1^4/9 = \text{Bi}^2$ and obtain

$$\lambda_1^2 = 3\,\text{Bi}\left[1 - \frac{\text{Bi}}{5} + \mathcal{O}(\text{Bi}^2)\right] \tag{77}$$

$$\begin{aligned} A_1 &= \frac{6\,\text{Bi}^2}{3\,\text{Bi}\,[1-(\text{Bi}/5)]\{3\,\text{Bi}\,[1-(\text{Bi}/5)] + \text{Bi}\,(\text{Bi}-1)\}} + \mathcal{O}(\text{Bi}^2) \\ &= 1 + \mathcal{O}(\text{Bi}^2) \end{aligned} \tag{78}$$

Consequently for small Bi we may represent $\bar{\theta}$ of (74) for any value of τ by the first term of the series. The error of the approximation is $\mathcal{O}(\text{Bi}^2)$.

$$\bar{\theta}(\tau) \sim A_1 \exp(-\lambda_1^2\tau) \sim \exp\left[-\frac{3h}{\rho c_p R}\left(1 - \frac{\text{Bi}}{5}\right)t\right] \tag{79}$$

We next define a one-dimensional model in the same way as in sections 1.2 and 1.3. Integrate (71) over the volume of the sphere to obtain

$$\frac{d\bar{\theta}}{dt} = 3\frac{k}{\rho c_p R^2}\left(\frac{\partial\theta}{\partial x}\right)_{x=1} = -3\frac{h}{\rho c_p R}\theta_{x=1} \tag{80}$$

$\theta_{x=1}$ is unknown but it will not be much different from $\bar{\theta}$ if Bi is small. Our one-dimensional model is consequently defined by

$$\frac{d\bar{\theta}}{dt} = -3\frac{\bar{h}}{\rho c_p R}\bar{\theta} \tag{81}$$

The "effective" heat transfer coefficient $\bar{h}$ of the one-dimensional model is smaller than h since $\theta_{x=1} < \bar{\theta}$. $\bar{h}$ will be represented by the "adding of resistances" principle that is frequently used in engineering practice:

$$\frac{1}{\bar{h}} = \frac{1}{h} + \alpha\frac{R}{k} = \frac{1}{h}(1 + \alpha \text{ Bi}) \tag{82}$$

where α is an empirical constant. Evidently $\bar{h} \to h$ when Bi $\to 0$.

We have seen that (79) with λ_1 given by (77) and A_1 by (78) represents the correct solution (74) for $\bar{\theta}$ with an error $\mathcal{O}(\text{Bi}^2)$ for any value of t.

If we wish the solution of the one-dimensional model (81) to represent $\bar{\theta}$ to the same degree of accuracy, we must calculate $\bar{h}$ by

$$-3\frac{\bar{h}}{\rho c_p R} = -\left(1 - \frac{\text{Bi}}{5}\right)\frac{3h}{\rho c_p R} = -\frac{3h}{[1 + (\text{Bi}/5)]\rho c_p R} + \mathcal{O}(\text{Bi}^2)$$

or

$$\frac{1}{\bar{h}} = \frac{1}{h}\left[1 + \frac{1}{5}\text{Bi} + \mathcal{O}(\text{Bi}^2)\right]$$

and we have determined $\alpha = \frac{1}{5}$ in (82).

The flaw of the present method is that a closed-form expression (76) for the eigenvalues of the complete model must be available to obtain the constant α in (82). A simple technique that does not require a closed-form solution of the full model is presented next. It is similar to that used in our analysis of the accuracy of Galerkin's method in section 2.5, and we develop the technique further in section 9.1.

The eigenfunctions F_j of (71) are solutions of

$$\lambda F = \frac{1}{x^s}\frac{d}{dx}\left(x^s\frac{dF}{dx}\right) \tag{83}$$

$$\frac{dF}{dx} = 0 \quad \text{at } x = 0$$

$$\frac{dF}{dx} = -\text{Bi}\, F \quad \text{at } x = 1 \tag{84}$$

$$\theta = \sum_1^\infty a_j F_j(x) \exp\left(-\lambda_j \frac{k}{\rho c_p R^2} t\right)$$

For the trivial problem (71) the eigenfunctions are of course known [$F_j = \sin \lambda_j x$, where λ_j is a solution of (76)], but here we present the first eigenfunction and its corresponding eigenvalue in form of infinite series in increasing powers of Bi:

$$\lambda_1 = q_0 + q_1 \text{Bi} + q_2 \text{Bi}^2 + \dots$$

$$F_1 = f_0 + f_1 \text{Bi} + f_2 \text{Bi}^2 + \dots \tag{85}$$

The zero-order perturbation terms q_0 and f_0 are known from the solution of (71) with Bi = 0: $q_0 = 0$ and $f_0 = 1$.

Higher-order perturbation terms are found by inserting (85) into (83) and (84) and collecting terms that are multiplied by Bi, $\text{Bi}^2, \ldots$. From (83) a chain of differential equations in $f_1, f_2, \ldots$ results:

$$q_0 f_1 + q_1 f_0 = \frac{1}{x^s}\frac{d}{dx}\left(x^s \frac{df_1}{dx}\right) \tag{86}$$

$$q_0 f_2 + q_1 f_1 + q_2 f_0 = \frac{1}{x^s}\frac{d}{dx}\left(x^s \frac{df_2}{dx}\right) \tag{87}$$

or, in general,

$$\sum_{k=0}^{N} q_k f_{N-k} = \nabla^2 f_N \tag{88}$$

Boundary conditions for (88) are obtained from (84):

$$\frac{df_N}{dx} = 0 \quad \text{at } x = 0 \quad \text{and} \quad \frac{df_N}{dx} + f_{N-1} = 0 \quad \text{at } x = 1 \tag{89}$$

An extra side condition $f_N(0) = 0$ may be imposed upon all f_N ($N > 0$) since F_1 can be normalized to 1 at $x = 0$ and $f_0(0) = 1$.

Integrating (86) and using the three side conditions (89) gives

$$q_1 = -(s+1) \quad \text{and} \quad f_1 = \frac{1}{2(s+1)} q_1 x^2 = -\frac{1}{2}x^2$$

This result is inserted into (87), which on integration gives

$$q_2 = \frac{s+1}{s+3} \quad \text{and} \quad f_2 = \frac{s+1}{8(s+3)}x^4 + \frac{1}{2(s+3)}x^2$$

Higher-order perturbations can be found when q_1, q_2, f_1, and f_2 are inserted into (88) with $N = 3, 4, \ldots$ but we stop here since terms up to and including Bi^2 in the perturbation series for λ_1 have now been obtained.

$$\lambda_1 \frac{k}{\rho c_p R^2} = \lambda_1 \frac{h}{\rho c_p R \,\text{Bi}} = \frac{h}{\rho c_p R}\left[-(s+1) + \frac{s+1}{s+3}\text{Bi}\right] + \mathcal{O}(\text{Bi}^2)$$

Returning to our definition of $\bar{h}$ in (81) and repeating the argument that was used to obtain α in (82) from the analytical solution of (71), we obtain

$$-(s+1)\frac{\bar{h}}{\rho c_p R} = -(s+1)\frac{h}{\rho c_p R}\left(1 - \frac{1}{s+3}\text{Bi}\right)$$

or

$$\bar{h} = h\left(1 - \frac{1}{s + 3}\text{Bi}\right) = h\left\{\frac{1}{1 + [\text{Bi}/(s + 3)]}\right\} + \mathcal{O}(\text{Bi}^2) \tag{90}$$

which means that $\alpha = 1/(s + 3)$ for the three geometrics $s = 0$, 1, and 2. The approximation error is $\mathcal{O}(\text{Bi}^2)$.

In the case $s = 2$ we have seen that the present approximation for the average temperature by the solution of a one-dimensional model matches the preexponential factor A_1 as well as the exponent up to and including the first-order term in Bi. Using the orthogonality properties of the eigenfunctions, this can be shown to be true for any Sturm–Liouville problem with uniform initial distribution of the dependent variable.

The best one-dimensional homogeneous reactor flow model is discussed in section 1.2. Formula (1.27) gives the best choice of an effective wall heat transfer coefficient $\bar{U}$ when radial gradients are neglected by the averaging process (1.17). The value $\alpha = \frac{1}{4}$ for a flat velocity profile is now well understood since it results from (90) with $s = 1$—the flow problem in a reactor is, of course, identical to (71) with $s = 1$ when the velocity is constant in the reactor cross section. For a laminar velocity profile $v_z = 2v_{\text{av}}(1 - x^2)$ the parent eigenvalue problem cannot be solved analytically, but the technique described in this chapter is applied just as easily to give

$$\frac{1}{\bar{U}} = \frac{1}{U} + \frac{11}{24}\frac{R}{k^r} \tag{91}$$

We finally see how a result similar to (90) can be derived by one-point collocation. The discrete version of (71) is

$$\frac{2(s + 1)}{1 - u_1}(\theta_{x=1} - \theta_1) = \frac{d\theta_1}{d\tau} \tag{92}$$

where θ_1 is the value of θ at $u = x^2 = u_1$.

$\theta_{x=1}$ is eliminated by means of the discrete version of the boundary condition at $u = 1$:

$$2\left(\frac{\partial\theta}{\partial u}\right)_{u=1} = \frac{2}{1 - u_1}(\theta_{x=1} - \theta_1) = -\text{Bi}\,\theta_{x=1} \tag{93}$$

or

$$\frac{d\theta_1}{d\tau} = \frac{\rho c_p R^2}{k}\frac{d\theta_1}{dt} = -\frac{(s + 1)\,\text{Bi}}{1 + [\text{Bi}\,(1 - u_1)/2]}\theta_1$$

The average temperature is found by a Gauss–Legendre quadrature

$$\bar{\theta} = \int_0^1 \theta\,dx^{s+1} \sim \theta(u = u_1)$$

$\bar{h}$ is defined by the relation

$$\frac{d\bar{\theta}}{dt} \simeq \frac{(s + 1)h}{R\rho c_p} \frac{1}{1 + [\text{Bi}\,(1 - u_1)/2]}\theta(u = u_1) \equiv -\frac{(s + 1)\bar{h}}{R\rho c_p}\theta(u = u_1)$$

or

$$\frac{1}{\bar{h}} = \frac{1}{h} + \frac{1 - u_1}{2}\frac{R}{k} \tag{94}$$

We have seen in subsection 6.1.5 that provided the collocation abscissa u_1 was chosen at the zero of

$$P_1^{[0,(s-1)/2]}(u) = \frac{s + 3}{s + 1}u - 1$$

the terms up to and including u in the solution are correctly represented. Inserting $u_1 = (s + 1)/(s + 3)$ leads exactly to the choice $\alpha = 1/(s + 3)$ of formula (90) and a one-dimensional model that is correct up to and including the linear term in Bi. This result is expected since the development that we have given in this chapter is in essence the same as the one used in chapter 2 to prove that one particular choice of collocation points leads to the highest possible number of correct terms in $\eta(\Phi)$.

$u_1 = \frac{1}{2}$ appears to be the best choice of a collocation point for a tubular reactor (1.14) and (1.15) when the velocity profile is flat and the wall heat transfer coefficient U is small. For laminar flow in an empty tubular reactor, $u_1 = \frac{1}{2}$ yields the correct representative velocity [$v(u = \frac{1}{2}) = 2v_{\text{av}}(1 - u_1) = v_{\text{av}}$, but α of (90) is calculated to $\frac{1}{4}$ while the correct value of α should be $\frac{11}{24}$ (see Exercise 6)].

It should be noted that no one-point collocation method—not even the method of subsection 6.1.5—leads both to the correct value of v_{av} and of α in this case. Neither can the Taylor dispersion approach—the inclusion of radial gradients via an effective axial dispersion coefficient—that is used in table 1.4 to give an accurate one-dimensional approximation for (1.3.4) be found by one-point collocation. The perturbation method is still applicable, as seen in Exercise 7.

Finlayson (1971) proposes to use $u_1 = \frac{1}{3}$ in a one-dimensional approximation for the tubular reactor with radial concentration and temperature gradients but a flat velocity profile. This choice of u_1 that leads to $\alpha = \frac{1}{3}$ in (90) is less satisfactory than $u_1 = \frac{1}{2}$ when Bi is small. For large Bi the boundary condition at $x = 1$ becomes the appropriate one for collocation at the zeros of $P_N^{(1,0)}(u)$ in cylinder geometry but now a one-point collocation cannot suffice to represent the pronounced radial gradients in the reactor.

The question of when Bi is "small enough" to permit a fully satisfactory lumping of the two-dimensional uniform velocity reactor model may qualitatively be discussed on the basis of table 6.6 where $\lambda_1/2\,\mathrm{Bi}$ [λ_1 is the solution of $\lambda J_1(\lambda) - \mathrm{Bi}\,J_0(\lambda) = 0$] is taken to represent $\bar{u}/u$ from the full model while $1/(1 + 0.25\,\mathrm{Bi})$ is the result obtained from (90).

For small Bi ($\lesssim 0.2$), there is hardly any difference between the two models. For Bi $\lesssim 2$, there is still no more than a few percent difference but now, of course, the first term of the full solution does not adequately represent $\bar{\theta}$ for small ζ. The results of the table should be interpreted as accordance of the two models for large values of the axial coordinate ζ where only one term in the series is necessary to represent the solution.

TABLE 6.6
COMPARISON OF TWO-DIMENSIONAL AND ONE-DIMENSIONAL REACTOR MODELS

Bi	$(\bar{h}/h)_{\mathrm{exact}} \sim \lambda_1/2\,\mathrm{Bi}$	$(\bar{h}/h)_{\mathrm{appr.}} = 1/(1 + 0.25\,\mathrm{Bi})$
0.01	0.9975	0.9975
0.1	0.9755	0.9756
0.2	0.9517	0.9523
0.5	0.8851	0.8889
1	0.789	0.800
2	0.640	0.667
5	0.400	0.444

6.3 One-Point Collocation for Initial Value Problems

The collocation methods of section 4.2 for solution of single or coupled first-order differential equations are developed from considerations on simple linear problems just as the optimal collocation method of chapter 2 for boundary value problems with known ordinate at $x = 1$ is derived for the simplest possible two-point boundary value problem (34). For nonlinear boundary value problems we would still recommend the use of, e.g., zeros of $P_N^{(1,0)}(u)$ for collocation in cylinder geometry when a high accuracy of the first few power series coefficients is desired, since an optimal quadrature equivalent to Galerkin's method is obtained. We cannot, as in section 2.5, prove that a certain number of terms in a perturbation series (2.103) for the Fourier coefficients are correct.

The optimality of the stepwise collocation of section 4.2 does, however, carry over also for nonlinear equations. This is presently proved rigorously for one collocation point in each subinterval, but a general proof is outside the scope of the text.

We solve

$$\frac{dy}{dx} = f(x, y) \quad \text{with } y = y_0 = 0 \quad \text{at } x = x_0 = 0 \tag{95}$$

in a subinterval of size h using one collocation point $x_1 = h/3$, i.e., by the predictor-corrector methods (4.31) to (4.35) based on the zeros of $P_N^{(1,0)}(x)$.

First we develop a solution to (95) that is correct to $\mathcal{O}(h^4)$. The one-point collocation solution subsequently is compared with this solution to prove that the collocation solution has an error $\mathcal{O}(h^4)$.

Continued differentiation of (95) gives

$$d\left(\frac{dy}{dx}\right) = \frac{\partial f}{\partial x}\,dx + \frac{\partial f}{\partial y}\,dy = f_x\,dx + f_y\,dy$$

or

$$\frac{d^2y}{dx^2} = f_x + \frac{\partial f}{\partial y}\frac{dy}{dx} = f_x + f_y f$$

$$d\left(\frac{d^2y}{dx^2}\right) = \frac{\partial}{\partial x}(f_x + f_y f)\,dx + \frac{\partial}{\partial y}(f_x + f_y f)\,dy$$

or

$$\frac{d^3y}{dx^3} = f_{xx} + f_y f_x + (2f_{xy} + f_y^2 + f f_{yy})f$$

Now $y(h)$ can be obtained in terms of f and the partial derivatives of f—all taken at $x_0 = 0$:

$$\begin{aligned} y(h) &= y_0 + \left(\frac{dy}{dx}\right)_{x_0} h + \frac{1}{2}\left(\frac{d^2y}{dx^2}\right)_{x_0} h^2 + \frac{1}{6}\left(\frac{d^3y}{dx^3}\right)_{x_0} h^3 + \mathcal{O}(h^4) \\ &= hf + \frac{1}{2}h^2(f_x + f f_y) + \frac{1}{6}h^3[f_{xx} + f_x f_y + (2f_{xy} + f_y^2 + f f_{yy})f] \\ &\quad + \mathcal{O}(h^4) \end{aligned} \tag{96}$$

We shall now develop the collocation solution: The ordinate y_1 at the collocation ordinate $x_1 = h/3$ is given by

$$3(y_1 - y_0) = 3y_1 = hf\left(\frac{h}{3}, y_1\right) \tag{97}$$

$f(h/3, y_1)$ is found by a Taylor series from (x_0, y_0) and y_1 is inserted as a second-degree polynomial in h with coefficients a_1 and a_2 to be determined in the following:

$$f(x, y) = f + xf_x + yf_y + \tfrac{1}{2}x^2 f_{xx} + \tfrac{1}{2}y^2 f_{yy} + xyf_{xy} + \ldots \tag{98}$$

where f and the partial derivatives on the right-hand side of (98) are evaluated at $(x_0, y_0) = (0, 0)$.

$$y_1 = ha_1 + h^2 a_2 + \dots \tag{99}$$

$$f\left(\frac{h}{3}, y_1\right) = f + \frac{h}{3}f_x + (a_1 h + a_2 h^2)f_y + \frac{1}{18}h^2 f_{xx} + \frac{1}{2}(a_1 h + a_2 h^2)^2 f_{yy} + \frac{h}{3}(a_1 h + a_2 h^2)f_{xy} \tag{100}$$

Equations (99) and (100) are substituted into (97) and (a_1, a_2) are determined by equating coefficients to h and h^2 on both sides of the equation.

$$a_1 = \tfrac{1}{3}f \quad \text{and} \quad a_2 = \tfrac{1}{9}(f_x + ff_y) \tag{101}$$

Inserting these values into (99) gives

$$y(h) = 3y_1 - 2y_{x=0} = 3y_1 = fh + \tfrac{1}{3}(f_x + ff_y)h^2$$

and comparison with (96) shows that this preliminary $y(h)$ has an error $\mathcal{O}(h^2)$.

The corrected value of $y(h)$ is given by a quadrature

$$y_c(h) = y_0 + h\left[\frac{3}{4}f\left(\frac{h}{3}, y_1\right) + \frac{1}{4}f(h, 3y_1)\right] \tag{102}$$

$f(h/3, y_1)$ is given by

$$f\left(\frac{h}{3}, y_1\right) = f + \frac{h}{3}(f_x + ff_y) + h^2\left(\frac{1}{18}f_{xx} + f_y\frac{f_x + ff_y}{9} + \frac{f^2}{18}f_{yy} + \frac{ff_{xy}}{9}\right) + \mathcal{O}(h^3)$$

while

$$\begin{aligned} f(h, 3y_1) &= f + hf_x + 3(a_1 h + a_2 h^2)f_y + \frac{1}{2}h^2 f_{xx} + \frac{9}{2}(a_1 h + a_2 h^2)^2 f_{yy} \\ &\quad + 3h(a_1 h + a_2 h^2)f_{xy} + \dots \\ &= f + h(f_x + ff_y) + h^2\left(\frac{1}{2}f_{xx} + f_y\frac{f_x + ff_y}{3} + \frac{1}{2}f^2 f_{yy} + ff_{xy}\right) \\ &\quad + \mathcal{O}(h^3) \end{aligned} \tag{103}$$

$$\begin{aligned} y_c(h) &= \frac{h}{4}\left[3f\left(\frac{h}{3}, y_1\right) + f(h, 3y_1)\right] \\ &= hf + \frac{1}{2}h^2(f_x + ff_y) + \frac{1}{6}h^3[f_{xx} + f_y f_x + (2f_{xy} + f_y^2 + ff_{yy})f] \\ &\quad + \mathcal{O}(h^4) \end{aligned}$$

It is seen that the corrected y-value at $x = h$ is the same as the value obtained in (96).

EXERCISES

1. Use formulas (6.46) to (6.50) to calculate the effectiveness factor $\eta(\phi)$ for a first order isothermal reaction in plane parallel symmetry. Let A equal 5. Derive the corresponding analytical solution and compare the power series expansions to confirm that terms up to and including Φ^4 are correctly represented.

2. For a cube of edge length e, the volume to surface area ratio is $e^3/6e^2 = (e/2)/3$. The appropriate Thiele modulus Λ for a first-order isothermal reaction is $\Lambda = (e/2)/3\sqrt{k/D}$.

a. Derive a one-point collocation expression for $\eta(\Lambda)$ and use this to compute the first approximation η_1 to $\eta(\Lambda = 1)$.
b. How many collocation equations must be set up to obtain the approximation for η based on two interior points in each coordinate direction?
c. Derive the quadrature formula for this approximation.
d. Set up the collocation equations for $\Lambda = 1$, solve for the ordinates, and calculate η_2.

3. In section 6.1 an optimal A-dependent Galerkin collocation method was developed. Collocation at the zeros of $P_N^{1,(s-1)/2}(u)$ using the boundary condition (35) to eliminate $y(1)$ was inferior compared with a method of moments collocation method based on $P_N^{0,(s-1)/2}(u)$.
The boundary condition at $x = 1$ can be accounted for, however, by integration of the differential equation

$$\int_V \nabla^2 y\, dV = \nabla y|_{x=1} = A[1 - y(1)] = \Phi^2 \int_V R(y)\, dV \tag{1}$$

It may be shown that quadrature evaluation of the integral using a Radau formula gives a boundary equation of the same accuracy as the collocation equations. Consequently collocation at the zeros of $P_N^{[1,(s-1)/2]}(u)$ combined with elimination of $y(1)$ by (1) should have an accuracy of the same order as the A-dependent collocation method.

a. Solve $\nabla^2 y = \Phi^2 y$ with $y^{(1)} + Ay = A$ at $x = 1$ and $y^{(1)}(0) = 0$ by this method for $s = 0$ and $N = 1$. Compare the result with a power series expansion of the true solution for η and show that the coefficients to Φ^0, Φ^2, and Φ^4 are correct. Also compare the coefficients to Φ^6 to see how small the error of the first erroneous term is.
b. Solve $\nabla^2 y = \Phi^2 y^2$ for $A = 5$, $s = 1$, and $\Phi = 2$ by
   i. Method of moments collocation method with $N = 1$ and $N = 2$.
   ii. Simple collocation using zeros of $P_N^{(1,0)}(u)$ for $N = 1$.
   iii. Improved Galerkin collocation again using zeros of $P_N^{(1,0)}(u)$ for $N = 1$.
   iv. The method of subsection 6.1.5.

   Which result is probably most accurate?
   Answers for η:
   i. $N = 1$:: 0.4045
   $N = 2$: 0.4311 (or 0.4288, depending on whether η is found from slope at $x = 1$ or from a quadrature).

Why is there a difference in the two results for $N = 2$?
ii. 0.4581
iii. 0.43397
iv. 0.43559

c. Solve $\nabla^2 y = \Phi^2 y^2$ for $s = 1$ and $\Phi = 2$ with boundary condition $y^{(1)} = 5(1 - y^2)$ at $x = 1$ by the improved Galerkin collocation method. Compare accuracy/required effort for this method and for method of moments collocation in this example and in the example of part b with a linear boundary condition.
Answer: $\eta = 0.4980$

4. a. Find the steady state solution of (70) for $\Phi = 1.1$, $\gamma = 30$, $\beta = 0.15$, and spherical symmetry.
b. Calculate the stability limit $\text{Le}_{\min}$ for this set of parameters.
c. Draw a sketch of $\text{Le}_{\min}(\theta)$ and determine a value Φ^* below which the system is stable for all values of $\text{Le} \geq 0$.
d. The sketch of $\text{Le}_{\min}(\theta)$ is based on a low-order approximation for $\theta(x)$. Where do you suspect the sketch to give a qualitatively incorrect picture of the stability properties of the system?
The parameter values of this example were also used by Hellinckx, Grootjans, and van den Bosch (1972).
The true value of $\text{Le}_{\min}$ in part b is 0.39, while Φ^* of part c should be 1.014.

5. Derive a one-point collocation graphical method analogous to the Stewart–Villadsen method of subsection 6.1.2, but for the general boundary conditions

$$\frac{dy}{dx} = \text{Bi}_M(1 - y) \qquad \frac{d\theta}{dx} = \text{Bi}(1 - \theta) \quad \text{at } x = 1$$

Result:

$$-C_{11}(1 - y) = (\Phi^*)^2 y \exp\left[\gamma\left(1 - \frac{1}{\theta}\right)\right]$$

where $\theta = 1 + \beta^*(1 - y)$.

$$\Phi^* = \Phi\sqrt{1 + \frac{1}{\text{Bi}_M}} \qquad \beta^* = \beta\frac{1 + (1/\text{Bi})}{1 + (1/\text{Bi}_M)}$$

and (y, θ) equals the collocation point value of y and θ.
How would you choose the collocation point?

6. Show that for laminar flow in an empty tube the effective wall heat transfer coefficient $\bar{U}$ is given by

$$\frac{1}{\bar{U}} = \frac{1}{U} + \frac{11}{24}\frac{R}{k}$$

7. In Exercise 4.12 the full model for laminar flow with first-order isothermal reaction in a tubular reactor is solved by Nth-order collocation and it is shown by an analysis of the computer output that

$$\begin{aligned}\lambda_1 &= -\frac{kL}{v_{av}}\left[1 - \frac{1}{48}\frac{kR^2}{D^r} + \mathcal{O}\left(\frac{kR^2}{D^r}\right)^2\right] \\ &= -\text{Da}\left[1 - \frac{1}{48}\text{Da}\frac{R^2 v_{av}}{D^r L} + \mathcal{O}\left(\text{Da}\frac{R^2 v_{av}}{D^r L}\right)^2\right]\end{aligned} \tag{1}$$

a. Derive the result (1) by the perturbation method of subsection 6.2.2.
b. In the equivalent one-dimensional model, radial gradients are taken into account through an "effective" axial diffusion term (1.43):

$$\frac{dy}{d\zeta} + \frac{kL}{v_{av}}y - \frac{1}{\text{Pe}_{ef}}\frac{d^2y}{d\zeta^2} = 0 \tag{2}$$

The boundary conditions are given in (1.45):

$$\begin{aligned} y(0) - \frac{1}{\text{Pe}_{ef}}\left(\frac{dy}{d\zeta}\right)_{\zeta=0} &= 1 \\ \left(\frac{dy}{d\zeta}\right)_{\zeta=1} &= 0 \end{aligned} \tag{3}$$

Solve equations (2) and (3) and show that the exit concentration is

$$y(\zeta = 1) = \exp\left[-\frac{kL}{v_{av}}\left(1 - \frac{kL}{v_{av}\text{Pe}_{ef}}\right)\right] + \mathcal{O}\left(\frac{kL}{v_{av}\text{Pe}_{ef}}\right)^2$$

and consequently by comparison with (1) that the best choice of Pe_{ef} is

$$\text{Pe}_{ef} = 48\frac{D_r L}{v_{av}R^2} \tag{4}$$

c. Equation (2) can also be solved using so-called semiinfinite boundary conditions

$$y(0) = 1 \quad \text{and} \quad y \to \text{constant for } \zeta \to \infty \tag{5}$$

Show that the solution of (2) and (5) differs from the solution of (2) and (3) (or of the full model) only in terms of order $(\text{Da}/\text{Pe}_{ef})^2$. Consequently, no advantage is obtained by using the "complicated" boundary conditions (3) rather than (5) when the radial gradients are taken into account via the Taylor dispersion model approximation.

REFERENCES

1. STEWART, W. E., and VILLADSEN, J. *AIChE Journal* 15 (1969):28.
2. BISCHOFF, K. B. *Chem. Eng. Sci.* 22 (1967):525.
3. LUSS, D. *Chem. Eng. Sci.* 23 (1968):1249.
4. MICHELSEN, M. L., and VILLADSEN, J. *Chem. Eng. Sci.* 27 (1972):751.

5. Paterson, W. R., and Cresswell, D. L. *Chem. Eng. Sci.* 26 (1971):605.

6. Van den Bosch, B., and Padmanabhan, L. *Chem. Eng. Sci.* 29 (1974):1217.

7. Hatfield, B., and Aris, R. *Chem. Eng. Sci.* 24 (1969):1213.

8. Stewart, W. E., and Sørensen, J. P. Transactions 2nd ISCRE, (May, 1972), paper B8-(75-88), Amsterdam: Elsevier (1972).

9. Villadsen, J., and Stewart, W. E. *Chem. Eng. Sci.* 22 (1967):1483.

10. Sørensen, J. P., Guertin, E. W., and Stewart, W. E. *AIChE Journal* 19 (1973):969, 1286.

11. Aris, R. *Chem. Eng. Sci.* 6 (1957):262.

12. Hellinckx, L., Grootjans, J., and Van den Bosch, B. *Chem. Eng. Sci.* 27 (1972):644.

13. Finlayson, B. A. *Chem. Eng. Sci.* 26 (1971):1081.

14. Guertin, E. W., Sørensen, J. P., and Stewart, W. E. Paper 10a. *AIChE National Meeting*, Boston, Mass. (1975).

Global Spline Collocation 7

Introduction

Problems that are cataloged as *difficult* in some way or the other are best treated as highly individual species. This unfortunately makes a concise textbook approach to such problems less satisfactory than is the case for the simple problems that are taken up in chapter 4 to illustrate standard collocation procedures. Chapter 6, "One-Point Collocation," introduced the concept of a "burnt out" and a "reaction" zone (i.e., a model approximation in which the dependent variable was equated to zero in part of the range of the independent variable) to treat steep profiles in catalyst pellets. We now formalize this procedure and point to its application to continuous models with a wedge-shaped or otherwise strange looking solution but also to models that contain a discontinuity in the second derivative or in a higher derivative of the solution at one or more values of the independent variable. The isothermal catalyst effectiveness problem for very large Thiele modulus or the Weisz–Hicks problem with significant heat effects are mathematically speaking very well-behaved problems—$y(x)$ has continuous derivatives of any order—but they are very difficult to solve by a global approximation method based on polynomials due to the strange shape of the solution. The unsatisfactory accuracy obtained by global polynomial approximation for the inlet section of a fluid flow—heat transfer problem or for the start (small t) of a stationary diffusion problem is discussed in chapter 4. The wedge-shaped profile of the "penetration front" is as difficult to determine numerically as the very similar looking solution of the one-dimensional two-point boundary value problems.

True discontinuities of $y^{(2)}(x)$ or of some higher derivative can be found in eigenvalue problems as well as in two-point boundary value problems. Partial differential equations may also contain a discontinuity in the side conditions, i.e., a discontinuous initial profile.

A forward integration technique using N_i collocation points in integration step d_i is used in chapters 4 and 5 to produce very satisfactory approximations even when as in subsection 5.5.5 a solution by global collocation was impossible. The resulting approximate solution is spliced together from polynomial functions of degree N_i in integration step d_i using continuity of y and dy/dx at the spline points.

Here we introduce a combination of this spline method and global collocation using the name *global spline collocation* for the hybrid method: In each of M subintervals $d_1, d_2, \ldots, d_i, \ldots, d_M$ collocation is applied at N_i interior points. The collocation equations of subinterval d_i are set up using the collocation abscissas and the two subinterval end points as nodes. Continuity of $y(x)$ and of $y^{(1)}(x)$ is imposed at the junctions between intervals and this gives a sufficient number of equations to determine the collocation ordinates as well as the spline-point ordinates.

The first example is a boundary value problem with a discontinuity in $y^{(2)}(x)$ or of $y^{(3)}(x)$ at $x = x^*$. A model with a rapidly varying solution follows as the first (almost trivial) application of global spline collocation to penetration problems.

Eigenvalue problems are treated in nearly the same way as two-point boundary value problems as shown in the next examples—one for a one-dimensional eigenvalue problem and one that sets the scope for penetration front calculations in general. Finally a discontinuous initial profile for a partial differential equation is briefly discussed.

No selection of examples can illustrate the possibilities that are offered by the proposed relaxation of high-order continuity of the approximate profiles. Basically the method belongs in the numerical analyst's "bag of tricks" and each example is probably best treated on an individual basis, exploiting whatever is known semianalytically or otherwise about the solution.

7.1 Global Spline Collocation for Two-Point Boundary Value Problems

7.1.1 Definition of the method

Let y be given as the solution of

$$\nabla^2 y - f(x, y) = 0 \tag{1}$$

$$y^{(1)} + Ay = C_1 \quad \text{at } x = 0, \quad y^{(1)} + By = C_2 \quad \text{at } x = 1 \tag{2}$$

Divide the x-interval $[0, 1]$ into M-subintervals $d_1, d_2, \ldots, d_i, \ldots, d_M$ of not necessarily equal length and collocate at N_i interior points of subinterval d_i. Impose the conditions $y_-(x_{si}) = y_+(x_{si})$ and $y_-^{(1)}(x_{si}) = y_+^{(1)}(x_{si})$ at the junctions x_{si} between the subintervals. The continuity of $y^{(1)}$ is expressed in terms of the discretization matrices $\mathbf{A}_-$ and $\mathbf{A}_+$ based on the ordinates of the interval to the left and to the right of the junction x_{si} including both interval end points.

There results a set of algebraic equations $\mathbf{F}(\mathbf{y}) = \mathbf{0}$ in the $\sum_1^M N_i$ collocation ordinates, the $M - 1$ spline-point ordinates, and possibly the two boundary ordinates, unless one or both are given explicitly through the boundary conditions. If the left-hand boundary ordinate is unknown, it can be expressed as a linear combination of the collocation ordinates in d_1 and the first spline ordinate at x_{s1}. Likewise, the right-hand ordinate can be expressed in terms of the collocation ordinates in d_M and the spline ordinate at $x_{s,M-1}$.

Thus the total number of unknowns and equations can be reduced to $\sum_1^M N_i + M - 1$ and the structure of the system of algebraic equations is shown graphically in (3). Points marked + are obtained from the $\mathbf{B}$ and $\mathbf{A}$ matrices alone, while points marked 0 contain, e.g., $B_{jj}y_j$ modified by the nonlinearity $f(x_j, y_j)$ in (1).

$$
\begin{array}{ccccccccccc}
0 & + & + & + & + & & & & & & \\
+ & 0 & + & + & + & & & & & & \\
+ & + & 0 & + & + & & & & & & \\
+ & + & + & 0 & + & & & & & & \\
+ & + & + & + & + & + & + & + & + & & \\
 & & & & + & 0 & + & + & + & & \\
 & & & & + & + & 0 & + & + & & \\
 & & & & + & + & + & 0 & + & & \\
 & & & & + & + & + & + & + & + & + \\
 & & & & \Big| & & & & + & 0 & + \\
 & & & & \Big| & & & & + & + & 0 \\
 & & & & \downarrow & & & & \downarrow & & \\
 & & & & x_{s1} & & & & x_{s2} & &
\end{array}
\tag{3}
$$

If the problem is symmetric in x, polynomials in $u = x^2$ can be used in the innermost interval to eliminate the boundary condition at $x = 0$ immediately. Otherwise there is probably no particular advantage in choosing even polynomials and perhaps the best choice is to use $P_{N_i}^{(0,0)}(x)$ throughout the subintervals. The fifth line of (3) expresses that

$$
\left.\frac{dy}{dx}\right|_{x_{s1-}} \left(\text{or } 2\left.\frac{dy}{du}\right|_{x_{s1-}}\right) = \left.\frac{dy}{dx}\right|_{x_{s1+}} \tag{4}
$$

$$\frac{1}{d_1}\sum_{i=1}^{5} A^-_{5i}y_i\left(\text{or}\,\frac{2}{d_1}\sum_{i=1}^{5} A^-_{5i}y_i\right) = \frac{1}{d_2}\sum_{5}^{9} A^+_{1,k-4}y_k \tag{5}$$

where $\mathbf{A}^-$ is the discretization matrix for dy/dx in d_1 (based on $N_1 + 2$ or $N_1 + 1$ interpolation nodes) and $\mathbf{A}^+$ is the discretization matrix for dy/dx in d_2 (based on $N_2 + 2$ interpolation nodes).

In (3), line 9 is set up in the same way as line 5 except that now

$$\left.\frac{dy}{dx}\right|_{x_{s2-}} = \left.\frac{dy}{dx}\right|_{x_{s2+}} \tag{6}$$

is expressed in terms of the appropriate elements of $\mathbf{A}$ matrices to the left and to the right of x_{s2}.

The Jacobian matrix $\mathbf{J}$ is (as usual) constructed in parallel with $F_j(\mathbf{y})$ for a given estimate of the ordinate vector $\mathbf{y}$, and in each new iteration only the diagonal elements of $\mathbf{J}$ and $F_j(\mathbf{y})$ need to be recomputed.

The structure of (3) is quite sparse and this may if necessary be used in computer storage of the system of equations. The block structure of $F_j(y_i)$ is automatically utilized in a normal Gauss elimination routine with pivotation of rows within each block.

The spline ordinates may be eliminated just as the ordinates at the boundaries of the original interval. In this way the number of equations is further reduced at the expense of obtaining a full matrix. This may be a proper procedure if many spline points with small N_i are used.

To find the ordinate at a given abscissa using the spline points and the collocation points as nodes, the interpolation program of chapter 3 can be used exactly as described there. If a given x falls in subinterval d_i, only those ordinates that belong to d_i (that is, the interior collocation ordinates and the two end point ordinates) are used.

The integral of y or of some function of y [e.g., $f(x, y)$ in (1)] is found by Gauss–Jacobi quadrature on each subinterval.

7.1.2 Pore mouth poisoning

The global spline collocation procedure was tested on a model for pore mouth poisoning with diffusion restriction. The main reaction is isothermal, second order on cylindrical catalyst pellets, and the activity of the catalyst at a given time is described by the following expression in x:

$$\text{Rate constant} = \begin{cases} k_0 & \text{for } x \le x^* \\ k_0 \exp[-q(x - x^*)] & \text{for } x > x^* \end{cases} \tag{7}$$

This leads to the following differential equation for dimensionless concentration y.

$$\frac{d^2y}{dx^2} + \frac{1}{x}\frac{dy}{dx} - \Phi^2 y^2 = 0 \quad \text{for } x \le x^* \tag{8}$$

$$\frac{d^2y}{dx^2} + \frac{1}{x}\frac{dy}{dx} - \Phi^2 \exp[-q(x - x^*)]y^2 = 0 \quad \text{for } x^* < x \le 1$$

$$\frac{dy}{dx} = 0 \quad \text{at } x = 0 \quad \text{and} \quad y = 1 \quad \text{at } x = 1 \tag{9}$$

The effectiveness factor is defined as

$$\eta = \int_0^{x^*} y^2\,dx + \int_{x^*}^{1} \exp[-q(x - x^*)]y^2\,dx \tag{10}$$

Table 7.1 shows η as a function of N in a global collocation process for $\phi^2 = 4$ and $q = 10^6$ in (9), i.e., a situation where the rate of reaction abruptly drops to zero for $x > x^* = 0.7$.

TABLE 7.1
RESULTS OF GLOBAL COLLOCATION METHOD

N	6	7	8	9	10	11	12	13	15
$y(x = 0.7)$	0.837	0.803	0.833	0.806	0.831	0.808	0.829	0.810	0.811
η	0.225	0.278	0.232	0.273	0.236	0.270	0.238	0.268	0.266

The process does not reach a stable value of η or of $y(x = 0.7)$ for a reasonably small N. Exponential extrapolation of the six-digit upper sequence results for η yields a limit of 0.260783, while the lower sequence (N even) yields a different number, 0.246383. The true value of η is 0.24902 as found below by spline collocation, and it may be concluded that global collocation is unsuited for the present problem.

The profile is very well behaved: For $\phi^2 = 4$ and $q = 0$ (i.e., no poisoning), $N = 4$, 6, and 8 all yield $y(x = 0) = 0.56370$ and $\eta = 0.592214$. Hence collocation in each of the two intervals $d_1 = [0, 0.7]$ and $d_2 = [0.7, 1]$ is expected to give equally rapid convergence. This is confirmed by the results of table 7.2:

TABLE 7.2
GLOBAL SPLINE COLLOCATION. SPLINE POINT AT $x_{S1} = 0.7$

N_1, N_2	(6, 2)	(5, 3)	(4, 4)	(3, 5)
$y(0)$	0.60923	0.60924	0.60924	0.60921
$y(0.7)$	0.822346	0.8223634	0.8223634	0.8223636
η	0.249008	0.2490172	0.2490173	0.2490172

A total of eight collocation points is entirely satisfactory and different allocations of points in the two subintervals [0, 0.7] and [0.7, 1] give practically the same result. Three points in the inner interval [0, 0.7] is sufficient, while two points in [0.7, 1] is barely sufficient to represent the logarithmic concentration profile in the outer, reaction–free zone of the pellet.

The global spline collocation procedure was also used for an intermediate value of q in (9). With $q = 100$ the activity drops by a factor e^{10} when x increases from 0.7 to 0.8. $y(x)$ is still a smooth function ~0.8–0.9 in this interval where the rate of reaction [or $y^2(x)$] drops from $\sim 4 \cdot 0.8^2 = 2.6$ to practically zero. It may be conjectured that insertion of two spline points, one at $x = 0.7$ and one somewhat arbitrarily between 0.7 and 0.8, may lead to the most satisfactory allocation of collocation points. Table 7.3 shows the result of this investigation.

TABLE 7.3
SPLINE COLLOCATION FOR $R(x > 0.7, y) = \exp[-100(x - 0.7)]y^2$ AND $\phi = 2$.

Location of spline points	Number of collocation points in each subinterval	η (exact: 0.2560985)	$y(0.7)$ (exact: 0.817593)
—	4	0.24839	0.82055
—	8	0.25244	0.81928
—	12	0.25364	0.81886
—	16	0.25444	0.81849
0.7	4 4	0.25387	0.81928
0.7	3 5	0.25536	0.81819
0.7	4 6	0.25591	0.81776
0.7	4 7	0.25606	0.81762
0.7	5 7	0.25606	0.81762
0.7	4 8	0.25610	0.81759
0.7	4 9	0.25610	0.81759
0.7 0.73	4 4 4	0.25599	0.81765
0.7 0.75	4 3 3	0.25604	0.81763
0.7 0.75	4 4 4	0.25608	0.81760
0.7 0.75	4 5 3	0.25606	0.81761
0.7 0.75	4 5 4	0.25608	0.81759
0.7 0.77	4 6 2	0.25608	0.81759
0.7 0.77	4 7 2	0.25608	0.81759
0.7 0.77	4 7 3	0.25609	0.81759

The best estimate of $[\eta, y(0.7)]$ is obtained with $x_{s1} = 0.7$, $N_1 = 4$, and $N_2 = 8$ (or 9): $\eta = 0.2560986$ and $y(0.7) = 0.817593$.

Global collocation is again much less satisfactory than the spline collocation methods. $N_1 = 4$, $N_2 = 8$ yields seven correct digits of η, while $N = 16$ hardly leads to three correct digits. It should be noted that the failure of global collocation is not caused by rapid changes of y or $y^{(1)}$, the two quantities that can be used to judge ill-conditioning of a function by inspection of its graph. $y(x)$ and $y^{(1)}(x)$ are both perfectly smooth functions, but $y^{(2)}(x)$ varies rapidly with x in the x-interval $[0.7, 0.8]$ and $y^{(3)}$ is discontinuous at $x = 0.7$. The inability of a high-order polynomial to represent a function $y^{(3)}(x)$ that is discontinuous at a point in the approximation interval makes global collocation unsuitable for the present problem just as well as in the previous example where the discontinuity is in $y^{(2)}(x)$. The approximation error of, e.g., η decreases slowly (linear convergence) with N: Exponential extrapolation of the data for $N = 8$, 12, and 16 indicates an error of $\sim 2 \cdot 10^{-4}$ for the extremely high N value $N = 60$. Insertion of a spline point at $x = 0.7$ is obviously necessary. Four collocation points are sufficient for the interval [0, 0.7] as seen from the discussion of the previous example and confirmed by table 7.3 where four or five points in the inner interval always give the same result. Four points are now insufficient in the outer interval [0.7, 1] due to the rapid variation of $y^{(2)}(x)$ and of the higher derivatives in part of this interval.

For $0.7 \leq x \leq 0.73$, $y \sim 0.82$ and the rate of change of $y^{(2)}(x)$ is

$$y^{(3)}(x) \simeq \frac{(4 \cdot 0.82^2) - [(4 \cdot 0.82^2)/e^3]}{0.03} = \frac{2.68 - 0.13}{0.03} = 85$$

y is represented by a polynomial of degree $N_i + 1$ in this interval (N_i collocation points and the two end point ordinates). For $y = x^{N_i+1}$ the maximum value of

$$\frac{d}{dx}\left(\frac{d^2y}{dx^2} + \frac{1}{x}\frac{dy}{dx}\right) \quad \text{is} \quad (N_i + 1)^2(N_i - 1) \tag{11}$$

Consequently by the rule of thumb mentioned in subsection 2.3.5 the number of collocation points N_i in the x-interval $(0.7, 0.73)$ must be greater than or equal to 4 to obtain a sufficiently good representation for y. This is well confirmed by the results in table 7.3 and it is also seen that a substantial number of collocation points must be placed in the interval [0.73, 1] due to the rapid change of higher derivatives in this outer region where the rate of change of $y^{(3)}$ has become small.

In this example, it is interesting to note that insertion of an extra spline point at $x = 0.73$, 0.75, or 0.77 does not improve the accuracy for the same total number of collocation points: $N_1 = 4$, $N_2 = 8$ ($x_{s1} = 0.7$) is better than any distribution of eight collocation points between two outer subintervals with $x_{s2} = 0.73$, 0.75, or 0.77.

The first spline point was required by the discontinuity of $y^{(3)}(x)$ at $x = 0.7$, but the advantage of inserting a second or further spline points to separate regions of greatly different values of $y^{(2)}(x)$ is not nearly so obvious and much numerical experimentation is needed before general guidelines can be given. A large number of spline points will probably always be wasteful since the spline-point ordinates enter as extra unknowns without contributing significantly to the overall accuracy of the approximation.

7.1.3 Effectiveness factor for extremely high Thiele modulus

The isothermal effectiveness factor problem

$$\begin{aligned} &\frac{d^2y}{dx^2} - \Phi^2 y^n = 0 \qquad (n > 0) \\ &y^{(1)}(0) = 0 \qquad y(1) = 1 \\ &\eta = \int_0^1 y^n\, dx \end{aligned} \tag{12}$$

is treated by global collocation in chapter 4. For $n = 2$, this procedure breaks down if Φ exceeds 75 to 100.

TABLE 7.4
CATALYST EFFECTIVENESS η FOR LARGE THIELE MODULUS; GLOBAL COLLOCATION

N/Φ	50	75	100	125	150
4	0.02702	—	—	—	—
8	0.01689	0.01214	0.0100	0.00894	—
12	0.01635	0.01099	0.00839	0.00691	0.00597
Exact	0.01633	0.01089	0.00816	0.00653	0.00544

When Φ increases, the number of iterations in the Newton algorithm for solution of the collocation equations increases rapidly, especially for small N. Finally, at $\Phi = 75$, convergence fails for $N = 4$, while $N = 8$ breaks down at $\Phi = 150$. The reason is simply that the collocation equations have no real-valued solution $\mathbf{y}$. Even in problems where a solution may be obtained, e.g., $N = 8$ $\Phi = 50$, the very poor determination of the concentration profile is quite unacceptable if a companion reaction whose rate of reaction may be strongly influenced by the species described by (12) takes place in the pellet.

The last line of table 7.4 is obtained from the penetration solution that appears when (12) is solved by quadrature as described in Aris (1975), p. 109:

$$\frac{dy}{dx} = \Phi\sqrt{\frac{2}{n+1}}(y^{n+1} - y_0^{n+1})^{1/2} \tag{13}$$

where $y_0 = y(x = 0)$ is given by

$$\Phi\sqrt{\frac{2}{n+1}}y_0^{(n-1)/2} = \int_1^{1/y_0} \frac{d\xi}{(\xi^{n+1} - 1)^{1/2}} \tag{14}$$

and

$$\eta = \frac{1}{\Phi}\sqrt{\frac{2}{n+1}}(1 - y_0^{n+1})^{1/2} \tag{15}$$

For $\Phi = 1000$ and $n = 2$, the solution of (14) is $y_0 = 0.8804 \cdot 10^{-5}$ (see Exercise 3.6). y_0 may be neglected in (13) for $y \gtrsim 20y_0$, and if $y_0 \sim 0$, the profile is easily found by integration of (13):

$$1 - x = \sqrt{6} \cdot 10^{-3}(y^{-1/2} - 1) \tag{16}$$

$y > 20y_0$ for $x > 0.818$ by equation (16); for $0.818 < x < 1$, (16) can be taken to represent the exact solution of (12) with $n = 2$ and $\Phi = 1000$.

A concentration profile as extreme as that obtained from (12) with $(\Phi, n) = (1000, 2)$ is a typical case for spline collocation and we use different criteria for selecting one spline point x_s.

The first three columns of table 7.5 provide a systematic choice of x_s based on the penetration solution with $y_0 \sim 0$.

TABLE 7.5
SYSTEMATIC CHOICE OF SPLINE-POINT ABSCISSA x_s FOR A SECOND-ORDER REACTION, $n = 2$ IN (12)

Φ	$y_s = 0.01$	$R_s = 0.5$	$R_s = 1$	$1 - [2\Phi^2 \int_0^1 R(y)\,dy]^{-1/2}$
10	—	0.3237	0.4703	0.8775
20	—	0.4711	0.5747	0.9387
50	0.5590	0.6370	0.7026	0.9755
100	0.7795	0.7332	0.7795	0.9877
500	0.9560	0.8746	0.8954	0.9975
1000	0.9779	0.9103	0.9250	0.9987

We have seen that (12) with $\Phi = 1$, i.e., with $\max(\Phi^2 y) = 1$, can be solved with a satisfactory accuracy using one or at most two collocation points. Thus if x_s is chosen such that $R = R_s = 1$ at x_s, it is sufficient to allocate one or two collocation points to the interval $[0, x_s]$, while the remaining points can be used in $[x_s, 1]$ to give an accurate representation of the rapidly decreasing part of the profile.

Another possibility is to chose x_s such that y is a small fraction, e.g., 1%, of $y(x = 1)$ when $x = x_s$. This is in close analogy to the "burnt-out," "reaction zone" concept of Paterson and Cresswell where y is taken to be zero at $x = x_s$.

Finally the last column of table 7.5 is a very generally applicable choice of x_s based upon integration of the differential equation

$$\left.\frac{dy}{dx}\right|_1 = \left[2\Phi^2 \int_{y_0}^{1} R(y)\,dy\right]^{1/2} \sim \left[2\Phi^2 \int_0^1 R(y)\,dy\right]^{1/2} \tag{17}$$

$y_0 \sim 0$ for large Φ and the integral can be evaluated by quadrature for any rate expression that depends on y only.

Table 7.6 shows the results for 12 collocation points in [0.9779, 1] and 1 point in [0, 0.9779]. The collocation abscissas of the table are at x_i ($i = 11, 8, 5, 2$) in [0.9779, 1] and at x_1 in [0, 0.9779]. Finally, the collocation ordinates have been inserted in (16) to give the true abscissas that correspond to the ordinates in the first line of the table.

TABLE 7.6
SPLINE COLLOCATION SOLUTION OF $y^{(2)} - 10^6 y^2 = 0$ USING 1 COLLOCATION POINT IN [0, 0.9779] AND 12 POINTS IN [0.9779, 1]

y	0.48720	0.06764	0.02049	0.01465	0.01444	0.00021
x	0.99894	0.99301	0.98488	0.97895	0.9779	0.56459
x by (16)	0.99894	0.99303	0.98533	0.98221	0.98206	0.8334

The collocation results are apparently very accurate, at least for $y > 0.02$, while the last few ordinates are much too high—the true value of $y(0.9779) = 0.01$ from table 7.5.

The relatively high error of the small ordinates is, however, of small importance for the computation of η, which is found almost exclusively from the large ordinates—and these are determined well enough, as seen in table 7.6.

The profile for $N = 9$ is almost the same as that for $N = 12$. Accurate results are found for $y > 0.1$ and $y(x_s) = 0.01442$, which is very close to the $N = 12$ result. $\eta(N = 9) = 0.0008077$, while

$\eta(N = 12) = 0.0008159$; both are satisfactory approximations to $\eta(\text{exact}) = 0.0008165$.

It appears that

1. The error of the small ordinates is of no great consequence.
2. The value of including a spline point in $[0,\ 0.9779]$ is fictitious.

Since it is immaterial whether $y(0.9779)$ is 0 {as in a generalized Paterson and Cresswell method with $N = 9$ or 12 points in $[x_s, 1]$}, 0.01, or 0.01444, it seems reasonable to represent the solution in $[0, x_s]$ simply by the condition $dy/dx = 0$ for $x < x_s$. This might be called a zero-order spline method and it is a conceptional improvement on Paterson's and Cresswell's method that hardly costs any extra computation since the ordinate $y(x_s)$ is eliminated before the calculations are started. Equation (12) is rewritten in terms of $(x - x_s)/(1 - x_s) = v$:

$$\frac{d^2y}{dv^2} - (1 - x_s)^2\Phi^2 y^n = 0$$
$$\frac{dy}{dv} = 0 \quad \text{at } v = 0 \qquad y = 1 \quad \text{at } v = 1 \tag{18}$$

First we use (17) to obtain $x_s = 0.9987$ for $\Phi = 1000$. Now $(1 - x_s)^2\Phi^2 = 1.7$ for $\Phi = 10^3$ and two to four points are certainly enough to represent $y(v)$.

The solution is shown on the upper part of figure 7-1 and $10^4 \cdot \eta$ is given in the first line of table 7.7.

TABLE 7.7

SPLINE COLLOCATION WITH NO COLLOCATION POINTS IN $[0, x_s]$

x_s/N	1	2	3	4	8	12
0.9987	6.275	7.085	7.158	7.155	7.155	7.155
0.9779	—	—	—	11.74	8.230	8.169

As expected, a very small N is sufficient for an adequate representation of $y(v)$: Collocation ordinates for $N = 2$ are almost on the curve drawn through the $N = 4$ points. It is seen, however, that any N gives a large deviation of the collocation profile from the penetration profile (16) except close to $x = 1$. The slope at $x = 1$ is quite close to the correct slope, which means that η is tolerably well determined.

Also shown in figure 7-1 are solutions of (18) with $x_s = 0.9779$. Now $(1 - x_s)^2\Phi^2 = 448$ and $N = 12$ is just sufficient to obtain 0.05% accuracy for η, while $N = 4$ is much too small to represent $y(v)$: The $N = 4$ curve wriggles around the $N = 12$ curve with a large difference in slope at $x = 1$, which leads to an appreciable error in η.

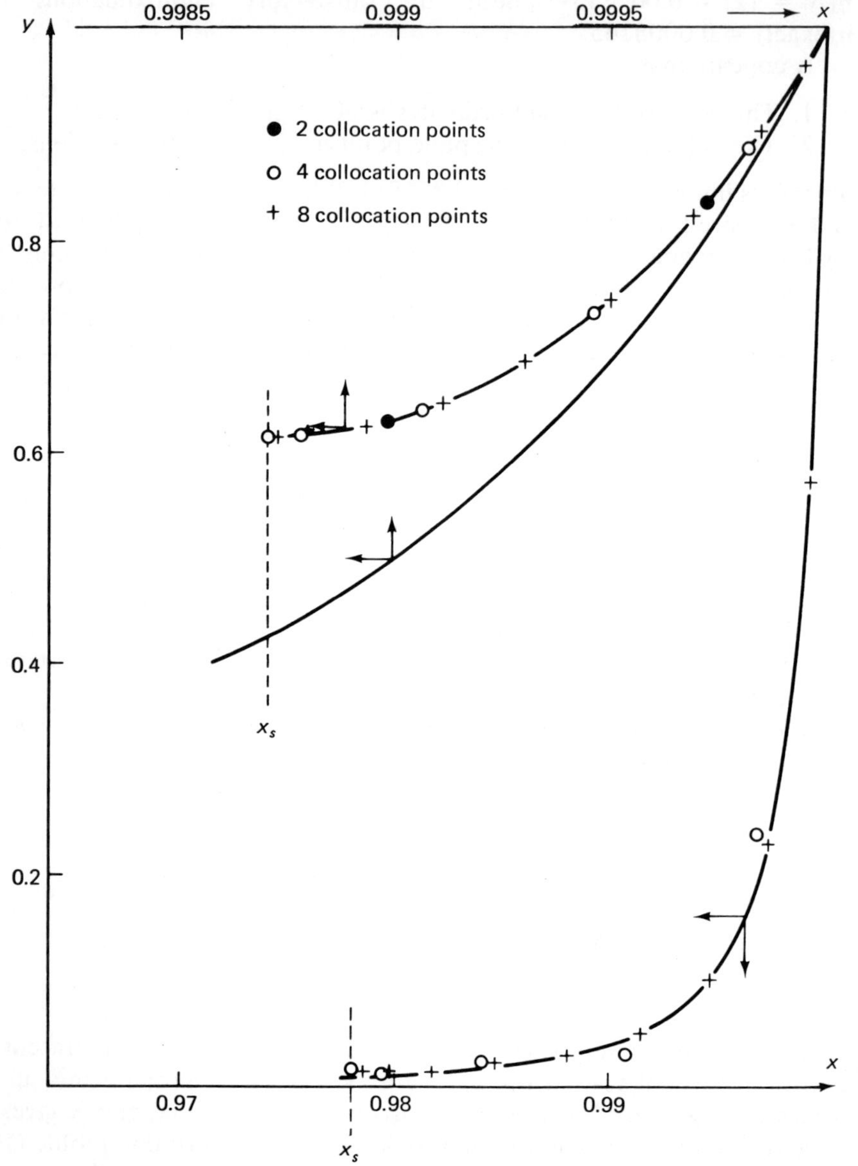

Figure 7-1. Zero-order spline collocation with two different spline abscissas. Solution of equation (7.16) in [0.9985, 1] shown for comparison.

The following conclusion is apparent: If the interval $1 - x_s$ is large enough to include the major part of the concentration drop, the true solution in $[x_s, 1]$ can be obtained but only for a large N. On the other hand, if x_s is closer to 1, the exact solution can never be found even for large N since the boundary condition $dy/dx = 0$ at $x = x_s$ does not hold with sufficient accuracy. However, a stationary value of η that is reasonably correct is found with a small N even though the profile is considerably in error.

For practical calculations the last situation is probably to be preferred, and if necessary improved results can be obtained by including a "buffer interval" $[x_{s2}, x_{s1}]$ with a small number of collocation points.

Table 7.8 gives results in which two spline points $x_{s2} = 0.991$ and $x_{s1} = 0.9987$ have been chosen. dy/dx is equated to zero at x_{s2}.

TABLE 7.8
SPLINE POINTS AT $x_{s2} = 0.991$ AND $x_{s1} = 0.9987$;
N_2 COLLOCATION POINTS IN $[x_{s2}, x_{s1}]$, N_1 POINTS IN $[x_{s1}, 1]$

N_2	4	4	3	2	2	2	1	1
N_1	4	3	3	4	3	2	4	3
$\eta \cdot 10^4$	8.155	8.152	8.120	8.007	8.004	7.925	7.721	7.718

Figure 7-2 illustrates how the profile approaches the correct profile not only in $[x_{s1}, 1]$ but also in $[x_{s2}, x_{s1}]$ when N_2 is increased from 2 to 3. $N_1 = 2$ and 3 give almost identical results.

The ideas presented here can easily be organized into a general algorithm for computation of steep profiles by collocation: The first spline point x_{s1} is chosen from (17) using a numerical quadrature if necessary. The solution of (18) with a small N indicates whether $1 - x_{s1}$ is large enough. If $y(x_{s1}) < 0.1$, the computed profile covers the penetration region and a larger N can be used to improve the accuracy of η. If $y(x_{s1}) > 0.1$, a second spline point is chosen. The length of a suitable interval $[x_{s2}, x_{s1}]$ can be estimated (automatically) from $R(y)$ or arbitrarily chosen as, e.g., $10(1 - x_{s1})$. One or two collocation points in $[x_{s2}, x_{s1}]$ will probably be enough to obtain a sufficiently accurate profile in $[x_{s1}, 1]$ and η.

7.2 Eigenvalues and Entry Length Problems by Spline Collocation

7.2.1 One-dimensional eigenvalue problems

The structure of a global spline collocation formulation of an eigenvalue problem is not much different from that used for two-point boundary value problems in section 7.1.

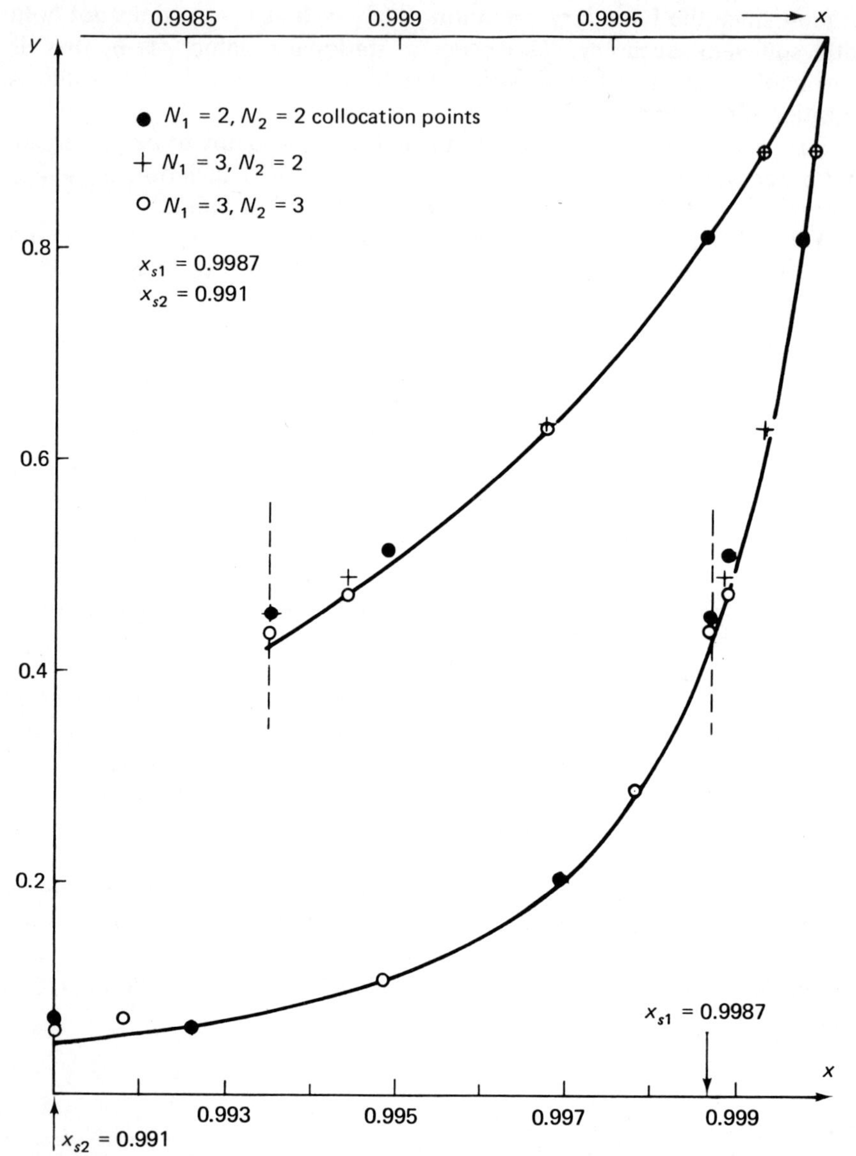

Figure 7-2. Effect of one extra spline point. Curves show solution of equation (7.16).

Let a linear eigenvalue problem be given by

$$\nabla^2 y - \lambda f(x) y = 0 \tag{19}$$

$$y^{(1)}(0) + Ay(0) = 0 \qquad y^{(1)}(1) + By(1) = 0 \tag{20}$$

The collocation formulation of (19) and (20) with one spline point x_s, N_1 collocation points in $[0, x_s]$, and N_2 collocation points in $[x_s, 1]$ is

$$\mathbf{U}\mathbf{y} + \mathbf{V}\begin{pmatrix} y(x = 0) \\ y(x_s) \\ y(x = 1) \end{pmatrix} = \lambda \mathbf{y} \tag{21}$$

where $\mathbf{y}$ is the vector of $N = N_1 + N_2$ collocation ordinates.

$$\mathbf{y} = (y_1, y_2, \ldots, y_{N_1}, y_{N_1+1}, \ldots, y_{N_1+N_2}) \tag{22}$$

and $\mathbf{V}$ is an $(N \times 3)$ matrix that contains the elements of the discretization matrix for $\nabla^2 y/f(x)$ that correspond to the three ordinates $y(0)$, $y(x_s)$, and $y(x = 1)$.

The boundary conditions are written

$$\mathbf{V}_1\mathbf{y} + \mathbf{V}_2\begin{pmatrix} y(x = 0) \\ y(x_s) \\ y(x = 1) \end{pmatrix} = 0 \tag{23}$$

The $(3 \times N)$ matrix $\mathbf{V}_1$ and the (3×3) matrix $\mathbf{V}_2$ are obtained by discretization of the boundary conditions and the spline condition

$$\left.\frac{dy}{dx}\right|_{x_s-} = \left.\frac{dy}{dx}\right|_{x_s+} \tag{24}$$

Combination of (23) and (21) yields an $(N_1 + N_2)$ eigenvalue problem

$$(\mathbf{U} - \mathbf{V}\mathbf{V}_2^{-1}\mathbf{V}_1)\mathbf{y} = \lambda \mathbf{y} \tag{25}$$

which is solved in the usual way.

The global spline collocation procedure was tested on several problems of the form (19) and (20) with discontinuities in $f(x)$ or in $f^{(1)}(x)$. One example is

$$f(x) = \begin{cases} 4 & 0 \le x \le \dfrac{1}{2} \\[2ex] \dfrac{2}{x} & \dfrac{1}{2} \le x \le 1 \end{cases} \tag{26}$$

$$y(0) = y^{(1)}(1) = 0 \tag{27}$$

Equations (19), (26), and (27) could be the mathematical model for a column that is kept in vertical position at $x = 1$ while a normal force P is applied at $x = 0$. $f(x) = [1/EI(x)]$. E is the modulus of elasticity and $I(x)$ is the moment of inertia, which is proportional to R^4 where R is the radius of the column. When $P \ge \lambda_1$, the smallest eigenvalue of (19), the column will collapse. Thus (26) describes a column with constant cross section for $0 \le x \le \frac{1}{2}$ and a widening base from $x = \frac{1}{2}$ to $x = 1$.

Table 7.9 states the very disappointing result of a global N-point collocation using zeros of $P_N^{(0,0)}(x)$. The last row shows that λ_1 does converge to its correct value (0.842215) determined by global spline collocation. As seen in table 7.10, the rate of convergence is no better than for a straightforward finite difference method based on second-order differences and N subintervals.

TABLE 7.9
SLOW CONVERGENCE OF THE FIRST EIGENVALUE BY GLOBAL COLLOCATION

Collocation points N	4	10	20	22	24	26
$10 \cdot \lambda_1$	8.4682	8.4422	8.4238	8.4235	8.4233	8.4231
$\lvert\lambda_1 - \lambda_{1ex}\rvert \cdot 10^6$	46,021	7003	1677	1379	1154	979
$\lvert\lambda_1 - \lambda_{1ex}\rvert \cdot N^2$	0.736	0.700	0.670	0.667	0.664	0.662

TABLE 7.10
CONVERGENCE TO λ_1 BY A STANDARD FINITE DIFFERENCE METHOD

Subintervals N	8	16	32	64	128	256
$\lvert\lambda_1 - \lambda_{1ex}\rvert \cdot 10^6$	4632	1179	295	73.2	18.5	3.7
$\lvert\lambda_1 - \lambda_{1ex}\rvert \cdot N^2$	0.296	0.302	0.302	0.300	0.303	0.242

In both methods, λ_1 converges to λ_{1ex} at a rate inversely proportional to N^2 and the computational work for fixed N is much greater in the global collocation method than in a finite difference method where the tridiagonal structure of the matrix can be utilized.

Table 7.11 shows the first four eigenvalues by global spline collocation. x_s is chosen at 0.5 and N collocation points [zeros of $P_N^{(0,0)}(x)$] are used in each of the two subintervals.

TABLE 7.11
SPLINE COLLOCATION WITH N COLLOCATION POINTS IN EACH SUBINTERVAL

N	1	2	4	6
λ_1	0.880	0.8418	0.84221511	0.84221510985
λ_2	6.41	7.191	6.8098	6.8093835
λ_3		19.98	18.50	18.4794
λ_4		34.45	35.54	36.389

A total of $2N = 12$ collocation points is sufficient to obtain machine accuracy of the first eigenvalue. It may safely be concluded that linear eigenvalue problems with a discontinuity at a given x in $y^{(2)}$ or in higher derivatives can be solved by the proposed global spline collocation procedure.

7.2.2 Entry length problems in partial differential equations

The linear partial differential equation of sections 4.3 and 4.4,

$$(1 - x^2)\frac{\partial y}{\partial z} = \frac{\partial^2 y}{\partial x^2} \tag{28}$$

$$y(z, 0) = y^{(1)}(z, 1) = 0$$

$$y(0, x) = 1$$

has previously been solved by global collocation to obtain the Sherwood number defined by (4.68) or (4.69). Global collocation works well for $z > 0.02$ to 0.05 but the discussion of subsection 4.4.3 (table 4.5) has revealed that the method is unable to produce accurate enough results for $z < 0.005$ to elucidate the form of $\text{Sh}(z)$ in the penetration region.

Now we show how spline collocation can be used to obtain a solution that is accurate for small z-values.

A spline point x_s is inserted between $x = 0$ and $x = 1$. For $x_s < x < 1$ the boundary condition $y^{(1)}(z, x) = 0$ is assumed to hold and the collocation procedure is applied only to the small interval $(0, x_s)$, the penetration region.

Inserting $v = x/x_s$ in (28) yields

$$x_s^2[1 - (vx_s)^2]\frac{\partial y}{\partial z} = \frac{\partial^2 y}{\partial v^2} \tag{29}$$

$$y(z, v = 0) = y^{(1)}(z, v = 1) = 0$$

$$y(0, v) = 1$$

Equation (29) is collocated at the zeros of $P_N^{(0,1)}(v)$ just as in subsection 4.3.5; the whole algorithm of that section can in fact be used with hardly any modifications.

The Sherwood number is calculated as

$$-\frac{2}{3}\frac{d/dz(\int_0^1 \frac{3}{2}y(1-x^2)\,dx]}{\int_0^1 \frac{3}{2}y(1-x^2)\,dx}$$

$$= \frac{-\frac{2}{3}d/dz\int_0^{x_s}\frac{3}{2}y(1-x^2)\,dx}{1-\frac{3}{2}[x_s-(x_s^3/3]+\int_0^{x_s}\frac{3}{2}y(1-x^2)\,dx} \tag{30}$$

$$= -\frac{2}{3}\frac{d/dz(I)}{I_k+I}, \text{ where } I_k = 1-\frac{3}{2}\left(x_s-\frac{1}{3}z_s^3\right)$$

and

$$I = \int_0^1 \frac{3}{2}x_s[1-(vx_s)^2]y\,dv = \mathbf{q}^T\mathbf{y} = \sum_1^N a_i \exp(\lambda_i z)$$

as in (4.92) and (4.93) with

$$q_j = \frac{3}{2}x_s[1-(v_jx_s)^2]w_j \tag{31}$$

and w_j = weights of the Radau quadrature formula.

$$\frac{dI}{dz} = \sum_1^N \lambda_i a_i \exp(\lambda_i z)$$

as in (4.94). By analogy with the effectiveness factor problem of subsection 7.1.3, the spline collocation solution for Sh (z) is expected to be satisfactory as long as the deviation of the $x = x_s$ ordinate from its initial value 1 is small.

Table 7.12 shows the results for $N = 8$ and 10 collocation points and different choices of x_s.

TABLE 7.12
PENETRATION FRONT CALCULATIONS BY SPLINE COLLOCATION

		Sh − Sh (N, x_s)			
	Exact value of Sh	$x_s = 0.05$ $N = 8$	$x_s = 0.1$ $N = 8$	$x_s = 0.1$ $N = 10$	$x_s = 0.2$ $N = 8$
0.00005	80.75	−0.028	0.370	0.052	−3.35
0.0001	57.39	0.0011	0.114	−0.013	−2.29
0.00015	47.04	$-4.6 \cdot 10^{-6}$	−0.031	0.0033	−0.451
0.0002	40.87	$2.9 \cdot 10^{-4}$	−0.015	$4.9 \cdot 10^{-4}$	0.186
0.0003	33.56	0.016	$7.5 \cdot 10^{-4}$	$-8.0 \cdot 10^{-5}$	0.210
0.0004	29.20	0.114	$5.6 \cdot 10^{-4}$	$-2.1 \cdot 10^{-6}$	0.058
0.0005	26.22	0.356	$1.0 \cdot 10^{-4}$	$2.6 \cdot 10^{-6}$	−0.0036
0.0006	24.03	0.750	$9.8 \cdot 10^{-7}$	$4.4 \cdot 10^{-6}$	−0.015
0.0008	20.95	1.85	$1.6 \cdot 10^{-4}$	$1.6 \cdot 10^{-4}$	−0.0075
0.001	18.85	3.12	0.0018	0.0018	−0.0012
0.003	11.34	9.11	0.826	0.826	$4.1 \cdot 10^{-5}$
0.005	9.04	8.73	2.49	2.49	0.0068

Choosing x_s close to zero ($x_s = 0.05$) leads to accurate results for small $z(z \sim 0.00015)$ and poor results for large $z(z \sim 0.001)$. Contrariwise, $x_s = 0.2$ leads to accurate results for $z \sim 0.003$ and to poor results when z is small (<0.00015). $N = 10$, $x_s = 0.1$ is accurate in the large z-interval $0.0001 < z < 0.001$. Oscillations around the true values of Sh are found for all choices of x_s.

These results are easily explained by means of the concentration profiles of figure 7-3. The curves are drawn through the collocation

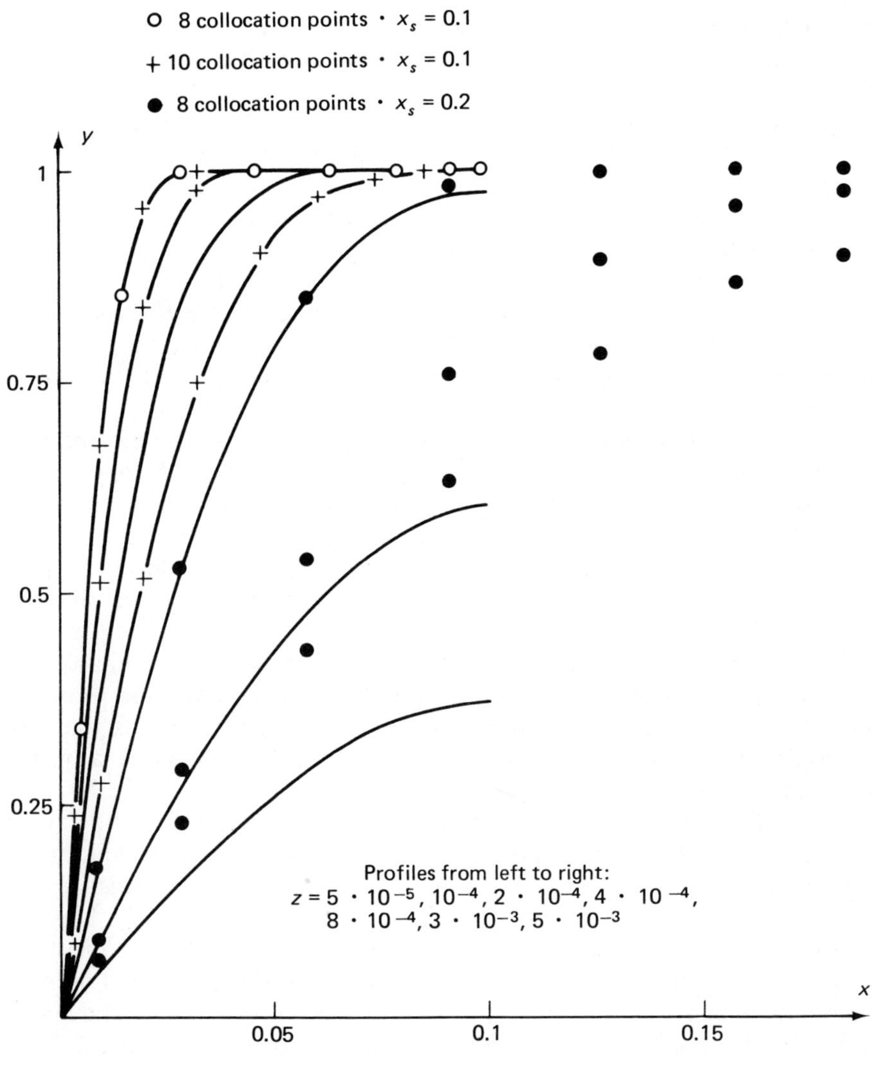

Figure 7-3. Penetration front.

ordinates ($N = 8$, $x_s = 0.1$) that are marked by open circles. The smooth profiles for $z > 10^{-4}$ are well represented by eight collocation points. Hence there is practically no difference between results for $N = 8$ and $N = 10$ when $z > 0.0001$ as seen from the + marked points for $N = 10$.

Close inspection of the results for $z = 0.00005$ indicates that the $N = 8$ profile wriggles somewhat around the $N = 10$ profile with, as usual, a very small deviation at the circled node points. For x smaller than the abscissa of the first collocation point, the $N = 8$ profile is below the more accurate $N = 10$ profile, giving a small but noticeable error of the $N = 8$ Sherwood number. The poor result for $z = 0.00005$ and $x_s = 0.2$ is also obvious. This extremely steep profile can hardly be represented by eight collocation points when $x_s = 0.1$ and much less so when $x_s = 0.2$.

Points obtained with $N = 8$, $x_s = 0.2$ are marked by closed circles for $z = 8 \cdot 10^{-4}$, $3 \cdot 10^{-3}$, and $5 \cdot 10^{-3}$. For these z-values, $N = 8$ is sufficient to represent the relatively smooth profiles in $[0, 0.1]$ or $[0, 0.2]$ and curves through the $N = 8$, $x_s = 0.2$ points can be used as "exact profiles" just as the penetration-solution curves of figures 7-1 and 7-2.

It is seen that the $x_s = 0.1$ profiles are considerably in error for $z > 0.001$ and $x \sim x_s$. This is of no consequence for the Sherwood number calculation, which is done mainly on the basis of the small x ordinates and these are determined well enough with $x_s = 0.1$ for z up to 0.003.

There is a striking resemblance between the present example and that of subsection 7.1.3: For $x_s = 0.1$ and a sufficiently large N, any profile with $0.00005 < z < 0.0008$ can be determined with an arbitrary accuracy. For $z > 0.001$, a small N gives the same result as a larger N but the result is not satisfactory since the slope of the true solution at x_s is comparable with the slope at $x = 0$ and the approximate solution assumes that the slope is zero at x_s.

The zero-order spline collocation method is especially suitable for high accuracy computations in the penetration zone. Pentration solutions are at best difficult to obtain by the semi-analytical methods that are exemplified in subsection 4.4.2, and the present method is a viable alternative to these methods.

A solution that is valid for both large and (reasonably) small z can be obtained as in the example above but including three or four collocation points in the inner zone. These points are quite useless for small z, but they enable the computation to be carried on to arbitrarily large z. Conversely, the collocation points in the outer zone are worthless for large z. Alternative variants of spline collocation where the total number of points is kept reasonably small are proposed in Exercises. The spline collocation procedure is used again in section 9.1 on a much more difficult example.

7.2.3 A discontinuous initial value profile

Discontinuities in the high-order derivatives—or near discontinuities, as in the previous example—are taken care of by spline collocation as described above.

In some problems the discontinuity lies in the initial value profile $y(x, z = 0)$. The thesis work of Venkata Subramanian (1974) deals with radial dispersion of a tracer that decays by a first-order reaction after having been injected into a laminar-flow Newtonian liquid at $z = 0$ through a thin-bore tube placed in the axis of the main stream:

$$(1 - x^2)\frac{\partial y}{\partial z} = \frac{\partial^2 y}{\partial x^2} + \frac{1}{x}\frac{\partial y}{\partial x} - \alpha y \tag{32}$$

$$y^{(1)}(1, z) = y^{(1)}(0, z) = 0 \tag{33}$$

$$\begin{aligned} y(x, 0) &= 1 \quad \text{for } 0 \le x \le \beta \\ y(x, 0) &= 0 \quad \text{for } \beta < x \le 1 \end{aligned} \tag{34}$$

A suitable approach to this problem is immediately suggested. A small interval $[\beta - \delta_1, \beta + \delta_2]$ is considered and (32) is only solved in this interval until y at $\beta - \delta_1$ is considerably smaller than $1 - e^{-\alpha z}$ or y at $\beta + \delta_2$ considerably larger than 0.

The required substitution is

$$x = (\delta_2 + \delta_1)v + \beta - \delta_1 \tag{35}$$

$$\begin{aligned} &\{1 - [(\delta_2 + \delta_1)v + \beta - \delta_1]^2\}\frac{\partial y}{\partial z} \\ &\qquad = \frac{1}{(\delta_1 + \delta_2)^2}\left(\frac{\partial^2 y}{\partial v^2} + \frac{\delta_2 + \delta_1}{(\delta_2 + \delta_1)v + \beta - \delta_1}\frac{\partial y}{\partial v}\right) - \alpha y \end{aligned}$$

$$y^{(1)}(v = 1, z) = y^{(1)}(v = 0, z) = 0$$

The transformed equation is easily solved by standard collocation taking, e.g., the zeros of $P_N^{(0,0)}(v)$ as collocation points. As z increases, the penetration zone is broadened by choosing a larger $\delta_1 - \delta_2$ and rediagonalizing the collocation matrix.

Finally when $\delta_1 = \beta$ and $\delta_2 = 1 - \beta$, global collocation using zeros of $P_N^{(0,0)}(x^2)$ is used to continue the calculations until the final z-value is reached.

EXERCISES

1. The rate of convergence of global collocation is dependent on the severity of the discontinuity that appears in the profile.

Solve the model of equation (7) by global N–point collocation for the following rate expressions:

$$R(x > 0.7, y) = 0, \qquad 4\exp[-100(x - 0.7)]$$

$$4\exp[-10^3(x - 0.7)^2] \quad \text{and} \quad 4\exp[-10^4(x - 0.7)^3]$$

$$R(x < 0.7, y) = 4$$

Compare the values of η [equation (10)] for different N with the true η value obtained by a suitable spline collocation procedure.

2. Solve equation (12) for $n = 0.2$ by global collocation and by a suitable spline collocation method. Φ should be chosen as 1, 1.75, 1.93, and 2.5.
Your first attempt to solve the equation may give either a solution with $y < 0$ for some x or no convergence at all. What happens?

3. The following apparently quite simple equation describes a zero-order reaction that approaches equilibrium in the key component. It is desired to find $y(x)$ with a given (a) absolute accuracy, and (b) relative accuracy.

$$\frac{d^2y}{dx^2} + \frac{2}{x}\frac{dy}{dx} - 80\left(1 - \frac{10^{-6}}{y}\right) = 0$$

$$y^{(1)}(0) = 0, \qquad y(1) = 1$$

What difficulties do you anticipate?
Solve the equation by global collocation and by a suitable spline collocation method.

4. It is desired to make a study of the Weisz–Hicks problem with film resistance by zero-order spline collocation.
Use the data of Hatfield and Aris [*Chem. Eng. Sci.* 24 (1969):1213] or Van den Bosch and Padmanabhan [*Chem. Eng. Sci.* 29 (1974):1217]: plane parallel symmetry, $\beta = \frac{1}{3}$, $\gamma = 27$, $\mathrm{Bi} = 100$, $\mathrm{Bi}_M = 300$ [equation (1.69)—the references use ν and σ for these quantities].
How would you choose x_s on the computer?
Note that the profile eventually becomes quite smooth inside the pellet when $\Phi \to \infty$ [$y(x = 1) \ll 1$].

5. Equation (26) is modified to

a. $$f(x) = \begin{cases} 6 & 0 \le x < \dfrac{1}{2} \\[2ex] \dfrac{2}{x} & \dfrac{1}{2} \le x \le 1 \end{cases}$$

b. $$f(x) = \begin{cases} 4 & 0 \le x \le \dfrac{1}{2} \\[2ex] \dfrac{1}{(x - \frac{1}{2})^2 + \frac{1}{4}} & \dfrac{1}{2} \le x \le 1 \end{cases}$$

Compare the rate of convergence of λ_1 to its true value (obtained by spline collocation) in a global collocation process for part a, equation (26), and part b.

6. Solve $\partial^2 y/\partial x^2 = \partial y/\partial t$,

$$\left.\frac{dy(t)}{dx}\right|_{x=1} + \text{Bi}\, y(1, t) = y^{(1)}(0, t) = 0 \quad \text{and} \quad y(x, 0) = 1$$

by spline collocation and t in the range 10^{-5} to 1.
Show that

$$1 - \int_0^1 y\, dx \sim 2\sqrt{\frac{t}{\pi}}$$

for t sufficiently small and $\text{Bi} \to \infty$. Repeat the calculations for

$$\frac{\partial^2 y}{\partial x^2} - \Phi^2 y = \frac{\partial y}{\partial t}$$

and obtain

$$1 - \int_0^1 y\, dx$$

as function of Φ for small t.

7. Solve equations (32) to (34) for $\alpha = 10$, $\beta = 0.28$, and $z = [10^{-4}, 0.25]$. The desired quantity is

$$\bar{y} = 4\int_0^1 (1 - x^2)xy\, dx$$

At approximately what value of z does the penetration front reach $x = 0$ and $x = 1$?

REFERENCES

Carey and Finlayson (1975) and Finlayson (1974) have applied the methods of this chapter on two-point boundary value problems. They use the name *collocation on finite elements* to stress the resemblance to ordinary finite difference methods, while we have preferred to include a reference to spline functions in the name. A spline function is a piecewise polynomial of degree M joined smoothly so that it has $M - 1$ continuous derivatives. Spline functions were used for approximation purposes even in Newton's day, but recent development in this field of approximation theory is due to Shoenberg [see, e.g., Shoenberg (1964)]. Strictly speaking it is only the $N = 1$ global "spline" collocation version that is a true spline function approximation {y is approximated by a second-degree polynomial and $[y, y^{(1)}]$ are continuous at the joints}. So-called "weak-spline" functions have been investigated, however [Barrodale and Young (1966)]. The requirement that the spline has $M - 1$ continuous derivatives at the joints is recognized to be too stringent, and these variants of spline functions are thus quite analogous to our concept of spline collocation.

Carey and Finlayson have some interesting references to Douglas and Du Pont (1973) and to De Boor and Swartz (1973). Douglas and Du Pont found that the spatial discretization error is proportional to $(1/M)^4$ when $N = 2$ collocation points, chosen as zeros of Legendre polynomials, are used. De Boor and Swartz generalized these results to $N > 2$ for linear equations, and Carey and Finlayson use computer experiments to show that the error of a spline point decreases with M^{-2N} also for a nonlinear problem. Global orthogonal collocation has the fastest rate of convergence $(1/N)^{K \cdot N}$ where K is reported to be 3.75, but the trouble is, of course, that N may be quite large before the rate predicted by this asymptotic formula is reached.

J. Jensen has used spline collocation to compute theoretical time-dependent poisoning profiles in a number of situations. His results are given in Jensen (1975) and in Jensen et al. (1976).

The mechanical stability problem of subsection 7.2.1 was suggested to us in 1972 by Dr. H. Bruun Nielsen of Numerical Institute (*DtH*) who was exasperated by the slow rate of convergence of a global collocation method. It is probably the first example that has been solved by global spline collocation and several variants of the problem are described in student publications from our university.

No results on penetration-front calculations appear to have been published although the method of subsection 7.2.2 seems to be be a rather obvious extension of standard collocation. The problem of subsection 7.2.3 was solved in cooperation with C. V. Venkatasubramanian [Department of Chemical Engineering, University of Salford (England)]. M. Fleischer and F. Worley, at the University of Houston, are using a three-dimensional variant of the same method to predict the dispersion of pollutants from tall chimneys.

1. CAREY, G. F., and FINLAYSON, B. A. *Chem. Eng. Sci.* 30 (1975):587–96.

2. FINLAYSON, B. A. *Cat. Rev.-Sci. Eng.* 10 (1974):69–138.

3. SCHOENBERG, I. J. "On Interpolating by Spline Functions and Its Minimal Properties" from *On Approximation Theory*, edited by Butzer and Korevaar. Stuttgart: Birkhäuser (1964).

4. BARRODALE, I., and YOUNG, A. *Comp. Journal* 9 (3) (1966):318–20.

5. DOUGLAS, J., and DU PONT, T. *Math. Comp.* 27 (1973):17–28.

6. DE BOOR, C., and SWARTZ, B. *Siam J. Numer. Anal.* 10 (1973):582–606.

7. JENSEN, J. V. Ph.D. thesis. Instituttet for Kemiteknik (1975).

8. JENSEN, J. V., NEWSON, E., and VILLADSEN, J. *ISCRE IV*, Heidelberg (1976):Preprints, 250–59.

Coupled Differential Equations 8

Introduction

Coupled linear differential equations are treated at length in chapter 4 and the nonlinear differential equation in one dependent variable is the subject of chapter 5. If the methods presented in the text are to have any lasting value, they should compete favorably with conventional finite difference methods in the solution of realistic research and design problems, that is, coupled nonlinear ordinary and partial differential equations.

These complicated problems should not be attacked, however, without a careful preliminary analysis of their structure. Any numerical method may be exposed in a very unfavorable light if it is used in a main program that is carelessly constructed

The valuable features of a numerical method are only recognized if supporting algorithms—e.g., subroutines for solving algebraic equations or stepsize selection algorithms—are well designed. Also a numerical method that has been developed for one type of problem may sometimes prove efficient for quite different problems if combined with other methods. Orthogonal collocation is typical in this respect. It was first developed for solution of boundary value problems, specifically catalyst effectiveness problems. Quite a few authors have used it successfully for solution of parabolic partial differential equations, and in this chapter and in the next it will be applied in unison with an efficient algorithm for solution of coupled first order differential equations to solve a number of research-like problems of very different structure.

Thus this chapter, which serves as a link between the textbook approach of the first five or six chapters and the research problems discussed in the last chapter, is concerned with several rather incoherent aspects of large numerical problems.

The structure of coupled nonlinear initial value problems and boundary value problems is first considered in section 8.1 and it becomes apparent that even a rather simple model leads to an unwieldy numerical complex.

Any given problem, on closer inspection, contains certain redeeming features and these should invariably be used to full advantage in setting up its numerical version. We illustrate this aspect of numerical adroitness in the last paragraph of section 8.1 by a catalyst effectiveness problem for two independent reactions.

Section 8.2 is the initial value problem revisited. Even the trivial linear problem of section 4.2, $y^{(1)} = Ky$, is taken up and a really difficult problem of coupled nonlinear equations is only reached via a set of coupled linear equations that are also in principle treated in chapter 4.

The emphasis in our present treatment of the initial value problem is on accuracy after many integration steps, rather than on accuracy within a single step. The problem of stability of a given integration method is now of decisive importance and methods that have desirable long-range integration properties are to be recommended as standard routines. Thus some of the end point methods that are discarded in subsection 4.2.3 due to their single-step properties are now revived.

Furthermore, a hybrid between purely explicit and purely implicit methods—the so-called semi-implicit Runge–Kutta methods—are presented as perhaps the final solution to the problem of integrating really "stiff" differential equations that occur ever so often in chemical engineering practice. It appears that these methods supported by a suitable stepsize adjustment procedure combine accuracy of integration for both fast and slow components with a very modest computing effort.

Sensitivity analysis—that is, the analysis of how the solution of a certain problem is changed when side conditions or parameters in the equation are changed—is the subject of section 8.3. This has not in itself any direct connection with the techniques discussed otherwise in the chapter except that the sensitivity function is usually almost directly available either from the collocation solution of the basic problem or for initial value problems (also from the results of the semi-implicit Runge–Kutta integration). Sensitivity analysis is nonetheless of such importance in cutting down the computational work of complicated problems that we feel obliged to call attention to some techniques for this purpose—especially since numerical practice does not seem at all well established in the field.

The final section 8.4 on nonlinear partial differential equations is quite short since all important features of their solution by collocation are contained in sections 8.1 to 8.3 and in chapter 4. Several applications of the methods of this chapter to partial differential equations are taken up in the exercises. These exercises as well as those of chapter 9 are to be regarded as an extension of the text for research students.

8.1 On the Numerical Structuring of Coupled Differential Equations

8.1.1 The initial value problem

Consider the analysis of an isothermal one-dimensional fixed-bed reactor for a series of coupled reactions. It is desired to compute the conversion of reactants $y_1, y_2, \ldots, y_M$ as functions of axial distance x.

The concentration profiles are described by

$$\frac{d}{dx}\mathbf{y} = \mathbf{r}(\mathbf{y}) \tag{1}$$

where

$$\mathbf{y}^T = (y_1, y_2, \ldots, y_M) \tag{2}$$

and

$$\mathbf{r}^T = [r_1(y_1, y_2, \ldots, y_M), r_2(y_1, y_2, \ldots, y_M), \ldots, r_M(y_1, \ldots, y_M)] \tag{3}$$

A numerical solution by orthogonal collocation is possible by the methods of chapter 4 using stepwise integration in the x-direction.

Assume that the solution $\mathbf{y}_0$ at $x = x_0$ is known and that our objective is to calculate the solution at $x = x_0 + h$. Introduction of

$$x = x_0 + uh \tag{4}$$

yields

$$\frac{d}{du}(\mathbf{y}) = h\mathbf{r}(\mathbf{y}) \tag{5}$$

Discretization at properly chosen interior collocation points u_1, $u_2, \ldots, u_N$ yields the following $N \times M$ set of collocation equations:

$$\mathbf{AY} + \mathbf{A}_0 \cdot \mathbf{y}_0^T = h \cdot \mathbf{R}(\mathbf{Y}) \tag{6}$$

$\mathbf{Y}$ is an $(N \times M)$ matrix, the ijth element representing the concentration of the jth component at the ith collocation point, $Y_{ij} = y_j(u = u_i)$; $\mathbf{y}_0$ is the M-vector of inlet concentrations, and the ijth element of the matrix $\mathbf{R}$ is given by $R_{ij} = r_j(Y_{i1}, Y_{i2}, \ldots, Y_{iM})$.

From the solution matrix $\mathbf{Y}$ of (6) the vector of concentrations $\mathbf{y}_{x_0+h}$ may as discussed in chapter 4 be obtained by extrapolation, quadrature, or a combination of these, depending on the choice of collocation points.

The $N \times M$ set of algebraic equations (6) can be solved by various methods, of which the two most obvious are briefly described. Multiplication by $\mathbf{A}^{-1}$ yields

$$\begin{aligned} \mathbf{Y} &= -\mathbf{A}^{-1}\mathbf{A}_0\mathbf{y}_0^T + h \cdot \mathbf{A}^{-1} \cdot \mathbf{R}(\mathbf{Y}) \\ &= \mathbf{1} \cdot \mathbf{y}_0^T + h \cdot \mathbf{A}^{-1} \cdot \mathbf{R}(\mathbf{Y}) \end{aligned} \tag{7}$$

and the iteration process

$$\mathbf{Y}^{(n+1)} = \mathbf{1} \cdot \mathbf{y}_0^T + h \cdot \mathbf{A}^{-1} \cdot \mathbf{R}[\mathbf{Y}^{(n)}] \tag{8}$$

may be applied. The evaluation is straightforward and the computational load per iteration is modest, requiring only setting up the $N \times M$ matrix $\mathbf{R}$ and subsequent multiplication by the $N \times N$ matrix $\mathbf{A}^{-1}$. As discussed below, convergence is often slow.

Alternatively, the Newton–Raphson method may be used. Consider, e.g., the case of three components ($M = 3$) and two collocation points ($N = 2$):

$$\begin{aligned} &\begin{pmatrix} A_{11} & A_{12} \\ A_{21} & A_{22} \end{pmatrix} \begin{pmatrix} Y_{11} & Y_{12} & Y_{13} \\ Y_{21} & Y_{22} & Y_{23} \end{pmatrix} + \begin{pmatrix} A_{10} \\ A_{20} \end{pmatrix} (Y_{01} \quad Y_{02} \quad Y_{03}) \\ &\quad = h \cdot \begin{bmatrix} r_1(Y_{11}, Y_{12}, Y_{13}) & r_2(Y_{11}, Y_{12}, Y_{13}) & r_3(Y_{11}, Y_{12}, Y_{13} \\ r_1(Y_{21}, Y_{22}, Y_{23}) & r_2(Y_{21}, Y_{22}, Y_{23}) & r_3(Y_{21}, Y_{22}, Y_{23}) \end{bmatrix} \end{aligned} \tag{9}$$

or

$$\begin{aligned} A_{11}Y_{11} + A_{12}Y_{21} + A_{10}Y_{01} - h \cdot r_1(Y_{11}, Y_{12}, Y_{13}) &= 0 \\ A_{21}Y_{11} + A_{22}Y_{21} + A_{20}Y_{01} - h \cdot r_1(Y_{21}, Y_{22}, Y_{23}) &= 0 \\ A_{11}Y_{12} + A_{12}Y_{22} + A_{10}Y_{02} - h \cdot r_2(Y_{11}, Y_{12}, Y_{13}) &= 0 \\ A_{21}Y_{12} + A_{22}Y_{22} + A_{20}Y_{02} - h \cdot r_2(Y_{21}, Y_{22}, Y_{23}) &= 0 \\ A_{11}Y_{13} + A_{12}Y_{23} + A_{10}Y_{03} - h \cdot r_3(Y_{11}, Y_{12}, Y_{13}) &= 0 \\ A_{21}Y_{13} + A_{22}Y_{23} + A_{20}Y_{03} - h \cdot r_3(Y_{21}, Y_{22}, Y_{23}) &= 0 \end{aligned} \tag{10}$$

Let

$$\mathbf{z}^T = (Y_{11}, Y_{21}, Y_{12}, Y_{22}, Y_{13}, Y_{23})$$

and

$$r_{ijk} = \left(\frac{\partial r_i}{\partial y_j}\right) \quad \text{at } (Y_{k1}, Y_{k2}, Y_{k3})$$

The Jacobian takes the following form

$$\left\{\begin{array}{cc:cc:cc} A_{11} - hr_{111} & A_{12} & -hr_{121} & 0 & -hr_{131} & 0 \\ A_{21} & A_{22} - hr_{112} & 0 & -hr_{122} & 0 & -hr_{132} \\ \hdashline -hr_{211} & 0 & A_{11} - hr_{221} & A_{12} & -hr_{231} & 0 \\ 0 & -hr_{212} & A_{21} & A_{22} - hr_{222} & 0 & -hr_{232} \\ \hdashline -hr_{311} & 0 & -hr_{321} & 0 & A_{11} - hr_{331} & A_{12} \\ 0 & -hr_{312} & 0 & -hr_{322} & A_{21} & A_{22} - hr_{332} \end{array}\right\} \tag{11}$$

In the general case of M components, N collocation points, and

$$\mathbf{z}^T = (Y_{11}, Y_{21}, \ldots, Y_{N1}, ----, Y_{1M}, Y_{2M}, \ldots, Y_{NM})$$

the $(M \times N)$ $(M \times N)$ Jacobian matrix $\mathbf{J}$ may be written

$$\mathbf{J} = \mathbf{J}_A - h \cdot \mathbf{J}_r \tag{12}$$

Each of the matrices $\mathbf{J}_A$ and $\mathbf{J}_r$ are composed of $M \times M$ blocks $(\mathbf{J}_A)_{ij}$ and $(\mathbf{J}_r)_{ij}$ that are of size $N \times N$:

$$(\mathbf{J}_A)_{ij} = \begin{cases} \mathbf{A} & i = j \\ \mathbf{O} & i \neq j \end{cases} \qquad \begin{array}{l} i = 1, 2, \ldots, M \\ j = 1, 2, \ldots, M \end{array} \tag{13}$$

and

$$(\mathbf{J}_r)_{ij} = \left\{\begin{array}{cccc} r_{ij1} & & & 0 \\ & r_{ij2} & & \\ & & \ddots & \\ 0 & & & r_{ijN} \end{array}\right\} \qquad \begin{array}{l} i = 1, 2, \ldots, M \\ j = 1, 2, \ldots, M \end{array} \tag{14}$$

$\mathbf{J}$ is thus a fairly sparse matrix with a total of $M \times N \times (M + N - 1)$ nonzero elements, but in general there seems to be no way in which the sparsity of $\mathbf{J}$ can be advantageously utilized in the solution of the linear equations.

The Newton–Raphson method is normally rapidly convergent, but the computational effort per iteration, which involves solution of $N \times M$

algebraic equations, may be substantial. The combined use of Newton-Raphson convergence and a high approximation order is particularly expensive for large systems of equations, and for such systems a desired accuracy is most conveniently obtained using a low approximation order and a small stepsize h.

In conclusion we notice that Nth-order collocation solution of the M coupled nonlinear equations (1) leads to solution of $M \times N$ nonlinear algebraic equations per integration step. Although the iteration process may be skilfully programmed, the solution of an initial value problem by a truly implicit method may be very expensive, and the methods of subsection 8.2.4 are probably superior.

8.1.2 The boundary value problem

Let us next consider a series of reactions occurring isothermally in a porous catalyst particle.

The concentration profile is found from

$$\nabla^2 \mathbf{y} = \mathbf{r}(\mathbf{y}) \tag{15}$$

with boundary conditions

$$\nabla \mathbf{y} = \mathbf{0} \quad \text{at } x = 0 \tag{16}$$

and

$$\mathbf{y} = \mathbf{y}_{N+1} \quad \text{at } x = 1 \tag{17}$$

Discretization at N interior collocation points yields a system of equations similar to (6):

$$\mathbf{CY} + \mathbf{C}_{N+1}\mathbf{y}_{N+1}^T = \mathbf{R}(\mathbf{Y}) \tag{18}$$

where $\mathbf{C}$ is the discretization operator for the Laplacian.

Again, direct substitution is possible, yielding

$$\mathbf{Y}^{(n+1)} = \mathbf{1} \cdot \mathbf{y}_{N+1}^T + \mathbf{C}^{-1}\mathbf{R}(\mathbf{Y}^{(n)}) \tag{19}$$

There is, however, an important distinction between (19) and the corresponding equation (8) for the initial value problem. Convergence of (8) is always possible if a sufficiently small value of the stepsize h is used. A similar option is *not* available for the boundary value problem.

The Jacobian to be used in the Newton–Raphson method corresponds in structure to that shown for the initial value problem, and convergence is normally easily obtained. It should be noticed, however, that a desired accuracy for the initial value problem could preferably be obtained using a combination of low approximation order N and small stepsize h. For the boundary value problem a high accuracy necessarily involves using a large N with a corresponding increase in the computational effort.

One possible remedy, not to be discussed here, would be to search for better trial functions as suggested in chapter 6. The extra work involved in setting up discretization matrices, finding optimal collocation points, or evaluating Galerkin integrals may be worthwhile in cases like these, where the computational price for a high-order method is inordinately high.

A further possibility is to apply a combination of the Newton–Raphson method and successive approximation or to utilize existing linear ties between the dependent variables. These methods, which have the effect of drastically reducing the number of equations to be solved, are discussed in the following subsection.

8.1.3 Economic structuring of the numerical problem

The general problem of determining effectiveness factors for complex reactions occurring in a catalyst particle with internal and external transport resistances is again taken up, but now using an example to illustrate how a particular problem should be treated.

Let us consider two independent reactions with four components A, B, C, and D.

$$A + B \rightleftarrows C \tag{20}$$

$$2A + C \rightleftarrows D \tag{21}$$

with reaction rates r_1 and r_2 and heats of reaction $-\Delta H_1$ and $-\Delta H_2$. Physical properties (i.e., diffusion coefficients D, external mass transfer coefficients k, heat transfer coefficient h, and thermal conductivity λ) are assumed to be constant.

The transport equations are

$$D_A \nabla^2 C_A = r_1 + 2r_2 \tag{22a}$$

$$D_B \nabla^2 C_B = r_1 \tag{22b}$$

$$D_C \nabla^2 C_C = -r_1 + r_2 \tag{22c}$$

$$D_D \nabla^2 C_D = -r_2 \tag{22d}$$

$$\lambda \nabla^2 T = \Delta H_1 \cdot r_1 + \Delta H_2 \cdot r_2 \tag{22e}$$

The boundary conditions are

$$\nabla C_i = \nabla T = 0 \quad \text{at } x = 0 \tag{23}$$

and at the surface

$$D_A\left(\frac{\partial C_A}{\partial x}\right)_s = k_A(C_{Ab} - C_{As})$$
$$\vdots$$
$$D_D\left(\frac{\partial C_D}{\partial x}\right)_s = k_D(C_{Db} - C_{Ds}) \tag{24}$$
$$\lambda\left(\frac{\partial T}{\partial x}\right)_s = h(T_b - T_s)$$

with subscripts b and s referring to bulk and surface conditions, respectively. The bulk phase concentrations and temperature are known quantities.

This system of equations may, of course, be solved as described in section 8.1 in terms of the five dependent variables C_A, C_B, C_C, C_D, and T, using different discretization matrices for the Laplacian due to the different boundary conditions (24). The final numerical model for an Nth-order method is a set of $5N$ algebraic equations. Considerable simplification is possible as shown below.

We select B and D as key components, one for each independent reaction.

$$r_1 = D_B\nabla^2 C_B \quad \text{and} \quad r_2 = -D_D\nabla^2 C_D$$

Substitution into the remaining equations yields

$$\begin{aligned} D_A\nabla^2 C_A &= D_B\nabla^2 C_B - 2D_D\nabla^2 C_D \\ D_C\nabla^2 C_C &= -D_B\nabla^2 C_B - D_D\nabla^2 C_D \\ \lambda\nabla^2 T &= \Delta H_1 D_B\nabla^2 C_B - \Delta H_2 D_D\nabla^2 C_D \end{aligned} \tag{25}$$

and after integrating twice

$$D_A(C_A - C_{As}) = D_B(C_B - C_{Bs}) - 2 \cdot D_D(C_D - C_{Ds}) \tag{26}$$

Similarly, the boundary condition at the surface yields

$$k_A(C_{Ab} - C_{As}) = k_B(C_{Bb} - C_{Bs}) - 2k_D(C_{Db} - C_{Ds}) \tag{27}$$

Equations (26) and (27) are combined into

$$\begin{aligned} C_A - C_{Ab} = &\left[\frac{D_B}{D_A}(C_B - C_{Bs}) - \frac{k_B}{k_A}(C_{Bb} - C_{Bs})\right] \\ &- 2\left[\frac{D_D}{D_A}(C_D - C_{Ds}) - \frac{k_D}{k_A}(C_{Db} - C_{Ds})\right] \end{aligned} \tag{28}$$

Similar expressions are obtained for C_C and T in terms of C_B, C_D, C_{Bs}, and C_{Ds}.

Collocation equations are formulated in the N interior points for components B and D. Their values at the surface C_{Bs} and C_{Ds} are not eliminated, but two additional equations are formulated that incorporate boundary conditions at the surface.

There results a system of $2N + 2$ algebraic equations in the $2N + 2$ collocation ordinates C_B (N values), C_D (N values), and the surface concentrations C_{Bs} and C_{Ds}.

Further simplifications are often possible. Assume, for example, that the rate of the second reaction is small. The concentration of component D is set equal to its bulk-phase value throughout the pellet, and the resulting set of $N + 1$ equations is solved for the concentration profile of component B. Corrected values for the concentration profile of D are then evaluated, and if necessary the process is repeated.

8.2 The General Initial Value Problem—Accuracy, Convergence, and Stability Considerations

8.2.1 Accuracy within the single step, and convergence of single-step iterations

The general initial value problem is of the form

$$\frac{d}{dx}\mathbf{y} = \mathbf{f}(x, y) \tag{29}$$

As before, our objective is to arrive at the solution $\mathbf{y}$ at $x_0 + h$ from the known vector $\mathbf{y}_0$ at x_0.

Rewriting, we obtain

$$\frac{d}{du}\mathbf{y} = h \cdot \mathbf{f}(x_0 + uh, \mathbf{y}) \qquad 0 \le u \le 1 \tag{30}$$

where $x = x_0 + u \cdot h$.

It is shown in section 6.3 that the one-point Radau method for a single linear or nonlinear equation has an error of magnitude $\mathcal{O}(h^4)$; in chapter 4 the dominant error terms for coupled linear equations with constant coefficients are shown to be $\mathcal{O}(h^{2N+1})$, $\mathcal{O}(h^{2N+2})$, and $\mathcal{O}(h^{2N+3})$ for the Gauss method, the Radau method, and the Lobatto method, each with N interior points.

One may prove (following the procedure of section 6.3) that these results for the single-step accuracy are also true for the general nonlinear set of equations (29).

Formulas for the dominant error term are given in chapter 4 for the case of linear equations with constant coefficients. Similar results are not in general obtainable for (30). For many problems encountered in the chemical engineering literature, however, **f** is more strongly dependent on the dependent variable **y** than on the independent variable x:

$$\left(\frac{\partial f}{\partial y}\right)_x \gg \left(\frac{\partial f}{\partial x}\right)_y \tag{31}$$

In this case a first approximation to the integration error of (30) can be obtained from the integration error of the corresponding linearized problem:

$$\frac{d}{du}\mathbf{y} = h[\mathbf{f}_0 + \mathbf{Q}(\mathbf{y} - \mathbf{y}_0)] \tag{32}$$

where

$$\mathbf{f}_0 = \mathbf{f}(x_0, \mathbf{y}_0) \quad \text{and} \quad Q_{ij} = \left(\frac{\partial f_i}{\partial y_j}\right)_{x_0, \mathbf{y}_0}$$

Let λ be the largest modulus eigenvalue of **Q**. The integration error may then be estimated from, e.g., (4.32) or (4.41), substituting $h\lambda$ for k, provided $h\lambda \ll 1$. Integration over a finite x-interval with a given accuracy is obtainable either by using a small stepsize h and a small number of collocation points or by using a larger stepsize and a larger N.

The number of integrations to be performed is inversely proportional to the stepsize h, but the computational effort in each step increases with increasing N in a manner that depends on the method that is chosen for solution of the algebraic equations.

For a given accuracy and a selected solution method, an optimal combination of h and N exists, that is, a combination yielding the solution with the desired accuracy for the smallest computational effort. This optimal combination (and the choice of solution method) depends on the number of dependent variables, the desired accuracy, and the computational effort required in the evaluation of **f** and its derivatives relative to that of solving linear algebraic equations. Certain guidelines are presently given.

The Newton–Raphson method, which has been our preferred choice of algorithm for solving collocation equations in previous chapters, has the main advantage that with reasonably accurate starting values convergence to machine accuracy can be obtained in a very few, say two to four, iterations. The method is severely handicapped, however, when the number M of dependent variables is large. The dimension of the set of linear equations to be solved is $N \times M$, requiring a number of operations proportional to $(N \times M)^3$. Provided M is larger than 2, the combination of a small stepsize and a small N (say one to three) is usually to be preferred. The balance is shifted toward larger N in three cases: when the number of equations M is small, when a very high accuracy is desired, and when the evaluation of **f** is expensive.

For large systems of equations, collocation coupled to a Newton–Raphson algorithm becomes unattractive since by comparison the computational load of an explicit method such as the Runge–Kutta method is only proportional to M.

The direct substitution method may be an attractive alternative for large systems.

The collocation equations for the Gauss method can be written

$$\mathbf{Y} = \mathbf{1} \cdot \mathbf{y}_0^T + \mathbf{A}^{-1}\mathrm{h} \cdot \mathbf{f}(\mathbf{u}, \mathbf{Y}) \tag{33}$$

suggesting the iteration scheme

$$\mathbf{Y}^{(k+1)} = \mathbf{1} \cdot \mathbf{y}_0^T + \mathbf{A}^{-1}h \cdot \mathbf{f}[\mathbf{u}, \mathbf{Y}^{(k)}] \tag{34}$$

The effort in each iteration step is modest, the number of operations being proportional to $M \times N^2$, but the process is only linearly convergent with a rate of convergence,

$$\frac{e_{k+1}}{e_k} \sim q = \frac{h \cdot |\lambda_Q|_{\max}}{|\lambda_A|_{\min}}$$

where $|\lambda_Q|$ is the modulus of the numerically largest eigenvalue of $\mathbf{Q}$, and $|\lambda_A|$ is the modulus of the numerically smallest eigenvalue of $\mathbf{A}$. e_k is the norm of $\mathbf{Y} - \mathbf{Y}^{(k)}$ where $\mathbf{Y}$ is the exact solution of (33).

The stepsize h must be chosen such that $h \cdot |\lambda_Q|_{\max}/|\lambda_A|_{\min} < 1$. The criterion for terminating the iteration sequence should be chosen in accordance with the accuracy expected from, e.g., (4.41).

Collocation coupled to a direct substitution algorithm for solution of the resulting algebraic equations has much in common with the Runge–Kutta method. The Gauss method with two interior collocation points corresponds in accuracy to the well-known fourth-order Runge–Kutta method, and in general a Gauss method with N interior points corresponds to a Runge–Kutta method of order $2N$. The computational effort for both methods is proportional to the number of dependent variables M.

8.2.2 Stability of integration

In the analysis of subsection 8.2.1, it is assumed that the stepsize h is chosen sufficiently small to make $h \cdot |\lambda_Q|_{\max} \ll 1$. Frequently systems of equations are encountered with large spread in the modulus of the eigenvalues. The eigenvalue with the smallest modulus determines the range of the independent variable—e.g., the "time" to reach a steady state—while the stepsize is determined by the eigenvalue of the largest modulus. The number of subintervals required becomes proportional to the ratio between the largest–modulus and smallest–modulus eigenvalues. Differential equations with a high value of this ratio are called *stiff* and a large number of subintervals are required for accurate integration of such equations.

It is therefore worthwhile considering whether a stepsize larger than $|\lambda_{\max}|^{-1}$ might be permissible. Let us as our first example consider single-step integration from $x = x_n$ to $x = x_n + h$ of the simple differential equation

$$\frac{dy}{dx} = \lambda y \quad \text{with } y(x_n) = y_n \tag{35}$$

We obtain

$$y_{n+1} = y(x_n + h) = \mu(h\lambda)y_n \tag{36}$$

where $\mu(h\lambda)$ for the Gauss, the Radau, and the Lobatto method is expressed as a ratio of two polynomials in $k = h\lambda$, the denominator polynomial being of degree N and the numerator polynomial of degree N, $N + 1$, and $N + 2$, respectively. For $N = 2$, we obtain, e.g.,

Gauss method: $$\mu(k) = \frac{k^2 + 6k + 12}{k^2 - 6k + 12}$$

Radau method: $$\mu(k) = \frac{k^3 + 9k^2 + 36k + 60}{3k^2 - 24k + 60}$$

Lobatto method: $$\mu(k) = \frac{k^4 + 12k^3 + 72k^2 + 240k + 260}{12k^2 + 120k + 360}$$

The multiplier $\mu(k)$ in the difference equation (36) is called the characteristic root of the integration method since continued integration of (35) leads to

$$y_{n+m} = y(x_n + mh) = y(x_n)[\mu(h\lambda)]^m \tag{37}$$

An integration method is said to be A-stable for the problem (35) provided $|\mu(h\lambda)| \leq 1$ for any value of λ with nonpositive real part.

A-stability implies that the integration error is limited, as the difference between the approximate and the exact solution at $x = x_n + mh$:

$$e(x_n + mh) = y_n\{[\mu(h\lambda)]^m - \exp(mh\lambda)\} \tag{38}$$

is finite for any value of λ with nonpositive real part. A method that leads to $|\mu| > 1$ will obviously by continued integration lead to an approximate solution with ever-increasing magnitude, while the true solution should decrease to zero (or remain of constant magnitude for purely imaginary λ) as $m \to \infty$.

Among the methods considered above, only the Gauss method is A-stable. As $\lambda \to \infty$, $\mu \to 1$. The μ of the Radau and the Lobatto methods increases proportionally to λ and λ^2, respectively, when $\lambda \to -\infty$. The standard fourth-order Runge–Kutta method is another non-A-stable method with

$$\mu(k) = 1 + k + \tfrac{1}{2}k^2 + \tfrac{1}{6}k^3 + \tfrac{1}{24}k^4$$

The error, $\mu(h\lambda) - \exp(h\lambda)$, is given as a function of $k = h\lambda$ for the three collocation methods ($N = 2$) and for the Runge–Kutta method in figure 8-1(a).

The fourth-order Runge–Kutta method, the Lobatto method, and the Radau method with $N = 2$ become unstable at $h\lambda = -2.8$, $h\lambda = -9.6$, and $h\lambda = -11.8$, respectively. The effective stability range of the two collocation methods is thus much larger than that of the Runge–Kutta method, but even so these methods are not very attractive for values of $|h\lambda|$ larger than say four to six. In spite of its stability, the Gauss method is not too attractive either since μ for large $|h\lambda|$ will either tend monotonously to 1 (N even) or to -1 (N uneven).

It is highly desirable to have an approximation technique that effectively damps the solution for $h\lambda \to -\infty$. The relative error of the solution is still very large, but at least the qualitative behavior is correctly represented. We compare two methods with this property. One is the end point collocation method of subsection 4.2.3, and the other is a semi-implicit Runge–Kutta method that is discussed in subsection 8.2.4.

When the right-hand interval end point is included as an extra collocation point, $y(1)$ is obtained directly from the collocation solution. With different choices of the interior collocation points [the zeros of $P_N^{(0,0)}$ or $P_N^{(1,0)}$], collocation methods corresponding to the Gauss or the Radau methods are obtained. They are characterized by a $\mu(k)$ with a numerator of degree N and a denominator of degree $N + 1$. Hence $\mu(k)$ tends to zero when $\lambda \to -\infty$, an important quality of these methods that may outweigh their smaller accuracy within a single step.

With N interior (i.e., $N + 1$ total) collocation points, the following expressions for $\mu(k)$ are obtained by solution of (35) in the form (36):

Gauss method:

$$N = 1: \quad \mu(k) = \frac{k + 4}{k^2 - 3k + 4}$$

$$N = 2: \quad \mu(k) = \frac{k^2 + 12k + 36}{-k^3 + 7k^2 - 24k + 36}$$

Radau method:

$$N = 1: \quad \mu(k) = \frac{2k + 6}{k^2 - 4k + 6}$$

$$N = 2: \quad \mu(k) = \frac{3k^2 + 24k + 60}{-k^3 + 9k^2 - 36k + 60}$$

Error curves for the end point methods for $N = 1$, 2, and 3 are given in figure 8-1(b). As discussed in subsection 4.2.3, the approximation order for the Gauss and Radau methods using N interior points and the end point equals that of the corresponding methods using interior points only, and no gain in the single-step accuracy is obtained by including the end point when the magnitude of k is small. The end point methods yield limited values of the error for any negative value of λ, however, and the error decreases asymptotically to 0 when $\lambda \to -\infty$.

For large negative λ, $\mu(h\lambda)$ converges to 0 as $(h\lambda)^{-1}$ for any Gauss method and as $(N + 1)(h\lambda)^{-1}$ for the Nth-order Radau method. The same amount of computation is required for an N-point Gauss method and an N-point Radau method. The former has the advantage of a smaller maximum error and a more favorable asymptotic behavior, whereas the Radau method is the more accurate for small values of $|h\lambda|$. The end point methods described require a total of at least two collocation points, and for most problems N would rarely be chosen larger than 2 or 3. The extra accuracy obtainable with a higher-order method does not compensate for the increased computational effort, and increased accuracy is most economically obtained by decreasing the stepsize h.

Let us compare the two collocation methods, first without and then with collocation at the right-hand end point, on the following linear problem

$$\begin{aligned} \frac{dy_1}{dt} &= -y_1 + 8.1y_2 \\ \frac{dy_2}{dt} &= y_1 - 9.1y_2 \end{aligned} \tag{39}$$

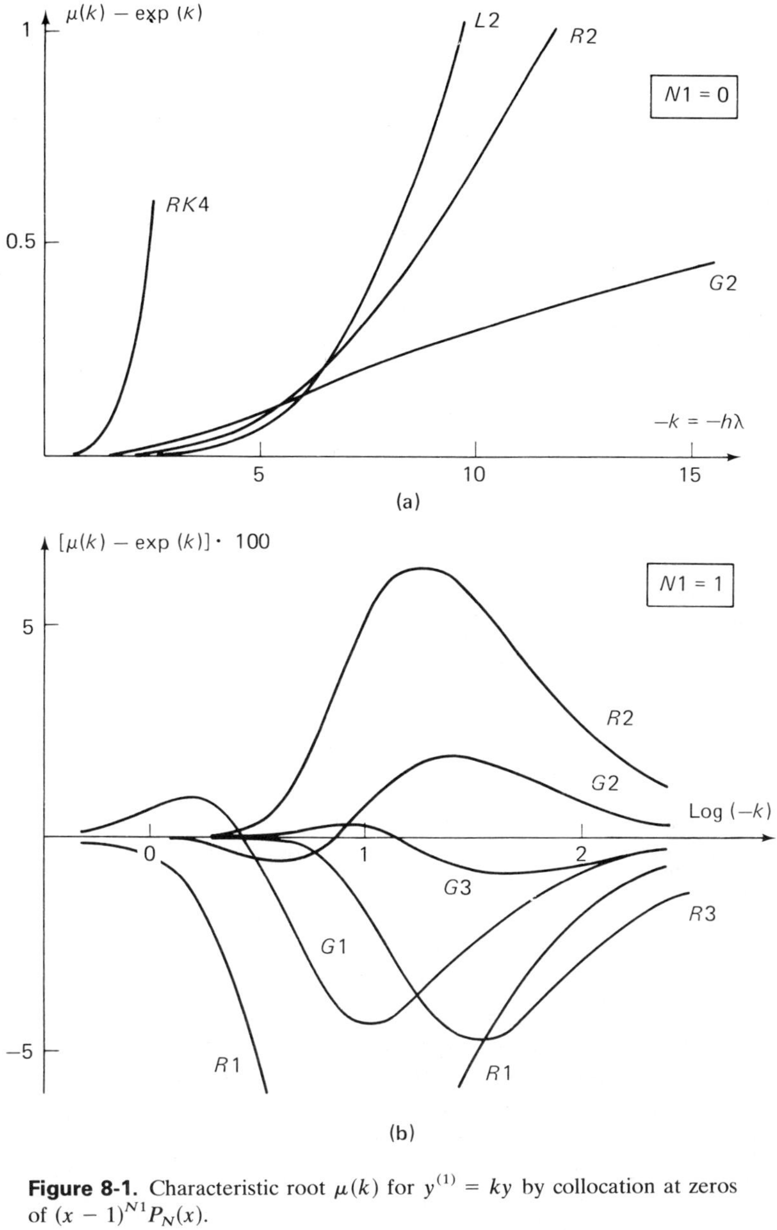

Figure 8-1. Characteristic root $\mu(k)$ for $y^{(1)} = ky$ by collocation at zeros of $(x - 1)^{N1}P_N(x)$.

with initial conditions $t = 0$: $y_1 = 1$ and $y_2 = 0$. Introducing

$$z_1 = y_1 + 0.9y_2$$

$$z_2 = y_1 - 9.0y_2$$

yields

$$\frac{dz_1}{dt} = -0.1z_1$$

$$\frac{dz_2}{dt} = -10.0z_2$$

with

$$z_1(0) = z_2(0) = 1$$

or

$$y_1(t) = \tfrac{10}{11}\exp(-0.1t) + \tfrac{1}{11}\exp(-10.0t)$$

$$y_2(t) = \tfrac{10}{99}\exp(-0.1t) - \tfrac{10}{99}\exp(-10.0t)$$

$y_2(t)$ increases rapidly from 0 to its maximum value 0.0955 at $t = 0.465$, after which it decreases exponentially following the first eigenfunction $\exp(-0.1t)$.

Let us integrate from $t = 0$ to $t = 25$, at which point y_2 has decreased to 0.00829.

The stability requirements for the fourth-order Runge–Kutta method (RK–4), the Radau, and the Lobatto methods are determined by the largest eigenvalue -10 of (39):

RK–4:	$10h < 2.8$	or	$h < 0.28$
Radau:	$10h < 11.8$	or	$h < 1.18$
Lobatto:	$10h < 9.8$	or	$h < 0.98$

$e_2(t)$, the difference between the approximate and the exact solution for $y_2(t)$ using two interior collocation points for the implicit methods, is given in table 8.1.

For $h < 0.5$, the Lobatto method is superior. The Gauss method is the only method that is stable for $h > 1.18$, but for these large h values the stiff component z_2 is very poorly represented, making the method unattractive in spite of its stability.

It is possible to increase the size of the stability region for the Radau and the Lobatto methods by increasing the number of collocation points N. The increase in the permissible value of h is only modest and the smaller number of steps does not compensate for the larger computational effort per step when a higher-order method is used. If the methods mentioned above are applied, one should use the lowest possible approximation order. For $N = 1$, we obtain for the Radau method

$$|h\lambda_{\max}| < 6 \quad \text{or} \quad h < 0.6$$

and for the Lobatto method

$$|h\lambda_{\max}| < 5.4\ldots \quad \text{or} \quad h < 0.54$$

TABLE 8.1
ACCUMULATED ERROR OF $y_2(t)$ FOR DIFFERENT STEPSIZE AND DIFFERENT METHODS

h	Method	$e_2(1)$	$e_2(10)$	$e_2(25)$
0.25	RK–4	$-1.8 \cdot 10^{-2}$	$-3.0 \cdot 10^{-9}$	$< 10^{-10}$
	Gauss	$-4.6 \cdot 10^{-6}$	$< 10^{-10}$	$< 10^{-10}$
	Radau	$1.1 \cdot 10^{-6}$	$< 10^{-10}$	$< 10^{-10}$
	Lobatto	$-3.9 \cdot 10^{-7}$	$< 10^{-10}$	$< 10^{-10}$
0.5	RK–4	Unstable	—	—
	Gauss	$-1.1 \cdot 10^{-3}$	$3.2 \cdot 10^{-10}$	$1.8 \cdot 10^{-10}$
	Radau	$-6.2 \cdot 10^{-4}$	$< 10^{-10}$	$< 10^{-10}$
	Lobatto	$-4.6 \cdot 10^{-4}$	$< 10^{-10}$	$< 10^{-10}$
1.0	Gauss	$-3.1 \cdot 10^{-2}$	$-6.5 \cdot 10^{-7}$	$2.9 \cdot 10^{-9}$
	Radau	$6.7 \cdot 10^{-2}$	$-1.8 \cdot 10^{-3}$	$-4.0 \cdot 10^{-6}$
	Lobatto	Unstable	—	—
2.5	Gauss		$-1.5 \cdot 10^{-2}$	$-8.3 \cdot 10^{-4}$
	Radau	Unstable	—	—

The Lobatto one-point method still permits twice as large a step as the Runge–Kutta method and it is somewhat more accurate. The error in each step is $\mathcal{O}(h^5)$ for both methods.

Using only one collocation point and a steplength $h = 0.2$, the following results in table 8.2 are obtained.

TABLE 8.2
ACCUMULATED ERROR OF $y_2(t)$ FOR $N = 1$ AND $h = 0.2$

	$t = 1$	$t = 10$	$t = 25$
Gauss	$4.0 \cdot 10^{-6}$	$1.2 \cdot 10^{-6}$	$6.9 \cdot 10^{-7}$
Radau	$2.8 \cdot 10^{-4}$	$< 10^{-9}$	$< 10^{-9}$
Lobatto	$2.8 \cdot 10^{-6}$	$< 10^{-9}$	$< 10^{-9}$

The small error at $t = 1$ for the Gauss method is coincidental, the reason being that the Gauss representation of $\exp(h\lambda)$, $(2 + h\lambda)/(2 - h\lambda)$, equates z_2 to 0 after the first step $x = h = 0.2$ when $\lambda = -10$. In contrast the Gauss method is not nearly so accurate at $t = 10$ and at $t = 25$ as the other two methods. This error is caused by integration of the slow component z_1 that is obtained with an insufficient accuracy for $N = 1$ and the Gauss method.

The conclusion of the previous paragraph is substantiated by the example that no method based on interior collocation points only is well suited for stiff systems. If the separation of the eigenvalues is large but not excessive, the best among the methods are probably the one-point Radau or Lobatto methods.

One possible remedy is to use a small stepsize initially until the stiff components have essentially vanished. From then on the stepsize may be increased provided a stable method, i.e., the Gauss method, is used. If an unstable method is used, the stiff component will eventually reappear and lead to erroneous results, while the Gauss method should be able to keep the errors at a small but nonzero level.

Next we consider the end point collocation methods. These methods are not only stable but they also provide a strong damping of the stiff components. However, the penalty for their use is either loss of accuracy in the integration of the slow components or an increase in the computational labor per step. The one-point Lobatto method has a local integration error of order $\mathcal{O}(h^5)$, and M algebraic equations must be solved at each step. In contrast, the Radau method with one interior point in addition to the end point has a local error $\mathcal{O}(h^4)$ and $2M$ algebraic equations must be solved per step. This shows that the end point methods can be unfavorable in both respects.

The simplest end point collocation method is collocation at the interval end point alone. The local error for this method is $\mathcal{O}(h^2)$, which probably makes the method too inaccurate for most purposes.

Here we are concerned mainly with the Gauss and Radau methods for which the errors are $\mathcal{O}(h^{2N+1})$ and $\mathcal{O}(h^{2N+2})$, respectively. The number N of interior collocation points is normally chosen as 1 or at most 2.

Values of $e_2(t)$ for various end point methods appear in table 8.3, which may be compared to table 8.1.

The results for large integration time indicate an increase in accuracy for the higher-order methods and a decrease in accuracy with increasing steplength. A given accuracy at large values of t is in the present case most economically obtained combining a small steplength and the one-point Radau method.

The maximum value of the error for these methods with constant h is always found after the first time step. The magnitude is normally smaller for the Gauss method than for the corresponding Radau method, but apart from this no systematic variation is seen.

8.2.3 Stepsize selection strategy

Although capable of yielding accurate results at large integration times the fairly large errors encountered for the end point methods after the first step indicates that for nonlinear problems a constant stepsize

TABLE 8.3

ACCUMULATED ERROR FOR $y_2(t)$ USING END POINT COLLOCATION

h	Method	Maximum value of e_2	$e_2(10)$	$e_2(25)$
0.5	G1	$3.0 \cdot 10^{-3}$	$3.7 \cdot 10^{-6}$	$2.1 \cdot 10^{-6}$
	R1	$8.6 \cdot 10^{-3}$	$6.4 \cdot 10^{-8}$	$3.6 \cdot 10^{-8}$
	G2	$4.6 \cdot 10^{-4}$	$1.7 \cdot 10^{-9}$	$< 10^{-9}$
	R2	$1.9 \cdot 10^{-3}$	$2.3 \cdot 10^{-9}$	$1.3 \cdot 10^{-9}$
1.0	G1	$4.5 \cdot 10^{-3}$	$1.4 \cdot 10^{-5}$	$8.0 \cdot 10^{-6}$
	R1	$9.7 \cdot 10^{-3}$	$5.0 \cdot 10^{-7}$	$2.8 \cdot 10^{-7}$
	G2	$8.1 \cdot 10^{-4}$	$3.3 \cdot 10^{-9}$	$1.9 \cdot 10^{-9}$
	R2	$5.2 \cdot 10^{-3}$	$< 10^{-9}$	$< 10^{-9}$
2.5	G1	$3.0 \cdot 10^{-3}$	$8.0 \cdot 10^{-5}$	$4.5 \cdot 10^{-5}$
	R1	$6.0 \cdot 10^{-3}$	$8.9 \cdot 10^{-6}$	$4.2 \cdot 10^{-6}$
	G2	$1.8 \cdot 10^{-3}$	$1.3 \cdot 10^{-7}$	$9.8 \cdot 10^{-7}$
	R2	$6.1 \cdot 10^{-3}$	$1.3 \cdot 10^{-6}$	$2.7 \cdot 10^{-9}$
5.0	G1	$2.0 \cdot 10^{-3}$	$2.4 \cdot 10^{-4}$	$1.5 \cdot 10^{-4}$
	R1	$3.5 \cdot 10^{-3}$	$1.8 \cdot 10^{-4}$	$3.2 \cdot 10^{-5}$
	G2	$1.4 \cdot 10^{-3}$	$2.0 \cdot 10^{-5}$	$9.8 \cdot 10^{-7}$
	R2	$4.3 \cdot 10^{-3}$	$1.8 \cdot 10^{-4}$	$6.9 \cdot 10^{-7}$

Nomenclature: G = Gauss, R = Radau method, G1 = one interior point ($N = 1$) Gauss method.

integration might lead to inaccurate results. For the linear problem (39) an initial error in the solution for the stiff component will not affect the solution for the slow component. This decoupling is not found for nonlinear problems and erroneous results initially may well propagate in time although exponential growth may not take place.

The stiff components should therefore be integrated correctly during the time period where they influence the solution, and in an effective integration procedure a small stepsize is normally called for initially.

Stepsize adjustment is conveniently obtained using the so-called full-step/half-step technique. Consider, e.g., the end point Radau method with one interior point. We know that the dominant single-step error is $\mathcal{O}(h^4)$, and our approximate solution at $x = x_n + h = x_{n+1}$ may be written

$$\mathbf{y}_{n+1}(h) = \mathbf{y}_{n+1}^{(ex)} + \mathbf{q} \cdot h^4 + \mathcal{O}(h^5) \tag{40}$$

where $\mathbf{y}_{n+1}^{(ex)}$ is the exact solution, and $\mathbf{q} \cdot h^4$ is the dominant (but unknown) error term.

Next we consider integration from x_n to $x_n + h$ in two steps, each of size $h/2$; we obtain

$$\mathbf{y}_{n+1}\left(\frac{h}{2}\right) = \mathbf{y}_{n+1}^{(ex)} + 2\mathbf{q}\left(\frac{h}{2}\right)^4 + \mathcal{O}(h^5)$$

where the factor 2 accounts for error accumulation in each of the two integration steps.

Subtraction yields

$$\mathbf{e} = \mathbf{y}_{n+1}\left(\frac{h}{2}\right) - \mathbf{y}_{n+1}(h) = -\frac{7}{8}\mathbf{q}h^4 + \mathcal{O}(h^5) \tag{41}$$

where the difference vector $\mathbf{e}$ between the two approximations is a measure of the accuracy obtained in the integration. Provided $\mathbf{e}$ is sufficiently small, the result is accepted and a refined estimate $\mathbf{y}_{n+1}^*$ is obtainable from

$$\mathbf{y}_{n+1}^* = \mathbf{y}_{n+1}\left(\frac{h}{2}\right) + \frac{1}{7}\mathbf{e} = \mathbf{y}_{n+1}^{(ex)} + \mathcal{O}(h^5) \tag{42}$$

We have thus obtained an increase in the approximation order by 1 and at the same time the accuracy of the integration process has been checked. Our criterion for acceptance is chosen as

$$e_i < \varepsilon_i \qquad i = 1, 2, \ldots, M \tag{43}$$

where ε_i is the selected tolerance for y_i. If this criterion is not satisfied, the stepsize h is halved and the integration is repeated.

When integrating a stiff problem, this procedure leads to small steps h_n whenever the solution changes rapidly, as is often the case at the start of the integration. As soon as the stiff component has faded away, one observes that the magnitude of $\mathbf{e}$ decreases rapidly and it becomes desirable to enlarge the stepsize. After a successful step with stepsize h_n we propose to adjust the stepsize h_{n+1} for the next integration by the following procedure: Choose h_{n+1} as the smaller of either

$$3h_n \tag{44a}$$

or

$$\left[4 \max\left(\frac{e_i}{\varepsilon_i}\right)\right]^{-0.25} h_n \tag{44b}$$

It frequently happens that e_i is much smaller than ε_i (see exercise 8.2 and 8.3). In this case the stepsize is increased by one of the two mechanisms (44). If $e_i < (\varepsilon_i/324)$ (44b) will increase h_n by more than a factor 3. If this happens there is a risk that in the next step a bisection by (43) would occur, leading to unnecessary computation. Consequently, we limit the stepsize increase by (44a). A stepsize decrease may also occur by (44b)—if $e_i > \frac{1}{4}\varepsilon_i$—but h_n can at most be reduced by a factor $\sqrt{2}$. Otherwise a bisection in (43) takes place. The factor 4 in (44b) and the factor 3 in (44a) are, of course, empirical but they are selected on the basis of large computing experience. The occurrence of $(\varepsilon_i/e_i)^{0.25}$ in (44b) is justified by the order of the error in (41).

Stepsize selection mechanisms quite different from that presented here are also used. One possibility is to run an integration method of order $\mathcal{O}(h^{k-1})$ in parallel with the method of order $\mathcal{O}(h^k)$. In this way an error estimate is also obtained.

In any case the extrapolation and steplength adjustment procedure (42) and (44) converts the end point Radau method into a method that is effectively fourth order for integration through many steps. When specifying tolerances ε_i, it should be taken into account that the actual error of the adjusted solution (42) is likely to be much smaller than ε.

Radau methods using a larger number of interior points would permit larger steps at the same accuracy, but it seems unlikely that this advantage could profitably counterbalance the increase in computational effort within each step.

8.2.4 Alternative methods for stiff systems

The Radau method with steplength adjustment is a versatile method that has been used for accurate integration of a number of notoriously stiff differential equations. It is, however, fairly expensive since for M coupled equations $2M$ usually nonlinear algebraic equations have to be solved in each integration step. We now introduce some equally reliable integration methods that require far less computational effort.

Let us first consider a scalar differential equation

$$\frac{dy}{dx} = f(y) \qquad y(x_n) = y_n \tag{45}$$

where $f(y)$ does not depend explicitly on x. A Taylor series from (x_n, y_n) is [see (6.96)]

$$y_{n+1} = y_n + hf + \tfrac{1}{2}h^2 f_y f + \tfrac{1}{6}h^3(f_{yy}f^2 + f_y^2 f) + \dots \tag{46}$$

where $f = f(y_n)$, $f_y = (\partial f/\partial y)_{y_n}$, etc.

Now a one-point Gauss method with collocation at $x = x_n + h/2$ and extrapolation to y_{n+1} gives

$$2(y_{1/2} - y_n) = hf(y_{1/2}) \qquad y_{n+1} = y(x_n + h) = y_n + 2(y_{1/2} - y_n) \tag{47}$$

The method is $\mathcal{O}(h^3)$ in y_{n+1} but one nonlinear equation—the first of (47)—must be solved for $y_{1/2}$.

If $f(y)$ is linearized from y_n, we obtain

$$2(y_{1/2} - y_n) = h[f(y_n) + f_y(y_{1/2} - y_n)] \quad \text{or} \quad y_{1/2} - y_n = (2 - hf_y)^{-1}hf$$

and finally

$$y_{n+1} = y_n + (1 - \tfrac{1}{2}f_y h)^{-1}hf \tag{48}$$

Expanding (48) yields

$$y_{n+1} = y_n + hf + \tfrac{1}{2}h^2 f_y f + \tfrac{1}{4}h^3 f_y^2 f$$

and comparison with (46) shows that terms up to and including h^2 are correctly represented.

Continuing next to M coupled differential equations, we obtain an expression similar to (48):

$$\mathbf{k}_1 = (\mathbf{I} - 0.5h\mathbf{f}_y)^{-1}h\mathbf{f}(\mathbf{y}) \qquad \mathbf{y}_{n+1} = \mathbf{y}_n + \mathbf{k}_1 \tag{49}$$

Equation (49) is written in the same terms as an explicit first-order Runge–Kutta method with Runge–Kutta constant $\mathbf{k}_1$.

It is an implicit method, however, since the first equation requires solution of M linear equations to obtain $\mathbf{k}_1$:

$$\begin{gathered}(\mathbf{I} - 0.5h\mathbf{f}_y)\mathbf{k}_1 = (\mathbf{I} - 0.5h\mathbf{J})\mathbf{k}_1 = h\mathbf{f}(\mathbf{y}) \\ J_{ij} = \left(\frac{\partial f_i}{\partial y_j}\right)_{\mathbf{y}_n} \quad \text{and} \quad \mathbf{f}(\mathbf{y}) = \mathbf{f}(\mathbf{y}_n)\end{gathered} \tag{50}$$

On the other hand, (50) is considerably more simple than the M nonlinear equations that have to be solved in a straightforward application of one-point Gauss collocation in (47), and we have seen that both (49) and (47) are accurate to $\mathcal{O}(h^3)$. For linear problems they, of course, give identical results and this means that (49) is A-stable just as the Gauss method and with characteristic root $\mu(\lambda h) \to -1$ when $\lambda \to -\infty$.

A number of methods similar to (49) but with a smaller value of $|\mu(\lambda h)|$ for $\lambda \to -\infty$—and thus more favorable damping properties—can be obtained with almost no extra computation.

The simplest device is to use (48) twice:

$$\begin{aligned}k_1 &= (1 - ahf_y)^{-1}hf \\ k_2 &= (1 - ahf_y)^{-1}k_1 \\ y_{n+1} &= y_n + R_1k_1 + R_2k_2\end{aligned} \tag{51}$$

The method is still required to be $\mathcal{O}(h^3)$ and the parameters a, R_1, and R_2 are determined such that the characteristic root μ has more suitable properties, i.e., such that the solution y_{n+1} fits the exponential better for large negative λ in (35).

$$\begin{aligned} k_1 &= h(f + ahf_y f + a^2h^2f_y^2 f + \dots) \\ k_2 &= h(f + 2ahf_y f + 3a^2h^2f_y^2 f + \dots) \\ y_{n+1} &= y_n + (R_1 + R_2)hf + (R_1 + 2R_2)h^2 a f_y f + \mathcal{O}(h^3) \end{aligned} \tag{52}$$

Fitting (52) to the power series (46) yields

$$R_1 + R_2 = 1, \qquad R_1 + 2R_2 = \frac{1}{2a}$$

$$R_1 = \frac{4a - 1}{2a}, \qquad R_2 = \frac{1 - 2a}{2a}$$

Inserting into (35) yields

$$k_1 = \frac{h\lambda y_n}{(1 - ah\lambda)} \qquad k_2 = \frac{h\lambda y_n}{(1 - ah\lambda)^2}$$

$$\begin{aligned} y_{n+1} = \mu(\lambda h)y_n &= \left[1 + \frac{4a - 1}{2a}\frac{h\lambda}{1 - ah\lambda} + \frac{1 - 2a}{2a}\frac{h\lambda}{(1 - ah\lambda)^2}\right]y_n \\ &= \frac{1 + (1 - 2a)h\lambda + (a^2 - 2a + \frac{1}{2})(h\lambda)^2}{(1 - ah\lambda)^2}y_n \end{aligned}$$

To obtain high accuracy for large $|\lambda h|$, we choose $\mu(-\infty) = 0$, which means that the coefficient to $(h\lambda)^2$ in the numerator is set equal to zero.

$$a^2 - 2a + \tfrac{1}{2} = 0 \quad \text{or} \quad a = \begin{cases} a_1 = 1 - \sqrt{\tfrac{1}{2}} \\ a_2 = 1 + \sqrt{\tfrac{1}{2}} \end{cases}$$

Both a_1 and a_2 lead to A-stable methods that effectively strangle stiff components since $\mu(-\infty) = 0$. The best fit to $\exp(\lambda h)$ for small negative λ is obtained with a_1, which is consequently to be preferred since both slow and fast components are then accurately integrated. Thus the constants of (51) are

$$a = 1 - \sqrt{\tfrac{1}{2}}, \qquad R_1 = 1 - \sqrt{\tfrac{1}{2}}, \qquad R_2 = \frac{\sqrt{2}}{2} \tag{53}$$

Higher-order methods can be constructed by including more constants. We have chosen a method proposed by Caillaud and Padmanabhan (1971), which has the same accuracy $\mathcal{O}(h^4)$ as the end point

Radau method. For a scalar equation the formulas for the Runge–Kutta constants and for y_{n+1} are

$$\begin{aligned}
k_1 &= (1 - ahf_y)^{-1}hf \\
k_2 &= (1 - ahf_y)^{-1}hf(y_n + b_2k_1) \\
k_3 &= (1 - ahf_y)^{-1}(b_{31}k_1 + b_{32}k_2) \\
y_{n+1} &= y_n + R_1k_1 + R_2k_2 + R_3k_3
\end{aligned} \tag{54}$$

$$\begin{aligned}
a &= 0.43586652 \quad \text{(the most appropriate zero of } a^3 - 3a^2 + \tfrac{3}{2}a - \tfrac{1}{6}) \\
b_2 &= \tfrac{3}{4} \\
b_{31} &= -\frac{1}{6a}(8a^2 - 2a + 1) = -0.63020209 \\
b_{32} &= \frac{2}{9a}(6a^2 - 6a + 1) = -0.24233789 \\
R_1 &= \tfrac{11}{27} - b_{31} = 1.037609496 \\
R_2 &= \tfrac{16}{27} - b_{32} = 0.83493048 \\
R_3 &= 1
\end{aligned} \tag{55}$$

In equations (54) it is assumed that the independent variable x does not appear explicitly (autonomous differential equations). If x or an explicit function of x should appear in a set of M coupled differential equations one may reformulate the set into $M + 1$ autonomous equations by introducing a new integration variable t letting x be the $M + 1$th component of the solution vector given by

$$\frac{dx}{dt} = \frac{dy_{M+1}}{dt} = 1$$

Equation (54) may also be modified to cope directly with $f(x, y)$ rather than $f(y)$ as the right hand side of equation (29). The result is (see also exercise 8.5):

$$\begin{aligned}
\mathbf{k}_1 &= (\mathbf{I} - ah\mathbf{J})^{-1}h[\mathbf{f}(x_n, \mathbf{y}_n) + ha\mathbf{f}_x] \\
\mathbf{k}_2 &= (\mathbf{I} - ah\mathbf{J})^{-1}h[\mathbf{f}(x_n + b_2h, \mathbf{y}_n + b_2\mathbf{k}_1) + ha\mathbf{f}_x] \\
\mathbf{k}_3 &= (\mathbf{I} - ah\mathbf{J})^{-1}[b_{31}(\mathbf{k}_1 + h^2a\mathbf{f}_x) + b_{32}(\mathbf{k}_2 + h^2a\mathbf{f}_x)] \\
\mathbf{y}_{n+1} &= \mathbf{y}_n + R_1\mathbf{k}_1 + R_2\mathbf{k}_2 + R_3\mathbf{k}_3
\end{aligned} \tag{56}$$

with a, b_2, b_3, R_1, R_2, and R_3 given by (55).

Equations (56) and (55) have been used with the stepsize adjustment algorithm of subsection 8.2.3 to solve various "difficult" stiff problems that have appeared in recent literature: the rate equations of Robertson (1967), a fluid bed reactor problem [Aiken and Lapidus (1974) and Luss and Amundson (1968)], the smog formation problem of Gelinas (1972), and certain variants of Van der Pol's equation that would be impossible to solve by explicit methods. Only the first and the last of the problems mentioned above are discussed here (subsection 8.2.5), while some of the other problems are given in Exercises.

Methods of the type (49) and (56) are called semi-implicit Runge–Kutta methods. Their general formulation for $\mathbf{y}^{(1)} = \mathbf{f}(\mathbf{y})$ is

$$\begin{aligned} \mathbf{k}_i &= (\mathbf{I} - a_i h\mathbf{J})^{-1} h\mathbf{f}\Big(\mathbf{y}_n + \sum_{j=1}^{i-1} b_{ij}\mathbf{k}_j\Big) \qquad i = 1, 2, \ldots, N \\ \mathbf{y}_{n+1} &= \mathbf{y}_n + \sum_{i=1}^{N} R_i \mathbf{k}_i \end{aligned} \tag{57}$$

They are formally quite similar to explicit Runge–Kutta methods:

$$\begin{aligned} \mathbf{k}_i &= h\mathbf{f}\Big(\mathbf{y}_n + \sum_{j=1}^{i-1} b_{ij}\mathbf{k}_j\Big) \qquad i = 1, 2, \ldots, N \\ \mathbf{y}_{n+1} &= \mathbf{y}_n + \sum_{i=1}^{N} R_i \mathbf{k}_i \end{aligned} \tag{58}$$

and also to implicit Runge–Kutta methods:

$$\begin{aligned} \mathbf{k}_i &= h\mathbf{f}\Big(\mathbf{y}_n + \sum_{j=1}^{N} b_{ij}\mathbf{k}_j\Big) \qquad i = 1, 2, \ldots, N \\ \mathbf{y}_{n+1} &= \mathbf{y}_n + \sum_{i=1}^{N} R_i \mathbf{k}_i \end{aligned} \tag{59}$$

Equation (58) is an explicit, sequential determination of the Runge–Kutta constants. Stabilization of the process is obtained through the first factor of (57); this is not a serious complication since with identical a_i and LU decomposition of the matrix $\mathbf{I} - ah\mathbf{J}$ all that is needed is the solution of M linear equations with N different right-hand sides and the process is again sequential. The implicit Runge–Kutta methods—which are reformulations of collocation methods as shown for the Gauss one-point method at the beginning of this subsection—are by comparison very difficult to apply. Here the N Runge–Kutta vectors $\mathbf{k}_i$ are determined simultaneously by solution of $N \times M$ (nonlinear) algebraic equations.

One possible advantage of the implicit methods is that they work almost equally well with an approximate Jacobi matrix that may if necessary be evaluated numerically and held constant through the iterations within an integration step. All that is required is that the iteration process converges to the solution of the algebraic equations. The semi-implicit methods require **J** to be evaluated accurately (e.g., by analytical differentiation) since the method has been constructed such that a certain number of terms in the Taylor series fit exactly. One other advantage of the implicit (collocation) methods might be the high order of the truncation error ($2N$) in each step. We have not made a thorough comparison, however, of the two types of methods in any examples.

As an example of program systems built on the principles of this section we have shown the implementation of the third-order semi-implicit method (54) and (55) in the Appendix, pp. A9–A11.

The subroutine call is

```
CALL STIFF 3 (N, ND, NPRINT, FUN, DFUN, OUT, XO, X1, HO, EPS, W, Y, YOLD,
YOLD1, IP, YA, YK1, YK2, YK3, DF, DFOLD, F, FOLD)
```

The parameter types are

INTEGER:	N, ND, NPRINT
REAL:	XO, X1, HO, EPS
INTEGER VECTOR (size ND):	IP*
REAL VECTOR (size ND):	W, Y, YOLD*, YOLD1*, YA*, YK1*, YK2*, YK3*, F*, FOLD*
REAL ARRAY (size ND × ND):	DF*, DFOLD*

Subroutine names:

FUN, DFUN, OUT

Parameters marked by a star are all internal quantities used by STIFF and its auxiliary subprograms. The user is only required to specify the following *input*:

N:	Number of equations to be integrated.
ND:	Main program array dimension.
NPRINT:	Printing interval. For NPRINT = K the solution Y is only printed at every Kth step. Y is always printed at X1.
XO, X1:	Limits of the independent variable between which the differential equation is solved.
HO:	Suggested initial half-steplength On exit HO contains suggested value for half-steplength for continued integration beyond X1.
EPS, **W**:	Tolerance parameters. ε_i of equation (44b) is computed internally as $(1 + \lvert Y(I)\rvert)\ EPS/W(I)$.

Y:	Vector of dependent variables at XO. On exit **Y** is vector of dependent variables at X1.
SUBROUTINE FUN (Y, F):	User supplied subprogram for function evaluation. F is the vector $\mathbf{F}(\mathbf{Y})$, the right-hand sides of the differential equations.
SUBROUTINE DFUN (Y, DF):	User supplied subprogram for evaluation of the Jacobian of **F**. **DF** is a matrix with elements $DF(I, J) = \partial F(I)/\partial Y(J)$.
SUBROUTINE OUT (X, Y, IH, Q):	User supplied subprogram for output.
X:	Current value of independent variable.
Y:	Current value of dependent variable vector.
IH:	Number of bisections (unsuccessful integrations) in the current step.
Q:	Steplength acceleration factor [e.g., 3 in equation (44a)].

Besides these user supplied routines, STIFF 3 operates with three *internal subroutines*:

SIRK 3:	Single-step semi-implicit integration.
LU:	Program for decomposing a matrix **A** to a lower and an upper triangular form. **A** = **LU**.
BACK:	Back substitution algorithm for solution to **LUx** = **b**.

STIFF 3 is called from a main program. If desired, a number NTAB of artificial end points XTAB for the integration may be inserted between XO and X1 via the main program. In this way, printout is assured at XTAB(1), XTAB(2), . . . , which may be convenient for comparison of integration results.

All input parameters are specified in the main program, and suitable table heads are printed.

FUN, DFUN, and OUT are declared EXTERNAL in the main program.

In Appendix A9, we have first listed STIFF 3 and its three internal subroutines, and also an implementation on the first problem of subsection 8.2.5 "Robertson's example" is shown in A21–23. The illustration includes a main program and three subroutines—FUN, DFUN, and OUT.

8.2.5 Numerical solution of two "stiff" problems

The semi-implicit method (56) with stepsize adjustment strategy (44) was used to solve two difficult "stiff" problems with three and two coupled nonlinear equations, respectively.

Our first example was originally proposed by Robertson (1967) and later discussed by Seinfeld, et al. (1970) in a review article on integration methods for stiff differential equations and by Caillaud and Padmanabhan in their paper on semi-implicit Runge–Kutta methods (1971).

$$\begin{aligned}\frac{dy_1}{dt} &= -0.04y_1 + 10^4 y_2 y_3\\ \frac{dy_2}{dt} &= 0.04y_1 - 10^4 y_2 y_3 - 3 \cdot 10^7 y_2^2\\ \frac{dy_3}{dt} &= 3 \cdot 10^7 y_2^2\end{aligned} \tag{60}$$

with $y_1 = 1$ and $y_2 = y_3 = 0$ for $t = 0$.

The Jacobian matrix is

$$\mathbf{J} = \begin{pmatrix} -0.04 & 10^4 y_3 & 10^4 y_2 \\ 0.04 & -10^4 y_3 - 6 \cdot 10^7 y_2 & -10^4 y_2 \\ 0 & 6 \cdot 10^7 y_2 & 0 \end{pmatrix}$$

At $t = 0$, the eigenvalues are $(-0.04, 0, 0)$. At $t \to \infty$ ($y_1 = y_2 = 0$, $y_3 = 1$), the eigenvalues are $[-(10^4 + 0.04), 0, 0]$. For intermediate t, one eigenvalue remains at zero ($y_1 + y_2 + y_3 = 1$ for all t). The other two eigenvalues are negative with one decreasing to -2400 in a very short time interval $(0, 0.02)$.

Computed results at $t = 10$ and the number of steps required to reach this t-value with different tolerance vectors $\boldsymbol{\varepsilon}$ are shown in table 8.4. Caillaud's and Padmanabhan's results [their table 9, method ISI3 $(-\infty)$] at $t = 10$, which were reached after 200 equal size steps of $h = 0.05$, are included in the last line of the table.

Even though our stepsize correction mechanism corresponds to three single steps, it is beyond doubt a highly useful device: The error at $t = 10$ is reduced by a factor 10^4 and only 20% of the steps required in a constant stepsize integration are necessary. Not only the large t results are excellent, but results at intermediate t (0.4, 1, and 4) are equally good or in even better agreement with the "exact" solution that is defined as the solution with $\boldsymbol{\varepsilon} = 10^{-5}\,(1, 10^{-4}, 1)$.

The initial stepsize proposed to the algorithm at $t = 0$ was chosen as $h_0 = 10^{-4}$. If h_0 is too small, mechanism (44a) rapidly increases h to an appropriate size. If on the other hand h_0 is chosen too large, e.g., $5 \cdot 10^{-3}$, the procedure might become unstable if the tolerance $\boldsymbol{\varepsilon}$ is large [e.g., $10^{-2}\,(1, 1, 1)$]. Since the large t results are almost independent of $\boldsymbol{\varepsilon}$, we do not wish to make our tolerance criterion too strict, and it is advisable to use an initial step size that is reasonably small (here chosen as $|1/\lambda_{\max}|$). The weight factor of y_2 has almost negligible influence on the accuracy with which this very small component is determined, and no influence at all on y_1 and y_3. It is worth noting that y_2, which is a factor 10^4 smaller than y_1 and y_3, is determined with a relative accuracy better than 0.1% even when the absolute tolerance $\boldsymbol{\varepsilon} = 10^{-2}$ (1, 1, 1). The reason is that a small error in y_2 gives a large error in y_1 and y_3 and the size of h is regulated by the (in absolute sense) larger errors on y_1 and y_3.

TABLE 8.4

INTEGRATION OF ROBERTSON'S PROBLEM (60) BY STIFF 3 $t = 10$, $h_0 = 10^{-4}$, $y_1(0) = 1$, $y_2(0) = y_3(0) = 0$

Tolerance $\boldsymbol{\varepsilon}$	Number of steps to $t = 10$	y_1	$y_2 \cdot 10^4$	y_3
10^{-2} (1, 1, 1)	15	0.8413610	0.162334	0.158622
10^{-3} (1, 1, 1)	16	0.8413684	0.162350	0.1586153
10^{-4} (1, 1, 1)	17	0.8413696	0.162348	0.1586141
10^{-5} (1, 1, 1)	22	0.8413699	0.162341	0.1586139
10^{-2} (1, 10^{-2}, 1)	15	0.8413652	0.162340	0.1586186
10^{-3} (1, 10^{-2}, 1)	16	0.8413684	0.162354	0.1586153
10^{-4} (1, 10^{-2}, 1)	19	0.8413696	0.162348	0.1586141
10^{-5} (1, 10^{-2}, 1)	26	0.8413699	0.162341	0.1586139
10^{-2} (1, 10^{-4}, 1)	18	0.8413686	0.162353	0.1586152
10^{-3} (1, 10^{-4}, 1)	24	0.8413698	0.162341	0.1586140
10^{-4} (1, 10^{-4}, 1)	40	0.8413699	0.1623397	0.1586139
10^{-5} (1, 10^{-4}, 1)	83	0.8413699	0.1623391	0.1586138
Caillaud and Padmanabhan	200	0.851	0.170	0.149

y_2 and y_3 are shown in figure 8-2 for $\boldsymbol{\varepsilon} = 10^{-2}(1, 10^{-4}, 1)$. Note the deceleration of the stepsize increase mechanism (44b) close to the maximum value of y_2 where the fast exponential is tackled. From $t = 0.08225$ to $t = 1$, $h_{n+1} = 3h_n$ while $h_{n+1} \sim 2$–$1.5h_n$ for $1 < t < 10$ where the slow exponential is being integrated. When this exponential has been quenched, the stepsize mechanism again speeds up and $h_{n+1} = 3h_n$ in the steady state region.

The maximum of y_2 is found at $y_2 = 3.649 \cdot 10^{-5}$ in close agreement with the solution (3.651) of the middle equation of (60) for $dy_2/dt = 0$, $y_1 \sim 1$, and $y_2 \sim y_3$ at the maximum.

Our second example is Van der Pol's equation, which is discussed by Davis (1962), chapter 12:

$$\frac{dy_1}{dt} = y_2 \tag{61a}$$

$$\frac{dy_2}{dt} = K(1 - y_1^2)y_2 - y_1 \tag{61b}$$

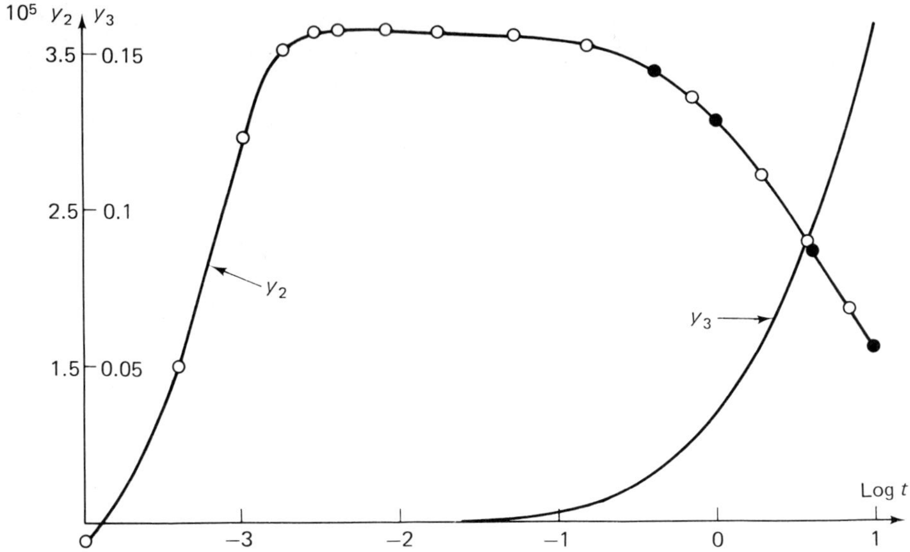

Figure 8-2. y_2 and y_3 for Robertson's problem; $\epsilon = 10^{-2}(1, 10^{-4}, 1)$. Eighteen steps from $t = 0$ to $t = 10$. Preset tabular points at $t = 0.4$, 1, 4, and 10, are shown with closed circles.

Figure 8-3 shows $y_1(t)$ for t close to zero and for large t when $K = 1000$ and $y_1(0) = y_2(0) = 1$.

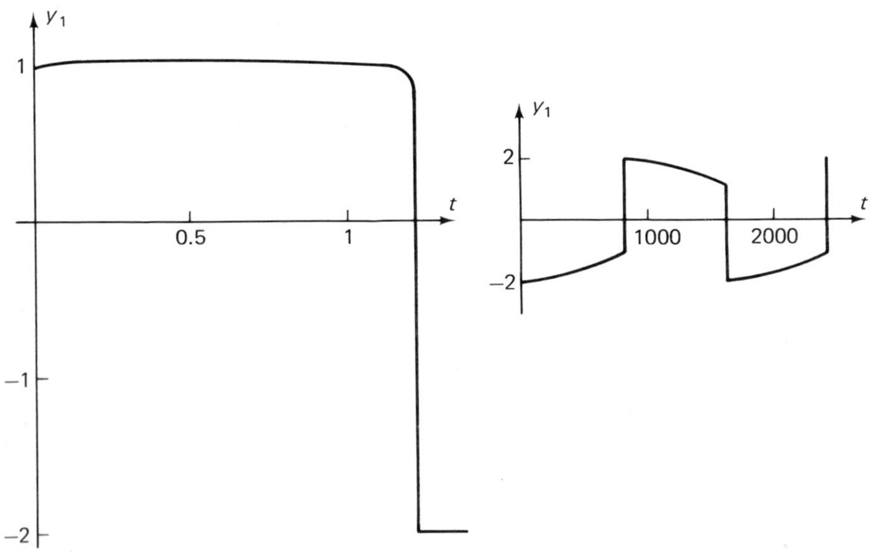

Figure 8-3. Solution of Van der Pol's equation (61) for $K = 1000$.

Initially both y_1 and $y_2 = dy_1/dt$ are positive. y_1 increases above 1, while y_2 decreases since both terms of (61b) are negative. y_1 passes a flat maximum at $(t, y_1) = (0.087, 1.03)$. For a short time after the maximum the two terms of (61b) are of opposite sign, but y_1 decreases by (61a); as soon as $y_1 < 1$, both terms of (61b) are again negative. The two equations now work together: y_1 decreases rapidly below 1 with a speed that is continuously increased by (61b), y_2 attaining a large negative value due to the large constant K. A minimum of y_2 is passed at $(y_1, y_2) = (-0.98, -1332)$. As soon as y_1 has passed -1 while y_2 is negative, the first term of (61b) becomes positive and works in cooperation with $-y_1$, which is positive. Consequently y_2 increases rapidly and becomes positive at $y_1 = -2.0000728$. The descent of y_1 from ~ 1 to -2 takes place in a time interval (1.21, 1.22). When $y_2 \sim 0$ but positive and y_1 is negative, the two terms of (61b) counterbalance with the result that y_2 remains virtually zero for a long time. From $t = 1.23$ to $t = 62.3$, y_2 increases from $6.66 \cdot 10^{-4}$ to $6.91 \cdot 10^{-4}$. At $t = 802$, y_2 is still only $6.51 \cdot 10^{-3}$ but over the long time interval y_1 has increased to -1.080. Soon after (at $t = 808.3$), y_1 passes -1 and y_2 increases tremendously fast, passing a maximum of 1333 at $y_1 = 0.994$.

A limit cycle has now been reached. y_1 oscillates between limits that are very close to -2 and 2. y_2 oscillates between 1333 and -1333, the minimum (maximum) being passed very close to $y = -1$ and 1. Close to the maximum (minimum) value of y_1, $y_2(t)$ has a point of inflexion with $|y_2| \sim -1/1500$. Most of the cycle time is spent near these inflexion points, y_1 working itself slowly away from either 2 or -2.

The computation was done by STIFF 3 with tolerance $\varepsilon_1 = \varepsilon_2 = 10^{-4}$ or 10^{-5}. For both values of $\boldsymbol{\varepsilon}$, identical results are obtained to eight digits.

The stepsize reduction mechanism (43) is very rarely activated—in more than 99% (98%) of the steps (44) predicts an acceptable h_{n+1} when $|\boldsymbol{\varepsilon}| = 10^{-5}$ (10^{-4}). When y_1 works itself away from 2 or -2, approximately 8 to 10 steps are taken with $h_{n+1} = 3h_n$; afterward strategy (44b) increases the stepsize more slowly until a maximum h of ~ 100 is reached. Thus the docile part of the cycle is traversed in less than 5% of the integration time while more than 80% of the integration time is used in the very short part of the cycle ($\Delta t < 0.1$) where $|y_2|$ increases from 0.1 to 1333 and goes back to 0.

The cycle time T is faithfully reproduced through many cycles. T is 1615.5 for $K = 1000$, a result that is confirmed by an analytical result cited by Davis (1962), p. 367, and derived by singular perturbation in Cole (1968), p. 55.

Davis claims incorrectly (p. 361) that the breadth $B = [y_1(\text{min}), y_1(\text{max})]$ of the limit cycle is 4 whatever the value of K in (61b). The numerical results show that $B = 2 \cdot 2.000073$ at $K = 1000$. B is only 4 for $K = 0$ or ∞, and a maximum of $B \sim 4.04$ is obtained for $K \sim 5$. The extrema of y_2 are close to but not exactly at $y_1 = \pm 1$. Inserting $y_1 = -1 + \varepsilon$ and $y_2 = 1333$ in (61b) yields

$$\frac{dy_2}{dt} \sim 0 \quad \text{for } K \cdot 2\varepsilon \cdot 1333 + 1 = 0 \quad \text{or} \quad \varepsilon = -3.710^{-6}$$

The computer results also indicate that for large $K \max|y_2| \sim \frac{4}{3}K$ and that the point of inflexion for y_2 is exactly at $y_1 = 2$ or -2 with a value of y_2 very close to $\frac{2}{3}K^{-1}$. We have not studied these details of the solution to the Van der Pol equation in any depth, however, since they are beside the point we want to prove here: that the proposed integration procedure for stiff differential equations is very reliable. But they do point to an exploration of purely mathematical problems by computer studies rather than by classical methods (e.g., perturbation theory) that may become of real importance once efficient and accurate numerical methods are available. Further examples in which this empirical line of attack is used are given in chapter 9.

8.3 Sensitivity Functions

The techniques used hitherto for solution of ordinary nonlinear single or coupled differential equations have mainly been based on a linearization of the nonlinear terms utilizing an approximate solution.

Consider, e.g., the simple initial value problem

$$\frac{dy}{dx} = f(x, y), \qquad y(0) = y_0 \tag{62}$$

and let an approximate solution $y^{(k)}(x)$ that satisfies the side condition at $x = 0$ be given in $x \in [0, 1]$.

$y^{(k)}(x)$ does not satisfy the differential equation, and we are looking for a correction term $\epsilon(x)$, which yields a better approximation:

$$y^{(k+1)}(x) = y^{(k)}(x) + \varepsilon(x) \tag{63}$$

Equation (45) is substituted into the differential equation:

$$\frac{d}{dx}y^{(k)}(x) + \frac{d}{dx}\varepsilon(x) = f[x, y^{(k)} + \varepsilon]$$

The term on the right-hand side is linearized at the approximate solution $y^{(k)}(x)$:

$$f[x, y^{(k)} + \varepsilon] = f[x, y^{(k)}] + \left(\frac{\partial f}{\partial y}\right)_{y^{(k)}} \cdot \varepsilon(x) + \mathcal{O}[\varepsilon(x)^2]$$

The correction term $\varepsilon(x)$ is now obtained as the solution of the *linear* problem

$$\begin{aligned}\frac{d}{dx}\varepsilon(x) - \left(\frac{\partial f}{\partial y}\right)_{y^{(k)}} \cdot \varepsilon(x) &= -\left[\frac{d}{dx}y^{(k)}(x)\right] - f[x, y^{(k)}] \\ &= -R^{(k)}(x)\end{aligned} \tag{64}$$

with initial condition $\varepsilon(0) = 0$.

The corrected solution $y^{(k+1)}(x) = y^{(k)}(x) + \varepsilon(x)$ may next be used as a basis for a renewed linearization, and the quadratically convergent process is repeated until the desired convergence is reached.

We shall now assume that a collocation solution of order N satisfies our accuracy criterion for the solution of (62) from $x = 0$ to $x = 1$. This means that the final result of the iteration process (62)–(64) $[y^{(k)}(x), k \to \infty]$ is adequately represented by an Nth order polynomial in x.

Hence the accepted solution is found by iterative solution of the N collocation equations

$$r_i = \sum_{j=0}^{N} A_{ij}y_j - f(x_i, y_i) = 0 \qquad i = 1, 2, \ldots, N \tag{65}$$

When (65) is solved by the Newton–Raphson method the kth and the $k + 1$ estimate of the solution is related by

$$\mathbf{y}^{(k+1)} = \mathbf{y}^{(k)} + \boldsymbol{\varepsilon}$$

where $\quad \mathbf{J}\boldsymbol{\varepsilon} = \mathbf{r} = \mathbf{r}(x, \mathbf{y}^{(k)})$

and element J_{ij} of the Jacobian matrix $\mathbf{J}$ is given by

$$J_{ij} = \left(\frac{\partial r_i}{\partial y_j}\right) = A_{ij} - \left(\frac{\partial f}{\partial y}\right)_{y=y_j} \delta_{ij} \tag{66}$$

We note that each step in the iteration process toward y is formally identical to solution of (64) by collocation

$$\sum_{j=0}^{N} A_{ij}\varepsilon_j - \left(\frac{\partial f}{\partial y}\right)_{y_j} \varepsilon_i = -\left[\sum_{j=0}^{N} A_{ij}y_j^{(k)} - f(x_i, y_i^{(k)})\right] \tag{67}$$

with $\varepsilon_0 = y^{(k+1)}(0) - y^{(k)}(0) = 0$, since all iterates to (62)–(63) are assumed to satisfy the side condition at $x = 0$.

The final solution of (67) is $\varepsilon_i = \varepsilon(x_i) = 0$. Since we assume that the N point collocation solution is a satisfactory approximation to the solution of (62) the analogy between the approximation sequence (62)–(63) that leads to $y^{(k+1)}(x) \equiv y^{(k)}(x)$ and the iterated solution of (65) is complete.

Frequently the right-hand side term of the differential equation contains a number of parameters,

$$f(x, y) = f(x, y, P_1, P_2, \ldots, P_M) = f(x, y, \mathbf{P}) \tag{68}$$

and it is of interest to investigate in which manner a change in the value of one of the parameters affects the solution of the differential equation.

Let us define the sensitivity function $z_m(x)$ by

$$z_m(x) = \frac{\partial}{\partial P_m}[y(k)]_{\mathbf{P}} \tag{69}$$

where $y(x)$ is the solution of

$$\frac{dy}{dx} = f(x, y, \mathbf{P}) \tag{70}$$

with given initial condition $y(0) = y_0$.

Differentiation of (70) with respect to P_m yields

$$\frac{\partial}{\partial P_m}\left(\frac{dy}{dx}\right) = \frac{\partial}{\partial P_m}[f(x, y, \mathbf{P})] \tag{71}$$

or

$$\begin{aligned} \frac{dz_m}{dx} &= \frac{\partial}{\partial P_m}[f(x, y, \mathbf{P})] \\ &= \left(\frac{\partial f}{\partial y}\right)_{\mathbf{P},y}\frac{\partial y}{\partial P_m} + \left(\frac{\partial f}{\partial P_m}\right)_{y,\mathbf{P}} \\ &= \left(\frac{\partial f}{\partial y}\right)_{\mathbf{P},y} z_m + \left(\frac{\partial y}{\partial P_m}\right)_{y,\mathbf{P}} \end{aligned} \tag{72}$$

where the partial derivatives are evaluated at the solution of (70) and hence become known functions of x when the solution of (70) has been evaluated.

The initial condition for z_m is $z_m(0) = \partial y(0)/\partial P_m = 0$ provided $y(0)$ does not depend on the parameter P_m.

Apart from the inhomogeneous terms, r in (64) and $\partial f/\partial P_m$ in (72), these equations are identical. In (67) $\mathbf{z}$ converges to $\mathbf{0}$ when $\mathbf{y}$ converges to the solution of the collocation equations (65); while in (72) $\mathbf{z}$ converges to the Nth order polynomial approximation for the desired sensitivity function when $\mathbf{y}$ converges to the solution of (65).

Thus, at the end of the iterations of (65), $\partial\mathbf{f}/\partial P_m$ in the collocation version of (72) is known at the accepted collocation point ordinates, and by solution of one extra set of linear equations [with the same coefficient matrix as in the last iteration of (65)] the sensitivity function is determined at the collocation points.

For a linear problem the process is even simpler. Let y be the solution of

$$\frac{d}{dx}y = f(x, \mathbf{P})y, \qquad y(0) = y_0 \tag{73}$$

The sensitivity functions are thus given by

$$\frac{d}{dx}z_m = f(x, \mathbf{P}) \cdot z_m + \left(\frac{\partial f}{\partial P_m}\right) \cdot y \tag{74}$$

Simultaneous collocation solution of the two sets of equations yields

$$\sum A_{ij}y_j - f(x_i, \mathbf{P}) \cdot y_i = -A_{i0} \cdot y_0 \tag{75}$$

and

$$\sum A_{ij}z_{m,j} - f(x_i, \mathbf{P}) \cdot z_{m,i} = \left(\frac{\partial f}{\partial P_m}\right)_{x_i} \cdot y_i \tag{76}$$

The matrix of coefficients for these two sets of linear algebraic equations are the same; however, evaluation of the right-hand side of (76) requires the solution of (75). The most economical method of solution is by triangular decomposition of the matrix of coefficients, back substitution in (75) to find the y_i, evaluation of the right-hand side(s) of (76), and solving for the z_m by back substitution. The alternatives, matrix inversion or solving the algebraic equations twice by Gaussian elimination, would require additional computation.

In the development presented here the parameters enter only through the term f. Completely analogous results are obtained, however, for parameters that enter into terms that contain derivatives or for parameters that occur in boundary conditions.

Sensitivity functions have been utilized twice in earlier chapters. In section 4.5 eigenvalues are determined by solution of an initial value problem, and the sensitivity functions with respect to the trial eigenvalues are utilized in a Newton method. Furthermore, sensitivity functions for a two-point boundary value problem are utilized in section 5.5 for solution of the Weisz–Hicks problems in terms of a tracing of the Φ^2–η curve. Other obvious applications would be in parameter estimation in systems described by differential equations. It should be noted that by the methods given above sensitivity functions are obtained essentially at no extra cost, whereas the price for their determination by, e.g., an explicit Runge–Kutta method would be that of solving a corresponding larger number of equations.

The semi-implicit Runge–Kutta methods are easily adapted for simultaneous integration of the sensitivity functions. The partial derivatives of the right-hand side of (70) with respect to y are equal to the partial derivatives with respect to z_m in (72) and the additional cost for the extra integration is only moderate.

8.4 Partial Differential Equations

Certain types of linear partial differential equations that by discretization in one coordinate yield a system of coupled linear ordinary differential equations with constant coefficients can be solved analytically by the matrix methods of chapter 4. Discretization of a general partial differential equation yields a set of coupled nonlinear differential equations where the number of independent variables is reduced by one. These equations are not normally amenable to analytic solution and certain structural properties of the equations that are pertinent to their numerical solution are now discussed.

Here we are only concerned with parabolic partial differential equations with two independent variables. Discretization in the "spatial" variable yields an initial value problem with the "time" variable as the remaining independent variable.

As the basis for our discussion, we first consider the simple linear equation:

$$\frac{\partial^2 y}{\partial x^2} = \frac{\partial y}{\partial t} \tag{77}$$

with boundary conditions

$$\begin{aligned} t &= 0: \quad & y &= 1 \quad & &\text{all } x \\ x &= 0: \quad & \frac{\partial y}{\partial x} &= 0 \quad & &\text{all } t > 0 \\ x &= 1: \quad & y &= 0 \quad & &\text{all } t > 0 \end{aligned} \tag{78}$$

The solution of (77) is

$$y(x, t) = \frac{4}{\pi} \sum_{i=1}^{\infty} \frac{(-1)^{i+1}}{(2i-1)} \cos\left[\frac{(2i-1)\pi}{2} x\right] \exp\left[\frac{-(2i-1)^2 \pi^2}{4} t\right] \tag{79}$$

For $t > 1$, the solution is well approximated by the first term in the series:

$$y(x, t) \simeq \frac{4}{\pi} \cos\left(\frac{\pi}{2} x\right) \exp\left(-\frac{\pi^2}{4} t\right) \tag{80}$$

For small values of t, however, the series is slowly convergent, and the discontinuity in the boundary conditions at $(t, x) = (0, 1)$ requires that a substantial number of terms are required to obtain a proper representation for small t.

To solve (77) by orthogonal collocation, one would introduce the new independent variable $u = x^2$ and discretize, e.g., at the zeros of $P_N^{(1,-1/2)}(u)$ or $P_N^{(0,-1/2)}(u)$. This yields the set of collocation equations

$$\frac{d}{dt}\mathbf{y} = \mathbf{C}\mathbf{y}, \qquad \mathbf{y}(t = 0) = \mathbf{1} \tag{81}$$

where

$$C_{ij} = 4u_i B_{ij} + 2A_{ij}$$

The solution of the discretized set (63) is

$$\mathbf{y} = \exp(\mathbf{C}t) \cdot \mathbf{1} \tag{82}$$

The eigenvalues λ_i of $\mathbf{C}$ approximate the first N eigenvalues of (77), and the eigenvectors of $\mathbf{C}$ represent the corresponding eigenfunctions at the collocation points.

The dominant eigenvalue $\lambda_1 = -(\pi^2/4)$ and the corresponding eigenfunction is correctly obtained using only two or three collocation points. An approximation of low order thus leads to accurate results for large values of t, but the approximation is poor close to $t = 0$ since a low-order polynomial is unable to represent the initially very steep profile $y(x)$.

Next, let us consider the nonlinear PDE:

$$\frac{\partial y}{\partial t} = \frac{\partial^2 y}{\partial x^2} - (1 - y)^2 \tag{83}$$

with boundary conditions:

$$\begin{aligned} t = 0&: \quad y = 1 \quad \text{all } x \\ x = 0&: \quad \frac{\partial y}{\partial x} = 0 \quad \text{all } t > 0 \\ x = 1&: \quad y = 0 \quad \text{all } t > 0 \end{aligned} \tag{84}$$

Discretization in the same manner yields

$$\frac{d}{dt}\mathbf{y} = \mathbf{f}(\mathbf{y}), \qquad \mathbf{y}(0) = \mathbf{1} \tag{85}$$

where

$$f_i = \sum_{j=1}^{N} C_{ij} y_j + (1 - y_i)^2$$

Equation (85) cannot be solved analytically but must be integrated numerically, e.g., by the methods described in section 8.2.

The Jacobian, $J_{ij} = \partial f_i/\partial y_j = C_{ij} - 2(1 - y_i)\,\delta_{ij}$, has eigenvalues that do not differ much from those of **C**: At $t = 0$, $y_i = 1$, and $\lambda_i = -[(2i - 1)/2]^2\pi^2$; as $t \to \infty$, the y_i converge to the same values at x_i that can be obtained by solution of the steady state equation

$$\frac{d^2y}{dx^2} - (1 - y)^2 = 0 \tag{86}$$

The eigenvectors of **J** are also quite similar to those of **C**, and it may be expected that a low-order discretization gives an excellent representation for large t.

Numerical integration of (85), using say $N = 3$, does not, however, yield an accurate solution for large but finite values of t. Unlike the linear problem (81) a fairly accurate representation of y in the initial phase is required since errors propagate into the solution for large t. To represent the initially steep profile, it is required that N is large.

Large values of N are expensive for two reasons. The number of coupled equations to be integrated increases proportional to N, and furthermore the resulting system becomes increasingly stiff, as seen from table 8.5, where the ratio between the largest and the smallest eigenvalue of **C** is given.

TABLE 8.5

RATIO BETWEEN LARGEST AND SMALLEST COLLOCATION EIGENVALUE FOR DIFFERENT N

N	1	2	3	4	5
λ_N/λ_1	1	10.3	35.6	89.1	188

The stiffness of the resulting set of coupled ordinary differential equations makes the A-stable methods of section 8.2 particularly attractive for integration of the collocation equations in the t–direction. The integration is conveniently performed with automatic stepsize adjustment, increasing the stepsize as the stiff components become of less importance.

The high approximation order (in x) is only required in the initial phase; as soon as a smoother profile is obtained, N may be decreased and starting values of y at the collocation points for the new N are found by interpolation. A measure of the smoothness of the profile is the maximum value of $\partial^2y/\partial x^2$, and N should be chosen at any step (except for the first, where a finite error is unavoidable) proportional to $\max\,(\partial^2y/\partial x^2)^{1/2}$, following the arguments of subsection 2.3.5. For large t, one would expect that N can be reduced to the value required to represent the steady state profile, that is, two or three for equation (86).

For small t, one may use the penetration front concept of subsection 7.2.2; consequently, by a suitable treatment of each regime of the solution a reasonably small computer expenditure is obtained even for quite complex problems.

Combination of orthogonal collocation in one (or more) spatial coordinates and an explicit integration method (e.g. fourth-order Runge–Kutta or a modified Euler method) has recently been reported by quite a few investigators to be an excellent method for solution of a single nonlinear PDE.

A satisfactory approximation is often obtained with a modest number of collocation points (e.g. $N < 5$) in which case the stiffness ratio, which determines the number of required integration steps, is not prohibitively large. The best results are frequently found with a low order integration method (e.g. the modified Euler method) and a steplength h close to the stability limit.

Explicit integration in the "time" direction may lead to serious problems when systems of nonlinear PDE are integrated. Here the local eigenvalues may differ by many orders of magnitude even for N small. One example is the transient catalyst pellet problem (1.62) to (1.63). If Le is close to 1 the dominant time constants for the mass and heat balances are of the same order of magnitude and explicit methods may be used. For Le $\ll 1$ or Le $\gg 1$ the temperature response is either much faster or much slower than the concentration response and explicit methods will fail due to a prohibitively large stiffness ratio even for low order collocation approximation.

Our computational experience is that the semi-implicit method of subsection 8.2.4 is far superior to explicit methods for this type of coupled equations and that it is at least as good as explicit methods for a single PDE. (See Exercise 9.6).

EXERCISES

1. Consider the end point Radau method with one interior collocation point for which $\mu(\lambda h)$ is given on p. 310.

a. Prove that $|\mu(h\lambda)| < 1$ for any purely imaginary $h\lambda = i\omega$ with $\omega \neq 0$.

b. We have seen in the text that $|\mu(h\lambda)| < 1$ for any negative $h\lambda$. $|\mu(h\lambda)|$ is also less than 1 for large, positive $h\lambda$. Hence by an inappropriate use of the method an increasing exponential will be strangled.

Show that the method is unstable [i.e., $\mu(\lambda h) > 1$] for $P = (a, b) \in \Omega$ where Ω is given by

$$y = \pm\sqrt{4x - x^2 + \sqrt{72x - 8x^2}}\,, \quad \text{and} \quad \lambda h = x + iy$$

c. Integrate

$$\frac{dy_1}{dt} = y_2$$
$$y_1 = 1, \quad y_2 = 0 \quad \text{at } t = 0$$
$$\frac{dy_2}{dt} = 6y_2 - 10y_1$$

by the Radau method of parts a and b using $h = 1$ and $h = 2$. Explain your results by comparison with the graph of Ω in part b. Is either of the two results close to the exact solution at $t = 1$? Try $h = 0.1$, 0.2, and 0.5.

2. We wish to study the improvement that can be obtained using the half-step, full-step correction technique (8.41). For the $N = 1$ end point Radau method, calculate the value of the following quantities:
a. $\mu(h\lambda)$
b. $\mu(h\lambda) - \exp(h\lambda)$
c. $e(h\lambda) = [\mu(h\lambda/2)]^2 - \mu(h\lambda)$
d. $y^*(h\lambda) = [\mu(h\lambda/2)]^2 + \frac{1}{7}e(h\lambda)$
e. $e^*(h\lambda) = y^*(h\lambda) - \exp(h\lambda)$
for $h\lambda = -0.05, -0.1, -0.25, -0.5, -1, -2, -5, -10, -50, -100$.
$e(h\lambda)$ is the difference between the half-step and the full-step solution, i.e., the quantity that decides whether the step $h\lambda$ is accepted. $e^*(h\lambda)$ is the error of the accepted solution that we hope is much smaller than $e(h\lambda)$.
What is the expected behavior of e^*/e for small $h\lambda$?
Check with the computed results for the 10 $h\lambda$ values given above.

3. We wish to solve $dy/dx = -y$, $y(0) = 1$ from $x = 0$ to x_s using the $N = 1$ end point Radau method with constant stepsize h. The correction technique (8.41) is used after each step.
Choose $x_s = 5$ and $h = 0.03125$, 0.0625, 0.125, 0.25, 0.5, 1, and 1.25; after each step determine the absolute deviation e between the full-step and the half-step integration result and also determine the relative deviation $e/y^*(x)$ where $y^*(x)$ is the corrected value of y after the step.
Approximately what is the relation between $e/y^*(x)$ and h?
Approximately what is the relation between the relative final error $Ef = (y^*(x_s) - \exp(-x_s))/\exp(-x_s)$ and h?
Note from your computed results how Ef varies with x_s.
How would you use the results of this exercise to choose the tolerance ε for a problem where a given accuracy of $y^*(x_s)$ is desired by the variable stepsize algorithm?

4. Investigate the region of instability Ω: $\mu(\lambda h) > 1$ for the third-order semi-implicit Runge–Kutta method (56). Tabulate the quantities a.–e. of Exercise 2 for this method, and verify that our choice of a in (55) is the best.
Hint: To obtain $\mu(h\lambda)$ as a function of $h\lambda$, note that it is the ratio of two third-degree polynomials, the denominator polynomial being $(1 - ah\lambda)^3$. $\mu(h\lambda)$ should correctly represent terms up to $(h\lambda)^3$ in a power series for $e^{h\lambda}$ and a should be chosen to make $\mu(-\infty) = 0$.

Next proceed as in Exercise 1 with $h\lambda = x + iy$ ($x = r\cos\Phi$, $y = r\sin\Phi$) to obtain the following relation between r and Φ for the boundary of Ω:

$$1 = \frac{q_1^2 r^4 + q_2^2 r^2 + 2q_1q_2r^3\cos\Phi + 2q_1r^2\cos 2\Phi + 2q_2 r\cos\Phi + 1}{(1 + a^2r^2 - 2ar\cos\Phi)^2}$$

where

$$q_1 = 3a^2 - 3a + \tfrac{1}{2} \quad \text{and} \quad q_2 = 1 - 3a$$

Additional hint: You should be especially careful when you investigate $\Phi \sim 0$!

5. In an autonomous system of differential equations, x does not appear explicitly. It is not difficult to generalize results like (54) for an autonomous equation to an equation of the form $y^{(1)} = f(x, y)$.
Derive (56) from (54) using the following technique: $\mathbf{z} = (z_1, z_2) = (y, x)$ is introduced in the equation and

$$\frac{dz_1}{dx} = f(z_1, z_2) \qquad \frac{dz_2}{dx} = 1$$

$$z_1(x_0) = y_0 \quad \text{and} \quad z_2(x_0) = x_0$$

The Jacobian matrix is

$$f_z = \mathbf{J} = \begin{pmatrix} f_y & f_x \\ 0 & 0 \end{pmatrix}$$

and this is inserted into (54).

6. Fourth-order semi-implicit Runge–Kutta methods are easily constructed. One example is

$$\begin{aligned}
\mathbf{k}_1 &= (\mathbf{I} - ah\mathbf{J})^{-1}h[f(\mathbf{y}_n) + ha\mathbf{f}_x] \\
\mathbf{k}_2 &= (\mathbf{I} - ah\mathbf{J})^{-1}h[f(\mathbf{y}_n + b\mathbf{k}_1, x_n + bh) + ha\mathbf{f}_x] \\
\mathbf{k}_3 &= (\mathbf{I} - ah\mathbf{J})^{-1}h[f(\mathbf{y}_n + b\mathbf{k}_2, x_n + bh) + ha\mathbf{f}_x] \\
\mathbf{k}_4 &= (\mathbf{I} - ah\mathbf{J})^{-1}[m_1(\mathbf{k}_1 + h^2a\mathbf{f}_x) + m_2(\mathbf{k}_2 + h^2a\mathbf{f}_x) + m_3(\mathbf{k}_3 + h^2a\mathbf{f}_x)] \\
\mathbf{y}_{n+1} &= \mathbf{y}_n + R_1\mathbf{k}_1 + R_2\mathbf{k}_2 + R_3\mathbf{k}_3 + \mathbf{k}_4
\end{aligned}$$

Determine $m_1, m_2, m_3, R_1, R_2, R_3$ as functions of a for $b = 0.75$, and find the characteristic root $\mu(\lambda h)$ for the method. The techniques of Exercises 4 and 5 should be used.

7. The following method is a hybrid between truly implicit methods and the semi-implicit method (56)

$$\begin{aligned}
\mathbf{k}_1 &= h\mathbf{f}(\mathbf{y}_n + \mathbf{k}_1 a) \\
\mathbf{k}_2 &= h\mathbf{f}(\mathbf{y}_n + b_2\mathbf{k}_1 + a\mathbf{k}_2) \\
\mathbf{k}_3 &= h\mathbf{f}(\mathbf{y}_n + c_3\mathbf{k}_1 + b_3\mathbf{k}_2 + a\mathbf{k}_3) \\
\mathbf{y}_{n+1} &= \mathbf{y}_n + R_1\mathbf{k}_1 + R_2\mathbf{k}_2 + R_3\mathbf{k}_3 + \mathcal{O}(h^4)
\end{aligned}$$

Each of the Runge–Kutta constants is determined by solution of one (nonlinear) equation. This must be done iteratively until convergence is reached, but presumably the same "Jacobian" **J** can be used [e.g., $\mathbf{k}_j^{(i)} = (\mathbf{I} - ha\mathbf{J})^{-1}\mathbf{k}_j^{(i-1)}$] throughout the iterations for all three Runge–Kutta constants. Discuss the advantages of this method compared to method (56) with stepsize adjustment (44)

a. When evaluation of $\mathbf{J}(\mathbf{y}_n)$ is difficult.
b. When h_n is rapidly increasing by (44).
c. When h_n is constant (integration of a single exponential).

How would you patch together an optimal third-order integration package from this method and methods (55) and (56)? Let $b_2 = b_3$ and $c_3 = 0$ and let $\mu(-\infty) = 0$. Calculate a, b, R_1, R_2, R_3. Use the Van der Pol equation to test the algorithm.

8. A photochemical smog model consisting of 64 independent reactions in 29 variables is discussed by R. J. Gelinas in *Journal of Computational Physics* 9 (1972):222.
Apply the third-order semi-implicit Runge–Kutta method on this example. You should make an input routine in which the reactants, the reactions, and the stoichiometric coefficients are designated in a suitable schematic manner. Take a second look at table I in the reference and select a much smaller set of equations that may presumably describe the system equally well. Carefully explain your reasons for neglecting each specific reaction.
Repeat the calculations and compare, e.g., the concentration of ozone as a function of time for the two models.

9. Aiken and Lapidus (1974) have presented a dynamic model for a fluid bed reactor:

$$\frac{d\mathbf{y}}{dt} = \begin{pmatrix} -1.30 & 10{,}400k & 1.30 & 0 \\ 0 & -1880(1+k) & 0 & 1880 \\ 267 & 0 & -269 & 0 \\ 0 & 320 & 0 & -321 \end{pmatrix}\mathbf{y} + \begin{pmatrix} 0 \\ 0 \\ 1752 \\ 0.1 \end{pmatrix}$$

$$\mathbf{y}(t=0) = (759.167, 0, 600, 0.1)$$

$$k = 0.0006 \exp\left(20.7 - \frac{15{,}000}{y_1}\right)$$

(y_1, y_2) are temperature and reactant partial pressure in the particle phase and y_3, y_4 are the same quantities in the fluid phase.
Aiken and Lapidus claim that the pellet temperature $y_1 \sim 700$ for $t \to \infty$. This result may qualitatively be checked by means of the equation for y_3 when $t \to \infty$. The exothermic reaction occurs on the solid, and the solid temperature y_1 must be above the fluid temperature y_3 for all t with the initial temperatures and the wall temperature as stated in the paper.
How does this agree with Aiken and Lapidus' result?

The example originates from a paper by Luss and Amundson[a] who used

$$\frac{dy_3}{dt} = 1752 + 266.7y_1 - 269.3y_3$$

rather than the rounded off values of Aiken and Lapidus.
Is Aiken and Lapidus' final value of y_1 in qualitative agreement with this equation for y_3?
What is your conclusion regarding the model?
Integrate both versions of the model to $t = 3000$ using STIFF 3. The tolerance vector $\boldsymbol{\varepsilon} = \varepsilon_0(760, 0, 0, 0)$, $\varepsilon_0 = 10^{-2}$, 10^{-3}, or 10^{-4}, and $HO = 10^{-3}$. When the weights of y_2, y_3, and y_4 are zero, the corresponding ε_i are infinity and the stepsize is regulated by the error of y_1 only.
The result at $t = 1000$ should be $y_1 = 702.92$ and $y_2 = 0.09075$ with Luss and Amundson's equation for y_3. Michelsen[b] obtained slightly different values since he erroneously took $y_3(0)$ to be 609.

10. Integrate the two examples of subsection 8.2.5 by method (8.55) and (8.56) and compute the eigenvalues of the Jacobian matrix as a function of the dependent variables. What is the variation of λ_1 and λ_2 through the limit cycle of the Van der Pol equation? Is there a qualitative difference between the eigenvalues of the Van der Pol equation and the eigenvalues of the Robertson problem?
Comment on the possible consequences for the solution of the two equations. In which part of the cycle is the Van der Pol equation a stiff problem?

11. In Exercise 4.12 a first-order isothermal reaction in a tube is considered. Repeat the calculations for a second-order reaction and investigate by an analysis of the computer output whether the radial concentration gradient can be accounted for by an axial dispersion term.
If your investigation shows that a model simplification is justified also in the nonlinear case, a numerical coefficient for the diffusion group should be extracted from the output.
Hint: Collocation in the radial direction is combined with STIFF 3 in the axial direction. Make a subroutine COLL (NCOLP) in which all collocation constants are generated. These data can be transferred to FUN and DFUN via COMMON statements. In this way almost nothing is changed in the main program shown in Appendix A21.

12. In many absorbents, surface diffusion is the governing mechanism of transport. The surface diffusion coefficient is strongly concentration dependent. Experimental values of the transport parameters are found by comparison of data with the solution of the diffusion model

$$\frac{\partial y}{\partial \tau} = \frac{1}{x^2}\frac{\partial}{\partial x}\left[x^2 D(y)\frac{\partial y}{\partial x}\right] \tag{1}$$

[a] Luss, D., and Amundson, N. R. *AIChE Journal* 14 (1968):211.
[b] Michelsen, M. L. *AIChE Journal* 22 (1976):594.

$$\begin{aligned}
&\tau = 0: && y = \frac{c}{c_0} = 1 \quad \text{and} \quad \tau = \frac{D_0 t}{R^2} \\
&x = 0: && \frac{\partial y}{\partial x} = 0 \qquad \text{and} \quad x = \frac{r}{R} \\
&x = 1: && y = 0 \quad \text{(or some more complicated side condition)}
\end{aligned} \tag{2}$$

$$\text{Diffusivity} = D_0 \cdot D(y) = D_0 \exp(ky)$$

Introduce $u = x^2$ and $v = \exp(ky)$ in (1). Show that the equation is transformed to

$$\frac{\partial v}{\partial \tau} = v\nabla^2 v = v\left(6\frac{\partial v}{\partial u} + 4u\frac{\partial^2 v}{\partial u^2}\right) \tag{3}$$

$$\begin{aligned}
&\tau = 0: && v = \exp(k) \\
&u = 1: && v = 1
\end{aligned} \tag{4}$$

Neretnieks[c] has solved (3) and (4) using collocation in the x-direction and Euler's method (or a fourth-order Runge–Kutta method) in the τ-direction.

a. Solve the collocation equations with STIFF 3 from $\tau = 0$ to 2 and $k = 5.3$ and 10.6 (the values used by Neretnieks). Compare your results for

$$\bar{y} = \int_0^1 y\,dx^3 = \frac{1}{k}\int_0^1 \ln v\,dx^3$$

with his figure 2.

b. It is of considerable interest to obtain a solution for small values of τ. Introduce variables similar to those of Exercise 4.9

$$\left.\begin{aligned} \eta &= (1 - x)\tau^{-1/2} \\ z &= \tau^{1/2} \end{aligned}\right\} \quad \text{or} \quad \left\{\begin{aligned} x &= 1 - \eta z \\ \tau &= z^2 \end{aligned}\right. \tag{5}$$

$$v = \sum z^i f_i \tag{6}$$

Show that equation (3) is transformed to

$$v\left(\frac{\partial^2 v}{\partial \eta^2} - \frac{2z}{1 - \eta z}\frac{\partial v}{\partial \eta}\right) = -\frac{1}{2}\eta\frac{\partial v}{\partial \eta} + \frac{1}{2}z\frac{\partial v}{\partial z} \tag{7}$$

Next show that the three first f_i of (6) are given by

$$\begin{aligned}
&f_0 f_0^{(2)} + \tfrac{1}{2}\eta f_0^{(1)} = 0 \\
&f_0 f_1^{(2)} + \tfrac{1}{2}\eta f_1^{(1)} - \tfrac{1}{2}f_1 + f_0^{(2)} f_1 - 2f_0^{(1)} f_0 = 0 \\
&f_0 f_2^{(2)} + \tfrac{1}{2}\eta f_2^{(1)} - f_2 + f_0^{(2)} f_2 - 2\eta f_0 f_0^{(1)} - 2f_0 f_1^{(1)} \\
&\qquad -2f_1 f_0^{(1)} + f_1 f_1^{(2)} = 0
\end{aligned}$$

$$\begin{aligned}
&\eta = 0: && f_0 = 1 && f_1 = f_2 \ldots = 0 \\
&\eta \to \infty: && f_0 = \exp(k), && f_1 = f_2 \ldots = 0
\end{aligned}$$

[c] Neretnieks, I. *Chem. Eng. Sci.* 31 (1976):465.

Finally show that

$$\bar{y}(\tau) = 1 - \frac{3}{k}\left[2\tau^{1/2}\left(\frac{df_0}{d\eta}\right)_{\eta=0} + \tau\left(\frac{df_1}{d\eta}\right)_{\eta=0} + \frac{2}{3}\tau^{3/2}\left(\frac{df_2}{d\eta}\right)_{\eta=0} + \ldots\right]$$

c. Compute the function $\bar{y}(\tau)$ for small τ by solution of the equations for f_0, f_1, f_2 either as described in Exercise 4.9 or using STIFF 3. In the latter case, some iterative scheme must be applied since the boundary conditions are given at $\eta = 0$ and at $\eta = \infty$.
For $k = 0$, $\bar{y}(\tau) = (6/\sqrt{\pi})\tau^{1/2}$ for small τ.
From the computed results, derive an empirical relation for the coefficient to $\tau^{1/2}$ as a function of k. This relation can immediately be used to find a value for k from experiments when D_0 is known.
Answer: For $k = 5.3$, the coefficient is 13.3.

13. The following linear partial differential equation [identical to eq. (4.65)] with a nonlinear boundary condition at $x = 1$ is important in the design of certain types of waste water treatment units.

$$(1 - x^2)\frac{\partial y}{\partial z} = \frac{\partial^2 y}{\partial x^2} \tag{1}$$

$$y = 1 \quad \text{for } z = 0 \quad \text{and } x \in [0, 1]$$

$$\left.\begin{aligned} &\frac{\partial y}{\partial x} = 0 \quad \text{for } x = 0 \quad \text{and } z > 0 \\ &\frac{\partial y}{\partial x} + Ky^n = 0 \quad \text{for } x = 1 \quad \text{and } z > 0 \end{aligned}\right. \tag{2}$$

According to Harremoës [Harremoës P.: *Journal Water Pollution Control Federation*, 48, 377, (1976)] the chemical reaction on the biological film at $x = 1$ has a reaction order $n = \frac{1}{2}$.

Collocation at the N zeros of $P_N^{(\alpha,\beta)}(u = x^2)$ yields N ordinary differential equations in $N + 1$ ordinates, and the boundary condition at $x = 0$ is automatically satisfied. Rather than solving N differential equations augmented by one nonlinear algebraic equation (the discretized boundary condition at $u = 1$) it is suggested to transform the algebraic equation into an ultra-stiff differential equation in y_{N+1} [$= y(u = 1)$] which is solved alongside the N "natural" differential equations in the collocation ordinates.

$$\varepsilon\frac{dy_{N+1}}{dz} = -\sum_{i=0}^{N+1} A_{N+1,i}y_i - \frac{K}{2}y_{N+1}^n \tag{3}$$

where ε is a small number (e.g. 10^{-5}).

a. Follow $y_i(z)$ and $\bar{y}(z)$ with $N = 2, 4, 6, 8$ until $\bar{y} < 0.01$. Different ε should be used. $K = 20$, $n = \frac{1}{4}, \frac{1}{2}$ and 1.
b. Use the standard method of section 4.3 to control the results for $n = 1$.

The approach suggested in the present exercise has many practical applications—e.g., in the study of the dynamics of absorption columns where the differential mass balance on the trays is augmented by a nonlinear equilibrium relationship. In exercise 9.7 a considerably more complicated model than that of the present exercise is solved by the same technique.

14. A two dimensional model is frequently required to give a realistic model for a strongly exothermic reaction in a fixed bed reactor. Froment [Froment, G. F., *Ind. Eng. Chem.*, 59 No. 2, (1967):18.] has used the following reaction scheme for oxidation of *o*-xylene (A) to phthalic anhydride (B) on vanadium catalyst:

$$A \xrightarrow{k_1} B \xrightarrow{k_2} C$$

$$A \xrightarrow{k_3} C$$

Waste materials (C = CO_2, CO, and H_2O) are formed as final products from both the consecutive and the parallel reaction path.

A number of assumptions are implicit in Froment's model. Transport restrictions inside the particles and between particles and fluid are neglected. Axial dispersion is likewise neglected and furthermore the volumetric flowrate and the oxygen concentration are treated as constants.

With these assumptions the steady state mass and energy balances are

$$\begin{aligned}\frac{\partial x}{\partial z} &= \alpha_M\left(\frac{\partial^2 x}{\partial r^2} + \frac{1}{r}\frac{\partial x}{\partial r}\right) + \beta_1 r_B \\ \frac{\partial w}{\partial z} &= \alpha_M\left(\frac{\partial^2 w}{\partial r^2} + \frac{1}{r}\frac{\partial w}{\partial r}\right) + \beta_1 r_C \\ \frac{\partial \theta}{\partial z} &= \alpha_H\left(\frac{\partial^2 \theta}{\partial r^2} + \frac{1}{r}\frac{\partial \theta}{\partial r}\right) + \beta_2 r_B + \beta_3 r_C\end{aligned} \tag{1}$$

x is the fractional conversion to B, w the fractional conversion to C, and $\theta = T/T_{\text{ref}}$

$$w = x = 0 \quad \text{and} \quad \theta = \theta_0 = \frac{T_0}{T_{\text{ref}}} \quad \text{at } z = 0$$

$$\frac{\partial x}{\partial r} = \frac{\partial w}{\partial r} = \frac{\partial \theta}{\partial r} = 0 \qquad \text{at } r = 0 \tag{2}$$

$$\frac{\partial x}{\partial r} = \frac{\partial w}{\partial r} = 0 \quad \text{and} \quad \frac{\partial \theta}{\partial r} = -\text{Bi}\,(\theta - \theta_w) \quad \text{at } r = 1$$

$$r_B = k_1(1 - x - w) - k_2 x$$

$$r_C = k_2 x + k_3(1 - x - w)$$

$$k_i = \exp\left[a_i + \gamma_i\left(1 - \frac{1}{\theta}\right)\right] \qquad i = 1, 2, 3$$

a. The following data are taken from Froment (other data are clearly stated in his paper): T_{ref} = 630 °K; inlet mole fraction of *o*-xylene = 0.00924;

inlet mole fraction of oxygen = 0.208; and a reactor length 3 m. The rate expressions are in units of k moles/kg catalyst/hr.
Derive the following values for the dimensionless groups:

$$\alpha_M = 5.76, \alpha_H = 10.97, \beta_1 = 5.106, \beta_2 = 3.144, \beta_3 = 11.16$$

$$\gamma_1 = 21.6, \gamma_2 = 25.1, \gamma_3 = 22.9$$

$$a_1 = -1.74, a_2 = -4.24, a_3 = -3.89$$

$$\text{Bi} = 2.5$$

b. Make a computer program for integration of (1) and (2). The desired output is radially averaged values of x, w and θ at axial positions $z = 0.2$, 0.4, 0.6, 0.8, and 1.
Let $\theta_w = \theta_0$ and investigate inlet temperature values in the range $0.98 < \theta_0 < 1.02$ to determine
1) the value of θ_0 that gives the highest yield of B.
2) the explosion limit.

c. The effective diffusivity of A and B in the catalyst pellets is assumed to be $2.5 \cdot 10^{-6}\ \text{m}^2/\text{s}$, and the catalyst bed porosity is $\varepsilon_b = 0.4$. Investigate whether the assumption of negligible intraparticle mass transport resistance is valid. If this is not the case formulate and solve a reactor model where intraparticle mass transport resistance is taken into account. External resistances to heat and mass transport are still assumed to be negligible and the pellets are of the same temperature as the bulk fluid phase. One-point collocation in the pellet phase is sufficient.

15. SNG is produced by methanation of carbonmonoxide on a Ni-based catalyst according to reaction (1):

$$3H_2 + CO \rightleftarrows CH_4 + H_2O \tag{1}$$

Some of the CO may be converted to CO_2 by the shift reaction (2):

$$CO + H_2O \rightleftarrows CO_2 + H_2 \tag{2}$$

The rate expressions are written as the product of a forward reaction rate R_{if} and an equilibrium correction factor $1 - K_i/K_{pi}$ where, e.g., $K_2 = p_{CO_2}p_{H_2}/p_{CO}p_{H_2O}$. The equilibrium constants are (Rostrup Nielsen)[d]

$$\log_{10} K_{p1} = \frac{11512}{T} - 12.905 \text{ for reaction (1)} \tag{3}$$

$$\log K_{p2} = \frac{1978}{T} - 1.851 \text{ for reaction (2)} \tag{4}$$

R_{1f} can be taken from Schoubye[e]:

$$R_{1f} = \frac{0.47 \cdot 10^9 \exp(-11760/T) p_{H_2}^{0.15}}{[1 + 10^{-5} \exp(8380/T)(p_{CO}/p_{H_2})]^{0.5}} \tag{5}$$

$$R_{2f} = 4.33 \cdot 10^6 \exp(-8810/T) \tag{6}$$

[d] Rostrup Nielsen, J. *Steam Reforming Catalysts*, Copenhagen: Teknisk Forlag (1975).
[e] Schoubye, P. *Journal of Catalysis* 14 (1969):238.

The last expression is an approximate version of rate expressions given by Bohlbro[f], e.g., his table 23 for iron oxide based shift catalyst. We use (6) in the absence of better published information for Ni-based shift catalysts.

We wish to calculate the effectiveness factor of a catalyst at the reactor entrance where $T = 300\,°C$. The total pressure is 30 atm and the pellet surface gas composition is

$$\begin{aligned} \text{mol \%}\quad H_2 &= 0.38 \\ CO &= 0.08 \\ CH_4 &= 0.40 \\ H_2O &= 0.12 \\ CO_2 &= 0.02 \end{aligned}$$

For simplicity, the effective diffusivity is taken to be

a. $3 \cdot 10^{-4}\,\text{cm}^2/\text{s}$, or

b. $5 \cdot 10^{-3}\,\text{cm}^2/\text{s}$

for all components. The pellet is considered isothermal (an inappropriate assumption: $-\Delta H_1 = 8.88$ kcal/mol and $-\Delta H_2 = 53.07$ kcal/mol and the relatively high pressure of 30 atm does give a rather high value of the heat generation parameter β).

Use global collocation as well as spline collocation for spherical pellets (0.5-cm diameter) and spline point, e.g., $r = 0.6$. CO and CO_2 are used as key components.

For the large diffusivity $5 \cdot 10^{-3}\,\text{cm}^2/\text{s}$, the effectiveness factor for the methanation reaction η_1 may be larger than 1.

For the small diffusivity the answer is $\eta_1 = 0.86$ and $\eta_2 = 0.18$. In both cases p_{CO_2} has a maximum close to $r = 1$.

The Newton–Raphson routine should be used in the following form:

$$\mathbf{z}^{(n+1)} = \mathbf{z}^{(n)} \exp\left[\frac{\Delta\mathbf{z}^{(n)}}{\mathbf{z}^{(n)}}\right] \qquad [\Delta\mathbf{z}^{(n)} = -[\mathbf{J}^{(n)}]^{-1}\mathbf{RES}^{(n)}]$$

for small values of $\mathbf{z}^{(n)}$ in order to avoid negative partial pressures of the key components.

Computations with approximately the gas composition of this exercise were made by Kinoshita.[g]

REFERENCES

There is not much published information on the difficulties involved in numerical solution of the complex models of section 8.1. Van den Bosch

[f] Bohlbro, H. *An Investigation on the Kinetics of the Conversion of CO with H_2O over Iron Oxide Catalysts.* Copenhagen: Gjellerup (1966).

[g] Kinoshita, G. "Simulation of a Fixed Bed Methanation Reactor," M.Sc. Thesis, Cambridge, Mass.: MIT (August 1973).

and Hellinckx (1974) solved a rather simple example of the type discussed in subsection 8.1.1 and obtained results concerning the optimal number N of collocation points per subinterval. They did not look for an optimal solution method for the algebraic equations and no conclusion of general validity can be drawn from their results, except perhaps that N should be chosen quite small.

The economic structuring of large reaction schemes is discussed in elementary textbooks on reaction engineering, but the optimal combination of numerical methods for solution of specific problems such as in Exercise 15 is often of proprietory nature and is not available in journals.

Contrariwise the amount of published information on "general" integration techniques for simple initial value problems as discussed in section 4.2 and again in section 8.2 is overwhelming.

Implicit Runge–Kutta methods such as (8.59) were probably first proposed by Butcher (1964), who also used the somewhat more appropriate expression "semi-explicit" for methods (8.57) that we have called "semi-implicit" in accordance with the nomenclature of, e.g., Rosenbrock (1963) and Seinfeld, Lapidus, and Hwang (1970). Many variants of semi-implicit methods have been proposed [e.g., Calahan (1968) and Caillaud and Padmanabhan (1971)] and many more could be constructed by the methods proposed in the text and in Exercise 4.6.

The close relation between implicit Runge–Kutta methods and the collocation methods of chapter 4 has been pointed out in many papers in *BIT*, e.g., Wright (1970) and Chipman (1973), who constructed implicit Runge–Kutta methods of the Gauss collocation type. Wright's paper is especially worth studying in this respect. The definition of A-stability orginates with Dahlquist (1963) and several papers from the Stockholm School have references to A-stable integration methods that are more or less the same as those discussed in section 8.2 [e.g., Ehle (1968), Axelsson (1969)]. Methods for which $|\mu(-\infty)| = 0$ are called strongly A-stable by Chipman (1971).

Test examples for different integration methods have, until recently, been in high demand. A paper in *BIT* by Enright, Hull, and Lindberg (1975) lists 25 carefully selected stiff problems divided into five different categories. The relative cost and reliability of several popular routines is compared for these examples.

Aiken and Lapidus (1974) have cited several chemical engineering examples of stiff equations, of which we have used one in Exercise 8.

Robertson's problem is found in many references, e.g., 4 and 13. Its original version is Robertson (1967). Van der Pol's equation is treated in some detail by Davis (1962) and by Cole (1968). A rather "mild" version (with $K = 5$) is included in ref. 13.

Nonlinear partial differential equations have been integrated by double collocation as described in section 8.4 by Villadsen (1970) and by

Sørensen and Stewart (1972)—both using the *o*-xylene oxidation as a test example. Comparisons between collocation with an explicit Runge–Kutta process, double collocation, and finite difference techniques are made by Finlayson (1971).

1. VAN DEN BOSCH, B., and HELLINCKX, L. *Chem. Eng. Journal* 7 (1974):73.
2. BUTCHER, J. C. *Mathematics of Computation* 18 (1964):50.
3. ROSENBROCK, H. H. *Computer Journal* 5 (1963):320.
4. SEINFELD, J. H., LAPIDUS, L., and HWANG, M. *Ind. Eng. Chem. Fund.* 9 (1970):266.
5. CALAHAN, D. A. *Proc. IEEE* (Letters) 56 (1964):744.
6. CAILLAUD, J. B., and PADMANABHAN, L. *Chem. Eng. Journal* 2 (1971):227.
7. WRIGHT, K. *BIT* 10 (1970):217.
8. CHIPMAN, F. H. *BIT* 13 (1973):391.
9. DAHLQUIST, G. G. *BIT* 3 (1963):27.
10. EHLE, B. L. *BIT* 8 (1968):276.
11. AXELSSON, O. *BIT* 9 (1969):185.
12. CHIPMAN, F. H. *BIT* 11 (1971):384.
13. ENRIGHT, W. H., HULL, T. E., and LINDBERG, B. *BIT* 15 (1975):10.
14. AIKEN, R. C., and LAPIDUS, L. *AIChE Journal* 20 (1974):368.
15. ROBERTSON, H. H. "Solution of a Set of Reaction Rate Equations," p. 178 in *Numerical Analysis*, ed. J. Walsh. Washington, D.C.: Thompson Book Co., (1967).
16. DAVIS, H. T. *Introduction to Non-Linear Differential and Integral Equations.* New York: Dover (1962).
17. COLE, J. D. *Perturbation Methods in Applied Mathematics.* Waltham, Mass.: Blaisdell (1968).
18. VILLADSEN, J. "Selected Approximation Methods for Chemical Engineering Problems," Instituttet for Kemiteknik, Denmark (1970).
19. SØRENSEN, J., and STEWART, W. E. *Proceedings 2nd ISCRE,* Amsterdam (1972):(B8) 75–88.
20. FINLAYSON, B. A. *Chem. Eng. Sci.* 26 (1971):1081.

Selected Research Problems 9

Introduction

In the introductory chapters 2 to 4 fairly simple examples are consistently chosen to illustrate the numerical methods. The purpose is, of course, to avoid overburdening the reader with physical detail that might obscure the exposition of the methods. Only in chapter 8, and possibly in chapter 5, have we deviated somewhat from this course, and the problems discussed in these chapters might have some independent research value.

Those computational problems that today have a significant research interest cannot be solved, however, just by applying one standard method—otherwise they would have been solved several years ago.

The purpose of this last chapter is to show how all the various methods of the previous chapters are used together to tackle problems that by current standards would be recognized as difficult.

The extended Graetz problem of section 9.1 is attacked first by global collocation and then by spline collocation to obtain a solution in the entry region where a standard collocation procedure breaks down. Finally, perturbation solutions are developed to extract the asymptotic Nusselt number without solving the full problem.

The stability problem for a catalyst particle is treated in section 9.2. The few existing publications have only scratched the surface of this obfuscate but highly interesting mathematical research subject. We have endeavored to give a fairly complete picture of the eigenvalue spectrum for a first-order irreversible reaction—fully recognizing that other and perhaps more interesting rate mechanisms would give a different picture.

In putting together the solution of our chosen problem, information derived by many different methods have been used: a Galerkin method combined with collocation using different sets of expansion functions; forward integration combined with a sensitivity analysis as described in chapters 4 and 8, and perturbation series developed from three simpler problems.

The steady state model for this system is treated in chapter 5, and one may conceive that a dynamic model based on the often quite extreme steady state profiles that are obtained in chapter 5 will give rise to unusually difficult numerical problems. All the standard methods will run into difficulties for some values of the parameters, and we shall have the occasion to show the benefits of a thoughtful combination of methods under unusual circumstances.

The fixed bed reactor problem of section 9.3 has been included to emphasize the usefulness of collocation in process control studies. Transfer functions can be difficult to obtain by standard "mixed cell" techniques, and it seems that a very significant improvement can be gained in the analysis when a reasonably low-order collocation process is used instead. Even though we have given only one example, it is hoped that this is general enough to illustrate the potential of collocation for this kind of problem where according to a recent review [Finlayson (1974)] the methods might find their most promising applications.

9.1 The Graetz Problem with Axial Conduction

The standard Graetz problem is concerned with laminar heat transfer to a fluid. The system is semi-infinite in the axial direction and finite in the other direction. Heat is transferred from wall to fluid by conduction and axially by convection.

An adequate discussion of this problem as stated mathematically in equations (4.65) and (4.66) has been given in section 4.3, and further computations for the entry region are developed in section 7.2.

Here we present results on the related but considerably more complicated problem where axial transport of heat occurs by conduction as well as by convection. The mathematical model for this system has been presented in equations (1.48) to (1.52), and we shall be specifically interested in computing the local Nusselt number defined by equations (1.53) and (1.54).

$$(1.49){:}\ (1 - x^2)\frac{\partial\theta}{\partial y} = \frac{1}{x}\frac{\partial}{\partial x}\left(x\frac{\partial\theta}{\partial x}\right) + \frac{1}{\mathrm{Pe}^2}\frac{\partial^2\theta}{\partial y^2} \tag{1}$$

$$(1.50)\text{–}(1.52){:}\quad \theta = 1 \quad \text{for } y \to -\infty, \text{ and } \theta \text{ finite } (\to 0) \text{ for } y \to \infty \tag{2}$$

$$\frac{\partial \theta}{\partial x} = 0 \quad \text{for } x = 0, \text{ and } -\infty < y < \infty \tag{3}$$

$$\frac{\partial \theta}{\partial x} = 0 \quad \text{for } x = 1, \text{ and } y < 0 \tag{4}$$

$$\theta = 0 \quad \text{for } x = 1, \text{ and } y \geq 0 \tag{5}$$

The specific situation considered in (4) and (5) is a tube insulated upstream ($y < 0$) and with a constant wall temperature downstream from $y = 0$ to ∞. Other types of boundary conditions at the wall are

1. constant but different wall temperature in the two sections of the tube.
2. constant but different heat flux at the tube wall in the two sections of the tube.
3. either constant wall temperature or constant wall flux for $y < 0$ and a film boundary condition at the wall for $y \geq 0$.

Some references to earlier work are given at the end of the chapter, but we are only concerned with boundary conditions (4) and (5) that were used by Michelsen and Villadsen (1974), the paper on which our presentation here is based.

(1.53):

$$\text{Nu}\,(y) = -\frac{2(\partial\theta/\partial x)|_{x=1}}{\bar{\theta}} = \frac{-\frac{1}{2}(d\bar{\theta}/dy) + (2/\text{Pe}^2)(d^2/dy^2)\int_0^1 x\theta\,dx}{\bar{\theta}} \tag{6}$$

where

$$\bar{\theta} = 4\int_0^1 x(1-x^2)\theta\,dx \quad \text{and} \quad \bar{\theta}_1 = \int_0^1 x\theta\,dx \tag{7}$$

A Fourier series solution of (1) is obtained by assuming $\theta = F(x)G(y) = F(x)\exp(\lambda y)$. Inserting this in (1) yields the following eigenproblem for $F(x)$:

$$\frac{d}{dx}\left(x\frac{dF}{dx}\right) - \left[\lambda x(1-x^2) - \frac{1}{\text{Pe}^2}\lambda^2 x\right]F = 0 \tag{8}$$

The solution of (8) $[F_i(x), \lambda_i]$ is different for $y > 0$ and for $y < 0$ since the boundary condition at $x = 1$ is different in the two tube sections:

$$\frac{dF}{dx} = 0 \quad \text{at } x = 1 \text{ for } y < 0 \quad \text{and} \quad F = 0 \quad \text{at } x = 1 \text{ for } y \geq 0$$

Thus (8) must be solved twice to obtain the solutions

$$[F_-(x), \lambda_-] \quad \text{for } y < 0 \quad \text{and} \quad [F_+(x), \lambda_+] \quad \text{for } y \geq 0 \tag{9}$$

For each region a numerical method, e.g., a Runge–Kutta method or, more conveniently, the forward integration method of section 4.5, must be used N times to obtain the first N eigenfunctions and eigenvalues. Equation (8) is nonlinear in λ, but to satisfy the side conditions at $y \to -\infty$ and $y \to \infty$ one can immediately exclude negative λ from the solutions $[F_-, \lambda_-]$ and positive λ from $[F_+, \lambda_+]$. Thus the additional difficulty due to extraneous solutions of (8) is easily circumvented in the search process for $[F, \lambda]$ by forward integration.

θ is now obtained as

$$\theta = \sum_1^\infty a_i F_i(x) \exp(\lambda_i y) \tag{10}$$

or in practice as

$$\theta \sim \sum_1^N a_i F_i(x) \exp(\lambda_i y) \tag{11}$$

A complete solution of (1) to (5) by the Fourier series method requires determination of N Fourier constants $\mathbf{a}_-$ for $y < 0$ (corresponding to $\lambda = 0$ and $N - 1$ positive eigenvalues $\boldsymbol{\lambda}_-$) and N Fourier constants $\mathbf{a}_+$ for $y > 0$ (corresponding to the N negative eigenvalues $\boldsymbol{\lambda}_+$).

If the members of each of the two sets of eigenfunctions F_- and F_+ had been mutually orthogonal, $2N$ orthogonality conditions would have been available for an easy determination of $\mathbf{a}_-$ and $\mathbf{a}_+$, one component at a time.

But (8) is not a Sturm–Liouville problem. Neither $\mathbf{F}_-$ nor $\mathbf{F}_+$ are mutually orthogonal and a set of auxiliary eigenfunctions that exhibit this property must be constructed from $\mathbf{F}_-$ and $\mathbf{F}_+$ by a Gram–Schmidt orthogonalization process before $\mathbf{a}_-$ and $\mathbf{a}_+$ can be computed.

The complexity of each step of the process (8) to (11), which is used by Hsu in several papers, e.g., Hsu (1971), casts some doubt on the value of a Fourier series approach for this type of problem.

9.1.1 Collocation solution of the model

Equation (1) is rewritten in terms of the two variables θ and $\phi = \partial\theta/\partial y$:

$$\frac{\partial \theta}{\partial y} = \phi$$

$$\frac{\partial \phi}{\partial y} = \mathrm{Pe}^2 (1 - x^2)\phi - \mathrm{Pe}^2 \frac{1}{x}\frac{\partial}{\partial x}\left(x \frac{\partial \theta}{\partial x}\right) \tag{12}$$

$$= \mathrm{Pe}^2 \left[(1 - u)\phi - 4\left(u \frac{\partial^2 \theta}{\partial u^2} + \frac{\partial \theta}{\partial u}\right)\right]$$

The collocation analog of (12) is

$$\begin{aligned} \frac{d\boldsymbol{\theta}}{dy} &= \boldsymbol{\phi} \\ \frac{d\boldsymbol{\phi}}{dy} &= \mathrm{Pe}^2\,\mathbf{V}\boldsymbol{\phi} - 4\,\mathrm{Pe}^2\,(\mathbf{UB} + \mathbf{A})\boldsymbol{\theta} \end{aligned} \tag{13}$$

A and **B** are discretization matrices for $\partial/\partial u$ and $\partial^2/\partial u^2$, respectively. **V** is a diagonal matrix $V_{ii} = 1 - u_i$, **U** is a diagonal matrix $U_{ii} = u_i$, and u_i are the collocation abscissas in $u = x^2$.

Thus with N collocation points (12) is transformed into $2N$ coupled first-order linear differential equations with constant coefficients. Introducing the $2N$ vector $\boldsymbol{\psi}$ in the collocation ordinates:

$$\boldsymbol{\psi} = (\theta_1, \theta_2, \ldots, \theta_N, \phi_2, \ldots, \phi_N)^T$$

makes it possible to write (13) in a compact form:

$$\frac{d\boldsymbol{\psi}}{dy} = \mathbf{Q}\boldsymbol{\psi} = \begin{pmatrix} \mathbf{0} & \mathbf{I} \\ -4\,\mathrm{Pe}^2\,(\mathbf{UB} + \mathbf{A}) & \mathrm{Pe}^2\,\mathbf{V} \end{pmatrix} \boldsymbol{\psi} \tag{14}$$

and the problem has been put into the standard form of subsection 4.3.3.

The difficulties due to different boundary conditions at $x = 1$ in the upstream and downstream regions have, of course, not been eliminated by the collocation approach. **A** and **B** are different in the two regions: For $y > 0$, $\theta(x = 1) = 0$ and **A** and **B**, as set up using $u_1, u_2, \ldots, u_N$ and $u_{N+1} = 1$ as interpolation points, can be used directly. For $y < 0$ the unknown boundary ordinate $\theta(x = 1)$ must be eliminated as described in subsection 4.3.3.

Thus (14) contains two $(2N \times 2N)$ sets of equations, one set for $y < 0$ and one for $y \geq 0$. The solution is

$$\boldsymbol{\psi} = \begin{cases} \exp(\mathbf{Q}_+ y)\boldsymbol{\psi}_0 = \mathbf{S}_+ \exp(\boldsymbol{\Lambda}_+ y)\mathbf{S}_+^{-1}\boldsymbol{\psi}_0 & \text{for } y \geq 0 \\ \exp(\mathbf{Q}_- y)\boldsymbol{\psi}_0 = \mathbf{S}_- \exp(\boldsymbol{\Lambda}_- y)\mathbf{S}_-^{-1}\boldsymbol{\psi}_0 & \text{for } y < 0 \end{cases} \tag{15}$$

$\boldsymbol{\Lambda}_+$ contains the $2N$ eigenvalues of $\mathbf{Q}_+$. N of these are positive and N are negative. $\boldsymbol{\Lambda}_-$ are the eigenvalues of $\mathbf{Q}_-$, and here N are positive, $N - 1$ are negative, and 1 is zero corresponding to the zero gradient of θ at $x = 1$ and at $x = 0$ for $y < 0$.

All collocation ordinates ψ_i must remain finite for $y \to \infty$ and for $y \to -\infty$. Consequently, N positive exponentials corresponding to positive eigenvalues of $\mathbf{Q}_+$ and $N - 1$ negative exponentials corresponding to negative eigenvalues of $\mathbf{Q}_-$ must be suppressed by orthogonality relations between the eigenrows $\mathbf{S}_+^{-1}$ (or $\mathbf{S}_-^{-1}$) of $\mathbf{Q}_+$ (or $\mathbf{Q}_-$) and the vector $\boldsymbol{\psi}_0$ that contains values of ψ at the collocation points for $y = 0$.

Thus in the computer diagonalization of $\mathbf{Q}$ using EISYS, as described in subsection 4.3.5, all rows of $\mathbf{S}^{-1}$ that correspond to nonadmissible eigenvalues are required to be orthogonal on $\boldsymbol{\psi}_0$. In this way $2N - 1$ linear algebraic equations in the $2N$ components of $\boldsymbol{\psi}_0$ are set up. Finally the requirement that $\theta \to 1$ for $y \to -\infty$ provides the $2N$th equation and $\boldsymbol{\psi}_0$ can be determined.

In summary the collocation solution of (1) is obtained through diagonalization of two $(2N \times 2N)$ matrices $\mathbf{Q}_+$ and $\mathbf{Q}_-$ followed by solution of $2N$ linear algebraic equations to obtain $\boldsymbol{\psi}_0$. The scalar quantities $\bar{\theta}$ of (7) and $\bar{\theta}_1 = \int_0^1 x\theta\, dx$ are obtained by Gauss quadrature, as described in subsection 4.3.5. Derivatives $d\bar{\theta}/dy$ and $d^2\bar{\theta}_1/dy^2$ that appear in (6) are obtained by analytic differentiation of the collocation series, as was described in subsection 4.3.5.

It is seen that the solution by collocation of equation (1) is only different from that of the standard problem in chapter 4 in some minor details. Thus the example may serve to illustrate how more complicated linear problems that even today have research interest can be attacked with the tools of chapter 4. Variants of the problem that are solved similarly include

1. film boundary condition at $x = 1$, $y > 0$.
2. a heat generation term $f(x)\theta$ included.
3. an insulated section for $y > y_1 > 0$.

In item 3 the solution is spliced together from three separate sections: $[-\infty, 0]$ with N admissible eigenvalues (from $\mathbf{Q}_-$), $[0, y_1]$ with $2N$ eigenvalues (from $\mathbf{Q}_+$), and finally $[y_1, \infty]$ with N admissible eigenvalues (the nonpositive eigenvalues of $\mathbf{Q}_-$). Continuity of $\boldsymbol{\psi}$ and $d\boldsymbol{\psi}/dy$ at 0 and y_1 provide equations for the additional $2N$ constants.

9.1.2 Comparison of Fourier series and collocation series for $y \sim 0$

A comparison of the collocation series and the Fourier series leads to exactly the same conclusions as in section 4.4: The first few collocation eigenvalues are the same as the corresponding Fourier eigenvalues computed by Hsu (1972), while eigenvalues with index $>N/2$ are numerically very much larger than the corresponding Fourier eigenvalues. Consequently the large y behavior of the two series is identical, while the collocation series can be used for a considerably smaller y than a Fourier series with the same number of terms.

For $y \to 0$, both series break down due to a discontinuity of $\partial\theta/\partial x$ at the wall $x = 1$: When y approaches zero from below, the wall flux is zero; but when y approaches zero from above, the wall flux tends to infinity.

It is difficult to give an unequivocal physical argument for this jump in $\partial\theta/\partial x|_{x=1}$ at $y = 0$, but there is ample numerical proof of its existence.

Figure 9-1 shows $\bar{\theta}$ as a function of y for discrete values of Pe. For Pe$\to\infty$ the upstream dissipation of heat disappears and the wall temperature is $\theta = 1$ throughout the $y < 0$ region. The jump of $\theta(x = 1)$ from 1 to 0 when crossing $y = 0$ clearly signifies that $\partial\theta/\partial x|_{x=1} = \frac{1}{4}\, d\bar{\theta}/dy$ is discontinuous at this y-value, a situation that has already been treated in chapter 4 for the basic Graetz problem. At a finite Pe, $d\bar{\theta}/dy$ is finite for all y, but the second term in the numerator of (6) and thus $\partial\theta/\partial x|_{x=1}$ itself may well be infinite at $y = 0$. The complexity of the definition of Nu(y) as the sum of convective and a conductive term is exactly what makes a physical argument concerning the behavior of Nu for $y \sim 0$ extremely precarious.

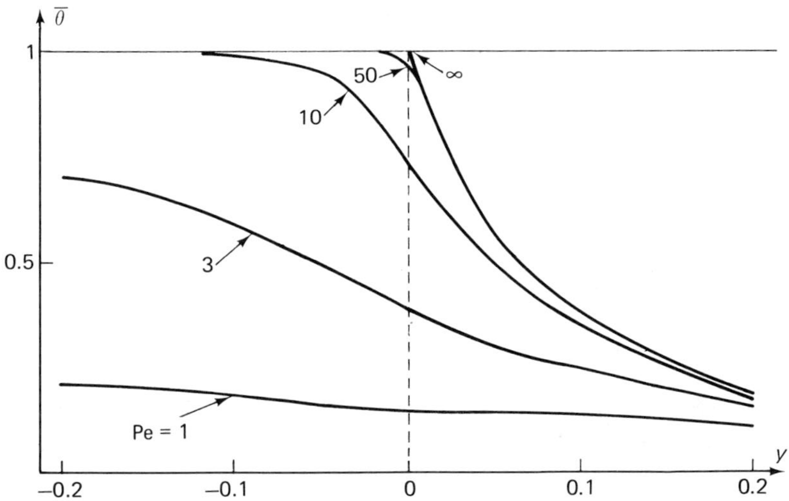

Figure 9-1. Mean temperature $\bar{\theta}$ equation (9.7) as a function of dimensionless axial position.

The numerical evidence for the discontinuity at $y = 0$ comes from several sources.

Table 9.1 shows computed values for $-\partial\theta/\partial u|_{u=1} = -\frac{1}{2}\,\partial\theta/\partial x|_{x=1}$ at $y = 0$ for Pe $= 10$ and various N. This quantity is, of course, readily available through an interpolation process based on the last N elements of $\boldsymbol{\psi}_0$ that contain $\partial\theta/\partial x$ at the interior interpolation nodes $u_1, u_2, \ldots, u_N$.

Also shown in the table are values of df/du at $u = 1$ for $f = (1 - u)^n$ using numerical differentiation formulas based on the same nodes and on the same approximation order N.

TABLE 9.1
THE GRADIENT OF θ AT $u = 1$, $y = 0$ DETERMINED BY NTH-ORDER COLLOCATION. COMPARISON WITH NUMERICAL DIFFERENTIATION OF $(1 - u)^n$

N	$-\partial\theta/\partial u\|_{u=1} = q_N$ (Pe = 10)	$\frac{N}{2}\ln\frac{q_N}{q_{N-1}}$	$(N/2)\ln(q_N/q_{N-1})$ for $(1-u)^n$				
			$n = \frac{1}{2}$	$\frac{1}{3}$	$\frac{1}{4}$	$\frac{1}{5}$	$\frac{1}{6}$
8	8.417	—					
9	9.411	0.5023	0.5038	0.6709	0.7543	0.8042	0.8377
10	10.404	0.5016	0.5033	0.6701	0.7536	0.8036	0.8369
11	11.396	0.5008	0.5026	0.6695	0.7529	0.8029	0.8362
12	12.389	0.5013	0.5023	0.6691	0.7524	0.8024	0.8358
13	13.382	0.5012	0.5017	0.6685	0.7522	0.8021	0.8354
14	14.374	0.5005	0.5016	0.6684	0.7519	0.8018	0.8350
15	15.367	0.5010	0.5014	0.6682	0.7515	0.8014	0.8349
16	16.359	0.5004	0.5013	0.6681	0.7513	0.8012	0.8346
18	18.344	—	0.5012	0.6677	0.7512	0.8011	0.8344
20	—	—	0.5006	0.6677	0.7509	0.8008	0.8341

It is obvious that the values of $\partial\theta/\partial u|_{u=1}$ do not converge when N is increased. In fact it appears that the ratio

$$\frac{N}{2}\ln\frac{q_N}{q_{N-1}} = \frac{N}{2}\ln\frac{(\partial\theta/\partial u)_{u=1,N}}{(\partial\theta/\partial u)_{u=1,N-1}} \to 0.5 \tag{16}$$

When differentiation formulas based on the same N interior nodes are applied to $f(u) = (1 - u)^n$, (16) seems to hold for $n = \frac{1}{2}$, a strong indication that $\theta(x, y = 0) \sim (1 - x^2)^{1/2}$ for $x \to 1$. For $N \lesssim 10$, (16) can be expressed equally well by the simpler formula

$$\frac{q_N}{q_{N-1}} = \frac{1}{2}\left[\left(\frac{N+1}{N}\right)^{2(n-1)} + \left(\frac{N}{N-1}\right)^{2(n-1)}\right]$$

or approximately

$$q_N = N^{2(1-n)}q_0 \tag{17}$$

It seems quite reasonable that the rate of divergence of q_N depends on n as shown in (17), but it is, of course, questionable to base any conclusions on the behavior of the divergent series for $\partial\theta/\partial u|_{u=1,y=0}$.

If the relationship between Nu and y can be derived for small y, much more conclusive evidence for the behavior of Nu at $y \to 0$ becomes available.

Figure 9-2 shows $J(y)$ for Pe = 0.2. This function is proportional to the total heat transported from the liquid to the wall between $y = 0$ and an arbitrary positive y:

$$J(y) = -4 \int_0^y \left.\frac{\partial \theta}{\partial x}\right|_{x=1} dy' \tag{18}$$

J is derived by integration of the collocation series for $\partial\theta/\partial x|_{x=1}$, which is given by the two terms in the numerator of (6).

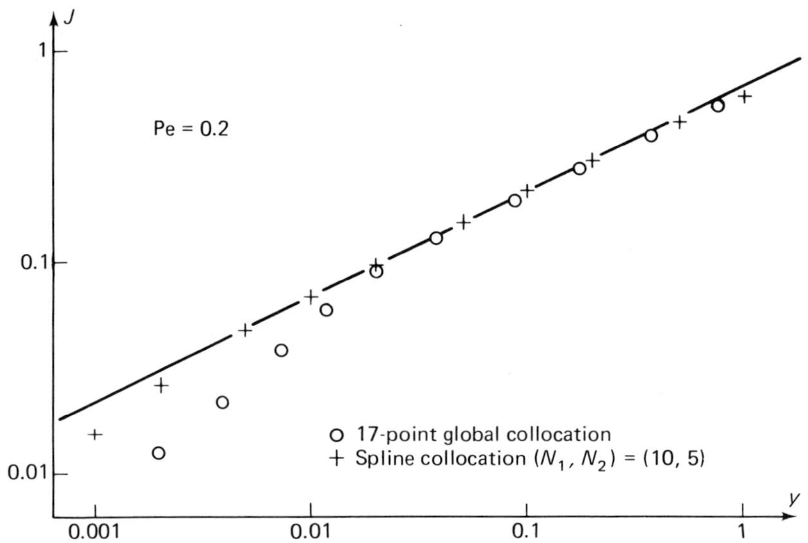

Figure 9-2. Total heat transfer from 0 to y.

The circled points mark results of a 17-point collocation process. In a region $0.02 < y < 0.2$, the values of J are clearly proportional to $y^{1/2}$. For $y \lesssim 0.005$, J is proportional to y, while $J \sim 1$ for $y \gtrsim 1$.

The last part of the curve is simply explained: For $y > 1$, the liquid has almost reached its final temperature, $\partial\theta/\partial x|_{x=1} \sim 0$ and no further increase in the upper limit of the integral in (18) leads to any increase in J. For $y < 0.005$, the 17-point collocation process yields a stationary value for $\partial\theta/\partial x|_{x=1}$ just as in table 9.1 for Pe = 10, and J becomes proportional to y. Clearly this result is due to the inadequacy of the finite-order polynomial approximation close to $y = 0$. The middle region where J is proportional to $y^{1/2}$ is significant, however, and may be used to approximate the behavior of J close to $y = 0$.

A confirmation of this last result is obtained by spline collocation. As discussed in chapter 7 the very rapid increase of θ from 0 at $x = 1$ to some value between 0 and 1 for $x = 1 - \varepsilon$ and y positive but close to 0 can be satisfactorily approximated if many collocation points are located close to $x = 1$.

Table 9.2 shows some computed values of J using a spline-point abscissa $u_s = 0.95$. $N_2 = 5$ or 6 points in $0 < u < 0.95$ give the same results except in a narrow zone around $y = 0.1$ where a five-term collocation series is barely enough to represent the solution. $N_1 = 9$ points in $0.95 < u < 1$ is clearly not enough to obtain convergent results when $y < 0.02$. For $N_1 = 10$ and $y < 0.0002$, J is proportional to y; but for $0.005 < y < 0.2$ the spline collocation values of J fall nicely on a straight line with slope $\frac{1}{2}$ in the log–log plot of figure 9.2. It is evident from the figure that spline collocation succeeds to widen the y-range by a factor 10 where the $J \propto y^{1/2}$ relationship is confirmed as compared to 17-point global collocation.

TABLE 9.2
COMPUTATION OF $J(y)$ FOR Pe = 0.2 BY SPLINE COLLOCATION.
$u_s = x_s^2 = 0.95$
N_1 POINTS IN $[u_s, 1]$ AND N_2 POINTS IN $[0, u_s]$

	$J(y) \cdot 10$ for (N_1, N_2)			
$y \cdot 10^5$	9, 5	10, 5	9, 6	10, 6
1	0.00166	0.00184	0.00166	
2	0.00332	0.00367	0.00332	
5	0.00827	0.00912	0.00827	
10	0.01641	0.01807	0.01642	
20	0.03232	0.03545	0.03232	0.03545
50	0.07715	0.08383	0.07715	0.08383
100	0.1434	0.1537	0.1434	0.1537
200	0.2513	0.2635	0.2514	0.2635
500	0.4637	0.4706	0.4637	0.4706
1000	0.6779	0.6809	0.6780	0.6811
2000	0.9686	0.9710	0.9690	0.9713
5000	1.534	1.536	1.536	
10,000	2.155	2.156	2.159	
20,000	3.005	3.005	3.011	
50,000	4.577	4.577	4.579	
100,000	6.099	6.099	6.099	

Hsu (1972), p. 2197, notes that "unsatisfactory results" are obtained for $Pe < 1.5$ due to a very slow increase of the numerical value of the negative Fourier eigenvalues λ_+ of (9). Michelsen and Villadsen (1974) reduced the data of Hsu's table 1 to obtain the Fourier eigenvalues λ_+ as

$$-\lambda_+ = \text{Pe}\,(\pi N - \text{constant})$$

with less than 0.005% error for $Pe < 1.5$ and $N =$ the eigenvalue number ≥ 4.

The global collocation eigenvalues are also proportional to Pe for $0.2 < \text{Pe} < 300$ and $N \gtrsim 8$, but they increase much more rapidly with N. They are approximated by

$$-\lambda_+ = 1.272\,\text{Pe}\,(N + 0.5139)^2$$

The very slow increase of λ_+ makes it virtually impossible to obtain convergence of the Fourier series for Pe = 0.2. Hsu's solution breaks down for $y \sim 0.05$ and Nu [figure 3 of Hsu (1972)] becomes constant for $y < 0.005$. In contrast the separation of collocation eigenvalues is sufficient to make a 17-point global collocation solution for Pe = 0.2 tolerably accurate for $y > 0.02$ and this makes it possible to detect the $J \propto y^{1/2}$ relation that only holds for $y < 0.1$. Spline collocation with 15 appropriately placed collocation points is even better. It is used here as in chapter 7 to obtain hard-to-derive solutions by numerical computation on an exponential series that in its ideal (Fourier) form is completely unsuited for the purpose.

For Pe $\rightarrow \infty$, J is given by the well-known Levèque solution, which was extended to three terms by Newman (1969) (see Exercise 4.9):

$$J(\text{Pe} \rightarrow \infty) = 4.0698y^{2/3} - 2.4y - 0.4454y^{4/3} + \mathcal{O}(y^{5/3}) \qquad (19)$$

He also considered large but finite Pe and found by a similar perturbation analysis inward (toward $y = 0$) from the Levèque solution that the final behavior of $J(y)$ for $y \rightarrow 0$ is

$$J(y) = 4.24\,\text{Pe}^{-1/4}y^{1/2} \quad \text{for Pe} \gg 1 \qquad (20)$$

Figure 9-3 and table 9.3 present numerical confirmation of (19) and (20) for Pe = 256 using spline collocation with $u_s = 0.98$, $N_1 = 10$, and $N_2 = 6$. Practically the same results are obtained when $u_s = 0.99$.

The two limiting solutions are easily separated on figure 9-3. For $0.00002 < y < 0.0002$, the points fall on the curve described by (20) with Pe = 256; for $0.002 < y < 0.05$, the Levèque solution (19) is seen to represent $J(y)$ extremely well.

The Pe-interval in which both limiting solutions can be confirmed by collocation is quite narrow. The Levèque solution (19) is located as shown in figure 9-3 for any value of Pe. Newman's solution (20) is shifted to the right (left) when Pe is increased (decreased). Thus when Pe is increased by a factor 4 from 256 to 1024, the point of intersection between the two limiting solutions is at $y \sim 10^{-4}$ and the collocation solution becomes too inaccurate to confirm (20). For Pe = 64 the intersection is at $y = 0.0065$ and Newman's solution will merge directly with the saturation solution for $y \gtrsim 0.1$.

Further results on the small y behavior of J are derived by collocation in Michelsen and Villadsen (1974) and collected in their figure 5.

TABLE 9.3

$J(y)$ COMPARED WITH TWO PENETRATION SOLUTIONS

Pe = 256, u_s = 0.98, N_1 = 10, N_2 = 6

y	Collocation	Equation (20)	Equation (19)
0.00001	0.00324	0.0035	
0.00002	0.00461	0.0047	
0.00005	0.00736	0.0075	0.0054
0.0001	0.0106	0.0106	0.0085
0.0002	0.0156	0.0150	0.0134
0.0005	0.0265	0.0237	0.0244
0.001	0.0403	0.0335	0.0383
0.002	0.0614	0.047	0.0600
0.005	0.1077		0.1066
0.01	0.1648		0.1639
0.02	0.2496		0.2494
0.05	0.4217		0.4242
0.1	0.6050		0.6161
0.2	0.8104		
0.5	0.9789		
1.	0.9994		

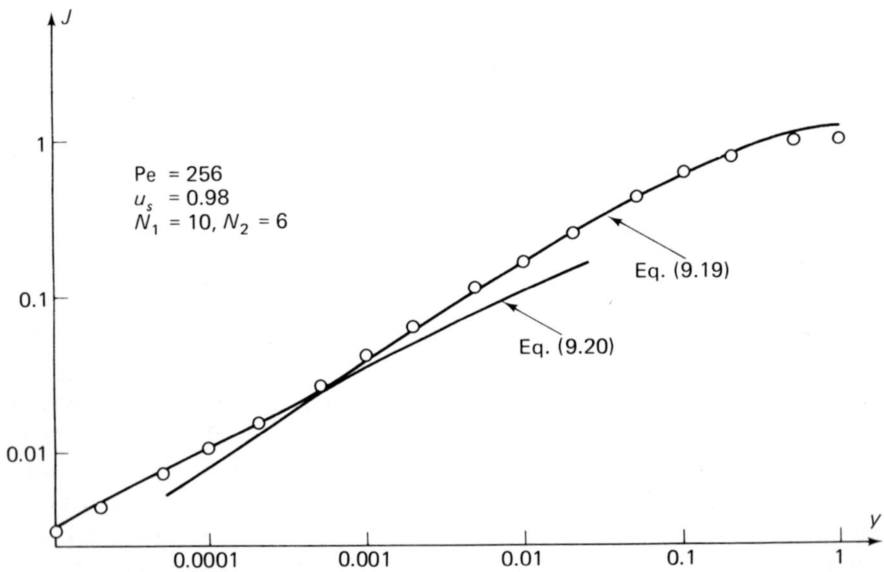

Figure 9-3. Spline collocation solution for J compared with two limiting solutions.

Newman's solution appears to be applicable even at Pe = 30 where the Levèque solution has disappeared. For decreasing Pe, the dependence of J upon Pe changes: In an intermediate region $3 < \text{Pe} < 20$, J is almost independent of Pe ($J \sim 1.7y^{1/2}$) while J is proportional to $\text{Pe}^{1/2}$ for $\text{Pe} < 0.5$ [$J = 1.5(\text{Pe}\, y)^{1/2} = 1.5(z/R)^{1/2}$].

9.1.3 Perturbation solutions for the far downstream Nu at large and small Pe

In practice, the Peclet number will always be fairly large, e.g., >20, and the contribution of axial conduction to the total heat transfer is small compared to the contribution from convection. Thus for practical purposes a relation between the far downstream Nusselt number and the Peclet number, Nu_{as} (Pe), for large Pe is sufficient. To obtain Nu_{as} for any given Pe the collocation procedure with diagonalization of $\mathbf{Q}$ must be carried through with a sufficiently high N to obtain an accurate representation of the first term in the Fourier series.

This is a complicated process, and it appears much more satisfactory to obtain Nu_{as} (Pe) as a perturbation series based on the solution of the simple Graetz problem with $\text{Pe} = \infty$ for which we know that $\text{Nu}_{as} = 3.657$. In the following this perturbation series is derived. For completeness we also derive a perturbation solution from another extreme case Pe = 0, and finally we come to the conclusion that Nu_{as} (Pe) can be expressed by a combination of the two series for any Pe. Thus (1) need not be solved at all to obtain the important quantity Nu_{as} (Pe). It is sufficient to solve two Sturm–Liouville problems [(1) with $\text{Pe}=\infty$ or 0] and manipulate these solutions.

When the Fourier series solution for θ is inserted into (6), we obtain

$$\text{Nu}_{as} = -\frac{dF_1/dx|_{x=1}}{2\int_0^1 x(1-x^2)F_1\,dx} \tag{21}$$

since only the first term in the exponential series is needed when $y \to \infty$. In (21), $[\lambda_1, F_1]$ are the first eigenvalue and eigenfunction of the set $[\lambda_+, F_+]$ of (9). Thus all we need are expressions for $F_1(x, \text{Pe})$ and $\lambda_1(\text{Pe})$.

Let us first consider the general case where a Sturm–Liouville problem is perturbed by $g(x, \lambda)F$:

$$\frac{d}{dx}\left[p(x)\frac{dF}{dx}\right] - [q(x) + \lambda r(x)]F + \alpha g(x, \lambda)F = 0 \tag{22}$$

p, q, and r are the general functions that appear in Sturm–Liouville problems and α is a small constant.

We call the eigenvalues and eigenfunctions of the unperturbed problem ($\alpha = 0$) Λ_i and $Y_i(x)$. These are supposed to be known for $i = 1, 2, \ldots$.

The first eigenfunction $F_1(x)$ and the first eigenvalue λ_1 of (22) with $\alpha \neq 0$ are now expanded in power series in the perturbation parameter α:

$$F_1(x) = \sum_{i=0}^{\infty} \alpha^i f_i(x) \quad \text{and} \quad \lambda_1 = \sum_{i=0}^{\infty} \alpha^i s_i \tag{23}$$

$f_0(x) = Y_1$ and $s_0 = \Lambda_1$.

When (23) is inserted into (22) and terms in α^M are collected, one obtains

$$\frac{d}{dx}\left(p\frac{df_M}{dx}\right) - (q + \Lambda_1 r)f_M - s_M r Y_1 + h_M(x) = 0 \tag{24}$$

where

$$h_M(x) = h_M(x, s_0, s_1, \ldots, s_{M-1}, f_0, f_1, \ldots, f_{M-1}) \tag{25}$$

is a known function when the perturbation functions f_j and the perturbation constants s_j with $j < M$ have been determined. Equation (24) is consequently a recurrence formula for f_M.

Next we expand f_M on the set Y_j, $j = 2, 3, \ldots, \infty$ (f_i is a linearly independent set and since $f_0 = Y_1$, no further contribution from Y_1 is found for $i > 0$).

$$f_M = \sum_2^{\infty} c_{j,M} Y_j \tag{26}$$

Equation (26) is inserted into (24) using (22) with $\alpha = 0$ to eliminate derivatives.

$$\sum_2^{\infty} c_{j,M}(\Lambda_j - \Lambda_1) r Y_j - s_M r Y_1 + h_M(x) = 0 \tag{27}$$

Equation (27) is multiplied by Y_1 and integrated over the interval of orthogonality {e.g., [0, 1]} to give

$$s_M = \int_0^1 Y_1 h_M \, dx \quad \text{if } \langle r Y_j Y_K \rangle = \delta_{jK} \tag{28}$$

Similarly, we multiply (27) by Y_K and integrate to obtain $c_{K,M}$:

$$(\Lambda_K - \Lambda_1)c_{K,M} + \int_0^1 Y_K h_M \, dx = 0 \tag{29}$$

Thus for any M the Mth perturbation constant s_M is calculated from (28) while $c_{K,M}$ can be calculated for $K = 2, 3, \ldots$ from (29), which does not depend on s_M.

For our present problem with Pe → ∞, by comparison of (8) and (22) we obtain

$$p(x) = x, \qquad q(x) = 0, \qquad r(x) = x(1 - x^2),$$
$$\alpha = \frac{1}{\text{Pe}^2}, \quad \text{and} \quad g(x, \lambda) = \lambda^2 x$$

Furthermore Y_1 is "the first Graetz function," which is available in the literature in form of a power series and $\Lambda_1 = -7.31359$.

When the zero-order representation $\lambda_1 = \Lambda_1$ and $F_1 = Y_1$ is inserted in $g(x, \lambda_1)F_1$, one obtains $h_1 = \Lambda_1^2 x Y_1$ and

$$s_1 = \int_0^1 h_1 Y_1 \, dx = \Lambda_1^2 \int_0^1 x Y_1^2 \, dx = 66.926$$

Consequently,

$$\lambda_1 \sim -7.31359 + 66.926 \, \text{Pe}^{-2} + \mathcal{O}(\text{Pe}^{-4}) \tag{30}$$

To obtain the first perturbation function f_1 to $Y_1(x)$, we have to determine

$$c_{K,1} = \Lambda_1^2 \frac{\int_0^1 x Y_1 Y_K \, dx}{\Lambda_1 - \Lambda_K} \quad \text{for } K = 2, 3, \ldots, \infty \tag{31}$$

Power series for higher eigenfunctions Y_K ($K > 1$) of the unperturbed Graetz problem are also available and using (31) the coefficients of f_1 in the series (26) can be determined. Finally (23) with only two terms in the series for F_1 yields

$$F_1(x) \sim Y_1(x) + \frac{1}{\text{Pe}^2} f_1(x) \tag{32}$$

$F_1(x)$ is differentiated and inserted with (30) into (21) to yield

$$\text{Nu}_{as} \sim -\frac{\Lambda_1}{2} + \frac{4.487}{\text{Pe}^2} = 3.6579 + \frac{4.487}{\text{Pe}^2} \tag{33}$$

The process leading to (33) via power series for $Y_K(x)$ is quite complicated and objectionable from a numerical point of view: F_1 is given by an infinite series in (23). Each of the terms $f_i(x)$ is given by another infinite series (26) and the coefficients of this last series are found by a complicated manipulation on power series in (31). The series (26) may be expected to converge reasonably fast but even so the differentiation of F_1 in (21) may necessitate that more than one term f_1 is retained in (23) to find an accurate value of Nu_{as}.

The perturbation process as described above is thus deemed to be unsound from a numerical standpoint and too complicated to use except for the lowest-order approximation.

Next we present a collocation analog of the method which works whether a series solution for Y_K is known or not and which requires almost no additional computational work besides solving the basic problem (22) with $\alpha = 0$ by collocation.

Consider the Nth-order collocation analog of the unperturbed Graetz problem:

$$\mathbf{CY} = \mathbf{\Lambda Y} \tag{34}$$

$\mathbf{C}$ is the discretization matrix for $[(1 - x^2)x]^{-1} d/dx(x\, d/dx)$, $\mathbf{\Lambda}$ is a diagonal matrix of the collocation eigenvalues, and we know from chapter 4 that the first $N/2$ eigenvalues (and especially the eigenvalues of smallest modulus) are very nearly the same as the Fourier eigenvalues when N collocation points are used.

Using the method of section 4.3 we obtain the eigenvector $\mathbf{Y}_i$ and the eigenrow $\mathbf{U}_i$ for each eigenvalue Λ_i. Similarly, for (22),

$$\mathbf{CF} + \alpha\mathbf{G}(\boldsymbol{\lambda})\mathbf{F} = \boldsymbol{\lambda}\mathbf{F} \tag{35}$$

Expand λ_1 and F_1 as in (23) and collect terms in α^M:

$$\mathbf{Cf}_M + \mathbf{h}_M = \Lambda_1\mathbf{f}_M + s_M\mathbf{Y}_1 \tag{36}$$

expand $\mathbf{f}_M$ on $\mathbf{Y}_K$, $K = 2, 3, \ldots, N$,

$$\mathbf{f}_M = \sum_2^N c_{j,M}\mathbf{Y}_j \tag{37}$$

and insert in (36):

$$\sum_2^N c_{j,M}(\Lambda_j - \Lambda_1)\mathbf{Y}_j + \mathbf{h}_M = s_M\mathbf{Y}_1 \tag{38}$$

Multiply by $\mathbf{U}_1^T$ or by $\mathbf{U}_K^T$ $(K > 1)$ to obtain

$$s_M = \mathbf{U}_1^T\mathbf{h}_M \tag{39}$$

$$(\Lambda_K - \Lambda_1)c_{K,M} = -\mathbf{U}_K^T\mathbf{h}_M \tag{40}$$

$\mathbf{h}_M$ is a known vector of values of $h_M(x)$ at the collocation points. $\mathbf{U}_K$, $K = 1, 2, \ldots, N$, is available when the unperturbed problem has been solved for a given N. Thus in a stepwise process s_M, $c_{K,M}$ $(K = 2, 3, \ldots, N)$, and $\mathbf{h}_{M+1}$ are computed simply as scalar products. The process can be extended to any M.

The values of s_i depend on the approximation order N of the unperturbed problem and it may be expected that the s_i converge simultaneously with the Λ_i to constant values when N is increased. Thus at the cost of computing a few matrix-vector or scalar products the solution F_1, λ_1 to (22) can be obtained simultaneously with Y_i, Λ_i.

Tables 9.4 and 9.5 show how well the process converges.

TABLE 9.4
COEFFICIENTS s_i IN $\lambda_1 = \sum_0^\infty (\mathrm{Pe}^{-2})^i s_i$ BY COLLOCATION

N	$s_0 = \Lambda_1$	s_1	s_2	$s_{11} \cdot 10^{-17}$
3	−7.3142	66.95	−1220	1.498
4	−7.31359	66.9265	−1218.65	1.4893
6	−7.31359	66.9262	−1218.64	1.48917
8	−7.31359	66.9263	−1218.64	1.48917

TABLE 9.5
COEFFICIENTS n_i IN $\mathrm{Nu}_{as} = \sum_0^\infty (Pe^{-2})^i n_i$

N	$n_o = (\Lambda_1/2)$	n_1	n_2	$n_{11} \cdot 10^{-15}$
3	3.657	4.4816	−78.87	0.8998
4	3.65679	4.48724	−78.9786	0.89812
6	3.65679	4.48731	−78.9802	0.89811
8	3.65679	4.48731	−78.9804	0.89812

It is interesting that s_i and n_i with a large index i that are almost impossible to obtain by algebraic manipulations on the power series of the exact eigenfunctions converge as fast as s_0 and n_0 to constant values.

A perturbation series in increasing powers of Pe is developed by the same technique: Substitute $\lambda = \mathrm{Pe}\,\lambda_*$ in (8) to obtain

$$\frac{d}{dx}\left(x\frac{dF}{dx}\right) + \lambda_*^2 xF - \lambda_* \,\mathrm{Pe}\, x(1 - x^2)F = 0 \tag{41}$$

A comparison with (22) shows that

$$p(x) = -r(x) = x, \qquad \alpha = \mathrm{Pe}, \qquad g(x, \lambda) = -\lambda_* x(1 - x^2)$$

The eigenfunctions Y_i for $\mathrm{Pe} = 0$ are Bessel functions $J_0(\Lambda_i x)$ and the corresponding eigenvalues $\Lambda_1, \Lambda_2, \ldots$ are negative zeros of $J_0(\rho)$ or $-2.4048256, -5.52, \ldots$, etc.

Michelsen and Villadsen (1974) used the properties of Bessel functions to obtain the value of Nu_{as} for $\mathrm{Pe} = 0$. Their result was $n_0 = \Lambda_1^4/8 = 4.180654$, but higher-order perturbations are as difficult to find analytically as in the preceding perturbation series from $\mathrm{Pe} \to \infty$.

The collocation analog (35) to (40) is easily applied, however, and tables 9.6 and 9.7 show perturbation constants in

$$\lambda_{1*} = \sum_0^\infty s_i \alpha^i = \sum_0^\infty s_i \,\mathrm{Pe}^i \tag{42}$$

$$\mathrm{Nu}_{as} = \sum_0^\infty n_i \,\mathrm{Pe}^i \tag{43}$$

TABLE 9.6
COEFFICIENTS IN $\lambda_{1*} = \lambda_1/\mathrm{Pe} = \sum_0^\infty s_i \,\mathrm{Pe}^i$ BY COLLOCATION

N	$s_0 = \Lambda_1$	s_1	s_2	$s_{11}\cdot 10^{12}$
3	−2.404826	0.390969	−0.30285	−0.87
4	−2.404826	0.390972	−0.30280	−0.508
6	−2.404826	0.390972	−0.30280	−0.49678
8	−2.404826	0.390972	−0.30280	−0.49678

TABLE 9.7
COEFFICIENTS IN $\mathrm{Nu}_{as} = \sum_0^\infty n_i \,\mathrm{Pe}^i$

N	n_0	n_1	n_2	$n_{11}\cdot 10^{10}$
3	4.18065	−0.18354	0.03368	−0.65
4	4.18065	−0.183554	0.033645	−0.6263
6	4.18065	−0.183554	0.033645	−0.62748
8	4.18065	−0.183554	0.033645	−0.62748

There is no way to determine in advance the radius of convergence for a perturbation series. The ratio n_i/n_{i-1} is, however, empirically found to approach a constant value (~33 for table 9.5 and ~0.15 for table 9.7) even for $i = 3$–5. Thus the series of table 9.5 is assumed to be convergent for $\mathrm{Pe} \geq 5.5$, while the series of table 9.7 is assumed to be convergent for $\mathrm{Pe} \leq 6.5$. For $\mathrm{Pe} = 6$, both series can be used to predict Nu_{as} to four-digit accuracy.

The overlap between the range of applicability of the two perturbation series makes it unnecessary to integrate the full equation (1) even once to obtain Nu_{as} (Pe). All that is needed is a collocation solution for the two limiting Sturm–Liouville problems $\mathrm{Pe} = \infty$ and $\mathrm{Pe} = 0$ combined with one of the two perturbation series.

In general the numerical situation may not be as favorable as in the present example, but the technique can always be used close to limiting solutions and it hardly requires any additional computational work besides solving the basic Sturm–Liouville problems.

9.2 Asymptotic Stability of a Catalyst Particle

In section 1.3 and again in subsection 6.2.2 we have studied a model that describes the concentration and temperature inside a catalyst particle as functions of time t. Only small deviations $(\hat{y}, \hat{\theta})$ from a steady state (y_{ss}, θ_{ss}) were considered and the rate expression $R(y, \theta)$ was linearized around (y_{ss}, θ_{ss}).

The two dependent variables are determined from the coupled linear partial differential equations (1.82):

$$\frac{\partial \hat{y}}{\partial t} = \nabla^2 \hat{y} - R_y \hat{y} - R_\theta \hat{\theta} \tag{44}$$

$$\mathrm{Le}\frac{\partial \hat{\theta}}{\partial t} = \nabla^2 \hat{\theta} + \beta(R_y \hat{y} + R_\theta \hat{\theta}) \tag{45}$$

Initial conditions are $\hat{y} = \hat{y}_0$ and $\hat{\theta} = \hat{\theta}_0$ (both close to 0), while the boundary conditions at $x = 0$ and at the particle surface $x = 1$ are chosen as in chapter 6:

$$\frac{\partial \hat{\theta}}{\partial x} = \frac{\partial \hat{y}}{\partial x} = 0 \quad \text{at } x = 0 \tag{46}$$

$$\hat{\theta} = \hat{y} = 0 \quad \text{at } x = 1 \tag{47}$$

We substitute $\hat{y}(t, x) = \exp(\lambda t) y(x)$ and $\hat{\theta}(t, x) = \exp(\lambda t)\theta(x)$ into (44) and (45) to obtain the following eigenvalue problem:

$$\lambda y = \nabla^2 y - R_y y - R_\theta \theta \tag{48}$$

$$\mathrm{Le}\,\lambda\theta = \nabla^2\theta + \beta(R_y y + R_\theta \theta) \tag{49}$$

with the homogeneous boundary conditions (46) and (47) for (y, θ).

A given steady state (y_{ss}, θ_{ss}) is asymptotically stable if all eigenvalues λ of (48) and (49) have negative real part.

In the following we make a detailed analysis of the eigenvalues of (48) and (49) as functions of the capacitance parameter Le. The critical value of Le for which occurrence of eigenvalues with positive real part is first observed is of special interest. We note that an estimate of Le_{cr} has already been found in subsection 6.2.2 by one-point collocation. Here much more accurate numerical methods for determining the eigenvalues are presented: A combination of Galerkin's method and high-order collocation, forward integration, and perturbation methods from limiting solutions.

We recognize these methods from chapter 4 and also from the example in section 9.1. In fact the linear problem (48) and (49) is not different in principle from the extended Graetz problem of section 9.1.

Neither is a Sturm–Liouville problem except for exceptional values of the parameter that are investigated [Pe in section 9.1 and Le in (48) and (49)] and the eigenfunctions cannot be found analytically.

The present problem is more difficult in several aspects, however: the partial derivatives R_y and R_θ are complicated functions of x. They are calculated from the steady state solution that is given by a nonlinear equation except for trivial rate expressions. The steady state solution depends on at least three parameter groups (Φ, β, and γ for an Nth-order irreversible reaction) and in the analysis of (44) and (45) these parameters appear besides the dynamic parameter Le. The large number of parameters makes it a difficult task to obtain a clear picture of how the eigenvalues depend on Le.

9.2.1 Solution for Le=1

For one specific value Le = 1, (48) and (49) is easily reduced to a Sturm–Liouville problem. Multiply (48) by β and add to (49) to obtain

$$\lambda(\beta y + \theta) = \nabla^2(\beta y + \theta) \tag{50}$$

or

$$\lambda z = \nabla^2 z \quad \text{with } z = \beta y + \theta$$

with boundary conditions $z = 0$ at $x = 1$ and $dz/dx = 0$ at $x = 0$. For spherical geometry the eigenfunctions and eigenvalues are

$$z_k(x) = \frac{\sin k\pi x}{x}, \qquad \lambda_k = -k^2\pi^2, \qquad k = 1, 2, \ldots \tag{51}$$

Equation (51) represents one set of solutions to (48) and (49) for Le = 1. It can never give rise to instability since all eigenvalues are negative. The other solution is obtained when $z = 0$—the trivial solution of (50)—is inserted in (48) or (49):

$$\lambda y = \nabla^2 y - R_y y + \beta R_\theta y \tag{52}$$

with $y(1) = 0$ and $dy/dx = 0$ at $x = 0$.

Equation (52) is the same as (5.82) except that the Thiele modulus has now been incorporated into the rate expression R. In (5.82) θ has been eliminated from R using the relation $\theta + \beta y = 0$, which holds for all values of x when Le = 1. In (52) both derivatives are formally present but

$$\left(\frac{\partial R(y, x)}{\partial y}\right)_x\bigg|_{y_s} = \left(\frac{\partial R(y, \theta, x)}{\partial \theta}\right)_{x,y}\bigg|_{y_s,\theta_s}\frac{\partial \theta}{\partial y} + \left(\frac{\partial R(y, \theta, x)}{\partial y}\right)_{x,\theta}\bigg|_{y_s,\theta_s} \tag{53}$$

$$= -\beta R_\theta + R_y \quad \text{when } \theta + \beta y = 0$$

The eigenfunctions and eigenvalues of (52) may be found one by one by integration of (52) from $x = 0$ to $x = 1$ with $y(0) = 1$ and $y^{(1)}(0) = 0$ and iterating on λ until $y(1) = 0$. From the discussion in subsection 5.5.6 we know that all eigenvalues are negative if (5.85) or (52) with zero on the left-hand side has a solution $y(x)$ that in this integration process remains positive for all $x \in [0, 1]$.

Unfortunately the results for Le = 1 yield little information about the behavior of (48) and (49) for Le $\neq$ 1. One important result has been proved, however, for any geometry and for any rate expression:

1. If (52) and (53) have an odd number of positive eigenvalues for Le = 1, the investigated steady state is unstable for any other value of Le.
   Furthermore all available computational experience points to the validity of the following theorems.
2. If (52) and (53) have any number (even or odd) of positive eigenvalues at Le = 1, the investigated steady state is unstable for any value of Le.
3. When the rate constant is described by an Arrhenius expression with a positive product $\gamma\beta$ and the reaction order is positive, then a given steady state is unstable for any value of Le in $[0, \text{Le}_{cr}]$ if it is unstable for Le = Le_{cr}.
4. As a possible generalization of No. 3, there are no disconnected Le regions of instability and vice versa; there are no disconnected Le-regions of stability; or an eventual Le_{cr} is unique.

9.2.2 Eigenvalues for Le $\neq$ 1 using Galerkin's method and trigonometric trial functions

Luss and Lee (1970) computed eigenvalues of (48) and (49) for a first-order irreversible reaction:

$$R_y = \Phi^2 \exp\left[\gamma\left(1 - \frac{1}{\theta_{ss}}\right)\right] \tag{54}$$

$$R_\theta = \Phi^2 y_{ss} \frac{\gamma}{\theta_{ss}^2} \exp\left[\gamma\left(1 - \frac{1}{\theta_{ss}}\right)\right] \tag{55}$$

They chose $(\gamma, \beta, \Phi) = (30, 0.15, 1.1)$, for which the steady state solution (y_{ss}, θ_{ss}) is unique. Their computations were based on Galerkin's method with trial functions equal to $z_k(x)$, the solution of (50). Each of the trial functions satisfies the homogeneous boundary conditions (46) and (47), and each of the trial solutions (56) and (57) is a consequently valid representation for y and θ.

$$y = \sum_1^N a_k \frac{\sin k\pi x}{x} \tag{56}$$

$$\theta = \sum_1^N b_k \frac{\sin k\pi x}{x} \tag{57}$$

Equations (56) and (57) are inserted into (48) and (49) to yield the following residuals:

$$\sum_1^N \{a_k[(-k^2\pi^2 - \lambda) - R_y] - b_k R_\theta\} \frac{\sin k\pi x}{x} \tag{58}$$

$$\sum_1^N \left[a_k \frac{\beta}{\text{Le}} R_y + b_k \left(-\frac{k^2\pi^2}{\text{Le}} + \frac{\beta}{\text{Le}} R_\theta - \lambda \right) \right] \frac{\sin k\pi x}{x} \tag{59}$$

Each residual (58) or (59) is made orthogonal on the N trial functions $(\sin j\pi x)/x$, $j = 1, 2, \ldots, N$ over the system volume. There results a $2N \times 2N$ eigenvalue problem

$$\begin{pmatrix} \mathbf{S}_1 & \mathbf{S}_2 \\ \dfrac{1}{\text{Le}}\mathbf{S}_3 & \dfrac{1}{\text{Le}}\mathbf{S}_4 \end{pmatrix} \begin{pmatrix} \mathbf{a} \\ \mathbf{b} \end{pmatrix} = \frac{1}{2}\lambda \begin{pmatrix} \mathbf{a} \\ \mathbf{b} \end{pmatrix} \tag{60}$$

where the $N \times N$ matrices $\mathbf{S}_1$, $\mathbf{S}_2$, $\mathbf{S}_3$, and $\mathbf{S}_4$ are given by

$$\begin{aligned} (S_1)_{kj} &= (S_1)_{jk} = -\tfrac{1}{2}k^2\pi^2\delta_{kj} - \int_0^1 \sin k\pi x \sin j\pi x R_y \, dx \\ (S_2)_{kj} &= (S_2)_{jk} = -\int_0^1 \sin k\pi x \sin j\pi x R_\theta \, dx \\ (S_3)_{kj} &= (S_3)_{jk} = \beta \int_0^1 \sin k\pi x \sin j\pi x R_y \, dx \\ (S_4)_{kj} &= (S_4)_{jk} = -\tfrac{1}{2}k^2\pi^2\delta_{kj} + \beta \int_0^1 \sin k\pi x \sin j\pi x R_\theta \, dx \end{aligned} \tag{61}$$

Each of the integrals in (61) can be evaluated by quadrature from the steady state profiles (y_{ss}, θ_{ss}). Once the elements of $\mathbf{S}_1$ to $\mathbf{S}_4$ have been calculated for a given N, they can be used for any higher N—the extra rows and columns are filled up by supplementary computation of integrals.

For the case $(\gamma, \beta, \Phi) = (30, 0.15, 1.1)$, Luss and Lee obtained satisfactory results with $N = 7$. The system was found to be stable for $\text{Le} > 0.4$ and unstable for smaller values of Le. Some numerical difficulties were reported for small values of Le ($\text{Le} < 0.1$).

Luss (1974) has also investigated the case $(\gamma, \beta, \Phi) = (40, 0.5, 0.264)$ for which five steady states exist. The middle one is known to give two positive eigenvalues of (52) and Luss found that they remained positive even for very large Le (the suggested Theorem 2 in subsection 9.2.1). In the present context it is noteworthy that he had to use 38 terms in the expansions (56) and (57) and thus had to solve a (76×76) eigenvalue problem for each investigated value of Le.

It is to be expected from the discussion in chapter 4 that a collocation analog of this procedure would give approximately the same accuracy for about the same work—perhaps with the small advantage that the integrals in (61) need not be computed in the collocation method, but by far the heaviest work lies in solving the high-order eigenvalue problem.

The $(2N \times 2N)$ collocation eigenvalue problem is

$$\begin{pmatrix} \mathbf{C} - \mathbf{R}_y & -\mathbf{R}_\theta \\ \dfrac{1}{\text{Le}}\mathbf{R}_y & \dfrac{1}{\text{Le}}(\mathbf{C} + \beta\mathbf{R}_\theta) \end{pmatrix} \begin{pmatrix} \mathbf{y} \\ \boldsymbol{\theta} \end{pmatrix} = \lambda \begin{pmatrix} \mathbf{y} \\ \boldsymbol{\theta} \end{pmatrix} \tag{62}$$

$\mathbf{C}$ is the discretization matrix for the Laplacian, $(\mathbf{y}, \boldsymbol{\theta})$ is the $2N$ vector of collocation ordinates, and $\mathbf{R}_y$ and $\mathbf{R}_\theta$ are diagonal matrices.

We were able to pursue the $(\gamma, \beta, \Phi) = (30, 0.15, 1.1)$ example for $N = 7$ to at least $\text{Le} = 10^{-6}$ without any numerical difficulties. The reason is probably not that the collocation method is better than Galerkin's method but rather that the *QR* algorithm of our EISYS program (Appendix A4) is better than the one used by Luss and Lee.

9.2.3 Expansion on two sets of trial functions

In either the Galerkin method or its collocation analog the bulk of computational work lies in solving a large-order eigenvalue problem. An obvious improvement is to include both sets of eigenfunctions from (52) and from (50) in the trial functions—the sine series from (50) are apparently very slowly convergent in some cases.

Let the eigenfunctions and eigenvalues of (52) and (50) be, respectively, $[v_i(x), \lambda_i^{(v)}]$ and $[z_i(x), \lambda_i^{(z)}]$. Expand (y, θ) on the two series:

$$y = \sum_{i=1}^{N} a_i z_i + \sum_{i=1}^{N} b_i v_i \tag{63}$$

$$\theta = \sum_{i=1}^{N} c_i z_i + \sum_{i=1}^{N} d_i v_i \tag{64}$$

These are substituted into (48) and (49) and derivatives are eliminated using the definition of z_i and v_i:

$$\nabla^2 z_i = \lambda_i^{(z)} z_i \quad \text{and} \quad \nabla^2 v_i = [\lambda_i^{(v)} + R_y - \beta R_\theta] v_i$$

Next the residuals are orthogonalized over the system volume on each of the N first z_j and v_i trial functions. In this way nine ($N \times N$) matrices appear:

$$\begin{aligned}
A_{ij} &= \int_0^1 x^2 z_i z_j \, dx = \frac{1}{2}\delta_{ij} \\
(A_1)_{ij} &= \int_0^1 x^2 z_i z_j R_y \, dx, \qquad (A_2)_{ij} = \int_0^1 x^2 z_i z_j R_\theta \, dx \\
B_{ij} &= \int_0^1 x^2 v_i v_j \, dx = \frac{1}{2}\delta_{ij} \\
(B_1)_{ij} &= \int_0^1 x^2 v_i v_j R_y \, dx, \qquad (B_2)_{ij} = \int_0^1 x^2 v_i v_j R_\theta \, dx \\
G_{ij} &= \int_0^1 x^2 z_i v_j \, dx \\
(G_1)_{ij} &= \int_0^1 x^2 z_i v_j R_y \, dx, \qquad (G_2)_{ij} = \int_0^1 x^2 z_i v_j R_\theta \, dx
\end{aligned} \tag{65}$$

These matrices are combined into three ($4N \times 4N$) matrices:

$$\mathbf{L} = \begin{Bmatrix} \mathbf{I} & \mathbf{0} & \mathbf{0} & \mathbf{0} \\ \mathbf{0} & \mathbf{I} & \mathbf{0} & \mathbf{0} \\ \mathbf{0} & \mathbf{0} & \mathrm{Le}\,\mathbf{I} & \mathbf{0} \\ \mathbf{0} & \mathbf{0} & \mathbf{0} & \mathrm{Le}\,\mathbf{I} \end{Bmatrix} \tag{66}$$

$$\mathbf{M} = \begin{Bmatrix} \mathbf{A} & \mathbf{G} & \mathbf{0} & \mathbf{0} \\ \mathbf{G}^T & \mathbf{B} & \mathbf{0} & \mathbf{0} \\ \mathbf{0} & \mathbf{0} & \mathbf{A} & \mathbf{G} \\ \mathbf{0} & \mathbf{0} & \mathbf{G}^T & \mathbf{B} \end{Bmatrix} \tag{67}$$

$$\mathbf{R} = \begin{Bmatrix} (\mathbf{A}\mathbf{\Lambda}^z - \mathbf{A}_1) & (\mathbf{G}\mathbf{\Lambda}^v - \beta\mathbf{G}_2) & -\mathbf{A}_2 & -\mathbf{G}_2 \\ (\mathbf{G}^T\mathbf{\Lambda}^z - \mathbf{G}_1^T) & (\mathbf{B}\mathbf{\Lambda}^v - \beta\mathbf{B}_2) & -\mathbf{G}_2^T & -\mathbf{B}_2 \\ \beta\mathbf{A}_1 & \beta\mathbf{G}_1 & (\mathbf{A}\mathbf{\Lambda}^z + \beta\mathbf{A}_2) & (\mathbf{G}\mathbf{\Lambda}^v + \mathbf{G}_1) \\ \beta\mathbf{G}_1^T & \beta\mathbf{B}_1 & (\mathbf{G}^T\mathbf{\Lambda}^z + \beta\mathbf{G}_2^T) & (\mathbf{B}\mathbf{\Lambda}^v + \mathbf{B}_1) \end{Bmatrix} \tag{68}$$

where

$$\mathbf{\Lambda}^z = \begin{pmatrix} \lambda_1^{(z)} & & 0 \\ & \lambda_2^{(z)} & \\ 0 & & \lambda_N^{(z)} \end{pmatrix} = \begin{pmatrix} -\pi^2 & & 0 \\ & -4\pi^2 & \\ 0 & & -N^2\pi^2 \end{pmatrix} \tag{69}$$

$$\mathbf{\Lambda}^v = \begin{pmatrix} \lambda_1^{(v)} & & 0 \\ & \lambda_2^{(v)} & \\ 0 & & \lambda_N^{(v)} \end{pmatrix} \tag{70}$$

The final formulation of our eigenvalue problem is now

$$\mathbf{Q}\begin{pmatrix} \mathbf{a} \\ \mathbf{b} \\ \mathbf{c} \\ \mathbf{d} \end{pmatrix} = \lambda \begin{pmatrix} \mathbf{a} \\ \mathbf{b} \\ \mathbf{c} \\ \mathbf{d} \end{pmatrix} \tag{71}$$

where $\mathbf{Q} = \mathbf{L}^{-1}\mathbf{M}^{-1}\mathbf{R}$.

$\mathbf{M}^{-1}\mathbf{R}$ is independent of Le and it has only to be evaluated for each steady state that is investigated. The premultiplication of $\mathbf{M}^{-1}\mathbf{R}$ by $\mathbf{L}^{-1}$ from (66) implies that for each Le the eigenproblem (71) appears when the last $2N$ rows of $\mathbf{M}^{-1}\mathbf{R}$ are divided by Le.

The present expansion is much more consistent than that used in subsection 9.2.2. Here we utilize not only terms that arise from the diffusion part of the problem—the trigonometric series—but also probably the much more critical terms that represent the chemical reaction. The result is that accurate values of the eigenvalues of (48) and (49) are determined with a much smaller N.

Table 9.8 shows that even with $N = 1$ the two first eigenvalues are determined quite satisfactorily over a wide range of Le-values.

For $N = 2$ the relative error of λ_1 and λ_2 is always $<10^{-5}$; in general, for systems with one or three steady states it was found that $4N - 2$ of the $4N$ computed eigenvalues were determined qualitatively correct.

In the computations we use collocation with $N = 10$ [at the zeros of $P_{10}^{(1,1/2)}(x^2)$] to determine the eigenvalues $\lambda_i^{(v)}$ and eigenfunctions $v_i(x)$ of (52). The integrals of (65) were evaluated by Radau quadrature. The "exact" eigenvalues here and in the following tables are obtained by forward integration as described in subsection 9.2.5.

For higher-approximation order (even at $N = 2$ and 3) the matrix $\mathbf{M}$ in (67) becomes nearly singular, which shows that the higher eigenfunctions z_i and v_j become nearly identical. Figure 5 of Michelsen and Villadsen (1972) (with a considerably greater influence of the reaction

TABLE 9.8
LOWEST-ORDER APPROXIMATION FOR λ_1 AND λ_2
BY THE MODIFIED GALERKIN METHOD
$(\gamma, \beta, \Phi) = (30, 0.15, 1.1)$

	Approximate values		Exact values	
Le	λ_1	λ_2	λ_1	λ_2
0.01	886.8	3.75467	906.442	3.74496
0.50	-2.30524 $+i7.54641$	-2.30524 $-i7.54641$	-2.31344 $+i7.56524$	-2.31344 $-i7.56524$
1	-3.12196	-9.86960	-3.12196	-9.86960
2	-0.948047	-16.1884 $\pm i0.8579$	-0.948035	-15.0428 $\pm i0.6990$
100	-0.0144551	-20.90	-0.0144514	-20.5920

term) illustrates this point: Even for the large $\lambda\beta = 28$ of that example, v_8 is probably indistinguishable from the corresponding function in the z_i-function set: $(\sin 8\pi x)/x$. Fortunately this is of small practical consequence since the similarity of the higher eigenfunctions indicate that accurate results are found with a small N: It is necessary to have the first different looking eigenfunction v_1 of (52) represented in the expansion but the higher eigenfunctions v_i contain much less additional information.

The modified Galerkin method was also used on the five steadystates example of Luss (1974). As might be expected, the $N = 1$ approximation is unable to give qualitatively correct results for the dynamics of the middle steady state in this rather extreme example. For Le = 1, the first and second eigenvalues are determined as, respectively, $\lambda_1 = 1653.3$ and $\lambda_2 = -1.8$. λ_1 is, of course, represented correctly since it is the first eigenvalue of (52), the eigenvalues of which are found exactly by forward integration as discussed below, but λ_2 is of the wrong sign—its correct value is $\lambda_2 = 2.7323$. When $N = 2$ is used, we include in our basis both eigenfunctions of (52) that give rise to positive eigenvalues at Le = 1, and now qualitatively correct results are obtained.

Table 9.9 illustrates that the two positive eigenvalues remain positive also for Le = 10^4. Calculations for even larger Le show that they approach 0 through positive values and are inversely proportional to Le. N-independent results are obtained even with $N = 3$ to 4. The $N = 2$ case, corresponding to an 8×8 eigenvalue problem, gives results as accurate as the 76×76 eigenvalue problem that Luss had to solve using only the z_i trial functions.

The modified Galerkin method can, of course, be used with an arbitrary number of terms from either of the two series. Luss's choice of only z_i functions is not optimal, while the opposite choice of only v_i functions is much better, as seen in table 9.10, a continuation of table 9.9.

TABLE 9.9
FIRST TWO EIGENVALUES $(\gamma, \beta, \Phi, \text{Le}) = (40, 0.5, 0.264, 10^4)$
BY MODIFIED GALERKIN METHOD

N	λ_1 Le	λ_2 Le
1	829.4	−1.78
2	593.4	2.849
3	571.3	2.845
4	555.0	2.845
Exact	554.5	2.845

TABLE 9.10
FIRST TWO EIGENVALUES $(\gamma, \beta, \Phi, \text{Le}) = (40, 0.5, 0.264, 10^4)$
USING ONLY "REACTION" EIGENFUNCTIONS IN THE EXPANSION

N	Le λ_1	Le λ_2
3	618.5	2.90
4	610.0	2.878
Exact	554.5	2.845

The form of the first eigenfunction $v_1(x)$ is so peculiar in this example that it has to be found by forward integration of (52) using 200 subdivisions and three collocation points in each subinterval. The large number of subintervals is necessary due to the rapid variation of R_y and R_θ, and it is obvious that a large number of z_i-functions have to be used in order to represent $v_1(x)$.

Whereas it is impossible to obtain the correct shape of the steady state profiles and thus R_y and R_θ by global collocation or by Galerkin's method directly used on the steady state equations, the coordinate transformation of subsection 5.5.4

$$u = \frac{(a + 1)x}{1 + ax}$$

with a fairly large value of a (e.g., $a = 10$) makes it possible to use global collocation on (48) and (49) to obtain the eigenvalues. Thus at Le = 1 and $(\gamma, \beta, \Phi) = (40, 0.5, 0.264)$ the dominant eigenvalue was found with an accuracy better than 10^{-5} with $N = 12$, but each new value of Le requires the solution of a $2N \times 2N$ eigenvalue problem with $N \sim 12$ and in general it is not possible in advance to decide on the best choice of coordinate transformation.

In conclusion it appears that the modified Galerkin method is the most favorable both with respect to accuracy and general applicability.

9.2.4 Further results of a trigonometric expansion

In the preceding subsections we have compared two methods both based on Galerkin's method, but using different trial functions. The expansion in eigenfunctions v_i from (52) and z_i from (50) was shown to be the most economical.

The eigenvalues of (50) are immediately available, however, and we may expect that the large eigenvalues of (48) and (49), that according to the discussion of subsection 9.2.3 are not strongly dependent on the reaction eigenvalue problem (52), can be constructed from the solution of the diffusion eigenvalue problem (50). In fact it turns out that this idea is very fruitful if the steady state solution is not too extreme. Not only do we obtain accurate approximations to the high eigenvalues that are the most difficult to evaluate by a global method, but also the first eigenvalues are obtained qualitatively correct and with hardly any computational work.

First let us consider the reaction eigenproblem (52) with eigenfunctions $v_i(x)$:

$$\lambda v = \nabla^2 v - R_y v + \beta R_\theta v$$

Using Galerkin's method and trial functions $(\sin k\pi x)/x$, we obtain

$$\mathbf{Mv} = \lambda \mathbf{v}$$

with

$$M_{jk} = M_{kj} = -k^2\pi^2\delta_{kj} - S_{kj} + T_{kj} \tag{72}$$

$$S_{kj} = 2\int_0^1 \sin k\pi x \sin j\pi x R_y \, dx \tag{73}$$

$$T_{kj} = 2\beta \int_0^1 \sin k\pi x \sin j\pi x R_\theta \, dx \tag{74}$$

S_{kj} and T_{kj} are both finite for any k and, for large (k, j), $\mathbf{M}$ is diagonal dominant due to the large term $-k^2\pi^2$ in the diagonal. Consequently, for large k,

$$\lambda_k \sim M_{kk} = -k^2\pi^2 - S_{kk} + T_{kk} \tag{75}$$

and furthermore

$$S_{kk} \sim \int_0^1 R_y \, dx, \qquad T_{kk} \sim \beta \int_0^1 R_\theta \, dx \tag{76}$$

since

$$2\int_0^1 \sin^2 k\pi x f(x) \, dx \to \int_0^1 f(x)\, dx \quad \text{for } k \to \infty$$

Using (75) we completely bypass the matrix diagonalization process. Once the two integrals in (76) have been found by quadrature from the steady state solution, the eigenvalues can be written directly. Table 9.11 shows that even the second eigenvalue of Luss and Lee's example $(\gamma, \beta, \Phi, \text{Le}) = (30, 0.15, 1.1, 1)$ is well determined by (75) and (76) and if the final—and perhaps somewhat unnecessary—simplification of S_{kk} and T_{kk} is left out, the first eigenvalue is also reasonably accurate.

TABLE 9.11
SOLUTION OF REACTION EIGENVALUE PROBLEM WITHOUT DIAGONALIZATION

	λ_k		
k	Eq. (75), (76)	Eq. (75), (73), (74)	Exact
1	−6.2	−3.41	−3.12
2	−35.9	−35.53	−35.48
3	−85.2	−84.92	−84.94
4	−154.3	−154.1	−154.0
5	−243.1	−243.0	−242.8

Next we use the same technique for the full problem (48) and (49). Here reasonably accurate approximations for the large eigenvalues are presumably found from

$$\det \begin{vmatrix} (-k^2\pi^2 - S_{kk}) - \lambda_k & -T_{kk} \\ \dfrac{1}{\text{Le}} S_{kk} & \dfrac{1}{\text{Le}}(-k^2\pi^2 + T_{kk}) - \lambda_k \end{vmatrix} = 0 \tag{77}$$

or

$$\lambda_k^2 + \left(k^2\pi^2 + S_{kk} + \frac{k^2\pi^2 - T_{kk}}{\text{Le}}\right)\lambda_k + \frac{k^2\pi^2}{\text{Le}}(k^2\pi^2 + S_{kk} - T_{kk}) = 0 \tag{78}$$

For Le = 1, (78) simplifies to

$$\lambda_k^2 + (2k^2\pi^2 + S_{kk} - T_{kk})\lambda_k + k^2\pi^2(k^2\pi^2 + S_{kk} - T_{kk}) = 0$$

with solution $\lambda_k = -k^2\pi^2$ and $\lambda_k = -k^2\pi^2 - S_{kk} + T_{kk}$ as expected from the previous discussion.

Equally simple results are found for Le → 0 and Le → ∞.

1. Le → 0

We neglect the small terms of (78) and obtain

$$\lambda_k^2 + \frac{k^2\pi^2 - T_{kk}}{\text{Le}}\lambda_k + \frac{k^2\pi^2}{\text{Le}}(k^2\pi^2 + S_{kk} - T_{kk}) = 0 \tag{79}$$

One solution with λ_k inversely proportional to Le is obtained by neglecting the last term of (79):

$$\text{Le}\,\lambda_k = -k^2\pi^2 + T_{kk} \tag{80}$$

Another solution with λ_k independent of Le is found by neglecting the first term of (79):

$$\lambda_k = \frac{-k^2\pi^2 - S_{kk} + T_{kk}}{1 - (T_{kk}/k^2\pi^2)} \sim -k^2\pi^2 - S_{kk} \tag{81}$$

since k is assumed to be large.

2. Le $\to \infty$

The eigenvalues that do not depend on Le are immediately extracted from (78):

$$\lambda_k = -k^2\pi^2 - S_{kk} \tag{82}$$

Another set of eigenvalues that are inversely proportional to Le also appear:

$$\text{Le}\,\lambda_k = \frac{-k^2\pi^2 - S_{kk} + T_{kk}}{1 + (S_{kk}/k^2\pi^2)} \sim -k^2\pi^2 + T_{kk} \tag{83}$$

There is a striking similarity between the behavior of the large eigenvalues for Le $\to 0$ and Le $\to \infty$. For a given (large) k the eigenvalue either approaches a constant value given by (82) and (81) or the product of Le and the eigenvalue approaches a constant value given by (83) and (80). Since T_{kk} and S_{kk} are finite, the eigenvalues become negative and real for large enough k. Nothing can be said about the low-order eigenvalues since they are not covered by the present treatment: They may be real or complex with positive or negative real part when Le $\to 0$ or Le $\to \infty$.

Returning to (78) with a value of Le that is neither 0 nor ∞, we find a double zero if

$$\left(k^2\pi^2 + S_{kk} + \frac{k^2\pi^2 - T_{kk}}{\text{Le}}\right)^2 = \frac{4k^2\pi^2}{\text{Le}}(k^2\pi^2 + S_{kk} - T_{kk}) \tag{84}$$

For a given k, (84) determines two values Le_+ and Le_-, the solution of

$$(T_{kk} - k^2\pi^2)^2\left(\frac{1}{\text{Le}} - 1\right)^2 - 2\left(-S_{kk} - T_{kk} + \frac{2S_{kk}T_{kk}}{T_{kk} - k^2\pi^2}\right)(T_{kk} - k^2\pi^2) \times \left(\frac{1}{\text{Le}} - 1\right) + (S_{kk} - T_{kk})^2 = 0$$

$$\left.\begin{matrix}\dfrac{1}{\text{Le}_+}\\[2ex] \dfrac{1}{\text{Le}_-}\end{matrix}\right\} = 1 + \frac{Q_{kk} \pm \sqrt{Q_{kk}^2 - (S_{kk} - T_{kk})^2}}{T_{kk} - k^2\pi^2} \tag{85}$$

where

$$Q_{kk} = -(S_{kk} + T_{kk}) + \frac{2S_{kk}T_{kk}}{T_{kk} - k^2\pi^2} \tag{86}$$

Corresponding to the given k there is a pair of complex eigenvalues when $\mathrm{Le}_- < \mathrm{Le} < \mathrm{Le}_+$.

For large k we may furthermore simplify Q_{kk}:

$$Q_{kk} \sim -(S_{kk} + T_{kk})$$

$$\left.\begin{array}{c} \dfrac{1}{\mathrm{Le}_+} \\ \dfrac{1}{\mathrm{Le}_-} \end{array}\right\} = 1 + \frac{S_{kk} + T_{kk} \pm 2\sqrt{S_{kk}T_{kk}}}{-(k\pi)^2} = 1 + \left(\frac{\sqrt{S_{kk}} \mp \sqrt{T_{kk}}}{k\pi}\right)^2 \tag{87}$$

The existence of real zeros in (85) for $k \to \infty$ requires that S_{kk} and T_{kk} are of the same sign. This is true when $\gamma \cdot \beta$ is positive and the reaction is first order.

Using the approximation (76) for S_{kk} and T_{kk} and the parameter values of Luss and Lee (1970), the Le range in which a given high-order eigenvalue is complex is determined by the following inequality:

$$\frac{1}{\mathrm{Le}_+} = 1 + \frac{0.022}{k^2} < \frac{1}{\mathrm{Le}} < 1 + \frac{6.1}{k^2} = \frac{1}{\mathrm{Le}_-}$$

$$\frac{0.022}{k^2}\,\mathrm{Le} < 1 - \mathrm{Le} < \frac{6.1}{k^2}\,\mathrm{Le}$$

or for $\mathrm{Le} \sim 1$

$$\frac{0.022}{k^2} < 1 - \mathrm{Le} < \frac{6.1}{k^2} \tag{88}$$

Equation (88) shows that an increasing number of eigenvalues are complex when the interval $1 - \mathrm{Le} \to 0$. When $\mathrm{Le} = 1 - 10^{-3}$, inequality (88) holds for $k = 5$ to 78 and the corresponding eigenvalues form complex pairs. For $\mathrm{Le} = 1$ to 10^{-6} eigenvalues of index 149 to 2469 form complex pairs. It is an interesting feature of the problem that the number of complex eigenvalue pairs increases to infinity when $\mathrm{Le} \to 1$—but when $\mathrm{Le} = 1$, all eigenvalues are known to be real since (50) and (52) are both Sturm–Liouville problems.

Table 9.12 illustrates the use of formulas (80) to (83) for Le at either 0 or infinity, and table 9.13 gives the Le range in which eigenvalues of index k from the diffusion—and reaction—part of the eigenvalue problem (48) and (49) collapse to a complex pair.

It is quite obvious that the trigonometric expansion as used in this subsection is of considerable value. The first few eigenvalues may not be too well determined, but the higher eigenvalues are estimated with sufficient accuracy to provide a qualitative understanding of the whole eigenvalue spectrum for any value of Le between 0 and ∞. Also we have seen in table 9.11 that all except the first one or two eigenvalues of the reaction part of the problem are determined as the sum of the diffusion eigenvalues and a constant. This strongly suggests that only one or two v_i functions should be used in the method of subsection 9.2.3. The remaining trial functions can be taken from either set (and we naturally prefer the simple trigonometric functions).

TABLE 9.12
EIGENVALUES AT Le $\to$ 0 AND Le $\to \infty$ BY FORMULAS (80) TO (83)
$(\gamma, \beta, \Phi) = (30, 0.15, 1.1)$

	Le = 0				Le = ∞			
k	(80)–(81)		(Exact)		(83)–(82)		(Exact)	
	Le λ_k	λ_k	Le λ_k	λ_k	Le λ_k	λ_k	Le λ_k	λ_k
1	8.83	3.81	9.38	3.63	−1.52	−22.1	−1.44	−20.68
2	−22.2	−63.2	−22.4	−54.3	−26.6	−52.8	−26.8	−53.3
3	−71.6	−105.3	−71.7	−108.8	−73.9	−102.1	−74.2	−102.4
4	−140.8	−168.2	−140.7	−173.8	−142.1	−171.2	−142.3	−171.3
5	−229.7	−261.0	−229.5	−261.5	−230.6	−260.0	−230.6	−260.1

TABLE 9.13
REGION IN WHICH COMPLEX EIGENVALUES ARE FORMED
$(\gamma, \beta, \Phi) = (30, 0.15, 1.1)$

k	Approximate region using (85) and (86) and (73) and (74)	Exact region of complex eigenvalues
1	0.17–0.90	0.18–0.88
2	0.18–0.993	0.13–0.993
3	0.49–0.997	0.53–0.997
4	0.68–0.9984	0.69–0.9985
5	0.78–0.9990	0.78–0.9988

9.2.5 Eigenvalues by forward integration

In the preceding subsections, results obtained by various methods have been compared with so-called exact values.

These "exact" results are calculated by forward integration, as described in section 4.5. The eigenvalues and eigenfunctions are found one set at a time by a time-consuming procedure that can be continued, however, to arbitrary accuracy for any given eigenvalue.

A trial value of λ is inserted into (48) and (49) and the two equations are integrated by forward integration first with initial condition (89) and next with initial condition (90):

$$\text{At } x = 0: \qquad y = 1, \qquad \theta = \frac{dy}{dx} = \frac{d\theta}{dx} = 0 \tag{89}$$

$$\text{At } x = 0: \qquad \theta = 1, \qquad y = \frac{dy}{dx} = \frac{d\theta}{dx} = 0 \tag{90}$$

Let the resulting values of y and θ at $x = 1$ be

$$\text{For (89):} \qquad y_1(1, \lambda), \qquad \theta_1(1, \lambda) \tag{91}$$

$$\text{For (90):} \qquad y_2(1, \lambda), \qquad \theta_2(1, \lambda) \tag{92}$$

The trial value of λ is an eigenvalue of (48) and (49) if and only if

$$\det \begin{pmatrix} y_1(1, \lambda) & y_2(1, \lambda) \\ \theta_1(1, \lambda) & \theta_2(1, \lambda) \end{pmatrix} = 0 \tag{93}$$

The eigenvalues are found iteratively by simultaneous solution for $dy/d\lambda$ and $d\theta/d\lambda$ and a Newton–Raphson correction algorithm as described in section 4.5. The trial value for λ may, of course, be complex and for this reason the previously used method of checking the stability of the steady state by a single integration using $\lambda = 0$ cannot be used. This simple device is applicable only for Sturm–Liouville problems with real eigenvalues as in the reaction eigenvalue problem of subsection 9.2.1.

Highly accurate λ values require a large number of subdivisions or a large number of collocation points in each subinterval, especially for difficult problems. Thus with five steady states 200 equidistant subintervals (in x) and three interior collocation points were used for the middle steady state.

The computations of the forward integration procedure are greatly reduced if reasonably accurate initial estimates of the eigenvalues (e.g., from a low-order Galerkin method for the first eigenvalues and from the method of subsection 9.2.4 for the high-order eigenvalues) are used. The method has the important advantage over any of the previously discussed methods that it can be used regardless of how strange the steady state profiles look and also in those cases where suitable trial functions cannot be found for the Galerkin method.

An illustration is provided by the case of boundary conditions of the radiation type at $x = 1$ and with different Bi_M and Bi [equation (1.83)]. When the Biot number for mass and heat transfer are different, (48) and (49) do not reduce to two Sturm–Liouville problems even when Le = 1. Global collocation as in (62) still works, of course, since it is based on a standard set of expansion functions (polynomials), but the simple version of Galerkin's method in subsection 9.2.3 fails. The forward integration method is easily modified to cope with this set of boundary conditions at $x = 1$:

$$\left(\frac{dy}{dx} + \text{Bi}_M\, y\right)_{1,\lambda} \quad \text{and} \quad \left(\frac{d\theta}{dx} + \text{Bi}\,\theta\right)_{1,\lambda}$$

are simply substituted for y and θ in (93).

9.2.6 The limiting eigenvalue problems at Le → 0 and Le → ∞

We have used the Sturm–Liouville problems that can be extracted from (48) and (49) when Le = 1 to construct approximation methods for the eigenvalues when Le ≠ 1, and these methods have been shown to be applicable throughout the entire Le-range $10^{-6} < \text{Le} < 10^4$. It is tempting, however, to investigate the two limiting values Le → 0 and Le → ∞ in order to obtain expansions based on the solution at these particular parameter values. It is especially the prospect of obtaining perturbation series for λ based on Le = ∞, Le = 1, and Le = 0 and spanning the whole Le-range $[0, \infty]$ that makes this investigation interesting.

As we have already seen for Le = 1, the eigenvalues appear in two separate groups also for any other value of Le. For Le → 0 we can rescale (48) to obtain

$$(\text{Le}\,\lambda)y = \text{Le}\,\nabla^2 y - \text{Le}\,R_y y - \text{Le}\,R_\theta \theta$$
$$(\text{Le}\,\lambda)\theta = \nabla^2\theta + \beta R_y y + \beta R_\theta \theta$$

The right-hand side of the rescaled (48) is zero for Le = 0 and if we focus our attention on the group of eigenvalues that are inversely proportional to Le, the equation has the solution $y = 0$. Consequently, the eigenvalues that we have found approximately in (80) appear from the Sturm–Liouville problem

$$(\text{Le}\,\lambda)\theta = \nabla^2\theta + \beta R_\theta \theta \tag{49a}$$

The eigenvalues of (49a) are easily determined by forward integration or in not too extreme cases by high-order collocation:

$$(\text{Le}\,\lambda)\boldsymbol{\theta} = (\mathbf{C} + \beta\mathbf{R}_\theta)\boldsymbol{\theta} \tag{94}$$

The second group of eigenvalues is obtained directly from (44) and (45) using the quasi-stationarity principle that was discussed in section (1.3). θ is given by the steady state equation

$$\nabla^2\theta + \beta R_y y + \beta R_\theta \theta = 0 \tag{95}$$

and it follows slavishly the concentration profile given by (44).

Thus the second eigenvalue problem is (48) with θ given by (95). The collocation version of (48), (95) is

$$\begin{aligned} \lambda \mathbf{y} &= (\mathbf{C} - \mathbf{R}_y)\mathbf{y} - \mathbf{R}_\theta \boldsymbol{\theta} \\ \mathbf{0} &= (\mathbf{C} + \beta \mathbf{R}_\theta)\boldsymbol{\theta} + \beta \mathbf{R}_y \mathbf{y} \end{aligned} \tag{96}$$

or

$$\begin{aligned} \boldsymbol{\theta} &= -\beta(\mathbf{C} + \beta \mathbf{R}_\theta)^{-1}\mathbf{R}_y \mathbf{y} \\ \lambda \mathbf{y} &= [\mathbf{C} - \mathbf{R}_y + \beta \mathbf{R}_\theta (\mathbf{C} + \beta \mathbf{R}_\theta)^{-1}\mathbf{R}_y]\mathbf{y} \end{aligned} \tag{97}$$

$$= \mathbf{C}(\mathbf{C} + \beta \mathbf{R}_\theta)^{-1}(\mathbf{C} - \mathbf{R}_y + \beta \mathbf{R}_\theta)\mathbf{y} \tag{98}$$

It is interesting to note that the discretization matrices that appear in (98) arise from discretization of (49a), (50), and (52). The eigenvalues of (96) are the "constant group" eigenvalues estimated by (81).

For Le → ∞, the eigenvalues are also separated into two groups. One group of constant eigenvalues [corresponding to (82)] is obtained from

$$\lambda y = \nabla^2 - R_y y \tag{48a}$$

since the right-hand side of (49) is finite, leading to the solution $\theta = 0$ for Le → ∞.

The group of eigenvalues that were estimated by (83) is derived by a quasi-steady state assumption for y:

$$\begin{aligned} \nabla^2 y - R_y y - R_\theta \theta &= 0 \\ (\text{Le}\,\lambda)\theta &= \nabla^2\theta + \beta R_y y + \beta R_\theta \theta \end{aligned} \tag{99}$$

The discrete version of (48a) and (99) is

(48a): $$\lambda \mathbf{y} = (\mathbf{C} - \mathbf{R}_y)y \tag{100}$$

(99): $$(\text{Le}\,\lambda)\boldsymbol{\theta} = \mathbf{C}(\mathbf{C} - \mathbf{R}_y)^{-1}(\mathbf{C} - \mathbf{R}_y + \beta \mathbf{R}_\theta)\boldsymbol{\theta} \tag{101}$$

These two eigenvalue problems are very similar to (94) and (98). Equations (49a) and (48a) are both Sturm–Liouville problems, but their counterparts [(48) and (95)] and (99) are not. We know from subsection 9.2.4 that the high-order eigenvalues at Le = 0 and ∞ are real, and usually this is also the case for the first eigenvalues. But exceptions exist, especially for large γ and β values. Thus at $(\gamma, \beta, \Phi) = (28, 1, 0.1908)$, a case studied by Michelsen and Villadsen (1972), the first few eigenvalues at Le = 0 in the "constant eigenvalue group" are −11.36, −28.58 ± 30.19i, −78.43, −161.74, while the corresponding eigenvalues of the Sturm–Liouville problem are, as expected, all found to be real. Le λ → 1362, 8.72, −21.48, −70.62, −142.6 for Le → 0.

The form of the eigenvalue problem (49a) strongly suggests why a steady state may be stable at Le = 1 but unstable at smaller values of Le. Except for the scalar factor Le, the difference between (52) and (49a) is that the latter lacks the stabilizing term $-R_y y$, which tends to shift the eigenvalues to the left since $R_y y > 0$ for an irreversible nth-order ($n > 0$) reaction.

9.2.7 Perturbation methods

We have shown that the non-self-adjoint eigenvalue problem (48) and (49) is reduced to simpler problems, with separation of the eigenvalues into two distinct groups, when Le is 1, 0, or ∞.

The eigenfunctions at the limiting solutions are now used to construct perturbation series for λ when Le is close to 1, 0, or ∞, much in the same manner as the Sturm–Liouville problems at Pe = 0 and ∞ are used in the previous section.

First we study the eigenvalues for small Le, remembering that each of the two eigenvalue groups (the "Le-number dependent" and the "constant" group) must be treated separately.

In subsection 9.2.6, Equations (48) and (49) were written

$$(\text{Le}\,\lambda)y = \text{Le}\,\nabla^2 y - \text{Le}\,R_y y - \text{Le}\,R_\theta \theta \tag{102}$$

$$(\text{Le}\,\lambda)\theta = \nabla^2\theta + \beta R_y y + \beta R_\theta \theta \tag{103}$$

and the right-hand side of (102) is equated to 0 to give the eigenproblem (49a) for the (Le λ) group at Le = 0.

Here we expand the eigenvalues and eigenfunctions of (102) and (103) in the following series for small Le:

$$\text{Le}\,\lambda = s_0 + s_1\,\text{Le} + s_2\,\text{Le}^2 + \ldots + s_k\,\text{Le}^k + \ldots \tag{104}$$

$$y = y_0 + y_1\,\text{Le} + y_2\,\text{Le}^2 + \ldots + y_k\,\text{Le}^k + \ldots \tag{105}$$

$$\theta = \theta_0 + \theta_1\,\text{Le} + \theta_2\,\text{Le}^2 + \ldots + \theta_k\,\text{Le}^k + \ldots \tag{106}$$

Inserting into (102) and (103) and collecting terms in Le^k yields

$k = 0$:

$$s_0 y_0 = 0 \tag{107}$$

$$s_0\theta_0 = (\nabla^2 + \beta R_\theta)\theta_0 + \beta R_y y_0 \tag{108}$$

Since $s_0 \neq 0$, we obtain $y_0 = 0$ and (108) is identical to (49a), yielding $s_0 = \underset{\text{Le}\to 0}{\text{Le}\,\lambda}$ and θ_0 = the eigenfunction $\theta(x)$ at Le = 0.

$k = 1$:

$$s_0 y_1 + s_1 y_0 = (\nabla^2 - R_y)y_0 - R_\theta\theta_0 \tag{109}$$

$$s_0\theta_1 + s_1\theta_0 = (\nabla^2 + \beta R_\theta)\theta_1 + \beta R_y y_1 \tag{110}$$

Inserting $y_0 = 0$, we obtain

(109):
$$y_1 = -\frac{1}{s_0} R_\theta \theta_0 \tag{111}$$

(110) and (111):
$$(\nabla^2 + \beta R_\theta - s_0)\theta_1 - s_1\theta_0 = \frac{\beta}{s_0} R_y R_\theta \theta_0 \tag{112}$$

where the right-hand side of (112) contains the known functions R_y, R_θ, and the lower index term θ_0. In general, for the Mth term of the expansions (104) and (106),

$$s_0 y_M + s_1 y_{M-1} + \ldots + s_M y_0 = (\nabla^2 - R_y) y_{M-1} - R_\theta \theta_{M-1}$$
$$s_0 \theta_M + s_1 \theta_{M-1} + \ldots + s_M \theta_0 = (\nabla^2 + \beta R_\theta)\theta_M + \beta R_y y_M$$

or

$$y_M = \frac{1}{s_0}\left[-\sum_{j=1}^{M-1} s_j y_{M-j} + (\nabla^2 - R_y) y_{M-1} - R_\theta \theta_{M-1} \right] \tag{113}$$

$$(\nabla^2 + \beta R_\theta - s_0)\theta_M - s_M \theta_0 = \sum_{j=1}^{M-1} s_j \theta_{M-j} - \beta R_y y_M = q_M \tag{114}$$

When (113) is inserted into (114), the right-hand side q_M is seen to contain only perturbation constants s_j and perturbation functions θ_j, y_j of index less than M.

Here we use the collocation version of the perturbation method in section 9.1 to obtain s_M and expansion coefficients of θ_M in terms of the eigenfunctions θ_0 for Le $= 0$.

The basic problem to be solved by the Nth-order collocations is (108) or

$$s_0 \boldsymbol{\theta}_0 = (\mathbf{C} + \beta \mathbf{R}_\theta)\boldsymbol{\theta}_0 \tag{115}$$

Eigenvalues and eigenvectors of (115) are denoted $s_0^{(j)}$ and $\boldsymbol{\theta}_0^{(j)}$, $j = 1, 2, \ldots, N$.

Each eigenvalue of (102) and (103) may be treated by the method of section 9.1, but here we only consider the first (largest) eigenvalue λ_1 since this has immediate interest in a stability analysis. Each of the vectors $\boldsymbol{\theta}_1, \boldsymbol{\theta}_2, \ldots$, the perturbation functions of (106) corresponding to λ_1 and taken at the collocation points of (115), are expanded on the last $N - 1$ eigenvectors $\boldsymbol{\theta}_0^{(2)}, \boldsymbol{\theta}_0^{(3)}, \ldots, \boldsymbol{\theta}_0^{(N)}$ of (115) as in (37):

$$\boldsymbol{\theta}_M = \sum_{j=2}^{N} c_{j,M} \boldsymbol{\theta}_0^{(j)}, \qquad M = 1, 2, \ldots \tag{116}$$

Equation (116) is inserted into the discrete version of (114) utilizing

$$(\mathbf{C} + \beta\mathbf{R}_\theta)\boldsymbol{\theta}_0^{(j)} = s_0^{(j)}\boldsymbol{\theta}_0^{(j)}$$

to give

$$\sum_{j=2}^{N} c_{j,M}[s_0^{(j)} - s_0^{(1)}]\boldsymbol{\theta}_0^{(j)} - s_M\boldsymbol{\theta}_0^{(1)} = \mathbf{q}_M \tag{117}$$

The eigenrow of (115) corresponding to $\boldsymbol{\theta}_0^{(j)}$ is $\mathbf{U}_0^{(j)}$. Equation (117) is multiplied by $\mathbf{U}_0^{(1)}$ and $\mathbf{U}_0^{(K)}$, respectively, to give

$$s_M = -[\mathbf{U}_0^{(1)}]^T\mathbf{q}_M \tag{118}$$

$$c_{K,M}[s_0^{(K)} - s_0^{(1)}] = [\mathbf{U}_0^{(K)}]^T\mathbf{q}_M \tag{119}$$

Recursive application of (118) and (119) starting with $M = 1$ to compute s_1, $c_{K,1}$, $K = 2, \ldots, N$, and q_2 yields the terms in (104) to (106) at almost no extra cost once (115) has been solved by collocation.

Turning next to the group of constant eigenvalues small Le we use (48) and (49) as they stand and expand

$$\lambda = s_0 + s_1 \,\mathrm{Le} + s_2 \,\mathrm{Le}^2 + \ldots \tag{120}$$

$$y = y_0 + y_1 \,\mathrm{Le} + y_2 \,\mathrm{Le}^2 + \ldots \tag{121}$$

$$\theta = \theta_0 + \theta_1 \,\mathrm{Le} + \theta_2 \,\mathrm{Le} + \ldots \tag{122}$$

where the s_k and the perturbation functions are naturally different from those just found from (104) to (106). The first few perturbation equations become

$$s_0 y_0 = \nabla^2 y_0 - R_y y_0 - R_\theta \theta_0 \tag{123}$$

$$0 = \nabla^2 \theta_0 + \beta R_\theta \theta_0 + \beta R_y y_0 \tag{124}$$

and

$$s_0 y_1 + s_1 y_0 = \nabla^2 y_1 - R_y y_1 - R_\theta \theta_1 \tag{125}$$

$$s_0 \theta_0 = \nabla^2 \theta_1 + \beta R_\theta \theta_1 + \beta R_y y_1 \tag{126}$$

or, for $k = M$,

$$s_0 y_M + s_1 y_{M-1} + \ldots + s_M y_0 = \nabla^2 y_M - R_y y_M - R_\theta \theta_M \tag{127}$$

$$s_0 \theta_{M-1} + s_1 \theta_{M-2} + \ldots + s_{M-1} \theta_0 = \nabla^2 \theta_M + \beta R_\theta \theta_M + \beta R_y y_M \tag{128}$$

The discrete version of (128) is solved for $\boldsymbol{\theta}_M$:

$$\boldsymbol{\theta}_M = (\mathbf{C} + \beta\mathbf{R}_\theta)^{-1}\left(\sum_{j=0}^{M-1} s_j \boldsymbol{\theta}_{M-j-1} - \beta\mathbf{R}_y\mathbf{y}_M\right) \tag{129}$$

and $\boldsymbol{\theta}_M$ is substituted into (127) giving

$$[\mathbf{C} - \mathbf{R}_y + \beta\mathbf{R}_\theta(\mathbf{C} + \beta\mathbf{R}_\theta)^{-1}\mathbf{R}_y]\mathbf{y}_M - s_0\mathbf{y}_M - s_M\mathbf{y}_0$$
$$= \sum_{j=1}^{M-1} s_j\mathbf{y}_{M-j} + \mathbf{R}_\theta(\mathbf{C} + \beta\mathbf{R}_\theta)^{-1}\sum_{j=0}^{M-1} s_j\boldsymbol{\theta}_{M-j-1} = \mathbf{q}_M \tag{130}$$

where again $\mathbf{q}_M$ contains only terms of index less than M.

The expansion is here based on the eigenfunctions and eigenvalues of (123) and (124) or in the discrete version on the eigenvectors $\mathbf{y}_0^{(j)}$ and eigenvalues $s_0^{(j)}$ of

$$[\mathbf{C} - \mathbf{R}_y + \beta\mathbf{R}_\theta(\mathbf{C} + \beta\mathbf{R}_\theta)^{-1}\mathbf{R}_y]\mathbf{y}_0 = s_0\mathbf{y}_0 \tag{131}$$

Equations (123) and (124) are, as we have already seen in subsection 9.2.6, not a Sturm–Liouville problem, and the eigenvalues of (131) may consequently be complex. In all cases we have studied the first eigenvalue has been real [e.g., the example $(\gamma, \beta, \Phi) = (28, 1, 0.1908)$ of subsection 9.2.6], but a modification of the expansion technique has had to be used in cases where complex $s_0^{(j)} (j > 1)$ occurred.

The two expansions from Le $= \infty$ are developed in exactly the same manner as described above using series of the form

$$\lambda = s_0 + s_1\,\mathrm{Le}^{-1} + s_2\,\mathrm{Le}^{-2} + \dots \tag{132a}$$

or

$$\mathrm{Le}\,\lambda = s_0 + s_1\,\mathrm{Le}^{-1} + s_2\,\mathrm{Le}^{-2} + \dots \tag{132b}$$

The third, and perhaps most important, perturbation series is that based on Le $= 1$. Rewrite (48) and (49) as

$$\lambda\begin{pmatrix} y \\ \theta \end{pmatrix} = \begin{pmatrix} \nabla^2 - R_y & -R_\theta \\ \dfrac{\beta R_y}{\mathrm{Le}} & \dfrac{\nabla^2 + \beta R_\theta}{\mathrm{Le}} \end{pmatrix}\begin{pmatrix} y \\ \theta \end{pmatrix}$$
$$= \begin{pmatrix} \nabla^2 - R_y & -R_\theta \\ \beta R_y & \nabla^2 + \beta R_\theta \end{pmatrix}\begin{pmatrix} y \\ \theta \end{pmatrix} + \frac{1 - \mathrm{Le}}{\mathrm{Le}}\begin{pmatrix} 0 & 0 \\ \beta R_y & \nabla^2 + \beta R_\theta \end{pmatrix}\begin{pmatrix} y \\ \theta \end{pmatrix} \tag{133}$$

or, alternatively, rewrite (102) and (103) as

$$(\lambda\,\mathrm{Le})\begin{pmatrix} y \\ \theta \end{pmatrix} = \begin{pmatrix} \nabla^2 - R_y & -R_\theta \\ \beta R_y & \nabla^2 + \beta R_\theta \end{pmatrix}\begin{pmatrix} y \\ \theta \end{pmatrix} + (\mathrm{Le} - 1)\begin{pmatrix} \nabla^2 - R_y & -R_\theta \\ 0 & 0 \end{pmatrix}\begin{pmatrix} y \\ \theta \end{pmatrix} \tag{134}$$

In both cases the discrete version of the problem is

$$\lambda\mathbf{w} = (\mathbf{G} + \Delta\mathbf{H})\mathbf{w}$$

where $\mathbf{G}$ is the collocation matrix for the eigenproblem at Le $= 1$, Δ is $(1 - \mathrm{Le})/\mathrm{Le}$ in (133) and $(\mathrm{Le} - 1)$ in (134). $\mathbf{H}$ is constructed from either the θ or the y part of the Le $= 1$ matrix.

We expand in powers of the small quantity Δ:

$$\lambda = \sum_0^\infty s_k \Delta^k$$
$$\begin{pmatrix} y \\ \theta \end{pmatrix} = \sum_0^\infty \begin{pmatrix} y_k \\ \theta_k \end{pmatrix} \Delta^k \tag{135}$$

and proceed as before. Perturbation series based on two series of eigenfunctions were used for Le → 0 and Le → ∞, giving rise to the "constant" eigenvalue λ_1 [e.g., from (120) to (122)] and the Le-number dependent eigenvalue Le λ_1 from (104) to (106). Here the problem formulation is simpler but the computational work larger since a $2N \times 2N$ problem has to be treated. We denote the eigenvalues that appear as $\lambda_1^{(v)}$ and $\lambda_1^{(z)}$ to emphasize that they tend to the reaction and diffusion eigenvalues respectively, of subsection 9.2.3 when Le → 1.

9.2.8 Results for $(\gamma, \beta, \Phi) = (30, 0.15, 1.1)$

Table 9.14 shows coefficients s_i in the two expansions (104) and (120) for λ_1. s_0 is, respectively, the limit of Le λ_1 when Le → 0, i.e., the first eigenvalue of (94) and the first of the "constant group" eigenvalues of (98).

TABLE 9.14
COEFFICIENTS s_i IN (Le λ_1) = $\sum_0^\infty s_i$ Le^i AND $\lambda_1 = \sum_0^\infty s_i$ Le^i
EIGENVALUES AND EIGENFUNCTIONS FOR Le = 0 FOUND BY 10-POINT COLLOCATION

i	Variable group Le λ_1	Constant group λ_1
0	9.3809	3.6342
1	−31.771	10.670
2	130.14	359.50
3	−995.51	1338.3
4	402.18	5371
5	$-1.8 \cdot 10^5$	22,766
6	$1.1 \cdot 10^8$	$1.004 \cdot 10^5$
7	$-7.8 \cdot 10^{10}$	$4.567 \cdot 10^5$

The series based on (104) breaks down for $i > 3$. When 14 collocation points are used, $s_0 - s_3$ are unaltered to the number of digits shown while s_5 differs by an order of magnitude. The problem must be seriously ill-conditioned since identical results were obtained when 9-digit and 14-digit machine accuracy were used in the computation. Excellent results were obtained, however, for Le < 0.1, as seen in table 9.15, with the first three perturbation terms

$$\text{Le}\,\lambda_1 \simeq 9.3809 - 31.771\,\text{Le} + 130.14\,\text{Le}^2 - 995.51\,\text{Le}^3 \qquad (136)$$

The s_i of the perturbation series for the "constant" λ_1 are much better behaved and the same values of s_i were obtained with $N = 10$ and 14 collocation points. The ratio s_i/s_{i-1} approaches 4.8, indicating that the series is convergent for $\text{Le} < 0.21$. Note from table 9.13 that this is close to the lower bound for the Le-region where the first pair of eigenvalues (coming from ∞ and 3.63 at $\text{Le} = 0$) meet to form a complex pair of eigenvalues.

TABLE 9.15
FIRST EIGENVALUE BY (104) AND BY (120): $(\gamma, \beta, \Phi) =$ (30, 0.15, 1.1)

i		0	1	2	3	10	Exact
Le = 0.5	λ_1 (104)	187.62	155.85	156.50	156.25	—	156.24
	λ_1 (120)	3.6342	4.1692	4.2591	4.2758	4.2800	4.2800
Le = 0.1	λ_1 (104)	93.81	62.04	63.34	62.34	—	62.24
	λ_1 (120)	3.6342	4.7041	5.0636	5.1974	5.2916	5.2927

Very similar results are obtained (tables 9.16 and 9.17) for the series (132a) and (132b) based upon an expansion from $\text{Le} \to \infty$, but now the "constant group" λ_1 perturbation coefficients s_i start to behave erratically for $i > 3$. The ratio between successive terms s_i/s_{i-1} in the series for Le λ_1 approaches 0.80, indicating an astonishing large convergence range $0.80 < \text{Le} < \infty$ for this series. Subsection 9.2.9 shows that the first eigenvalue of the Le λ series does not form a complex pair with any other eigenvalue in $1 < \text{Le} < \infty$, and this is probably the reason for the remarkable usefulness of this perturbation series.

TABLE 9.16
COEFFICIENTS s_i IN $\lambda_1 = \sum_0^\infty s_i/\text{Le}^i$ AND IN $\lambda_1\,\text{Le} = \sum_0^\infty s_i/\text{Le}^i$
$N = 10$ FOR λ_1 SERIES AND $N = 10$ OR 14 FOR λ_1 Le SERIES

i	Constant group λ_1	Variable group Le λ_1
0	−20.6831	−1.4385
1	9.0341	−0.66401
2	1.2599	−0.35037
3	0.71440	−0.20520
4	65.403	−0.12970
5	$1.4 \cdot 10^4$	−0.086745
6	$4.0 \cdot 10^6$	−0.060403
7	$1.3 \cdot 10^9$	−0.043370

TABLE 9.17
FIRST EIGENVALUE BY (132a) AND BY (132b) $(\gamma, \beta, \Phi) =$ (30, 0.15, 1.1)

i		0	1	2	3	10	Exact
Le = 5	λ_1 (132a)	−20.68	−18.88	−18.83	−18.82	—	−18.82
	λ_1 (132b)	−0.2877	−0.3143	−0.3171	−0.3174	−0.3174	−0.3174
Le = 2	λ_1 (132a)	−20.68	−16.16	−15.85	−15.76	—	$-15.04 \pm i0.70$
	λ_1 (132b)	−0.7193	−0.8853	−0.9290	−0.9419	−0.9480	−0.9480

The results of the expansion from Le = 1 are quite disappointing. The series (135) for $\lambda_1^{(v)}$ and $\lambda_1^{(z)}$ converge in only a small interval around Le = 1. Thus for $(\gamma, \beta, \Phi) = (30,\ 0.15,\ 1.1)$ the following series in $\Delta = (1/\text{Le}) - 1$ is obtained:

$$\lambda_1^{(v)} = -3.12196 - 8.4449\Delta - 21.003\Delta^2 - 78.365\Delta^3 - 357.59\Delta^4$$
$$\lambda_1^{(z)} = -9.86960 + 16.710\Delta + 20.898\Delta^2 + 79.099\Delta^3 + 355.86\Delta^4 \tag{137}$$

There is no difficulty in obtaining higher-order expansion coefficients that soon become equal but of opposite sign. The ratio between successive terms approaches 6.6Δ, which means that the series is convergent when $|\Delta| = |(1/\text{Le}) - 1| < 0.15$ or for $0.87 < \text{Le} < 1.18$.

If $\Delta = \text{Le} - 1$ is used, one obtains

$$\text{Le}\,\lambda_1^{(v)} = -3.12196 + 5.3229\Delta - 21.003\Delta^2 + 99.368\Delta^3 - 535.32\Delta^4$$
$$\text{Le}\,\lambda_1^{(z)} = -9.86960 - 26.5792\Delta + 20.898\Delta^2 - 99.997\Delta^3 + 534.96\Delta^4 \tag{138}$$

In (138) the ratio between successive terms approaches $\sim 8\Delta$ and the series appears to be convergent for $0.875 < \text{Le} < 1.125$. For both series it is apparent that the range of convergence is determined by the value of Le (0.88 from table 9.13) where the two eigenvalues form a complex pair.

9.2.9 Qualitative description of the eigenvalue spectrum

In the previous subsections we have been collecting pieces of a puzzle that can now be put together to what can at least be called a rough sketch of the eigenvalue spectrum of (48) and (49) for different values of Le.

In subsection 9.1.1 the eigenvalues come from two groups—one is the reaction set of eigenfunctions (52) and the other is the diffusion set of eigenfunctions, the sine functions. Both groups of eigenfunctions should be represented in a series expansion of the eigenfunctions for Le $\neq$ 1 in order to obtain an accurate approximation with a small number of terms.

For large and small values of Le, the eigenvalues separate into two groups: one with eigenvalues proportional to Le^{-1} and one with constant eigenvalues. This is first seen in subsection 9.2.4, then it appears in subsection 9.2.6, and finally we have based our perturbation series in subsection 9.2.7 on this knowledge.

All eigenvalues are real at Le = 1 and the constant group (variable group) eigenvalues at Le = ∞ (Le = 0) are derived from Sturm–Liouville problems and are thus real. Often all eigenvalues at Le = 0 and ∞ are real, but this cannot be guaranteed.

In between these limiting values for Le any eigenvalue may become complex for large enough R_θ and R_y. The high-order eigenvalues of the "reaction" and "diffusion" groups are different by an almost constant amount determined in (75) and they become complex very close to Le = 1.

To understand the somewhat bewildering results of the perturbation analysis, a few more details on the range of Le in which the eigenvalues form complex pairs must be given in continuation of the approximate data of table 9.13.

In table 9.18 the first few eigenvalues of the Le = 1 problem are collected from tables 9.8 and 9.11.

TABLE 9.18
REACTION AND DIFFUSION EIGENVALUES FOR Le = 1.
$(\gamma, \beta, \Phi) = (30, 0.15, 1.1)$

λ_1	−3.122	$-9.870 = -\pi^2$
λ_2	−35.48	−39.48
λ_3	−84.94	−88.83
λ_4	−153.93	$-157.92 = -16\pi^2$

$\beta R_\theta - R_y$ is positive for all x when $(\gamma, \beta, \Phi) = (30, 0.15, 1.1)$ and the eigenvalues of (52) are consequently all situated to the right of the corresponding eigenvalues of (50). We see from table 9.11 that the difference between corresponding eigenvalues of the two sets quickly approaches $3.92 = T_{kk} - S_{kk}$ of (76).

Increasing Le from 1 to a value larger than 1 initially moves eigenvalues of the group connected with (52) to the right on the real axis and eigenvalues connected with (50) to the left. Conversely a decrease in Le from 1 moves the eigenvalues connected with (52) to the left and the other group of eigenvalues to the right—that is, toward each other. The rate of change of an eigenvalue with Le increases with increasing index of the eigenvalue.

The movement of the first pair of eigenvalues is illustrated in figure 9-4. The points marked A–P correspond to entries in table 9.19.

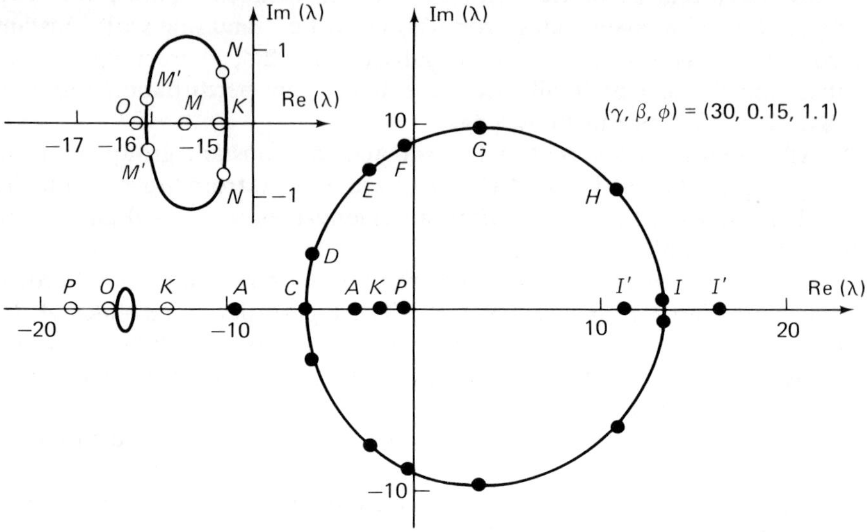

Figure 9-4. First eigenvalue pair and its interplay with higher eigenvalues.

TABLE 9.19
FIRST EIGENVALUE PAIR FOR Le < 1 AND ITS INTERPLAY WITH HIGHER EIGENVALUES FOR Le > 1. $(\gamma, \beta, \Phi) = (30, 0.15, 1.1)$

Code	λ	Le	Code	λ	Le
A	−3.122, −9.870	1	*J*	−2.485, −11.25	1.1
B	−4.568, −7.505	0.9	*K*	−1.806, −13.11	1.3
C	−5.593, −6.272	0.880	*L*	−1.297, −15.08	1.6
C	$-5.905 \pm 0.637i$	0.875	*M*	−1.240, −15.55(−16.74)	1.65
D	$-5.462 \pm 2.969i$	0.8	*M′*	$-1.212, -16.05 \pm 0.33i$	1.68
E	$-2.313 \pm 7.565i$	0.5	*N*	$-0.948, -15.04 \pm 0.70i$	2
F	$-0.168 \pm 8.873i$	0.4	*O*	−0.812, −16.18(−12.79) (−18.66)	2.25
G	$3.465 \pm 9.696i$	0.3	*P*	−0.408, −18.34 (−6.91)	4
H	$10.894 \pm 6.524i$	0.2			
I	$13.403 \pm 0.419i$	0.180			
I′	16.370, 11.143	0.178			

Once the complex pair has been formed at Le = 0.88, it moves toward the right in the complex plane, passes the imaginary axis at Le = 0.394, and disconnects to two real (positive) eigenvalues at Le = 0.179. One of these moves to the left and ends at 3.63 (table 9.12) for Le = 0, while the other wanders off to infinity.

The picture is much more complex for Le increasing from 1. The first eigenvalue of (52) moves steadily toward zero, without encountering or being bypassed by other eigenvalues. The first eigenvalue of (50) moves to the left, but at Le = 1.67 it meets the second eigenvalue of (52), which has moved from −35.48 to −16.74 when Le has increased from 1 to 1.65. The second eigenvalue of (52) separates out from the complex pair when Le = 2.1 and moves thereafter unmolested toward zero (−12.79 at Le = 2.25 and −6.91 at Le = 4), while the first eigenvalue of (50) forms a new complex pair with the third eigenvalue of (52) when Le has increased to 4.04. At Le = 4.07 this pair of eigenvalues splits up, the third eigenvalue of (52) decreases to zero, and the first eigenvalue of (50) moves toward the fourth eigenvalue of (52). This encounter does not take place, however. They come to within 0.1 of each other at Le = 7.3, but disengage, one continuing toward zero while the other bends back toward −20.68, its final destination when Le → ∞ (table 9.16).

A graphical representation of the movement of the first few eigenvalues when Le is varied from 1 is given in figure 9-5. Only the real part of the eigenvalue is shown. When two full-line curves meet, a complex pair is formed. It is seen that the eigenvalues for Le > 1 move in a very regular fashion even though the details are obscured by the formation of complex pairs: One group of eigenvalues decreases to zero roughly inversely proportional to Le and the other group tends to distinct, constant values for Le → ∞.

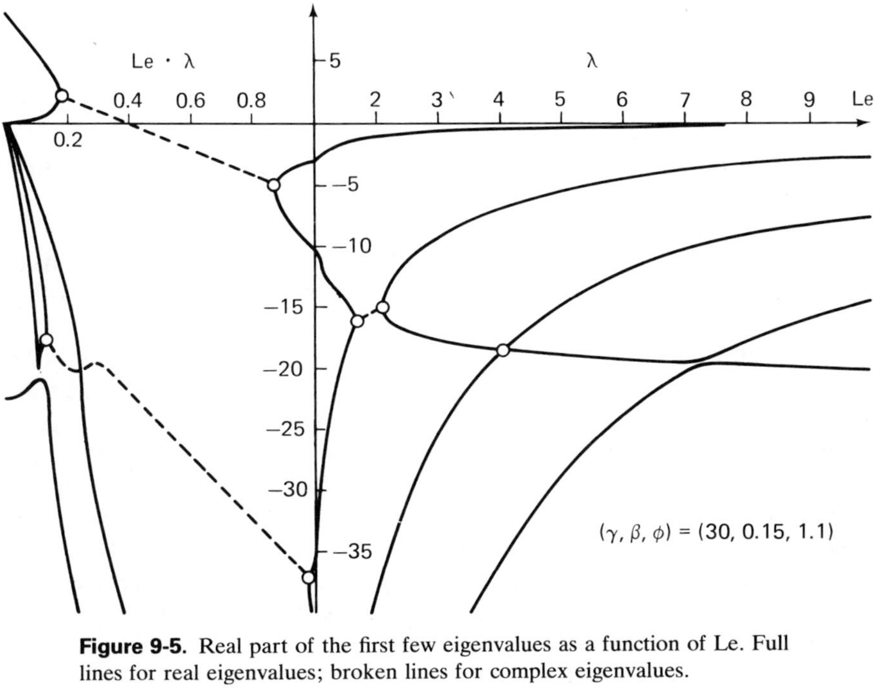

Figure 9-5. Real part of the first few eigenvalues as a function of Le. Full lines for real eigenvalues; broken lines for complex eigenvalues.

For $Le < 1$ we have shown $Le \cdot \lambda$ rather than λ as a function of Le. The first pair of eigenvalues form a complex pair in $0.179 < Le < 0.878$ and the second eigenvalue pair is complex in $0.126 < Le < 0.993$. In both cases the real part of the complex eigenvalue pair is a linear function of Le^{-1} as seen by the straight, broken lines. For the first eigenvalue pair that is unaffected by any other eigenvalues the straight line continues until the eigenvalues separate at $Le = 0.179$, while the interference with higher eigenvalues at $Le \sim 0.22$ is reflected in the last part of the broken line for the second eigenvalue pair.

We see in tables 9.14 and 9.16 that the perturbation series based upon the Sturm–Liouville problems (the variable group for $Le \to 0$ and the constant group for $Le \to \infty$) break down but also that the first few well-determined terms give a fairly accurate representation of the eigenvalues close to the limiting Le-values (the upper lines in tables 9.15 and 9.17).

Figure 9-5 may help to explain this result: For $Le \to 0$, all except the first (positive) eigenvalue of (49a) on their passage to $-\infty$ encounter an infinity of eigenvalues, solutions of (48) and (95), that move toward constant, negative final values for $Le = 0$. For $Le \to \infty$, the eigenvalues of (48a) that move toward constant, negative values (-20.68, $-53.3, \ldots$ from table 9.12) are encountered by an infinity of eigenvalues, solutions of (99), that move toward zero. Thus theoretically the convergence radius of the perturbation series for the Sturm–Liouville problems at either $Le = 0$ or $Le \to \infty$ should be zero, but since the interaction with other eigenvalues is very weak—formation of complex pairs takes place in a very narrow Le-region—fairly accurate results are obtained when the first terms of the divergent series are used.

At first glance it seems peculiar that the series for the positive eigenvalue in the variable group also breaks down, even though this particular eigenvalue does not encounter other eigenvalues during its passage from 13.5 (point I–I′ in table 9.19) to infinity, but the perturbation series is formally valid also for $Le < 0$, where crossings with other eigenvalues certainly take place.

The non-Sturm–Liouville problems (48) and (95) for $Le = 0$ and (99) for $Le \to \infty$ are well behaved and the perturbation series developed from these equations have a large radius of convergence that is in good agreement with the Le-value for which the particular eigenvalue first enters into a complex pair. Thus the series for the first constant group eigenvalue for small Le is convergent for $Le < 0.2$ and the series for the first variable group eigenvalue for $Le \to \infty$ is convergent for $Le \gtrsim 0.8$, although results for $Le > 0.179$ and $Le < 0.9$, respectively, are necessarily somewhat in error since the eigenvalues become complex beyond these limits.

The two series (137) and (138) from Le = 1 are convergent only in a short interval around Le = 1 dictated by the formation of a complex pair at Le = 0.88.

If we form the sum of the two series in either (137) or (138),

$$q = \lambda_1^{(z)} + \lambda_1^{(v)} = t_0 + \left(\frac{1}{\text{Le}} - 1\right)t_1 + \left(\frac{1}{\text{Le}} - 1\right)^2 t_2 + \ldots \tag{139}$$

$$\text{Le}\, q = \text{Le}\,[\lambda_1^{(z)} + \lambda_1^{(v)}] = t_0 + (\text{Le} - 1)t_1 + (\text{Le} - 1)^2 t_2 + \ldots \tag{140}$$

somewhat better results are obtained since the radius of convergence is now determined by the formation of a complex pair at Le = 1.67.

With the maximum $|\Delta|$ obtained from this value of Le, one obtains for series (139)

$$\left|\frac{\text{Le} - 1}{\text{Le}}\right| < \left|\frac{1.67 - 1}{1.67}\right| = 0.401$$

$$\left.\begin{array}{lll} \text{Le} > 1: & \dfrac{\text{Le} - 1}{\text{Le}} < 0.401 & \text{or} \quad \text{Le} < 1.67 \\ \text{Le} < 1: & \dfrac{\text{Le} - 1}{\text{Le}} > -0.401 & \text{or} \quad \text{Le} > 0.71 \end{array}\right\} 0.71 < \text{Le} < 1.67$$

while one obtains $0.33 < \text{Le} < 1.67$ directly for (140).

It is seen that with Le < 1 (140) can be used for a much smaller value of Le than (139). In particular, (140) with coefficients obtained by summation of coefficients in the two series (138) may be used to find the value for which the $(\gamma, \beta, \Phi) = (30, 0.15, 1.1)$ system becomes unstable:

$$\begin{aligned} \text{Le}_{cr}\,[\lambda_1^{(z)} + \lambda_1^{(v)}] = 0 &\sim \sum_{i=0}^{N} t_i(\text{Le}_{cr} - 1)^i \\ &= -12.9916 - 21.2563(\text{Le}_{cr} - 1) - 0.1047(\text{Le}_{cr} - 1)^2 \\ &\quad - 0.6289(\text{Le}_{cr} - 1)^3 \ldots \end{aligned} \tag{141}$$

For $N = 1$, (141) yields $\text{Le}_{cr} = 0.389$; for $N = 2$, $\text{Le}_{cr} = 0.387$; while, for any $N > 4$, $\text{Le}_{cr} = 0.393$ or 0.394 is obtained.

This result is, of course, in remarkable agreement with the exactly computed critical Le-value 0.3925, and the lowest-order perturbation solution, which is $\lambda_1^{(z)} + \lambda_1^{(v)} = (8.2647 - 21.2563\,\text{Le})/\text{Le}$ gives four- to five-digit accuracy for the sum of the real parts of $\lambda_1^{(z)}$ and $\lambda_1^{(v)}$ almost throughout the range of validity of (140).

The value of the computationally quite simple perturbation methods have again been demonstrated. The complexity of the present problem does require a much deeper analysis, however, than was the case for the comparatively trivial Graetz problem before the apparently disappointing results of a straightforward application of the series from Le = 1, 0, or ∞ can be explained. For more extreme values of (γ, β, Φ) than were used in this subsection, it may not be possible to use any perturbation series to obtain the critical value of Le below which instability occurs.

9.3 Fixed Bed Reactor Dynamics—Transfer Functions and State Space Formulation by Collocation

Let us consider an adiabatic tubular reactor filled with inert particles. A first-order exothermic, homogeneous reaction occurs in the fluid phase, and the solid particles act only as fluid distributors and as a heat capacitance. Plug flow and radial homogeneity are valid assumptions for the model, which has been shown to represent many laboratory experiments, e.g., the reaction between hydrogen peroxide and sodium thiosulfate. With minor modifications a heterogeneous reaction in the solid phase and heat loss through the wall of the reactor may be included. Axial dispersion may also be included at least for a gaseous fluid phase without further complications of the model. This last extension is discussed in subsection 9.3.4.

Stangeland and Foss (1970) and Michelsen, Vakil, and Foss (1973) have discussed the model in the following form:

$$\frac{\partial C}{\partial z} + \frac{\partial C}{\partial t} = -R(C, T) = -kC \exp\left(\frac{\gamma T}{T + T_r}\right) \tag{142}$$

$$\frac{\partial T}{\partial z} + \frac{\partial T}{\partial t} = R(C, T) + H(T_p - T) \tag{143}$$

$$\frac{\partial T_p}{\partial t} = H\beta(T - T_p) \tag{144}$$

The results of the present section are based on the paper mentioned last; and to make cross-reference easier we use the author's nomenclature rather than that of the present text. The similarity between (142) to (144) and our considerably more complicated model (1.87) does justify a brief comparison, however, between the two models.

The main difference between (142) to (144) and (1.87) is that in Stangeland's model the reaction takes place in the fluid phase, while it occurs on or inside pellets in (1.87). Stangeland also assumed that the linear velocity of reactants is constant through the reactor (ρ_b and v are temperature independent) and that there is no heat loss from the reactor wall ($H_W = 0$).

The variable t of (142) to (144) is time measured in units of fluid residence time $L\varepsilon_b/v_0$. H is the same as our H_b in (1.87) and β is the ratio of heat capacity of fluid to heat capacity of solid. Multiplication of H by β in (144) corresponds to the change of time scale between (1.87) and (144). When Stangeland rewrites his model into (155) to (157), the unit of time in the two models is again the same.

Other translations between variables of (1.87) and (142) to (144) are

$$C = \frac{c_b}{c_{b0}}, \qquad z = \zeta, \quad \text{and} \quad k = \text{Da} \tag{145}$$

The temperature T of (142) to (144) is the temperature rise above the steady state inlet temperature T_{b0} measured in units of the adiabatic temperature rise $\beta_{b0}T_{b0}$:

$$T = \frac{T_b - T_{b0}}{\beta_{b0}T_{b0}} \tag{146}$$

$$T_r = \frac{T_{b0}}{\beta_{b0}T_{b0}} = \frac{1}{\beta_{b0}} \tag{147}$$

$$\frac{T}{T + T_r} = \frac{T_b - T_{b0}}{T_b} = 1 - \frac{1}{\theta_b} \tag{148}$$

T_p is the solid temperature measured in the same units as T.

Stangeland's temperature function T is, in fact, preferable to our θ_b for computational purposes since more digits are preserved in $T/(T + T_r)$ than in θ_b.

The steady state model does not contain (144) and the two remaining equations in C_s, T_s for the adiabatic reactor

$$C_s + T_s = 1 \tag{149}$$

$$\frac{dT_s}{dz} = R(C_s, T_s) = R(1 - T_s, T_s) \tag{150}$$

with

$$C_s = 1 \quad \text{and} \quad T_s = 0 \quad \text{at } z = 0$$

are also slightly more convenient than the corresponding steady state equations in terms of (y_b, θ_b).

9.3.1 Numerical treatment of the linearized model

The steady state model can be solved to any degree of accuracy using any of the forward integration methods of section 8.2. The analysis of the transient model however, is just as in the previous section, based on discretization of (142) and (144), linearized from the steady state:

$$\frac{\partial c}{\partial z} + \frac{\partial c}{\partial t} = -R_c(z)c - R_\theta(z)\theta \tag{151}$$

$$\frac{\partial \theta}{\partial z} + \frac{\partial \theta}{\partial t} = R_c(z)c + R_\theta(z)\theta + H(\Phi - \theta) \tag{152}$$

$$\frac{\partial \Phi}{\partial t} = H\beta(\theta - \Phi) \tag{153}$$

with

$$c = C - C_s, \qquad \theta = T - T_s, \qquad \Phi = T_p - T_{ps} = T_p - T_s \tag{154}$$

It is now preferable to integrate the steady state equations by a global collocation method. Zeros of $P_N^{(0,0)}(z)$ are used as collocation points and $T_s(z = 0)$ is known (= 0). In this way R_c and R_θ are computed at the collocation points. $N = 6$ or at most 8 gives satisfactory results for the parameter values used in the following.

The time derivative is eliminated from (151) and (152) using the variable transformation $\tau = (t - z)\beta$:

$$\frac{\partial c}{\partial z} = -R_c c - R_\theta \theta \tag{155}$$

$$\frac{\partial \theta}{\partial z} = R_c c + R_\theta \theta + H(\Phi - \theta) \tag{156}$$

$$\frac{\partial \Phi}{\partial \tau} = H(\theta - \Phi) \tag{157}$$

τ is a characteristic time into which the transport delay between reactor inlet and position z is incorporated. The unit of the τ time scale is the thermal residence time, i.e., the time required for the thermal wave to pass through the system. β, the ratio between thermal capacity of fluid and solid, is eliminated by the transformation, and only one parameter H, which is proportional to the heat transfer coefficient between fluid and particles, remains besides the parameters of the kinetic expression. For a liquid-phase fluid, β is of the order of 1, whereas β has a much lower value (~0.001) for a gaseous fluid phase. In the last case the accumulation terms $\partial c/\partial t$ and $\partial T/\partial t$ can be neglected in (142) and (143) and axial dispersion can be included without any complications.

Discretization of (155) to (157) at the zeros of $P_N^{(0,0)}(z)$ yields

$$\mathbf{A}\mathbf{c} + \mathbf{A}_0 c_0 = -\mathbf{R}_c\mathbf{c} - \mathbf{R}_\theta\boldsymbol{\theta} \tag{158}$$

$$\mathbf{A}\boldsymbol{\theta} + \mathbf{A}_0\theta_0 = \mathbf{R}_c\mathbf{c} + \mathbf{R}_\theta\boldsymbol{\theta} + H(\boldsymbol{\Phi} - \boldsymbol{\theta}) \tag{159}$$

$$\frac{d\boldsymbol{\Phi}}{d\tau} = H(\boldsymbol{\theta} - \boldsymbol{\Phi}) \tag{160}$$

$\mathbf{A}$ is the matrix for the first derivative at the collocation points and $\mathbf{R}_c$, $\mathbf{R}_\theta$ are diagonal matrices. $2N$ algebraic equations (158) and (159) coupled to N ordinary differential equations (160) result for a single chemical reaction. For M independent reactions, (158) contains $M \times N$ algebraic equations rather than N, but except for the increase of size of the system of algebraic equations no further complications arise.

9.3.2 Transfer functions

The system of equations (158) to (160) is Laplace transformed with zero conditions and subsequently solved for the transformed variables $\bar{\boldsymbol{\theta}}(s)$, $\bar{\mathbf{c}}(s)$, and $\bar{\boldsymbol{\Phi}}(s)$:

$$\bar{\boldsymbol{\Phi}} = \frac{H}{s+H}\bar{\boldsymbol{\theta}} \tag{161}$$

$$\bar{\mathbf{c}} = -(\mathbf{A} + \mathbf{R}_c)^{-1}(\mathbf{R}_\theta\bar{\boldsymbol{\theta}} + \mathbf{A}_0\bar{c}_0) \tag{162}$$

$$\begin{aligned}\bar{\boldsymbol{\theta}} &= \left[\mathbf{A} + \mathbf{R}_c(\mathbf{A} + \mathbf{R}_c)^{-1}\mathbf{R}_\theta - \mathbf{R}_\theta + \frac{sH}{s+H}\mathbf{I}\right]^{-1}[-\mathbf{A}_0\bar{\theta}_0 \\ &\quad - \mathbf{R}_c(\mathbf{A} + \mathbf{R}_c)^{-1}\mathbf{A}_0\bar{c}_0] \\ &= (\mathbf{Q} + q\mathbf{I})^{-1}[-\mathbf{A}_0\bar{\theta}_0 - \mathbf{R}_c(\mathbf{A} + \mathbf{R}_c)^{-1}\mathbf{A}_0\bar{c}_0]\end{aligned} \tag{163}$$

where

$$q = \frac{sH}{s+H} \tag{164}$$

Equations (163) and (162) [with θ inserted from (163)] provide the transfer functions between the input $[\theta_0(\tau) = \theta_0(z = 0, \tau),\ c_0(\tau) = c_0(z = 0, \tau)]$ and (θ, c) at the collocation points.

The collocation ordinates can be used in an interpolation formula to find values of $\bar{c}$ and $\bar{\theta}$ at any other point; e.g.,

$$\bar{c}(z) = f_0(z)\bar{c}_0 + \mathbf{f}^T(z)\bar{\mathbf{c}} \tag{165}$$

where $f_j(z)$ is the Lagrange interpolation polynomial

$$f_j(z) = \frac{zP_N(z)}{(z - z_j)[zP_N(z)]_{z_j}^{(1)}} \tag{166}$$

We specifically wish to find the transfer functions between inlet and outlet $z = 1$. These are defined by

$$\begin{pmatrix} \bar{\theta} \\ \bar{c} \end{pmatrix}_{z=1} = \begin{pmatrix} G_{\theta\theta} & G_{\theta c} \\ G_{c\theta} & G_{cc} \end{pmatrix} \begin{pmatrix} \bar{\theta}_0 \\ \bar{c}_0 \end{pmatrix} \tag{167}$$

Each of the transfer functions can be found from (162), (163), and (165). The simplest to write up is $G_{\theta\theta}$, the transfer function between inlet and outlet temperature that is found directly from (163) and (155):

$$\begin{aligned} G_{\theta\theta} &= f_0(1) - \mathbf{f}^T(1)(\mathbf{Q} + q\mathbf{I})^{-1}\mathbf{A}_0 \\ &= f_0(1) - \mathbf{f}^T(1)\mathbf{U}(\mathbf{\Lambda} + q\mathbf{I})^{-1}\mathbf{U}^{-1}\mathbf{A}_0 \end{aligned} \tag{168}$$

Equation (168) is very similar to formulas we have used repeatedly in previous chapters [e.g., see equation (4.93), a mean value calculation]. Writing (168) in the form

$$G_{\theta\theta}(s) = f_0(1) + \sum_{i=1}^{N} \frac{\nu_i}{q + \lambda_i} \tag{169}$$

where $\nu_i = [-\mathbf{f}^T(1)\mathbf{U}]_i \cdot (\mathbf{U}^{-1}\mathbf{A}_0)_i$ brings the transfer function in the commonly used form—as the sum of partial fractions with q given by (164) and λ_i = eigenvalues of $\mathbf{Q}$. The poles of $G_{\theta\theta}$ are at

$$q_i = -\lambda_i \quad \text{or} \quad s_i = -\frac{\lambda_i H}{\lambda_i + H}$$

and these poles are of course identical for any of the four transfer functions between the inlet and any point in the reactor.

An approximate impulse response for the system in the form of a direct transmission term [from $f_0(1)\bar{\theta}_0$] and a sum of exponentials is immediately available after reintroducing s from (164) and inverse transformation of the partial fractions in the usual way.

Also (169) is directly amenable to frequency analysis. $s = i\omega$ is inserted with the N computed eigenvalues of $\mathbf{Q}$ and now amplitude and phase of $G_{\theta\theta}$ can be obtained numerically.

With the particular reactor model (155) to (157), we are in the fortunate position that expressions in closed form for the transfer functions have been derived [Stangeland and Foss (1970)], and certain important features of the collocation solution, e.g., (169) may be compared with results from an "exact" representation of the transfer functions. Insofar as these features are common for reactor control studies, one might arrive at some general conclusions as to the applicability of the collocation approach.

The most interesting transfer functions are $G_{\theta c}$ and $G_{\theta\theta}$ and these are given by

$$G_{\theta c}(q) = R(z = 1, T_s) \int_0^1 \frac{1}{C_s(y)} \exp(-qy)\, dy \tag{170}$$

$$G_{\theta\theta}(q) = \frac{R(z = 1, T_s)}{R(z = 0, T_s)} + \left[1 + \frac{q}{R_c(z = 1)}\right] G_{\theta c}(q) \tag{171}$$

The constant $R(z = 1, T_s)$ and the monotonically increasing positive function $1/C_s(z)$ are found from the steady state solution. Consequently when a value of $q = a + i\omega$ is inserted, the integral may be evaluated. Michelsen, Vakil, and Foss obtained identical results with sixth-order Gaussian quadrature applied to each of either 5 or 10 subdivisions of the interval [0, 1].

There are no simple poles of (170) but a pole of infinite order originating in the exponential term exists at $q \to -\infty$, i.e., when s tends to $-H$ from above as seen in (164). This immediately shows that a model analysis of the collocation model based on its poles, s_i, is at best unsound.

Equation (170) can have no real-valued zeros except the uninteresting zero at $q \to \infty$. It may have an infinite number of distinct complex zeros, however, and these may be compared to the zeros of the corresponding collocation $G_{\theta c}$ function.

A specific set of parameters $\gamma = 30$, $T_r = 7$, $k = 0.35$ were inserted into (149) and (150) to obtain the steady state solution, which is shown graphically in figure 9-6. The zeros q_k of $G_{\theta c}$ and $G_{\theta\theta}$ can be obtained as described above. The zeros $q_k = a_k + i\omega_k$ of (170) are marked in figure 9-7 as open circles, while the zeros of (171) are marked as full circles. Complex conjugate zeros $q_k = a_k - i\omega_k$ are not shown.

The transformation (164) maps the imaginary axis $s = i\omega$ of the complex s-plane into a circle with center at $(H/2, 0)$ and radius $H/2$. The positive half plane is mapped into the region within the circle that is also shown in figure 9-7 for $H = 17.5$.

With our choice of parameters no zeros q_k of (170) and three zeros ($q_k = 1.91$ and $q_k = 6.99 \pm i7.32$) of (171) are inside the circle. It is obvious that as H increases, an increasing number of the zeros of (170) and (171) (which are about evenly spaced with a vertical distance $\sim 2\pi$) are found within the circle of radius $H/2$.

Table 9.20 gives numerical values for the dominant zeros obtained from the collocation solution and from (170) and (171).

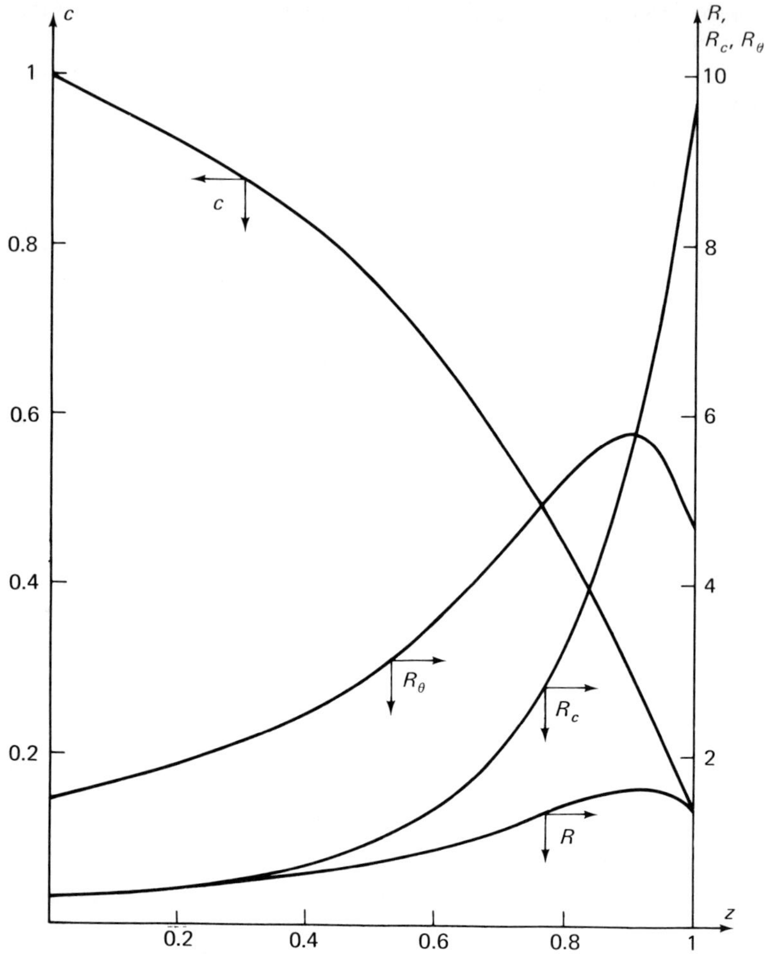

Figure 9-6. Steady state axial concentration, rate, and sensitivity profiles for fixed bed reactor; $\gamma = 30$, $T_r = 7$, and $k = 0.35$.

Bode plots of the frequency responses for $G_{\theta c}$ and $G_{\theta\theta}$ with $H = 17.5$ are shown in figure 9-8.

Each zero $q_k = Hs_k/(H + s_k)$ of $G_{\theta\theta}(q)$ inside the circle in figure 9-7 gives rise to a phase shift of $-180°$ when ω goes from 0 to ∞ in the frequency analysis $s = i\omega$. Any zero outside the circle gives an ultimate phase shift of zero when $\omega \to \infty$.

Figure 9-9 shows the magnitude of the transfer function error as a function of ω for $G_{\theta\theta}$ and various approximation order N. The error decreases rapidly with N, but ill-conditioning due to closely spaced eigenvectors of $\mathbf{Q}$ eventually sets a limit to the obtainable accuracy. The coefficients ν_i of (169) increase rapidly with N and are typically 10^7 for

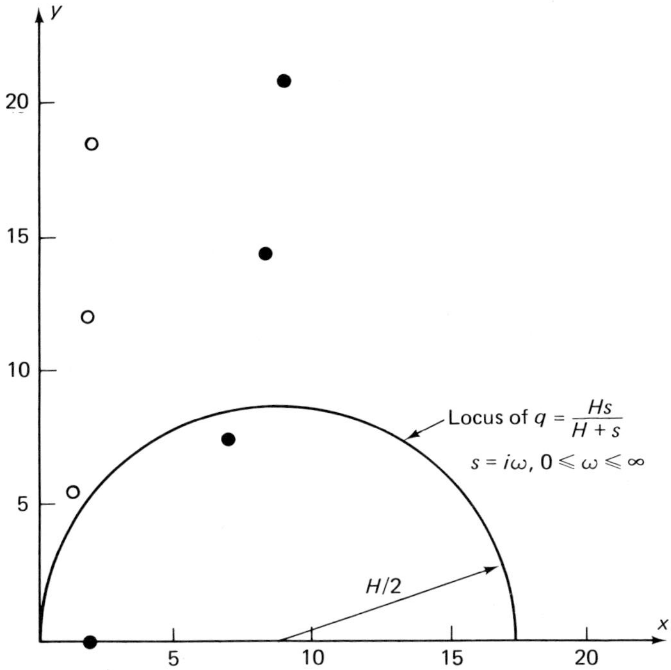

Figure 9-7. $q = x + iy$ plane with zeros of $G_{\theta c}$ (open circles) and zeros of $G_{\theta\theta}$ (full circles). The positive half plane Re $(s) > 0$ is mapped into the interior of the circle.

TABLE 9.20
DOMINANT ZEROS q_k OF TRANSFER FUNCTIONS BY Nth-ORDER COLLOCATION

N/k	1	2	3
	$G_{\theta c}(q)$		
8	1.4450 ± i5.5293	1.73 ± i12.00	
10	1.44434 ± i5.52879	1.772 ± i12.022	1.6 ± i18.7
12	1.444378 ± i5.528775	1.777007 ± i12.02075	1.881 ± i18.438
Exact	1.444376 ± i5.528776	1.777004 ± i12.02062	1.891 ± i18.432
	$G_{\theta\theta}(q)$		
8	1.9147	6.998 ± i7.3262	8.0 ± i13.0
10	1.91443	6.99578 ± i7.32611	8.02 ± i14.36
12	1.914412	6.995848 ± i7.326076	8.174 ± i14.292
Exact	1.914413	6.995854 ± i7.326081	8.170 ± i14.290

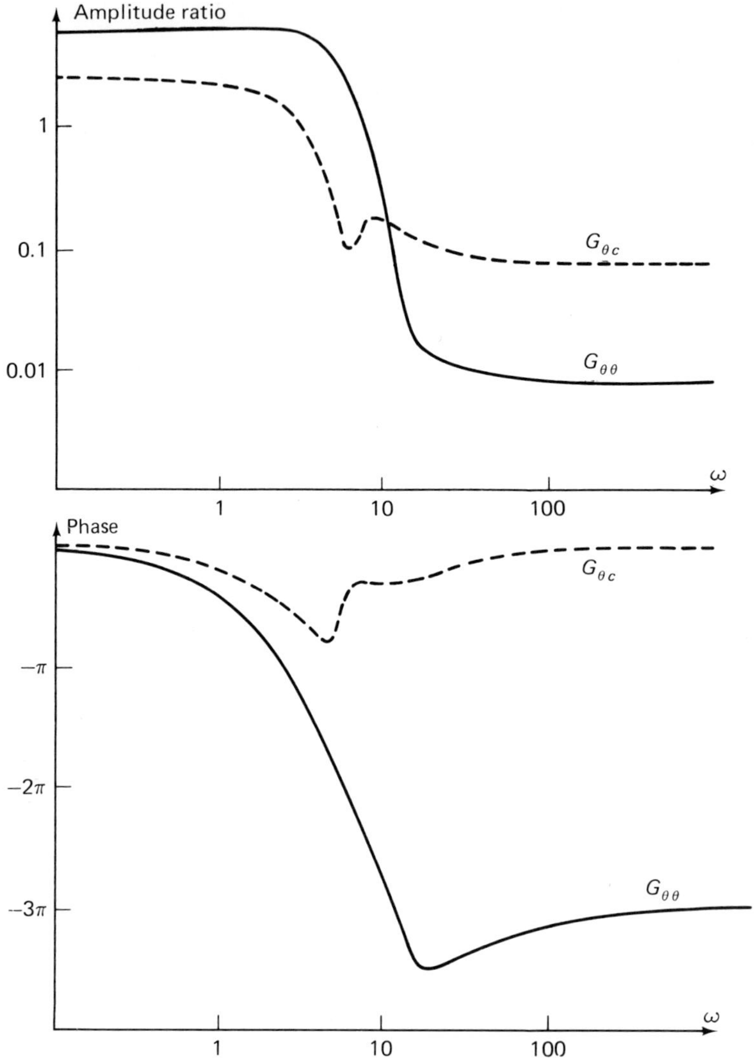

Figure 9-8. Bode plot for transfer functions.

$N = 12$. Even the $N = 6$ result is satisfactory for practical control purposes, however, and a similar accuracy could only be obtained in a tank-in-series model with about 1000 mixing cells, an indication of the power of the method as compared to conventional techniques.

In general, zeros close to or inside the circle on figure 9.7 (as explained above) will have a pronounced influence on the frequency response, and table 9.20 demonstrates that the location of these zeros is extremely well predicted in the approximate model. This does lend

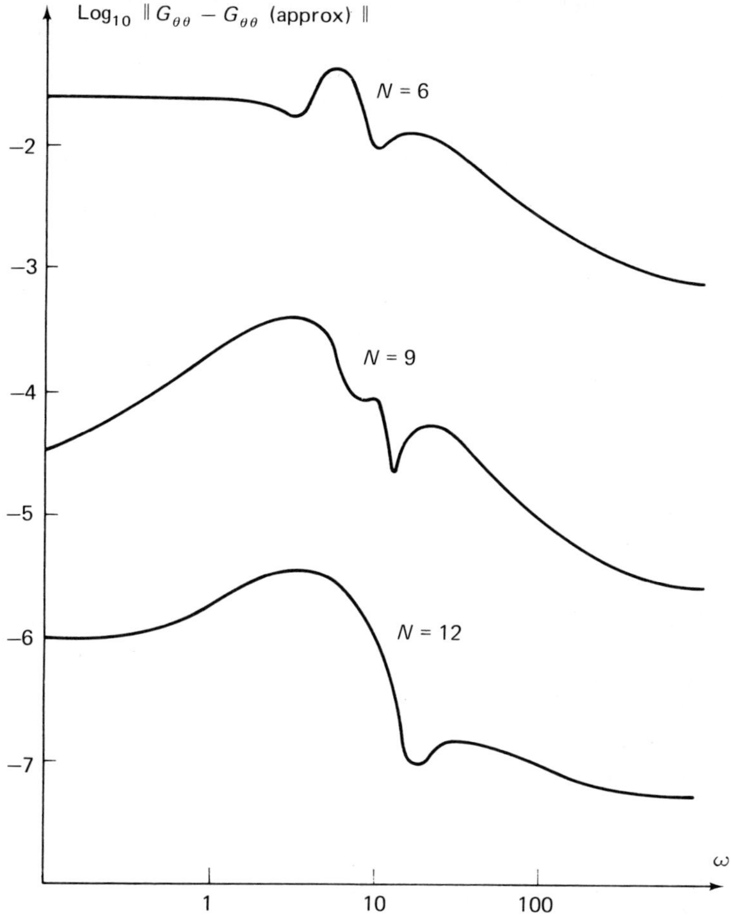

Figure 9-9. Approximation error for $G_{\theta\theta}$ using Nth-order collocation.

credibility to the collocation process as a means of obtaining transfer functions for dynamic reactor studies, but it should of course be remembered that as H increases and an increasing number of zeros fall inside the circle, the collocation method will eventually break down and an incorrect phase shift curve is calculated for large ω.

9.3.3 State-space equations

An alternative representation of the dynamic behavior of the reactor is the so-called state-space form in which the state of the system is given as a single set of coupled differential equations in the state variable while other dependent variables are expressed in terms of the state variable.

The most convenient choice of state variable is the particle temperature $\mathbf{\Phi}$ at the collocation points since it leads to a standard form directly amenable to control system analysis. Fluid temperature or concentration could also have been used as state variable, but the resulting model would involve time derivatives of input variables c_0 and θ_0.

We solve the algebraic equations (158) and (159) for $\mathbf{c}$ and $\boldsymbol{\theta}$ in terms of $\mathbf{\Phi}$, c_0, and θ_0 and insert into (160):

$$\begin{aligned} \frac{d\mathbf{\Phi}}{d\tau} &= -H\mathbf{\Phi} + H\boldsymbol{\theta} \\ &= [-H\mathbf{I} + H^2(\mathbf{Q} + H\mathbf{I})^{-1}]\mathbf{\Phi} + \mathbf{Q}_1\begin{pmatrix}\theta_0\\ c_0\end{pmatrix} \end{aligned} \tag{172}$$

with $\mathbf{Q}_1 = H(\mathbf{Q} + H\mathbf{I})^{-1}[-\mathbf{A}_0, -\mathbf{R}_c(\mathbf{A} + \mathbf{R}_c)^{-1}\mathbf{A}_0]$.

The solution of (158) and (159) at any point z can be obtained in terms of the state vector $\mathbf{\Phi}$ by interpolation:

$$\begin{pmatrix}\theta\\ c\end{pmatrix}_z = \mathbf{Q}_2(z)\mathbf{\Phi} + \mathbf{Q}_3(z)\begin{pmatrix}\theta_0\\ c_0\end{pmatrix} \tag{173}$$

The term involving θ_0 and c_0 in (173) represents a direct transmission of the input variables that arises owing to the use of characteristic time τ as the independent variable.

A control system analysis based on this representation of Φ, c, and θ using Kalman filtering for state estimation and a quadratic objective function has been described by Vakil, Michelsen, and Foss (1973).

9.3.4 Reactor dynamics with axial dispersion and heat loss included

For a gaseous fluid phase the parameter β of (144) is sufficiently small to make the accumulation terms in (142) and (143) negligible in comparison with the convective terms, and in that case (1.87) can be treated in the same way as Stangeland's model also with axial dispersion of heat and mass included. In the nomenclature of the present section one obtains

$$-\frac{1}{\mathrm{Pe}_M}\frac{\partial^2 c}{\partial x^2} + \frac{\partial c}{\partial x} = -R(c, \Phi) \tag{174}$$

$$-\frac{1}{\mathrm{Pe}_H}\frac{\partial^2 \theta}{\partial x^2} + \frac{\partial \theta}{\partial x} = H(\Phi - \theta) - b(\theta - \theta_W) \tag{175}$$

$$\frac{1}{\beta}\frac{\partial \Phi}{\partial t} = R(c, \Phi) + H(\theta - \Phi) \tag{176}$$

These equations can be solved in the same manner as (155) to (157) and qualitatively similar results are obtained. The profiles are often quite steep when realistic values of Pe_M, Pe_H, H are inserted and typically $N = 12$ to 18 collocation points are needed to obtain accurate results.

EXERCISES

1. Pirkle and Sigillito (1972) have found the eigenvalues for the $y > 0$ region of the extended Graetz problem using hypergeometric series. They expand θ as

$$\theta = \sum_1^\infty c_i F_i(x) \exp(-\Phi_i^2 y) \qquad (y > 0)$$

where

$$\frac{d^2F}{dx^2} + \frac{1}{x}\frac{dF}{dx} + \left[\frac{\Phi^4}{\text{Pe}^2} + \Phi^2(1 - x^2)\right]F = 0 \tag{1}$$

and $F(x = 1) = 0$. We have used the exponentials $\exp(\lambda_i y)$ with $\lambda_i = -\Phi_i^2 < 0$ in our expansion (9.10).

Show that (1) can be transformed to

$$\eta\frac{d^2M}{d\eta^2} + (1 - \eta)\frac{dM}{d\eta} - aM = 0 \quad \text{with } M(\Phi) = 0 \tag{2}$$

$$M = F\exp\left(\frac{\Phi x^2}{2}\right), \qquad \eta = \Phi x^2$$

$$a = \frac{1}{2} - \frac{\Phi}{4} - \frac{\Phi^3}{4}\alpha \quad \text{and } \alpha = \frac{1}{\text{Pe}^2}$$

By reference to Morse and Feshbach, show that the solution of (2) is

$$M = 1 + a\eta + \frac{a(a + 1)}{(2!)^2}\eta^2 + \frac{a(a + 1)(a + 2)}{(3!)^2}\eta^3 + \ldots \tag{3}$$

Show that (3) correctly degenerates to the series expansion for the Bessel function J_0 when $\text{Pe} \to 0$, and confirm that the eigenvalues $\Lambda = \lambda/\text{Pe}$ appear as zeros of $J_0(\rho)$ for $\text{Pe} = 0$.

2. Make a computer program for calculation of the eigenvalues Φ from (2) using Newton's method and analytical differentiation of M with respect to Φ. The following may be helpful:

$$\frac{\partial M}{\partial \Phi} = \sum_1^\infty \left[\frac{df_i(a)}{da}\left(-\frac{1}{4} - \frac{3\Phi^2}{4\text{Pe}^2}\right)\Phi^i + f_i(a)i\Phi^{i-1}\right]$$

$$f_1(a) = a, \qquad f_2(a) = \frac{a(a + 1)}{(2!)^2}, \qquad f_3(a) = \frac{a(a + 1)(a + 2)}{(3!)^2}, \qquad \text{etc.}$$

The calculation of df_i/da is made recursively using df_{i-1}/da and f_{i-1}.

Next calculate $\partial M/\partial \alpha$ and obtain $d\Phi/d\alpha$, the rate of change of the eigenvalue with the parameter α from

$$dM = 0 = \frac{\partial M}{\partial \alpha} d\alpha + \frac{\partial M}{\partial \Phi} d\Phi$$

Now from the solution of $M(\Phi) = 0$ at one α value (e.g., $\alpha = 0$), it is possible to find $\Phi(\alpha)$ by forward integration.
Extend the previous program with this feature, and check the result by comparison with the true solution of $M(\Phi) = 0$ for $\alpha = 0.01, 0.1, 1, 4, 25$, and each of the first three eigenvalues.

3. The first eigenfunction $M_1(\eta)$ is obtained simultaneously with the computation of Φ_1.
Show that Nu_{as} in (9.21) is obtained from

$$\text{Nu}_{as}(\alpha) = \frac{2\Phi_1 \exp[-(\Phi_1/2)](\partial M_1/\partial \eta)|_{\eta=\Phi_1}}{\int_0^1 (1-u)\exp[-(u\Phi_1/2)]M_1(\Phi_1 u)\,du} \tag{4}$$

The numerator of (4) is easily obtained from the series for M, but analytic integration of the series for the integrand in the denominator is cumbersome. Thus it is proposed to evaluate the integral by Gauss quadrature using $N = 1, 2, \ldots, M$ points until sufficiently accurate results are obtained.
Continue the computer exercise by calculation of Nu_{as} for the α values mentioned above. Check with the perturbation series of subsection 9.1.3.

4. In section 9.2, we have seen that the breakdown of the perturbation series was caused by the confluence of two real eigenvalues—one from each of two parent eigenvalue problems—into a complex pair.
The perturbation series for the extended Graetz problem also breaks down when Pe decreases below ~ 6 in one of the series or increases above ~ 6 in the other.
The computed eigenvalues for the $y > 0$ tube section have been real for any Pe in $[0, \infty]$ and collapse of eigenvalues can apparently not take place for real Pe.
The reason for the breakdown of the perturbation series of, e.g., table 9.4 must be sought in the behavior of the series for purely imaginary Pe—i.e., $\alpha < 0$ in (2).
Continue the integration of $d\Phi_1/d\alpha$ backward from $\alpha = 0$ until a vertical asymptote is reached at $\alpha \sim -1/6^2$ (i.e., Pe $\sim i6$). Show that the vertical asymptote corresponds to a double root of $M(\Phi) = 0$ and collapse of real eigenvalues at this α-value.

5. In Exercise 4 we have explained the failure of the perturbation series $\lambda_1(\text{Pe})$ at Pe ~ 6 by the behavior of the series for physically nonrealizable values of Pe in much the same way as the failure of the perturbation series for the first eigenvalue of the Le $\to$ 0 problem was explained in section 9.2.
There is no reason to believe that the first eigenvalue will enter into a complex pair for any positive value of Pe—this would lead to oscillatory heating far downstream where the first eigenfunction dominates, and this is unrealistic from a physical viewpoint.

Some of the higher eigenvalues of the non-Sturm–Liouville problem (9.8) could well be complex, however, without disturbing our picture of the physical situation.

All computational evidence points to the result that any eigenvalue for the $y > 0$ region is real for any real value of Pe. To our knowledge, this has not been explicitly proved in any publication on the subject.

Can you prove that the eigenvalues $\lambda_i(\text{Pe})$ are all real? The material of Exercises 1–4 *might* be helpful.

6. A large part of chapter 9 is devoted to an analysis of the linearized transient equations for the catalyst pellet. In this exercise a few features of the nonlinear pellet dynamics problem will be studied. The topic contains such a large variety of interesting phenomena that the exercise can serve only as an appetizer for a much more thorough study.

a. Write subprograms FUN, DFUN, and COLL for solution of the pellet dynamics problem following the guidelines of Exercise 8.11.

b. Test the program on Lee and Luss's example $\beta = 0.15$, $\gamma = 30$, and $\Phi = 1.1$. The following values of Le should be used: 50, 2, 1, 0.6, 0.4, 0.3, 0.1. The case Le = 0.1 is shown in figure 7-12 of Aris (1975).
The program should be run with different number of collocation points and different tolerances ε. All ordinates (N concentrations and N temperatures) should have the same weight in the tolerance vector. Note the influence of the tolerance. If ε is too large, the response is out of phase but the cycle time is constant and no numerical instability occurs. Even $N = 2$ gives a good estimate of the cycle time, but $N = 4$ to 6 must be used to obtain the finer details of the transients.

c. For Le = 50, the simplified model (1.85) and (1.86) should be compared with the full model.

d. Modify COLL to account for a film boundary resistance. How does this added resistance influence the stability limits, the maximum over-temperature of the pellet and the cycle time?

e. Change (γ, β) to the values of figure 5-4: $\gamma = 20$, $\beta = 0.3$.
Study the transients for Φ values between points B and C on the figure. Try different initial profiles to see which steady state is finally reached. Compare your numerical results with the theory of Aris (1975), chapter 7.

7. Breakthrough curves for fixed bed adsorption under constant pattern conditions are investigated by Fleck et al. [Fleck R. D., Kirwan D. J. and Hall K. R., *Ind. Eng. Chem. Fundam.* 12 (1973):95]. Transport equations for a Langmuir adsorption isotherm are summarized in their Table I. An external mass transfer resistance and two intraparticle transport mechanisms, pore diffusion and solid phase diffusion, are taken into account.

An instantaneous equilibrium exists between (dimensionless) concentration X of adsorbate in the fluid contained in the pores and concentration of adsorbate Y on the surface at the same position in the adsorbent

$$Y_{eq} = \frac{X}{R + (1 - R)X}$$

When axial dispersion is neglected and $R < 1$ (constant pattern conditions) the following relation holds between the average concentration $\bar{X}$ in the external fluid phase and the average concentration $\bar{Y}$ in the solid phase:

$$\bar{X} = \bar{Y}$$

The six cases investigated by Fleck et al. are

Model	Pore diffusion	Solid diffusion	External resistance
1	+	−	−
2	−	+	−
3	+	−	+
4	−	+	+
5	+	+	−
6	+	+	+

In Models 1 and 3 the transport equations are formulated in terms of the fluid concentration X and the average solid concentration $\bar{Y}$ is found from

$$\bar{Y} = \frac{1}{V}\int_V Y_{eq}(x)\,dV$$

a. Solve Models 1 and 3 for the parameters of figure 6 in the reference ($R = 0.2$). The calculations are started with a small uniform concentration (e.g. $X = 10^{-5}$) and the time variable is normalized as shown in the reference [eq. (16) and (17)].
b. Solve Models 2. and 4. for the parameters of figure 7 in the reference. The boundary condition at the particle surface is nonlinear for Model 4. The approach of Exercise 8.13 is used to reformulate this boundary condition to

$$\varepsilon\frac{dY_s}{d(N_sT)} = -\frac{RY_s}{(1 + (R - 1)Y_s)} + \bar{Y} - \frac{(N_S/N_F)}{5}\left(\frac{\partial Y}{\partial x}\right)_s$$

c. Formulate a general program which may handle any of the Models 1 to 6.
d. Axial dispersion in the fluid phase is easily included in the models. The constant pattern assumption now leads to

$$\frac{1}{N_D}\frac{d\bar{X}}{dT} = \bar{Y} - \bar{X}$$

where N_D is the number of dispersion transfer units.
Solve Model 3 with axial diffusion included and plot your results using

$$\frac{1}{N_{\text{total}}} = \frac{1}{N_F} + \frac{1}{N_P} + \frac{1}{N_D}$$

The time scale is here normalized such that

$$\int_0^\infty (1 - \bar{Y})\, dT = 1$$

N_F, N_S and N_P are defined in the reference.

8. A fixed bed of inert material is used as a heat exchanger between two gas streams.
Initially the bed temperature is uniform and equal to T_2, the temperature of the hot gas stream. The cold gas with constant inlet temperature $T_1 = T_1(0, t)$ is sent through the bed and exits with temperature $T_1(1, t) = T_1^*$. At time $t = t_{sw}$ the flow is reversed and hot gas is sent through the bed with temperature $T_2 = T_2(1, t)$ for another period t_{sw}. It exits at $z = 0$ with temperature $T_2(0, t) = T_2^*$. At time $t = 2t_{sw}$ the flow is again reversed. We wish to study the steady state cyclic operation of the heat exchanger.
Assume that the flowrate $G\,kg/h$ of the two gas streams is the same and that their heat capacity is the same and independent of temperature. The heat capacity of the gas is negligible compared to that of the solid. Intraparticle temperature gradients and axial heat dispersion are neglected. The gas-solid heat transfer coefficient is temperature independent.

a. Show that the gas temperature T and the solid temperature T_p are described by the differential equations (156, 157) with $R = 0$:

$$\frac{\partial \theta}{\partial z} = H(\Phi - \theta)$$

$$\frac{\partial \Phi}{d\tau} = H(\theta - \Phi)$$

$$\theta = \frac{T - T_1}{T_2 - T_1} \qquad \Phi = \frac{T_p - T_1}{T_2 - T_1}$$

τ is the ratio of actual time t to the thermal residence time $t_H = (WC_p)_p/GC_p$ where W is the weight of solid in the bed, and $(C_p)_p$ the specific heat of the solid.

b. The initial condition for the solid is $\Phi = 1$ for all z.
The inlet condition for the gas is $\theta = 0$ at $z = 0$ for $0 < \tau < t_{sw}/t_H$.
Define the initial thermal effectiveness η_1 as the ratio of heat transferred from solid to gas during the first period, to the heat that would be transferred if the outlet gas temperature had been T_2 (rather than T_1^*) during the whole period $\tau < t_{sw}/t_H$.

$$\eta_1 = \frac{1}{\tau_{sw}} \int_0^{\tau_{sw}} \theta(\tau, 1)\, d\tau$$

Calculate η_1 using collocation in z at the zeros of $P_N^{(0,0)}(z)$.

c. The effectiveness during the kth period is defined by

$$\eta_k = \frac{1}{\tau_{sw}} \int_{(2k-2)\tau_{sw}}^{(2k-1)\tau_{sw}} \theta(\tau, 1)\, d\tau$$

For large k this quantity approaches a final value η_{as}. Show by symmetry considerations that for large k the solid temperature profile Φ will cycle between limits Φ_1 (at $\tau = (2k - 2)\tau_{sw}$) and Φ_2 (at $\tau = (2k - 1)\tau_{sw}$) and that Φ_1 and Φ_2 are related by

$$\Phi_2(z) = 1 - \Phi_1(1 - z)$$

d. Show that

$$\eta_{as} \cdot \tau_{sw} = \int_0^1 (\Phi_1 - \Phi_2)\, dz$$

Compute η_{as} for various values of τ_{sw} and H and compare the results with table 35-5 in Jakob (1957).[a] The parameters Λ and Π of the reference are H and $\tau_{sw}H$, respectively.

e. Let the entry point of both the cold and the hot gas be at $z = 0$. For this cocurrent heat exchanger show that

$$\Phi_1(z) = 1 - \Phi_2(z) \quad \text{when } k \to \infty.$$

Determine η_{as} and compare with table 35-7 in Jakob (1957). Find the optimum value of τ_{sw} and the corresponding η_{as} for $H = 30$. Compare the results with the corresponding η_{as} for countercurrent operation and

a. the same value of τ_{sw}
b. $\tau_{sw} = 1$
c. $\tau_{sw} \to 0$

9. Steady state multiplicity for two parallel irreversible first order reactions ($A \to B, A \to C$) in a stirred tank reactor was studied by Luss and Chen (1975)[b] and by Michelsen (1977).[c] Here we consider the same system of reactions on catalyst particles in planar geometry.

a. In the absence of external transport resistances show that the steady state mass and energy balances are

$$\frac{d^2y}{dx^2} - \Phi_1^2 R_1(y, \theta) - \Phi_2^2 R_2(y, \theta) = 0$$
$$\frac{d^2\theta}{dx^2} + \beta_1 \Phi_1^2 R_1(y, \theta) + \beta_2 \Phi_2^2 R_2(y, \theta) = 0 \qquad (1)$$

$$\frac{d\theta}{dx} = \frac{dy}{dx} = 0 \quad \text{at } x = 0$$

$$\theta = y = 1 \quad \text{at } x = 1$$

where

$$R_1(y, \theta) = y \exp\left(\gamma_1\left(1 - \frac{1}{\theta}\right)\right)$$

$$R_2(y, \theta) = y \exp\left(\gamma_2\left(1 - \frac{1}{\theta}\right)\right)$$

[a] Jakob, M. "Heat Transfer," Vol. II, New York: Wiley (1957).
[b] Luss, D., Chen, G. T. *Chem. Eng. Sci.* 30, 1483 (1975).
[c] Michelsen, M. L. *Chem. Eng. Sci.* (1977, in press).

b. It is desired to investigate the multiplicity pattern as a function of Φ_1^2 for given values of γ_1, γ_2, β_1, β_2 and $\sigma = \Phi_2^2/\Phi_1^2$.
Rewrite (1) to

$$\begin{aligned} &\frac{d^2y}{du^2} - R_1(y, \theta) - \sigma R_2(y, \theta) = 0 \\ &\frac{d^2\theta}{du^2} + \beta_1 R_1(y, \theta) + \sigma\beta_2 R_2(y, \theta) = 0 \end{aligned} \tag{2}$$

where $u = \Phi_1 x$ and

$$\frac{dy}{du} = \frac{d\theta}{du} = 0 \quad \text{at } u = 0$$

$$y = \theta = 1 \quad \text{at } u = \Phi_1$$

An explicit relation between θ and y cannot be obtained, but the following inequalities can be derived

$$\begin{aligned} 1 + \beta_1(1 - y) < \theta < 1 + \beta_2(1 - y) \quad \text{for } \beta_2 > \beta_1 \\ 1 + \beta_2(1 - y) < \theta < 1 + \beta_1(1 - y) \quad \text{for } \beta_2 < \beta_1 \end{aligned} \tag{3}$$

To solve equation (2) the following procedure is recommended: Specify $y_0 = y(u = 0)$ and guess a value of $\theta_0 = \theta(u = 0)$. Now (2) can be integrated to $u = u_s$ where $y(u_s) = 1$. If θ_0 has been correctly guessed then $\theta(u_s) = 1$ and Φ_1 is determined ($\Phi_1 = u_s$).
Develop a quadratically convergent Newton iteration method for θ_0 using the sensitivity functions $z_y = (\partial y/\partial\theta_0)$ and $z_\theta = (\partial\theta/\partial\theta_0)$. θ_0 must satisfy the inequalities (3) at every iteration step. Multiple solutions θ_0 may occur.
Determine the multiplicity pattern for $\gamma_1 = 10$, $\gamma_2 = 50$, $\beta_1 = -0.4$, $\beta_2 = 0.8$, and $\sigma = 0.5$.
The results are plotted as Φ_1^2 vs. θ_0, and five solutions are possible in a certain Φ_1^2 range.

c. Consider the special case

$$\sigma\beta_2 + \beta_1 = 0 \tag{4}$$

and show that

$$\theta = 1 \quad \text{and } y = \frac{\cosh(\Phi_1 x\sqrt{1 + \sigma})}{\cosh(\Phi_1\sqrt{1 + \sigma})} \tag{5}$$

is a solution of (1).
Next consider the asymptotic stability problem for this particular steady state. Show that the eigenvalues of the linearized transient equations are given by

$$\lambda f = \frac{d^2 f}{dx^2} - \Phi_1^2(1 + \sigma)f \tag{6}$$

$$\text{Le}\,\lambda g = \frac{d^2 g}{dx^2} + \Phi_1^2(\gamma_1\beta_1 + \gamma_2\beta_2\sigma)\frac{\cosh(\Phi_1 x\sqrt{1 + \sigma})}{\cosh(\Phi_1\sqrt{1 + \sigma})} \tag{7}$$

where

$$f = g = 0 \quad \text{at } x = 1$$

$$\frac{df}{dx} = \frac{dg}{dx} = 0 \quad \text{at } x = 0$$

d. The eigenvalues of (6) are always negative (see subsection 5.5.6) while this is not necessarily the case for (7). Use (4) to rewrite (7)

$$\lambda' g = \frac{d^2g}{dx^2} + KA^2\frac{\cosh(Ax)}{\cosh A}g \tag{8}$$

$$A = \Phi_1\sqrt{1+\sigma} \qquad K = -\frac{\beta_1\beta_2}{\beta_2 - \beta_1}(\gamma_2 - \gamma_1) \qquad \text{and } \lambda' = \text{Le}\,\lambda$$

Let $\gamma_2 > \gamma_1$ and $\beta_2 > 0$. Now by (4) $\beta_1 < 0$ and $K > 0$. Provided K is sufficiently large (8) may have positive eigenvalues. We shall next determine the first eigenvalue of (8) as a function of K.
For $A < \pi/2$ we obtain the convergent expansion

$$\begin{aligned}\frac{\cosh(Ax)}{\cosh A} &= \frac{1 + \frac{1}{2}A^2x^2 + \frac{1}{24}A^4x^4 + \dots}{1 + \frac{1}{2}A^2 + \frac{1}{24}A_4 + \dots} \\ &= 1 + \tfrac{1}{2}A^2(1 - x^2) + \tfrac{1}{24}A^4(-5 + 6x^2 - x^4) + \dots \\ &= 1 + \alpha f_1(x) + \alpha^2 f_2(x) + \dots\end{aligned}$$

Now (8) is written

$$\lambda' g = \frac{d^2g}{dx^2} + K_0(1 + \alpha f_1(x) + \alpha^2 f_2(x) + \dots)g \tag{9}$$

where $K_0 = KA^2$ and

$$\lambda' = -\frac{\pi^2}{4} + K_0 + \alpha K_1 + \alpha K_2 + \dots \tag{10}$$

Determine an analytical value for K_1 and determine K_2 by the method of section 9.1. Compare the results with a collocation solution of (8).

e. The eigenvalues of (8) increase monotonously with A and positive eigenvalues will only occur if (8) has positive eigenvalues when $A \to \infty$. This will not happen unless K is larger than a certain critical value.
Consequently we wish to find K such that (8) has an eigenvalue 0 when $A \to \infty$. Show that $\lambda = 0$ is an eigenvalue of (8) provided K is an eigenvalue of

$$\frac{d^2G}{dx^2} + KA^2\frac{\cosh(Ax)}{\cosh A}G = 0 \tag{11}$$

Show by a coordinate transformation that the critical K value for $A \to \infty$ is the smallest zero of $J_0(2\sqrt{K})$.
Compare the result with a collocation solution of (11) for large, but not too large (why?), values of A.

REFERENCES

The Graetz problem with axial conduction has been treated by Hsu and co-workers in a long series of papers starting with Hsu (1967). The difficulties of Fourier series representation of the solution is most clearly exposed in Hsu (1971) and in Tan and Hsu (1972). The paper mentioned last contains the tables of eigenvalues that we have compared to collocation eigenvalues in subsections 9.1.1 and 9.1.2.

The eigenvalues were determined by Pirkle and Sigillito (1972) who used hypergeometric series for the solution. Their line of attack is illustrated in Exercises 1 and 2.

Newman (1969a) has nicely illustrated how further terms in the Levêque solution for the simple Graetz problem can be obtained. In his UCRL report Newman (1969b) gives a comprehensive treatment of all possible facets of the simple Graetz problem and it also includes the perturbation solution (20) that we have used to check the spline-collocation results to $y = 0$.

Michelsen and Villadsen (1974) give a discussion of the advantages of collocation series over Fourier series especially close to $y = 0$. The paper does not contain the spline collocation extension of subsection 9.1.2, and although the perturbation series from limiting Sturm–Liouville problems is in principle taken from this paper, our present collocation-perturbation approach of subsection 9.1.3 is a much more practical application of the method.

First- and second-order perturbation functions f_1 and f_2 to Sturm–Liouville problems are found in Nayfeh (1973), p. 68ff, by the analytical method of subsection 9.1.3 and it appears that the method dates back to Schrödinger (1926). The collocation-perturbation method described in subsection 9.1.3 is probably new and it seems to be the only viable approach except for trivial problems where Y_n are given explicitly and the integrals (28) and (29) can be handled in a reasonably simple way.

In spite of its small importance to practical chemical engineering, the extended Graetz problem continues to attract theoretical investigations, probably because it is well suited as a test example for numerical methods. The latest contribution is probably Verhoff and Fisher's (1973) solution by transformation of the finite–double-infinite model into a model which is finite in the axial as well as the radial direction and which can be treated by relaxation techniques.

The standard (and often cited) reference to asymptotic stability for catalyst particles at $\mathrm{Le} \neq 1$ is Lee and Luss (1970), while Michelsen and Villadsen (1972) looked at the stability of all intermediate steady state solutions for $\mathrm{Le} = 1$. This latter case has been studied by many authors with or without surface resistances and for up to three steady states. Luss (1972) and Jackson (1973) are continuations of Michelsen and Villadsen

(1972), while Luss (1974) continues Lee and Luss (1970) for a case of five steady states to see whether the middle steady state is stabilized for $\mathrm{Le} \to \infty$. His conclusion is the same as shown in table 9.9, that no stabilization occurs.

Besides the long series of papers on asymptotic stability for Le equal to 1 (and in fewer cases $\mathrm{Le} \neq 1$) there are many papers that investigate more or less general stability features of the nonlinearized model. These papers are almost all cited in the second volume of Aris (1975).

An interesting recent paper with computational results is Bidner and Calvelo (1974). They analyze the nonlinear model for a first-order irreversible reaction with internal and external heat and mass transfer resistance. The destabilizing influence of low Lewis number and the existence of limit cycles are confirmed.

As yet no more general rate expressions than first-order irreversible reactions have been investigated, but from a practical point of view the most interesting phenomena are likely to occur for other rate expressions.

The computations in section 9.2 were started in the MSc thesis of Wedel (1975). He also worked with negative R_y (e.g., the reactant-poisoned CO oxidation). These results are only reported in his thesis, while a summary of the calculations in section 9.2 is given in Wedel, Michelsen, and Villadsen (1977).

The material of section 9.3 is based upon Michelsen, Vakil, and Foss (1973), that is again a continuation of Stangeland and Foss (1970) and Crider and Foss (1968). The state space collocation representation is used in Vakil, Michelsen, and Foss (1973). Wæde Hansen, and Jørgensen (1974) analyzed the dynamics of a fixed bed reactor with a model similar to that of subsection 9.3.4 [they used $R(c, \theta)$ rather than $R(c, \Phi)$].

1. HSU, C. J. *Appl. Sci. Res.* 17 (1967):359.

2. HSU, C. J. *AIChE Journal* 17 (1971):732.

3. TAN, C. W., and HSU, C. J. *Int. J. Heat Mass Transfer* 15 (1972):2187.

4. PIRKLE, J. C., and SIGILLITO, V. G. *Journal Computational Physics* 9 (1972):207.

5. NEWMAN, J. *J. Heat transfer* 91 (1969):177.

6. NEWMAN, J. "The Graetz Problem," *UCRL Report* (1969): 18,646.

7. MICHELSEN, M., and VILLADSEN, J. *Int. J. Heat Mass Transfer* 17 (1974):1391.

8. NAYFEH, A. H. *Perturbation Methods*. New York: Wiley Interscience texts (1973).

9. SCHRÖDINGER, E. *Ann. Phys.* 80 (1926):437.

10. VERHOFF, F. H., and FISHER, D. P. *J. Heat Transfer* 95 (1973):132.

11. LEE, J. C. M., and LUSS, D. *AIChE Journal* 16 (1970):620.

12. MICHELSEN, M. L., and VILLADSEN, J. *Chem. Eng. Sci.* 27 (1972):751.

13. LUSS, D. *Chem. Eng. Sci.* 27 (1972):2299.

14. JACKSON, R. *Chem. Eng. Sci.* 28 (1973):1355.

15. LUSS, D. *Chem. Eng. Sci.* 29 (1974):1832.

16. ARIS, R. *The Mathematical Theory of Diffusion and Reaction in Permeable Catalysts.* Oxford: Clarendon (1975).

17. BIDNER, M. S., and CALVELO, A. *Chem. Eng. Sci.* 29 (1974):1909.

18. WEDEL, S. M.Sc. Thesis. Instituttet for Kemiteknik (DtH) (1975) (in Danish) and Wedel, S., Michelsen, M. L., Villadsen, J. *Chem. Eng. Sci.* 32 (1977):179.

19. MICHELSEN, M. L., VAKIL, H. B., and FOSS, A. S. *Ind. Eng. Chem. Fundam.* 12 (1973):323.

20. STANGELAND, B. E., and FOSS, A. S. *Ind. Eng. Chem. Fundam.* 9 (1970):38.

21. CRIDER, J. E., and FOSS, A. S. *AIChE Journal* 14 (1968):77.

22. VAKIL, H. B., MICHELSEN, M. L., and FOSS, A. S. *Ind. Eng. Chem. Fundam.* 12 (1973):328.

23. WÆDE HANSEN, K., and BAY JØRGENSEN, S. *Advances in Chemistry Series* 133 (1974):505. (Transactions 3rd ISCRE, Chicago, August 1974.)

24. FINLAYSON, B. A. *Catalysis Reviews Sci.-Eng.* 10 (1974):69.

Appendix A: Computer Programs with Test Examples

Examples illustrating use of program package

A1

```
      SUBROUTINE JCOBI (ND,N,N0,N1,AL,BE,DIF1,DIF2,DIF3,ROOT)
      IMPLICIT REAL*8 (A-H,O-Z)
      DIMENSION DIF1(ND),DIF2(ND),DIF3(ND),ROOT(ND)
C
C      EVALUATION OF ROOTS AND DERIVATIVES OF JACOBI POLYNOMIALS
C      P(N) (AL,BE);  MACHINE ACCURACY 16 D;
C
C      FIRST EVALUATION OF COEFFICIENTS IN RECURSION FORMULAS
C      RECURSION COEFFICIENTS ARE STORED IN DIF1 AND DIF2
C
      AB=AL+BE
      AD=BE-AL
      AP=BE*AL
      DIF1(1)=(AD/(AB+2)+1)/2
      DIF2(1)=0.
      IF (N.LT. 2) GO TO 15
      DO 10 I=2,N
      Z1=I-1
      Z=AB+2*Z1
      DIF1(I)=(AB*AD/Z/(Z+2)+1)/2
      IF (I .NE. 2) GO TO 11
      DIF2(I)=(AB+AP+Z1)/Z/Z/(Z+1)
      GO TO 10
   11 Z=Z*Z
      Y=Z1*(AB+Z1)
      Y=Y*(AP+Y)
      DIF2(I)=Y/Z/(Z-1)
   10 CONTINUE
C
C     ROOT DETERMINATION BY NEWTON METHOD WITH SUPPRESSION
C      OF PREVIOUSLY DETERMINED ROOTS
C
   15 X=0.
      DO 20 I=1,N
   25 XD=0.
      XN=1.
      XD1=0.
      XN1=0.
      DO 30 J=1,N
      XP=(DIF1(J)-X)*XN-DIF2(J)*XD
      XP1=(DIF1(J)-X)*XN1-DIF2(J)*XD1-XN
      XD=XN
      XD1=XN1
      XN=XP
   30 XN1=XP1
      ZC=1.
      Z=XN/XN1
      IF (I .EQ. 1) GO TO 21
      DO 22 J=2,I
   22 ZC=ZC-Z/(X-ROOT(J-1))
   21 Z=Z/ZC
      X=X-Z
      IF (DABS(Z) .GT. 1.D-09) GO TO 25
      ROOT(I)=X
      X=X+.0001
   20 CONTINUE
C
C      ADD EVENTUAL INTERPOLATION POINTS AT X=0 OR X=1
C
      NT=N+N0+N1
      IF (N0 .EQ. 0) GO TO 35
      DO 31 I=1,N
      J=N+1-I
   31 ROOT(J+1)=ROOT(J)
      ROOT(1)=0
   35 IF (N1 .EQ. 1) ROOT(NT) = 1.
C
C      NOW EVALUATE DERIVATIVES OF POLYNOMIAL
C
      DO 40 I=1,NT
      X=ROOT(I)
      DIF1(I)=1.
      DIF2(I)=0.
      DIF3(I)=0.
      DO 40 J=1,NT
      IF (J .EQ. I) GO TO 40
      Y=X-ROOT(J)
      DIF3(I)=Y*DIF3(I) + 3*DIF2(I)
      DIF2(I)=Y*DIF2(I) + 2*DIF1(I)
      DIF1(I)=Y*DIF1(I)
   40 CONTINUE
      RETURN
      END
```

A2

```
      SUBROUTINE DFOPR(ND,N,N0,N1,I,ID,DIF1,DIF2,DIF3,ROOT,VECT)
      IMPLICIT REAL*8 (A-H,O-Z)
      DIMENSION DIF1(ND),DIF2(ND),DIF3(ND),ROOT(ND),VECT(ND)
C
C      SUBROUTINE EVALUATES DISCRETIZATION MATRICES AND
C      GAUSSIAN QUADRATURE WEIGHTS, NORMALIZED TO SUM 1
C      ID = 1 :     DISCRETIZATION MATRIX FOR Y(1) (X)
C      ID = 2 :     DISCRETIZATION MATRIX FOR Y(2) (X)
C      ID = 3 :     GAUSSIAN QUADRATURE WEIGHTS(
C
      NT=N+N0+N1
      IF (ID .EQ. 3) GO TO 10
      DO 20 J=1,NT
      IF (J .NE. I) GO TO 21
      IF (ID .NE. 1) GO TO 5
      VECT(I)=DIF2(I)/DIF1(I)/2
      GO TO 20
    5 VECT(I)=DIF3(I)/DIF1(I)/3
      GO TO 20
   21 Y=ROOT(I)-ROOT(J)
      VECT(J)=DIF1(I)/DIF1(J)/Y
      IF (ID .EQ. 2) VECT(J)=VECT(J)*(DIF2(I)/DIF1(I)-2/Y)
   20 CONTINUE
      GO TO 50
   10 Y=0.
      DO 25 J=1,NT
      X=ROOT(J)
      AX=X*(1-X)
      IF (N0 .EQ. 0) AX=AX/X/X
      IF (N1 .EQ. 0) AX=AX/(1-X)/(1-X)
      VECT(J)=AX/DIF1(J)**2
   25 Y=Y+VECT(J)
      DO 60 J=1,NT
   60 VECT(J)=VECT(J)/Y
   50 RETURN
      END
```

```
      SUBROUTINE RADAU(ND,N,N0,N1,ID,AL,BE,ROOT,DIF1,VECT)
      IMPLICIT REAL*8(A-H,O-Z)
      DIMENSION ROOT(ND),DIF1(ND),VECT(ND)
C
C     RADAU AND LOBATTO QUADRATURE WEIGHTS
C       ID = 2 : RADAU QUADRATURE WITH X = 0
C       ID = 1 : RADAU QUADRATURE WITH X = 1
C       ID = 3 : LOBATTO QUADRATURE WITH BOTH ENDPOINTS
C
      S=0.
      NT=N+N0+N1
      DO 40 I=1,NT
      X=ROOT(I)
      IF (ID-2) 10,20,30
   10 AX=X
      IF (N0 .EQ. 0) AX=1/AX
      GO TO 40
   20 AX=1-X
      IF (N1 .EQ. 0) AX=1/AX
      GO TO 40
   30 AX=1.
   40 VECT(I)=AX/DIF1(I)**2
      IF (ID .NE. 2) VECT(NT)=VECT(NT)/(1+AL)
      IF (ID .GT. 1)VECT(1)=VECT(1)/(1+BE)
      DO 50 I=1,NT
   50 S=S+VECT(I)
      DO 60 I=1,NT
   60 VECT(I)=VECT(I)/S
      RETURN
      END
```

A3

```
      SUBROUTINE INTRP(ND,NT,X,ROOT,DIF1,XINTP)
      IMPLICIT REAL*8 (A-H,O-Z)
      DIMENSION ROOT(ND),DIF1(ND),XINTP(ND)
C
C      EVALUATION OF LAGRANGIAN INTERPOLATION COEFFICIENTS
C
      POL=1.
      DO 5 I=1,NT
      Y=X-ROOT(I)
      XINTP(I)=0.
      IF (Y .EQ. 0.D0) XINTP(I)=1.
    5 POL=POL*Y
      IF (POL .EQ. 0.D0) GO TO 10
      DO 6 I=1,NT
    6 XINTP(I)=POL/DIF1(I)/(X-ROOT(I))
   10 RETURN
      END
```

```
      SUBROUTINE GAUSL(ND,NCOL,N,NS,A)
      IMPLICIT REAL*8 (A-H,O-Z)
      DIMENSION A(ND,NCOL)
C
C      GAUSL SOLVES A*X=B, WHERE A IS N*N AND B IS N*NS, BY GAUSSIAN
C      ELIMINATION WITH PARTIAL PIVOTING. THE MATRIX (OR VECTOR) B
C      IS PLACED ADJACENT TO A IN COLUMNS N+1 TO N+NS.
C      A IS DESTROYED, AND THE RESULTING MATRIX X REPLACES B
C
      N1=N+1
      NT=N+NS
      IF (N .EQ. 1) GO TO 50
C      START ELIMINATION
      DO 10 I=2,N
      IP=I-1
      I1=IP
      X=DABS(A(I1,I1))
      DO 11 J=I,N
      IF (DABS(A(J,I1)) .LT. X) GO TO 11
      X=DABS(A(J,I1))
      IP=J
   11 CONTINUE
      IF (IP .EQ. I1) GO TO 13
C     ROW INTERCHANGE
      DO 12 J=I1,NT
      X=A(I1,J)
      A(I1,J)=A(IP,J)
   12 A(IP,J)=X
   13 DO 10 J=I,N
      X=A(J,I1)/A(I1,I1)
      DO 10 K=I,NT
   10 A(J,K)=A(J,K) - X*A(I1,K)
C      ELIMINATION FINISHED, NOW BACKSUBSTITUTION
   50 DO 20 IP=1,N
      I=N1-IP
      DO 20 K=N1,NT
      A(I,K) = A(I,K)/A(I,I)
      IF (I .EQ. 1) GO TO 20
      I1=I-1
      DO 25 J=1,I1
   25 A(J,K) = A(J,K) - A(I,K)*A(J,I)
   20 CONTINUE
      RETURN
      END
```

A4

```
AJOB     WATFIV  A102021
      SUBROUTINE EISYS(ND,NCOL,N,INDEX,EPS,NC,A)
      IMPLICIT REAL*8 (A-H,O-Z)
      DIMENSION NC(NCOL)
      DIMENSION A(ND,NCOL)
C
C     EISYS FINDS THE EIGENVALUES, EIGENVECTORS AND EIGENROWS OF AN N*N
C     SQUARE MATRIX, A
C
C     ND IS THE ROW DIMENSION OF A, DECLARED IN THE MAIN PROGRAM
C     NCOL IS THE COLUMN DIMENSION OF A AND THE DIMENSION OF NC
C
C     N IS THE SIZE OF THE ACTUAL MATRIX
C
C     INDEX IS AN INDICATOR
C     INDEX = -1 GIVES EIGENVALUES ONLY, AND THESE ARE STORED IN THE
C                FIRST COLUMN OF A, REAL AND IMAGINARY PARTS OF EVENTUAL
C                COMPLEX EIGENVALUES BEING IN ADJACENT POSITIONS
C     INDEX = 0 GIVES EIGENVALUES AND EIGENVECTORS. THEN EIGENVALUES ARE
C               ON THE MAIN DIAGONAL OF A, COMPLEX PAIRS AS A 2*2 BLOCK
C               WITH THE REAL PARTS (INDENTICAL) ON THE DIAGONAL, AND
C               THE IMAGINARY PARTS (IDENTICAL, BUT OF OPPOSITE SIGN)
C               AS ADJACENT OFF-DIAGONAL ELEMENTS
C               THE MATRIX OF EIGENVECTORS, Q, IS STORED IN ROWS N+1 -2N
C     INDEX = 1 GIVES EIGENVALUES, EIGENVECTORS AND EIGENROWS, EIGENVALU
C               -ES AND EIGENVECTORS BEING STORED AS FOR INDEX = 0
C               THE EIGENROWS, QINV, ARE STORED IN ROWS 2N+1 - 3N
C     ON EXIT INDEX GIVES THE NUMBER OF ITERATIONS IN THE QR-STEP
C     INDEX THEREFORE MUST BE A VARIABLE, NOT A CONSTANT
C
C     EPS IS THE TOLERANCE, AND A SUGGESTED VALUE IS 1.E-08
C
C     NC IS AN INDICATOR ARRAY FOR THE EIGENVALUES, NC(I)=0 INDICATES
C     THAT THE I'TH EIGENVALUE IS REAL, AND NC(I)=1 (AND NC(I+1)=2 )
C     INDICATES A COMPLEX PAIR OF EIGENVALUES ( WITH REAL PART IN
C     A(I,1) AND COMPLEX PART IN A(I+1,1) FOR INDEX=-1 )
C
C     STORAGE REQUIREMENTS
C     THE MINIMUM VALUE OF ND IS
C        N   FOR INDEX = -1  (EIGENVALUES ONLY)
C       2N FOR INDEX = 0     (EIGENVALUES AND -VECTORS)
C       3N FOR INDEX = 1     (EIGENVALUES, -VECTORS AND -ROWS)
C
C     THE DIAGONALISED MATRIX, D, THE MATRIX OF EIGENVECTORS, Q, AND
C     THE MATRIX OF EIGENROWS, QINV, HAVE THE PROPERTY, THAT
C           A  =  Q  *  D  *  QINV
C
      NS=2*N
      NT=NS
      IT=INDEX
      IF (INDEX .LE. 0) NS=N
      DO 10 I=1,N
      IF (INDEX .LT. 0) GO TO 10
      DO 11 J=1,N
   11 A(J+N,I)=0.
      A(I+N,I)=1.
   10 NC(I)=0
      CALL HOUS(ND,NCOL,N,A,INDEX)
      CALL QR(ND,NCOL,N,EPS,A,INDEX)
      IF (IT .LE. 0) GO TO 12


      DO 13 I=1,N
      DO 13 J=1,N
   13 A(I+2*N,J)=A(J+N,I)
   12 CALL ORDEN(ND,NCOL,N,IT,EPS,NC,A)
      IF (IT .GE. 0) CALL EIVEC(ND,NCOL,N,NT,NS,NC,A)
      RETURN
      END
```

A5

```
      SUBROUTINE HOUS(ND,NCOL,N,A,INDEX)
      IMPLICIT REAL*8 (A-H,O-Z)
      DIMENSION A(ND,NCOL)
      IF (N.LE. 2) GO TO 100
      NL=N-2
      NT=N+N
      IF (INDEX .LT. 0) NT=N
      DO 10 I=1,NL
      IPP=I+1
      IPA=I+2
      S=0.
      DO 11 J=IPA,N
   11 S=S+A(J,I)**2
      IF (S .EQ. 0.) GO TO 10
      S = DSQRT(S+A(IPP,I)**2)
      IF (A(IPP,I).LT. 0.) S=-S
      A(IPP,I)=A(IPP,I)+S
      SK=S*A(IPP,I)
      DO 15 J=IPP,N
      SX=0.
      DO 16 K=IPP,N
   16 SX=SX+A(K,I)*A(K,J)
      IF (SX .EQ. 0.) GO TO 15
      SX=SX/SK
      DO 17 K=IPP,N
   17 A(K,J) = A(K,J)-SX*A(K,I)
   15 CONTINUE
      DO 20 J=1,NT
      SX=0.
      DO 21 K=IPP,N
   21 SX=SX+A(K,I)*A(J,K)
      IF (SX .EQ. 0.) GO TO 20
      SX=SX/SK
      DO 23 K=IPP,N
   23 A(J,K)=A(J,K)-SX*A(K,I)
   20 CONTINUE
      A(IPP,I)=-S
      DO 22 J=IPA,N
   22 A(J,I)=0.
   10 CONTINUE
  100 RETURN
      END
```

A6

```
      SUBROUTINE QR(ND,NCOL,N,EPS,A,IT)
      IMPLICIT REAL*8 (A-H,O-Z)
      DIMENSION A(ND,NCOL)
      LOGICAL LO1
      MT=0
      IF (IT .LT. 0) MT=1
      IT=0
      NT=2*N
      SX=0.
      SY=0.
      I=N+1
   10 I=I-1
      IF (I .LE. 2) GO TO 100
   20 M=1
      IT=IT+1
      II=I
      DO 11 JA=2,II
      J=I+2-JA
      K=J-1
      X=DABS(A(J,K))/(DABS(A(J,J))+DABS(A(K,K))+EPS)
      IF (X.GT. EPS) GO TO 11
      M=J
      A(J,K)=0.
      IF (M.LE. I-2) GO TO 12
      IF (M .EQ. I-1) I=I-1
      GO TO 10
   11 CONTINUE
   12 IF (IT .EQ. 1) GO TO 30
      SX = A(I,I)+A(I-1,I-1)
      SY= A(I,I)*A(I-1,I-1)-A(I,I-1)*A(I-1,I)
      IF ((SX .EQ. 0.) .AND. (SY .EQ. 0.)) SX=A(I,I-1)
   30 X=A(M,M)/A(M+1,M)*(A(M,M)-SX)+A(M,M+1)+SY/A(M+1,M)
      Y=A(M,M)+A(M+1,M+1)-SX
      Z=A(M+2,M+1)
      IL=I-1
      NL1=N+MT*(I-N)
      NL2 = 1+(M-1)*MT
      NL3 = NT + (I-NT)*MT
      DO 15 J=M,IL
      LO1= (J.LT.IL)
      S=DSQRT(X*X+Y*Y+Z*Z)
      IF (X.LT. 0.) S=-S
      X=X+S
      Y=Y/X
      Z=Z/X
      X=X/S
      DO 16 K=J,NL1
      SX=A(J,K)+Y*A(J+1,K)
      IF (LO1) SX=SX+Z*A(J+2,K)
      SX=SX*X
      A(J,K)=A(J,K)-SX
      IF (LO1) A(J+2,K)=A(J+2,K)-Z*SX
   16 A(J+1,K)=A(J+1,K)-Y*SX
      DO 25 K=NL2,NL3
      IF (K.GT.N) GO TO 42
      IF ((K .GE. J+4) .OR. (K.GT. I)) GO TO 25
   42 SX = A(K,J)+Y*A(K,J+1)
      IF (LO1) SX=SX+Z*A(K,J+2)


      SX=SX*X
      A(K,J) =A(K,J)-SX
      A(K,J+1)=A(K,J+1)-Y*SX
      IF (LO1) A(K,J+2)=A(K,J+2)-Z*SX
   25 CONTINUE
      IF (J.LE. M) GO TO 26
      A(J,J-1)=-S
      A(J+1,J-1)=0.
      IF (J .LT. I-1) A(J+2,J-1)=0.
   26 X=A(J+1,J)
      Y=0.
      Z=0.
      IF (LO1) Y=A(J+2,J)
      IF (J .LT. I-2) Z=A(J+3,J)
   15 CONTINUE
      IF (IT .LT. 10*N) GO TO 20
  100 RETURN
      END
```

A7

```
      SUBROUTINE ORDEN(ND,NCOL,N,IT,EPS,NC,A)
      IMPLICIT REAL*8 (A-H,O-Z)
      DIMENSION NC(NCOL)
      DIMENSION A(ND,NCOL)
      NS=2*N
      NT=NS
      IF (IT .LE. 0) NS=N
      I=0
   15 I=I+1
      IF (I-N) 16,17,200
   16 X=DABS(A(I+1,I))/(DABS(A(I,I))+DABS(A(I+1,I+1))+EPS)
      IF (X .GT. EPS) GO TO 111
      A(I+1,I)=0.
   17 IF (IT .GE. 0) GO TO 15
      A(I,1)=A(I,I)
      GO TO 15
  111 Y = A(I,I) + A(I+1,I+1)
      Z = A(I,I)*A(I+1,I+1) - A(I,I+1)*A(I+1,I)
      X = Y*Y - 4.*Z
      IF (X .LT. 0.) GO TO 112
      R = (DABS(Y) + DSQRT(X))/2.
      IF (Y .LT. 0.) R=-R
      IF (IT .GE. 0) GO TO 120
      A(I,1) = R
      A(I+1,1) = Z/R
      GO TO 121
  120 Q = A(I,I)-R
      S = A(I,I)-Z/R
      IF (DABS(S) .GT. DABS(Q)) Q=S
      Q = A(I+1,I)/Q
      C = 1./DSQRT(Q*Q+1.)
      S = Q*C
      GO TO 113
  112 R = (A(I+1,I+1)-A(I,I))/(A(I+1,I) + A(I,I+1))
      IF (IT .GE. 0) GO TO 122
      A(I,1) = Y/2.
      A(I+1,1) = DSQRT(-X)/2.
      NC(I)=1
      NC(I+1)=2
      GO TO 121
  122 Q=1./DSQRT(1.+R*R)
      C=DSQRT((Q+1.)/2.)
      S=R*Q/C/2.
  113 DO 115 J=I,NS
      K1=I
      K2=J
      IF (J .GT. N) K1=I+2*N
      IF (J .GT. N) K2=J-N
      Q=A(K1,K2)
      A(K1,K2) = Q*C + S*A(K1+1,K2)
  115 A(K1+1,K2) = A(K1+1,K2)*C - S*Q
      DO 116 J=1,NT
      IF ((J.GE.I+2) .AND. (J.LE.N)) GO TO 116
      Q=A(J,I)
      A(J,I) = A(J,I)*C + A(J,I+1)*S
      A(J,I+1) = C*A(J,I+1) -Q*S
  116 CONTINUE
      IF (X .GE. 0.) GO TO 118

      NC(I)=1
      NC(I+1)=2
      R=DSQRT(DABS(A(I+1,I)/A(I,I+1)))
      DO 119 J=1,NT
      A(J,I+1)=A(J,I+1)*R
      IF (J .GT. N) GO TO 82
      A(I+1,J) = A(I+1,J)/R
      GO TO 119
   82 IF (J .GT. NS) GO TO 119
      A(I+2*N+1,J-N)=A(I+2*N+1,J-N)/R
  119 CONTINUE
      GO TO 121
  118 A(I+1,I)=0.
  121 I=I+1
      GO TO 15
  200 RETURN
      END
```

A8

```
      SUBROUTINE EIVEC(ND,NCOL,N,NVEC,NRO,NC,A)
      IMPLICIT REAL*8 (A-H,O-Z)
      DIMENSION NC(NCOL)
      DIMENSION A(ND,NCOL),X(2,2)
      IX=N
      NE=N+NRO
   10 IF (IX .LE. 0) GO TO 100
      C=A(IX,IX)
      D=0.
      I1=2
      IF (NC(IX) .LE. 0) GO TO 3
      I1=1
      D=A(IX-1,IX)
    3 JX=IX+I1-3
   60 IF (JX .LE. 0) GO TO 50
      AC=A(JX,JX)-C
      J1=2
      B=0.
      IF (NC(JX) .LE. 0) GO TO 6
      J1=1
      B=A(JX-1,JX)
    6 DO 8 J=1,2
      DO 8 I=1,2
    8 X(I,J)=0.
      DO 9 I=I1,2
      DO 9 J=J1,2
      X(J,I)=A(JX+J-2,IX+I-2)
    9 A(JX+J-2,IX+I-2)=0.
      U=(X(1,1)+X(2,2))/2.
      S=X(1,1)-U
      V=(X(1,2)-X(2,1))/2.
      T=X(1,2)-V
      DB=D-B
      DB=B-D
      DET=AC*AC+DB*DB
      R1=(U*AC+V*DB)/DET
      RI1=(V*AC-U*DB)/DET
      DB=D+B
      DET=AC*AC+DB*DB
      R2=(S*AC-T*DB)/DET
      RI2=(T*AC+S*DB)/DET
      X(1,1)=R1+R2
      X(2,2)=R1-R2
      X(1,2)=RI2-RI1
      X(2,1)=RI2+RI1
      DO 15 K=1,NE
      IF ((K .LE. N) .AND. (K .GE. JX+J1-2)) GO TO 15
      DO 20 I=I1,2
      DO 20 J=J1,2
      IF (K .LE. 2*N) A(K,IX+I-2)=A(K,IX+I-2)-X(I,J)*A(K,JX+J-2)
      IF (K .GT. 2*N) A(JX+J+2*N-2,K-2*N)=A(JX+J+2*N-2,K-2*N)
     1 + X(I,J)*A(IX+I+2*N-2,K-2*N)
   20 CONTINUE
   15 CONTINUE
      JX=JX+J1-3
      GO TO 60
   50 IX=IX+I1-3
      GO TO 10

  100 RETURN
      END
```

A9

```
      SUBROUTINE STIFF3(N,ND,NPRINT,FUN,DFUN,OUT,X0,X1,H0,EPS,W,Y,YOLD,
     XYOLD1,IP,YA,YK1,YK2,YK3,DF,DFOLD,F,FOLD)
      IMPLICIT REAL*8(A-H,O-Z)
      DIMENSION IP(ND),Y(ND),YOLD(ND),YOLD1(ND),YA(ND),YK1(ND),YK2(ND)
      DIMENSION YK3(ND),W(ND),F(ND),FOLD(ND),DF(ND,ND),DFOLD(ND,ND)
C
C     PARAMETER LIST OF STIFF 3
C     N       NUMBER OF EQUATIONS TO BE INTEGRATED
C     ND      DIMENSION OF VECTORS AND ARRAYS (FROM MAIN PROGRAM)
C     NPRINT  PRINT INTERVAL
C     FUN     USER SUPPLIED FUNCTION SUBPROGRAM
C     DFUN    USER SUPPLIED JACOBIAN SUBPROGRAM
C     OUT     USER SUPPLIED OUTPUT PROGRAM
C     X0      INITIAL VALUE OF INDEPENDENT VARIABLE
C     X1      FINAL VALUE OF INDEPENDENT VARIABLE
C     H0      INITIAL STEPSIZE: ON EXIT H0 CONTAINS THE SUGGESTED
C             STARTING VALUE FOR THE NEXT INTERVAL
C     EPS     TOLERANCE PARAMETER
C     W       VECTOR OF ERROR MULTIPLIERS FOR INDIVIDUAL COMPONENTS
C     IP      INTERNAL PIVOT VECTOR FOR LU-DECOMPOSITION
C     Y       VECTOR OF DEPENDENT VARIABLES, MUST BE DEFINED ON CALLING
C     YOLD    INTERNAL WORKING VECTOR OF DEPENDENT VARIABLES
C     YOLD1   INTERNAL WORKING VECTOR OF DEPENDENT VARIABLES
C     YA      INTERNAL WORKING VECTOR OF DEPENDENT VARIABLES
C     YK1     INTERNAL K1-VECTOR
C     YK2     INTERNAL K2-VECTOR
C     YK3     INTERNAL K3-VECTOR
C     DF      INTERNAL WORKING MATRIX FOR JACOBIAN
C     DFOLD   INTERNAL WORKING MATRIX FOR JACOBIAN
C     F       INTERNAL WORKING VECTOR FOR FUNCTION
C     FOLD    INTERNAL WORKING VECTOR FOR FUNCTION
C
      ICON = 0
C
C     ICON = 0 EXCEPT FOR LAST STEP WHICH ENDS EXACTLY AT X1
C
      NOUT = 0
      X = X0
      H = H0
      IF ( X0 + 2.*H .LT. X1) GO TO 1
C     LAST STEP - OR FIRST STEP LONGER THAN INTERVAL
C
C
    2 H = (X1 - X)/2
      ICON = 1
C
C     EVALUATE FUNCTION AND JACOBIAN
C
    1 IF (ICON .EQ. 0 .AND. X+4*H .GT. X1) H=(X1-X)/4
      CALL FUN(Y,F)
      CALL DFUN(Y,DF)
      IHA=-1
C
C     KEEP VALUES WHICH ARE USED IN HALF-STEP INTEGRATION
C
      DO 30 I = 1,N
      YOLD(I) = Y(I)
      FOLD(I) = F(I)
      DO 30 J = 1,N
   30 DFOLD(I,J) = DF(I,J)
C
C     PERFORM FULL-STEP INTEGRATION
C
   37 CALL SIRK3(N,ND,FUN,IP,F,Y,YK1,YK2,YK3,DF,2*H)
      DO 35 I=1,N
      YA(I) = Y(I)
      Y(I) = YOLD(I)
      F(I) = FOLD(I)
      DO 35 J = 1,N
   35 DF(I,J) = DFOLD(I,J)
   38 IHA = IHA + 1
C
C     FULL STEP FINISHED, START HALF-STEP INTEGRATION
C     IHA COUNTS NUMBER OF STEPLENGTH BISECTIONS
C
      CALL SIRK3(N,ND,FUN,IP,F,Y,YK1,YK2,YK3,DF,H)
      CALL FUN(Y,F)
      CALL DFUN(Y,DF)
      DO 40 I=1,N
   40 YOLD1(I) = Y(I)
      CALL SIRK3(N,ND,FUN,IP,F,Y,YK1,YK2,YK3,DF,H)
      E = 0.
C
C     HALF STEP INTEGRATION FINISHED
```

A10

```
C     COMPUTE DEVIATION AND COMPARE WITH TOLERANCE
C
      DO 41 I = 1,N
      ES = W(I) * DABS(YA(I) - Y(I))/(1. + DABS(Y(I)))
      IF (ES .GT. E) E = ES
   41 CONTINUE
      Q=E/EPS
      QA=(4.*Q)**.25
      IF (Q .LE. 1.) GO TO 48
C
C     DEVIATION TOO LARGE- RETURN TO HALF-STEP WITH SMALLER H
C
      DO 45 I = 1,N
      YA(I) = YOLD1(I)
      F(I) = FOLD(I)
      Y(I) = YOLD(I)
      DO 45 J = 1,N
   45 DF(I,J) = DFOLD(I,J)
      H = H/2
      ICON = 0
      GO TO 38
C
C     ADJUST Y-VECTOR
C
   48 DO 49 I = 1,N
   49 Y(I) = Y(I) + (Y(I) - YA(I))/7.D0
      X = X +2*H
C
C     COMPUTE NEXT STEPSIZE AND PRINT IF APPROPRIATE
C
      QA=1./(QA+1.D-10)
      IF (QA .GT. 3.) QA = 3.
      H = QA * H
      NOUT = NOUT + 1
      IF ( (NOUT/NPRINT)*NPRINT .EQ. NOUT .OR. ICON .EQ. 1) CALL OUT(X,Y


     X,IHA,QA)
      IF (ICON .EQ. 1) GO TO 187
      H0 = H
      IF (X + 2.*H .LT. X1) GO TO 1
      GO TO 2
  187 RETURN
      END
```

```
      SUBROUTINE BACK(ND,N,IPIV,A,V)
      IMPLICIT REAL*8(A-H,O-Z)
      DIMENSION IPIV(ND),A(ND,ND),V(ND)
C
C     SOLUTION OF LINEAR EQUATION
C     BY BACKSUBSTITUTION AFTER DECOMPOSITION
C
      N1=N-1
      DO 10 I=1,N1
      I1=I+1
      K=IPIV(I)
      IF (K.EQ.I) GO TO 11
      X=V(I)
      V(I)=V(K)
      V(K)=X
   11 DO 10 J=I1,N
   10 V(J)=V(J)+A(J,I)*V(I)
      V(N)=V(N)/A(N,N)
      DO 15 II=2,N
      I=N+1-II
      I1=I+1
      DO 16 J=I1,N
   16 V(I)=V(I)-A(I,J)*V(J)
   15 V(I)=V(I)/A(I,I)
      RETURN
      END
```

A11

```
      SUBROUTINE LU(ND,N,IPIV,A)
      IMPLICIT REAL*8(A-H,O-Z)
      DIMENSION IPIV(ND),A(ND,ND)
C
C     TRIANGULAR DECOMPOSITION BY GAUSSIAN ELIMINATION
C     WITH PARTIAL PIVOTING
C
      IPIV(N)=N
      N1=N-1
      DO 10 I=1,N1
      X=A(I,I)
      IF (X.LT. 0.) X=-X
      IPIV(I)=I


      I1=I+1
      DO 11 J=I1,N
      Y=A(J,I)
      IF (Y.LT.0.) Y=-Y
      IF (Y.LE.X) GO TO 11
      X=Y
      IPIV(I)=J
   11 CONTINUE
      IF (IPIV(I) .EQ. I) GO TO 14
      K=IPIV(I)
      DO 12 J=I,N
      X=A(I,J)
      A(I,J)=A(K,J)
   12 A(K,J)=X
   14 DO 10 J=I1,N
      X=-A(J,I)/A(I,I)
      A(J,I)=X
      DO 10 K=I1,N
   10 A(J,K)=A(J,K)+X*A(I,K)
      RETURN
      END
```

```
      SUBROUTINE SIRK3(N,ND,FUN,IPIV,F,Y,YK1,YK2,YK3,DF,H)
      IMPLICIT REAL*8(A-H,O-Z)
      DIMENSION F(ND),Y(ND),YK1(ND),YK2(ND),YK3(ND),IPIV(ND),DF(ND,ND)
      DIMENSION R(4)
      DATA A,R/.4358665215084589D0,1.0376094961318590D0,
     X.8349304838526377D0,-.6302020887244523D0,-.2423378912600452/
      DO 5 I = 1,N
      DO 6 J = 1,N
      DF(I,J) = -H * A * DF(I,J)
      IF (DABS(DF(I,J)) .LT. 1.D-12) DF(I,J) = 0.
    6 CONTINUE
    5 DF(I,I) = DF(I,I) + 1.
C
C     PERFORM TRIANGULAR DECOMPOSITION AND EVALUATE K1
C
      CALL LU(ND,N,IPIV,DF)
      CALL BACK(ND,N,IPIV,DF,F)
      DO 8 I = 1,N
      YK1(I) = H * F(I)
    8 YK2(I) = Y(I) + .75D0 * YK1(I)
      CALL FUN(YK2,F)
      CALL BACK(ND,N,IPIV,DF,F)
C
C     EVALUATE K2
C
      DO 9 I = 1,N
      YK2(I) = H * F(I)
      Y(I) = Y(I) + R(1) * YK1(I) + R(2) * YK2(I)
    9 YK2(I) = R(3) * YK1(I) + R(4) * YK2(I)
C
C     EVALUATE K3
C     FOR CONVENIENCE STORED IN YK2
C
      CALL BACK(ND,N,IPIV,DF,YK2)
      DO 10 I = 1,N
   10 Y(I) = Y(I) + YK2(I)
      RETURN
      END
```

A12

```
      IMPLICIT REAL*8(A-H,O-Z)
      DIMENSION DIF1(10),DIF2(10),DIF3(10),ROOT(10),V1(10),V2(10)
      DIMENSION XINTP(10),Y(10),BMAT(10,10)
    1 FORMAT(2I5,F10.5)
    3 FORMAT(1H1,'   GEOMETRY:   0=PLANAR, 1=CYLINDER, 2=SPHERE'//,' NUMBE
     XR OF COLLOCATION POINTS =',I5,/,' GEOMETRY FACTOR =',I5,/,' VALUE
     XOF THIELE MODULUS =',F10.5,//,' INTERPOLATION POINTS IN X**2',/)
    6 FORMAT(8F10.4)
   15 FORMAT(//' SOLUTION AT COLLOCATION POINTS ',/)
   19 FORMAT(//,' SOLUTION AT FIXED GRIDPOINTS',//,'     X            Y',/)
   17 FORMAT(F7.3,F19.9)
   50 FORMAT(//' EFFECTIVENESS FACTOR =',F15.9/' * * * * * * * * * * * * *
     1 * * * * * * * * * ',/)
  200 READ(5,1) N,IS,THM
      IF (N .EQ. 0) GO TO 100
      S=IS
      WRITE(6,3)N,IS,THM
      ALFA=1.
      BETA=(S-1)/2
      CALL JCOBI(10,N,0,1,ALFA,BETA,DIF1,DIF2,DIF3,ROOT)
      NT=N+1
      WRITE(6,6) (ROOT(I),I=1,NT)
C
C     SET UP LAPLACIAN MATRIX
C
      DO 10 I=1,NT
      CALL DFOPR(10,N,0,1,I,1,DIF1,DIF2,DIF3,ROOT,V1)
      CALL DFOPR(10,N,0,1,I,2,DIF1,DIF2,DIF3,ROOT,V2)
      DO 11 J=1,NT
   11 BMAT(I,J)=4*(ROOT(I)*V2(J)+(S+1)/2*V1(J))
      BMAT(I,I)=BMAT(I,I)-THM**2
   10 BMAT(I,NT)=-BMAT(I,NT)
C
C     SOLVE LINEAR EQUATIONS FOR Y
C
      CALL GAUSL(10,10,N,1,BMAT)
      Y(NT)=1.
C
C
C      EVALUATE EFFECTIVENESS FACTOR BY INTEGRATION
C
      CALL RADAU(10,N,0,1,1,0.D0,BETA,ROOT,DIF1,V1)
      ETA=V1(NT)
      DO 45 I=1,N
      Y(I)=BMAT(I,NT)
   45 ETA=ETA+V1(I)*Y(I)
      WRITE(6,15)
      WRITE(6,6)(Y(I),I=1,N)
      WRITE(6,19)
C
C      FIND SOLUTION AT X=0, X=.1, ETC.
C
      DO 40 I=1,11
      X=(I-1)/10.D0
      CALL INTRP(10,NT,X*X,ROOT,DIF1,XINTP)
      YV=0.
      DO 41 J=1,NT
   41 YV=YV+XINTP(J)*Y(J)
   40 WRITE(6,17) X,YV

      WRITE(6,50) ETA
      GO TO 200
  100 STOP
      END
```

A13

```
GEOMETRY:  0=PLANAR, 1=CYLINDER, 2=SPHERE

NUMBER OF COLLOCATION POINTS =     2
GEOMETRY FACTOR =     1
VALUE OF THIELE MODULUS =    2.00000

INTERPOLATION POINTS IN X**2

   0.1551     0.6449     1.0000

SOLUTION AT COLLOCATION POINTS

   0.5094     0.7706

SOLUTION AT FIXED GRIDPOINTS

   X            Y

 0.000          0.440000000
 0.100          0.444280000
 0.200          0.457280000
 0.300          0.479480000
 0.400          0.511680000
 0.500          0.555000000
 0.600          0.610880000
 0.700          0.681080000
 0.800          0.767680000
 0.900          0.873080000
 1.000          1.000000000

EFFECTIVENESS FACTOR =     0.697777778
* * * * * * * * * * * * * * * * * * * *


NUMBER OF COLLOCATION POINTS =     3
GEOMETRY FACTOR =     1
VALUE OF THIELE MODULUS =    2.00000

INTERPOLATION POINTS IN X**2

   0.0886     0.4095      0.7877      1.0000

SOLUTION AT COLLOCATION POINTS

   0.4784     0.6375      0.8585

SOLUTION AT FIXED GRIDPOINTS

   X            Y

 0.000          0.438652767
 0.100          0.443054144
 0.200          0.456389030
 0.300          0.479054737
 0.400          0.511730297
 0.500          0.555401714
 0.600          0.611397338
 0.700          0.681433326
 0.800          0.767669222
 0.900          0.872773631
 1.000          1.000000000

EFFECTIVENESS FACTOR =     0.697774659
* * * * * * * * * * * * * * * * * * * *
```

A14

```
NUMBER OF COLLOCATION POINTS =    4
GEOMETRY FACTOR =    1
VALUE OF THIELE MODULUS =   2.00000

INTERPOLATION POINTS IN X**2

   0.0571     0.2768     0.5836     0.8602     1.0000

SOLUTION AT COLLOCATION POINTS

   0.4641     0.5688     0.7345     0.9054

SOLUTION AT FIXED GRIDPOINTS

   X            Y

 0.000        0.438676539
 0.100        0.443074220
 0.200        0.456399640
 0.300        0.479054328
 0.400        0.511722328
 0.500        0.555393044
 0.600        0.611394414
 0.700        0.681437637
 0.800        0.767674965
 0.900        0.872772468
 1.000        1.000000000

EFFECTIVENESS FACTOR =     0.697774658
* * * * * * * * * * * * * * * * * * * *
```

```
NUMBER OF COLLOCATION POINTS =    6
GEOMETRY FACTOR =    1
VALUE OF THIELE MODULUS =   2.00000

INTERPOLATION POINTS IN X**2

   0.0293     0.1481     0.3370     0.5587     0.7692     0.9269     1.0000

SOLUTION AT COLLOCATION POINTS

   0.4516     0.5061     0.5994     0.7202     0.8468     0.9498

SOLUTION AT FIXED GRIDPOINTS

   X            Y

 0.000        0.438676280
 0.100        0.443074022
 0.200        0.456399583
 0.300        0.479054398
 0.400        0.511722427
 0.500        0.555393069
 0.600        0.611394354
 0.700        0.681437583
 0.800        0.767674997
 0.900        0.872772497
 1.000        1.000000000

EFFECTIVENESS FACTOR =     0.697774658
* * * * * * * * * * * * * * * * * * * *
```

A15

```
      IMPLICIT REAL*8(A-H,O-Z)
      DIMENSION DIF1(10),DIF2(10),DIF3(10),ROOT(10),V1(10),V2(10)
      DIMENSION XINTP(10),Y(10),BMAT(10,10)
      DIMENSION B(10,10)
    1 FORMAT(2I5,F10.5)
    3 FORMAT(1H1,'  GEOMETRY:  0=PLANAR, 1=CYLINDER, 2=SPHERE'//,' NUMBE
     XR OF COLLOCATION POINTS =',I5,/,' GEOMETRY FACTOR =',I5,/,' VALUE
     XOF THIELE MODULUS =',F10.5,//,' INTERPOLATION POINTS IN X**2',/)
    6 FORMAT(8F10.4)
   15 FORMAT(//,' CONVERGENCE IN ITERATION NO.',I4,' WITH RESIDUAL =',D1
     10.1//,' SOLUTION AT COLLOCATION POINTS ',/)
   19 FORMAT(//,' SOLUTION AT FIXED GRIDPOINTS',//,'    X            Y',/)
   17 FORMAT(F7.3,F19.9)
   50 FORMAT(//' EFFECTIVENESS FACTOR =',F15.9/' * * * * * * * * * * * *
     1 * * * * * * * * ',/)
  200 READ(5,1) N,IS,THM
      IF (N .EQ. 0) GO TO 100
      S=IS
      WRITE(6,3)N,IS,THM
      ALFA=1.
      BETA=(S-1)/2
      CALL JCOBI(10,N,0,1,ALFA,BETA,DIF1,DIF2,DIF3,ROOT)
      NT=N+1
      WRITE(6,6)  (ROOT(I),I=1,NT)
C
C     SET UP LAPLACIAN MATRIX
C
      DO 10 I=1,NT
      CALL DFOPR(10,N,0,1,I,1,DIF1,DIF2,DIF3,ROOT,V1)
      CALL DFOPR(10,N,0,1,I,2,DIF1,DIF2,DIF3,ROOT,V2)
      DO 11 J=1,NT
   11 BMAT(I,J)=4*(ROOT(I)*V2(J)+(S+1)/2*V1(J))
   10 CONTINUE
C
C       INITIAL ESTIMATE OF SOLUTION VECTOR
C
      DO 20 I=1,NT
   20 Y(I)=1.
      ITNO=0
C
C       SET UP JACOBIAN AND SOLVE ALGEBRAIC EQUATIONS
C
   30 DO 25 I=1,N
      B(I,NT)=BMAT(I,NT)*Y(NT)-THM**2*Y(I)**2
      DO 26 J=1,N
      B(I,J)=BMAT(I,J)
   26 B(I,NT)=B(I,NT)+BMAT(I,J)*Y(J)
   25 B(I,I)=B(I,I)-2*Y(I)*THM**2
      CALL GAUSL(10,10,N,1,B)
      RES=0.
      DO 28 I=1,N
      Y(I)=Y(I)-B(I,NT)
   28 RES=RES+B(I,NT)**2
      ITNO=ITNO+1
      IF (ITNO .GT. 10) GO TO 36
      IF (RES .GT. 1.D-16) GO TO 30
   36 WRITE(6,15) ITNO,RES
      WRITE(6,6)(Y(I),I=1,N)

C       EVALUATE EFFECTIVENESS FACTOR BY INTEGRATION
C
      CALL RADAU(10,N,0,1,1,0.D0,BETA,ROOT,DIF1,V1)
      ETA=V1(NT)
      DO 45 I=1,N
   45 ETA=ETA+V1(I)*Y(I)**2
      WRITE(6,19)
C
C       FIND SOLUTION AT X=0, X=.1, ETC.
C
      DO 40 I=1,11
      X=(I-1)/10.D0
      CALL INTRP(10,NT,X*X,ROOT,DIF1,XINTP)
      YV=0.
      DO 41 J=1,NT
   41 YV=YV+XINTP(J)*Y(J)
   40 WRITE(6,17) X,YV
      WRITE(6,50) ETA
      GO TO 200
  100 STOP
      END
```

A16

```
  GEOMETRY:  0=PLANAR, 1=CYLINDER, 2=SPHERE

NUMBER OF COLLOCATION POINTS =    2
GEOMETRY FACTOR =      1
VALUE OF THIELE MODULUS =   2.00000

INTERPOLATION POINTS IN X**2

   0.1551     0.6449     1.0000

CONVERGENCE IN ITERATION NO.   5 WITH RESIDUAL =   0.2D-28

SOLUTION AT COLLOCATION POINTS

   0.6151     0.8131

SOLUTION AT FIXED GRIDPOINTS

   X            Y

 0.000        0.566839428
 0.100        0.569741914
 0.200        0.578622598
 0.300        0.594001161
 0.400        0.616743736
 0.500        0.648062909
 0.600        0.689517720
 0.700        0.743013662
 0.800        0.810802680
 0.900        0.895483175
 1.000        1.000000000

EFFECTIVENESS FACTOR =     0.592379750
* * * * * * * * * * * * * * * * * * * * *


NUMBER OF COLLOCATION POINTS =    3
GEOMETRY FACTOR =      1
VALUE OF THIELE MODULUS =   2.00000

INTERPOLATION POINTS IN X**2

   0.0886     0.4095     0.7877     1.0000

CONVERGENCE IN ITERATION NO.   5 WITH RESIDUAL =   0.5D-28

SOLUTION AT COLLOCATION POINTS

   0.5926     0.7105     0.8829

SOLUTION AT FIXED GRIDPOINTS

   X            Y

 0.000        0.563461836
 0.100        0.566684838
 0.200        0.576447027
 0.300        0.593042115
 0.400        0.617006787
 0.500        0.649191464
 0.600        0.690859360
 0.700        0.743813848
 0.800        0.810554119
 0.900        0.894459156
 1.000        1.000000000

EFFECTIVENESS FACTOR =     0.592216302
* * * * * * * * * * * * * * * * * * * * *
```

A17

```
NUMBER OF COLLOCATION POINTS =    4
GEOMETRY FACTOR =     1
VALUE OF THIELE MODULUS =   2.00000

INTERPOLATION POINTS IN X**2

   0.0571     0.2768     0.5836     0.8602     1.0000

CONVERGENCE IN ITERATION NO.   5 WITH RESIDUAL =    0.7D-28

SOLUTION AT COLLOCATION POINTS

   0.5821     0.6590     0.7848     0.9211

SOLUTION AT FIXED GRIDPOINTS

   X             Y

  0.000         0.563710270
  0.100         0.566892989
  0.200         0.576552514
  0.300         0.593029622
  0.400         0.616916490
  0.500         0.649099992
  0.600         0.690836219
  0.700         0.743870172
  0.800         0.810618502
  0.900         0.894437187
  1.000         1.000000000

 EFFECTIVENESS FACTOR =     0.592214668
 * * * * * * * * * * * * * * * * * * * *


NUMBER OF COLLOCATION POINTS =    6
GEOMETRY FACTOR =     1
VALUE OF THIELE MODULUS =   2.00000

INTERPOLATION POINTS IN X**2

   0.0293     0.1481     0.3370     0.5587     0.7692     0.9269     1.0000

CONVERGENCE IN ITERATION NO.   5 WITH RESIDUAL =    0.1D-27

SOLUTION AT COLLOCATION POINTS

   0.5731     0.6128     0.6819     0.7736     0.8735     0.9578

SOLUTION AT FIXED GRIDPOINTS

   X             Y

  0.000         0.563694477
  0.100         0.566880941
  0.200         0.576549198
  0.300         0.593034131
  0.400         0.616922700
  0.500         0.649101418
  0.600         0.690832003
  0.700         0.743866552
  0.800         0.810621090
  0.900         0.894439281
  1.000         1.000000000

EFFECTIVENESS FACTOR =     0.592214656
* * * * * * * * * * * * * * * * * * * *
```

A18

```
      IMPLICIT REAL*8 (A-H,O-Z)
      DIMENSION DIF1(12),DIF2(12),DIF3(12),ROOT(12),Z1(12),Z2(12),Z3(12)
      DIMENSION V1(12),V2(12),A(30,10),NC(10)
      DIMENSION XTAB(13)
      DATA XTAB/1.D-4,2.D-4,5.D-4,1.D-3,2.D-3,5.D-3,1.D-2,2.D-2,5.D-2,
     X1.D-1,2.D-1,5.D-1,1.D0/
   80 FORMAT(F10.5,F20.6)
   70 FORMAT(/,'    DISTANCE       SHERWOOD NO. ')
   71 FORMAT(/' * * * * * * * * * * * * * * * *')
   26 FORMAT(F20.9,' * EXP(',D16.9,' Z)')
   20 FORMAT(/,' SOLUTION SERIES ',/)
   88 FORMAT(F20.9,F25.9)
    7 FORMAT(/' INTERPOLATION POINTS AND RADAU WEIGHTS '/)
    3 FORMAT(1H1,' APPROXIMATION ORDER =',I3/)
    4 FORMAT(I3)
    5 READ(5,4) N
      IF (N.EQ. 0) GO TO 100
      WRITE(6,3) N
      NT=N+2
C
C       SELECT THE APPROPRIATE POLYNOMIAL
C
      CALL JCOBI(12,N,1,1,0.D0,1.D0,DIF1,DIF2,DIF3,ROOT)
C
C       CONSTRUCT RADAU QUADRATURE WEIGHTS AND MULTIPLY BY 1.5*(1-X**2)
C
      CALL RADAU(12,N,1,1,2,0.D0,0.D0,ROOT,DIF1,V2)
      WRITE(6,7)
      WRITE(6,88)(ROOT(I),V2(I),I=1,NT)
      DO 12 J=1,N
   12 V2(J)=1.5*V2(J+1)*(1-ROOT(J+1)**2)
C
C       FIND VECTOR FOR FIRST DERIVATIVE AT X=1 AND FORM THE MODIFIED
C       LAPLACIAN
      CALL DFOPR(12,N,1,1,NT,1,DIF1,DIF2,DIF3,ROOT,V1)
      DO 10 I=1,N
      CALL DFOPR(12,N,1,1,I+1,2,DIF1,DIF2,DIF3,ROOT,Z2)
      DO 11 J=1,N
   11 A(I,J)=(Z2(J+1)-Z2(NT)/V1(NT)*V1(J+1))/(1-ROOT(I+1)**2)
   10 CONTINUE
C
C       FIND EIGENVALUES, -VECTORS AND -ROWS
C
      INDEX=1
      CALL EISYS(30,10,N,INDEX,1.D-12,NC,A)
C
C       NOW FIND VECTORS U AND V ( Z2 AND Z1 )
C       AND THE EXPANSION COEFFICIENTS (Z3)
C
      DO 14 I=1,N
      Z1(I)=0.
      Z2(I)=0.
      DO 15 J=1,N
      Z1(I)=Z1(I)+A(I+2*N,J)
   15 Z2(I)=Z2(I)+V2(J)*A(J+N,I)
   14 Z3(I)=Z1(I)*Z2(I)
      WRITE(6,20)
      DO 25 I=1,N
   25 WRITE(6,26) Z3(I),A(I,I)

C
C       FINALLY THE SHERWOOD NUMBER IS EVALUATED
C
      WRITE(6,70)
      WRITE(6,71)
      DO 75 I=1,13
      X=0.
      S=0.
      DO 76 J=1,N
      T=A(J,J)*XTAB(I)
      IF (T .LT. -60.D0) GO TO 76
      T=Z3(J)*DEXP(T)
      X=X+T
      S=S-T*A(J,J)
   76 CONTINUE
      S=2*S/X/3
   75 WRITE(6,80) XTAB(I),S
      WRITE(6,71)
      GO TO 5
  100 STOP
      END
```

A19

```
 APPROXIMATION ORDER =  2

INTERPOLATION POINTS AND RADAU WEIGHTS

         0.000000000                 0.111111111
         0.355051026                 0.512485826
         0.844948974                 0.376403063
         1.000000000                 0.000000000

SOLUTION SERIES

         0.807647988 * EXP(-0.513581570D 01 Z)
         0.025685345 * EXP(-0.346152954D 02 Z)

  DISTANCE       SHERWOOD NO.

* * * * * * * * * * * * * * *
  0.00010            4.027901
  0.00020            4.026178
  0.00050            4.021036
  0.00100            4.012560
  0.00200            3.995957
  0.00500            3.948829
  0.01000            3.878550
  0.02000            3.764476
  0.05000            3.565980
  0.10000            3.456603
  0.20000            3.425596
  0.50000            3.423877
  1.00000            3.423877

* * * * * * * * * * * * * * *
```

```
 APPROXIMATION ORDER =  4

INTERPOLATION POINTS AND RADAU WEIGHTS

         0.000000000                 0.040000000
         0.139759864                 0.223103901
         0.416409568                 0.311826523
         0.723156986                 0.281356015
         0.942895804                 0.143713561
         1.000000000                 0.000000000

SOLUTION SERIES

         0.000050314 * EXP(-0.448430809D 03 Z)
         0.055772257 * EXP(-0.923796259D 02 Z)
         0.094507452 * EXP(-0.401929540D 02 Z)
         0.789669977 * EXP(-0.512148582D 01 Z)

  DISTANCE       SHERWOOD NO.

* * * * * * * * * * * * * * *
  0.00010            9.198470
  0.00020            9.164826
  0.00050            9.065236
  0.00100            8.903598
  0.00200            8.595776
  0.00500            7.782284
  0.01000            6.725120
  0.02000            5.378527
  0.05000            3.939834
  0.10000            3.498585
  0.20000            3.416839
  0.50000            3.414324
  1.00000            3.414324

* * * * * * * * * * * * * * *
```

A20

```
 APPROXIMATION ORDER =  6

INTERPOLATION POINTS AND RADAU WEIGHTS

         0.000000000                  0.020408163
         0.073054329                  0.119613745
         0.230766138                  0.190474937
         0.441328481                  0.223554915
         0.663015310                  0.212351890
         0.851921400                  0.159102116
         0.970683573                  0.074494236
         1.000000000                  0.000000000
SOLUTION SERIES

         0.000000701 * EXP(-0.299940145D 04 Z)
         0.031502664 * EXP(-0.299498723D 03 Z)
         0.017030000 * EXP(-0.243946121D 03 Z)
         0.033780719 * EXP(-0.103916313D 03 Z)
         0.097371179 * EXP(-0.396740122D 02 Z)
         0.789702493 * EXP(-0.512166906D 01 Z)

  DISTANCE      SHERWOOD NO.

* * * * * * * * * * * * * * *
  0.00010          16.945327
  0.00020          16.696782
  0.00050          15.983968
  0.00100          14.897995
  0.00200          13.060787
  0.00500           9.477729
  0.01000           6.834489
  0.02000           5.160079
  0.05000           3.927955
  0.10000           3.503936
  0.20000           3.417278
  0.50000           3.414446
  1.00000           3.414446

* * * * * * * * * * * * * * *

 APPROXIMATION ORDER =  8

INTERPOLATION POINTS AND RADAU WEIGHTS

         0.000000000                  0.012345679
         0.044633955                  0.073827010
         0.144366257                  0.123594689
         0.286824757                  0.158421888
         0.454813315                  0.174136501
         0.628067835                  0.168846983
         0.785691521                  0.143193348
         0.908676392                  0.100276649
         0.982220085                  0.045357252
         1.000000000                  0.000000000
SOLUTION SERIES

         0.000000040 * EXP(-0.129408160D 05 Z)
         0.000233408 * EXP(-0.901318446D 03 Z)
         0.031395799 * EXP(-0.728240026D 03 Z)
         0.006239884 * EXP(-0.308849712D 03 Z)
         0.020695543 * EXP(-0.206976383D 03 Z)
         0.035958050 * EXP(-0.106202662D 03 Z)
         0.097256142 * EXP(-0.396608981D 02 Z)
         0.789702616 * EXP(-0.512166931D 01 Z)

  DISTANCE      SHERWOOD NO.

* * * * * * * * * * * * * * *
  0.00010          26.720290
  0.00020          25.654779
  0.00050          22.822743
  0.00100          19.115427
  0.00200          14.334396
  0.00500           9.003205
  0.01000           6.739859
  0.02000           5.187056
  0.05000           3.927075
  0.10000           3.503894
  0.20000           3.417281
  0.50000           3.414446
  1.00000           3.414446

* * * * * * * * * * * * * * *
```

A21

```
      IMPLICIT REAL*8(A-H,O-Z)
      DIMENSION XTAB(10),IP(30),METHOD(40),MPROB(40),MHEAD(40)
      DIMENSION Y(30),YOLD(30),YOLD1(30),YA(30),F(30),FOLD(30),YK1(30)
      DIMENSION YK2(30),YK3(30),DF(30,30),DFOLD(30,30),W(30)
      EXTERNAL FUN,DFUN,OUT
C
C     METHOD, MPROB AND MHEAD ARE HEADINGS
C     SUBPROGRAM NAMES DECLARED EXTERNAL
C
      READ(5,1) METHOD,MPROB,MHEAD
C
C
C     READ NUMBER OF EQUATIONS, NUMBER OF INTERVAL ENDPOINTS,
C     PRINTING INTERVAL
C     READ ERROR MULTIPLIERS AND DESIRED INTERVAL ENDPOINTS
C
      READ(5,2) NCOM,NTAB,NPRINT
      READ(5,3) (W(I),I=1,NCOM)
      READ(5,3) (XTAB(I),I=1,NTAB)
C
C     READ TOLERANCE, INITIAL VALUE OF INDEPENDENT VARIABLE
C     AND INITIAL HALFSTEP
C     INITIALIZE DEPENDENT VARIABLE
C
   10 READ(5,3) EPS,XST,H0
      IF (EPS.EQ.0.D0) GO TO 100
      READ(5,3)(Y(I),I=1,NCOM)
      WRITE(6,7)
      WRITE(6,1) METHOD
      WRITE(6,1) MPROB
      WRITE(6,4)(I,Y(I),W(I),I=1,NCOM)
      WRITE(6,5) EPS,H0
      WRITE(6,1) MHEAD
      WRITE(6,61)
      X1=XST
C
C     NOW SEQUENTIAL INTEGRATION OF INTERVALS
C
      DO 50 I=1,NTAB
      WRITE(6,6)
      X2=XTAB(I)
      CALL STIFF3(NCOM,30,NPRINT,FUN,DFUN,OUT,X1,X2,H0,EPS,W,Y,YOLD,
     XYOLD1,IP,YA,YK1,YK2,YK3,DF,DFOLD,F,FOLD)
   50 X1=X2
      WRITE(6,61)
      GO TO 10
    1 FORMAT(40A2)
    2 FORMAT(3I5)
    3 FORMAT(10F8.5)
    4 FORMAT(//,' INITIAL VALUES AND ERROR MULTIPLIERS'/,(/,I3,2F15.5))
    5 FORMAT(//,' TOLERANCE = ',D12.2,'  INITIAL HALF STEP =',D12.2//)
    6 FORMAT(' *')
    7 FORMAT(1H1)
   61 FORMAT(' * * * * * * * * * * * * * * * * * * * * * * * * * * * * * * *')
  100 STOP
      END
```

A22

```
      SUBROUTINE FUN(Y,F)
      IMPLICIT REAL*8(A-H,O-Z)
      DIMENSION Y(30),F(30)
      F(1)=-4.D-2*Y(1)+1.D4*Y(2)*Y(3)
      F(2)=-F(1)-3.D7*Y(2)**2
      F(3)=3.D7*Y(2)**2


      RETURN
      END
```

```
      SUBROUTINE OUT(X,Y,IH,Q)
      IMPLICIT REAL*8(A-H,O-Z)
      DIMENSION Y(30)
      Y2=1.D4*Y(2)
      WRITE(6,1)X,Y(1),Y2,Y(3),IH
    1 FORMAT(' *',F9.5,' *',3F12.6,9X,I3)
      RETURN
      END
```

```
      SUBROUTINE DFUN(Y,DF)
      IMPLICIT REAL*8(A-H,O-Z)
      DIMENSION Y(30),DF(30,30)
      DF(1,1)=-4.D-2
      DF(1,2)=1.D4*Y(3)
      DF(1,3)=1.D4*Y(2)
      DF(2,1)=4.D-2
      DF(2,2)=-1.D4*Y(3)-6.D7*Y(2)
      DF(2,3)=-1.D4*Y(2)
      DF(3,1)=0.
      DF(3,3)=0.
      DF(3,2)=6.D7*Y(2)
      RETURN
      END
```

A23

```
      IMPLICIT THIRD ORDER RUNGE KUTTA INTEGRATION
      ROBERTSON RATE EQUATIONS

 INITIAL VALUES AND ERROR MULTIPLIERS

  1          1.00000          1.00000

  2          0.00000        100.00000

  3          0.00000          1.00000

 TOLERANCE =       0.10D-01   INITIAL HALF STEP =     0.10D-02

     TIME            Y(1)        10**4*Y(2)       Y(3)          BISECTIONS
 * * * * * * * * * * * * * * * * * * * * * * * * * * *
 *
 *   0.00200 *     0.999920     0.467528     0.000033            0
 *   0.00348 *     0.999861     0.369046     0.000102            0
 *   0.00673 *     0.999731     0.364839     0.000232            0
 *   0.01646 *     0.999344     0.364033     0.000620            0
 *   0.04565 *     0.998190     0.361906     0.001774            0
 *   0.10000 *     0.996078     0.358044     0.003886            0
 *
 *   0.18759 *     0.992766     0.352057     0.007198            0
 *   0.40000 *     0.985172     0.338642     0.014794            0
 *
 *   0.66276 *     0.976530     0.323910     0.023438            0
 *   1.00000 *     0.966460     0.307466     0.033510            0
 *
 *   1.78827 *     0.946361     0.276962     0.053611            0
 *   4.00000 *     0.905517     0.224091     0.094461            0
 *
 *   6.36482 *     0.875153     0.191942     0.124828            0
 *  10.00000 *     0.841368     0.162357     0.158616            0
 * * * * * * * * * * * * * * * * * * * * * * * * * * *
```

```
      IMPLICIT THIRD ORDER RUNGE KUTTA INTEGRATION
      ROBERTSON RATE EQUATIONS

 INITIAL VALUES AND ERROR MULTIPLIERS

  1          1.00000          1.00000

  2          0.00000        100.00000

  3          0.00000          1.00000

 TOLERANCE =       0.10D-02   INITIAL HALF STEP =     0.10D-03

     TIME            Y(1)        10**4*Y(2)       Y(3)          BISECTIONS
 * * * * * * * * * * * * * * * * * * * * * * * * * * *
 *
 *   0.00020 *     0.999992     0.078746     0.000000            0
 *   0.00080 *     0.999968     0.257614     0.000006            0
 *   0.00188 *     0.999925     0.353926     0.000040            0
 *   0.00461 *     0.999816     0.364732     0.000148            0
 *   0.01046 *     0.999582     0.364465     0.000381            0
 *   0.02801 *     0.998886     0.363186     0.001078            0
 *   0.06400 *     0.997472     0.360589     0.002492            0
 *   0.10000 *     0.996078     0.358044     0.003886            0
 *
 *   0.20799 *     0.992011     0.350702     0.007954            0
 *   0.40000 *     0.985172     0.338641     0.014794            0
 *
 *   0.70000 *     0.975365     0.321968     0.024603            0
 *   1.00000 *     0.966460     0.307465     0.033510            0
 *
 *   1.90000 *     0.943824     0.273330     0.056148            0
 *   4.00000 *     0.905517     0.224086     0.094460            0
 *
 *   6.10871 *     0.878025     0.194738     0.121956            0
 *   8.05435 *     0.858040     0.176202     0.141942            0
 *  10.00000 *     0.841369     0.162344     0.158615            0
 * * * * * * * * * * * * * * * * * * * * * * * * * * *
```

Subject Index

Author Index

Note: Numbers in italic refer to pages on which the complete references are listed.